Understanding Marine Science

Understanding Marine Science

Edited by **Theodore Roa**

R CALLISTO REFERENCE

New York

Published by Callisto Reference,
106 Park Avenue, Suite 200,
New York, NY 10016, USA
www.callistoreference.com

Understanding Marine Science
Edited by Theodore Roa

International Standard Book Number: 978-1-63239-637-2 (Hardback)

Printed in the United States of America.

Contents

Preface

This is a research-focused book which presents theoretical and scientific studies in the field of marine science. It contains many exclusive researches contributed by eminent authors from all parts of the world. Marine science deals with the detailed study of the ocean, its ecosystems, and its life forms, as well as the study of coastal environments, oceanic currents, and the sea floor. Marine science is the amalgamation of marine biology, marine chemistry, marine geology, marine pollution, and physical oceanography. This book highlights the interdisciplinary nature of marine science, and covers a wide range of topics and is a quick and easy reference to a multitude of topics related to this subject. The extensive content of this book will provide the readers with a thorough understanding of the subject. Students, researchers, experts and all associated with marine science will benefit alike from this book.

Various studies have approached the subject by analyzing it with a single perspective, but the present book provides diverse methodologies and techniques to address this field. This book contains theories and applications needed for understanding the subject from different perspectives. The aim is to keep the readers informed about the progress in the field; therefore, the contributions were carefully examined to compile novel researches by specialists from across the globe.

Indeed, the job of the editor is the most crucial and challenging in compiling all chapters into a single book. In the end, I would extend my sincere thanks to the chapter authors for their profound work. I am also thankful for the support provided by my family and colleagues during the compilation of this book.

Editor

Increase in dimethylsulfide (DMS) emissions due to eutrophication of coastal waters offsets their reduction due to ocean acidification

Nathalie Gypens[1]* and Alberto V. Borges[2]

[1] Laboratoire d'Ecologie des Systèmes Aquatiques, Ecole Interfacultaire de Bioingénieurs, Université Libre de Bruxelles, Brussels, Belgium
[2] Unité d'Océanographie Chimique, Department of Astrophysics Geophysics and Oceanography, Institut de Physique (B5), Université de Liège, Liège, Belgium

Edited by:
Stelios Katsanevakis, Institute for Environment and Sustainability, Italy

Reviewed by:
Eric 'Pieter Achterberg, GEOMAR Helmholtz Centre for Ocean Research Kiel, Germany
Kevin John Flynn, Swansea University, UK
Thomas George Bell, Plymouth Marine Laboratory, UK

*Correspondence:
Nathalie Gypens, Laboratoire d'Ecologie des Systèmes Aquatiques, Ecole Interfacultaire de Bioingénieurs, Université Libre de Bruxelles, CP-221, Bd du Triomphe, 1050 Brussels, Belgium
e-mail: ngypens@ulb.ac.be

Available information from manipulative experiments suggested that the emission of dimethylsulfide (DMS) would decrease in response to the accumulation of anthropogenic CO_2 in the ocean (ocean acidification). However, in coastal environments, the carbonate chemistry of surface waters was also strongly modified by eutrophication and related changes in biological activity (increased primary production and change in phytoplankton dominance) during the last 50 years. Here, we tested the hypothesis that DMS emissions in marine coastal environments also strongly responded to eutrophication in addition to ocean acidification at decadal timescales. We used the R-MIRO-BIOGAS model in the eutrophied Southern Bight of the North Sea characterized by intense blooms of *Phaeocystis* that are high producers of dimethylsulfoniopropionate (DMSP), the precursor of DMS. We showed that, for the period from 1951 to 2007, eutrophication actually led to an increase of DMS emissions much stronger than the response of DMS emissions to ocean acidification.

Keywords: dimethylsulfide (DMS), eutrophication, ocean acidification, coastal waters, *Phaeocystis*, modeling

INTRODUCTION

Dimethylsulfide (DMS) is the largest natural source of sulfur to the atmosphere, and may influence climate regulation by affecting atmospheric chemistry and the heat balance of the atmosphere (Charlson et al., 1987) though the significance of this feedback still remains uncertain (Carslaw et al., 2010; Quinn and Bates, 2011). Dimethylsulfoniopropionate (DMSP), the precursor of DMS, is mainly produced by a limited number of marine microalgae; Haptophytes and dinoflagellates being usually characterized by higher cellular DMSP concentrations than diatoms (Keller et al., 1989; Stefels et al., 2007). The modification of microalgae dominance and production in response to climate warming (e.g., Blanchard et al., 2012), eutrophication (e.g., Mackenzie et al., 2011), or ocean acidification (e.g., Doney et al., 2009) is then expected to change DMS emissions in the future with a potential positive or a negative feedback on climate change (Bopp et al., 2003; Gabric et al., 2004; Kloster et al., 2007; Vallina et al., 2007; Cameron-Smith et al., 2011; Six et al., 2013).

Ocean acidification of surface waters corresponds to the increase of $[CO_2]$ and of $[H^+]$, the decrease of pH, of $[CO_3^{2-}]$, and of the saturation state of calcium carbonate, all related to the input of anthropogenic CO_2 from the atmosphere. These changes of the carbonate chemistry can alter the rates of primary production and calcification of numerous marine organisms and communities (Doney et al., 2009). Marine ecosystems also respond to other human pressures such as the increased delivery of nutrients from rivers to coastal waters mainly related to waste water discharge and use of fertilizers. This leads to eutrophication that corresponds to a general increase of primary production but also

a change in the phytoplankton community and structure that impacts the whole food-web and ecosystem functioning (Cloern, 2001).

Several experiments based on the manipulation of the seawater carbonate chemistry in micro- or mesocosms have shown that ocean acidification is expected to change the DMS(P) cycle (Vogt et al., 2008; Lee et al., 2009; Hopkins et al., 2010; Kim et al., 2010; Avgoustidi et al., 2012; Spielmeyer and Pohnert, 2012; Archer et al., 2013; Arnold et al., 2013). Six et al. (2013) used a simple relationship between DMS concentration and pH from a restricted set of experiments to predict the impact of ocean acidification on future climate through changes of the oceanic emission of DMS and its effect on atmospheric chemistry. However, in coastal environments, eutrophication is expected to counter to some extent the effect of ocean acidification on the dissolved carbonate chemistry of surface waters (Borges and Gypens, 2010). Further, DMS emissions would respond to the overall increase of primary production but also to the change of phytoplankton community structure related to eutrophication.

Here, we test the hypothesis that DMS emissions in coastal environments will also strongly respond to eutrophication in addition to ocean acidification at decadal timescales. We use as a case study the strongly eutrophied Southern Bight of the North Sea (SBNS) that is a hot spot of DMS emissions (Uher, 2006; Lana et al., 2011) because of the occurrence of intense blooms of *Phaeocystis* that are high DMSP producers. We use the R-MIRO-BIOGAS model that has been previously validated in the SBNS to which was added the parameterization of DMS change as a function of pH used in the global study of Six et al. (2013).

MATERIALS AND METHODS

The R-MIRO-BIOGAS model results of the coupling between the RIVERSTRAHLER model, a biogeochemical model of the river system (Garnier et al., 2002), and the MIRO-BIOGAS model. The coupled model describes the biological transformation of C, nitrogen (N), phosphorus (P), and silicate (Si) along the river-coastal continuum as a function of meteorological conditions and changing human activity on the watershed. C and nutrient river loads to the coastal zone are simulated by the RIVERSTRAHLER model from prescribed land use and human activities on the drainage basin (Billen et al., 2001, 2005; Passy et al., 2013). In the marine domain, MIRO-BIOGAS describes the dynamics of phytoplankton (diatoms, nanoflagellates and *Phaeocystis*), zooplankton and bacteria involved in the degradation of organic matter and the regeneration of inorganic nutrients in the water column and the sediment (Lancelot et al., 2005). The description of the carbonate system is based on the evolution of dissolved inorganic carbon (DIC) and total alkalinity (TA) and allows the calculation of partial pressure of CO_2 (pCO_2), pH and air-sea CO_2 fluxes (Gypens et al., 2004). The DMS(P) cycle describes the DMSP and DMS dynamics, including biological transformations by phytoplankton and bacteria, and physico-chemical processes (including photodegradation and DMS air-sea exchange) (Gypens et al., 2014).

To take account in the MIRO-BIOGAS model for the taxon-specific production of DMSP, the cellular production of DMSP is proportional to phytoplankton growth using a specific DMSP:C quota for each of the 3 phytoplankton groups described in the model (diatoms, nanoflagellates and *Phaeocystis*). The regulation of DMSP production by abiotic factors such as temperature, light, nutrients and salinity is not accounted. Indeed, at temperate latitudes, these effects are negligible (Stefels et al., 2007) and, as a first approximation, we consider that the cellular DMSP concentration of the different phytoplankton groups described in the model is fixed and specific (Gypens et al., 2014). The evolution of DMS concentration in the model results from its production by phytoplankton and bacterial enzymatic cleavage of DMSP in DMS and losses by bacterial consumption (assimilation and/or demethylation/methiolation), photochemical oxidation and emission to the atmosphere. The empirical relationship proposed by Six et al. (2013) modifies the DMS production (DMS_{prod}) as a function of pH change according to:

$$DMS_{prod\_pH} = F * DMS_{prod\_ref}$$

where DMS_{prod_pH} is the DMS production taking into account the effect of ocean acidification, DMS_{prod_ref} is the DMS production in the reference simulation (unperturbed by ocean acidification), and F is given by:

$$F = 1 + \gamma * \left(pH_{year} - pH_{1951}\right)$$

where γ corresponds to the value of the slope (0.58) of the linear regression of DMS concentration as function of pH in a selected number of mesocosm experiments that give a "medium" response according to Six et al. (2013), pH_{1951} corresponds to the pH mean

daily value in 1951, and pH_{year} is the corresponding pH mean daily value computed for each year of the model simulation.

Note that pH effects on model variables other than DMS are not included in the model such as pH effects on primary production, for which no consensual parameterization is available. Results amongst manipulative studies are indeed highly variable reflecting either inconsistencies in experimental approach (Richier et al., 2014), or taxon-specific differences of the physiology response to ocean acidification (Flynn et al., 2012).

For this application, the model is implemented in a 0D frame of three successive boxes delineated on the basis of the hydrological regime and river inputs (Figure 1 in Lancelot et al., 2005) from the English Channel to the Belgian coastal zone (BCZ), and model results are analyzed for the BCZ influenced by the inflowing of Atlantic waters from the French coastal zone and the Scheldt river. Model simulations were performed from 1951 to 2007 after a 10 year spin-up run by constraining the model by daily wind speed, sea surface temperature (SST), monthly atmospheric pCO_2, daily global solar radiation (climatology for 1989–1999), 10-days RIVERSTRAHLER simulations for C and nutrient river loads (Seine and Scheldt), and river DIC and TA inputs (Seine and Scheldt) computed as a function of freshwater discharge (Gypens et al., 2009).

Validation against field data and assessment of performance of the different components of the R-MIRO-BIOGAS model (RIVERSTRAHLER, R-MIRO, MIRO-CO2, MIRO-DMS) has been given and extensively discussed in other papers. Validation of the river loadings simulated by the RIVERSTRAHLER model for the Seine and Scheldt rivers during 1951–2007 was given by Billen et al. (2001, 2005) and Passy et al. (2013). Validation of the key biogeochemical variables simulated by R-MIRO such as nutrient concentration, phytoplankton composition and biomass (forced by the Seine and Scheldt river loadings simulated by RIVERSTRAHLER) in the BCZ was given by Lancelot et al. (2007) and Passy et al. (2013) for the *Phaeocystis* cells in particular. Validation of the seasonal and interannual variation of the seawater carbonate chemistry (DIC, TA, pCO_2) simulated by MIRO-CO2 was given by Gypens et al. (2004, 2011). The validation of DMS(P) dynamics for the year 1989 simulated by MIRO-DMS was given by Gypens et al. (2014).

In addition to the reference simulation, two other simulations described in **Table 1** were carried out to analyze model sensitivity to river inputs and atmospheric CO_2 forcings. The effect of changes in SST and light was not tested as it was previously shown to have little impact on the long-term trend of simulated net primary production (NPP) and phytoplankton community composition (Lancelot et al., 2007).

RESULTS AND DISCUSSION

DMS EMISSIONS IN THE BCZ

From 1951 to 2007, three periods can be distinguished in terms of river inputs of N and P, quality of nutrient enrichment defined by winter-time dissolved inorganic nitrogen to phosphate ($DIN:PO_4$) ratio, NPP, and pH (**Figures 1A,B,C**), as previously reported by Lancelot et al. (2007) and Borges and Gypens (2010). From 1951 to 1965, river nutrient inputs, NPP and pH remained stable. From 1966 to 1990, the river nutrient inputs increased and

Table 1 | Description of the R-MIRO-BIOGAS simulations.

Name	Setup	Application
Reference-Simulation	All forcings vary	Describes inter-annual variability and decadal changes of DMS fluxes from 1951 to 2007 in the SBNS
pH-Simulation	Includes the parameterization of DMS concentration change as a function of pH given by Six et al. (2013)	Allows to analyse the effect of pH change on DMS emissions from 1951 to 2007 in the SBNS
Ocean acidification only-Simulation	All forcings are maintained at the 1951 values except for atmospheric pCO_2 and includes the parameterization of DMS concentration change as a function of pH given by Six et al. (2013)	Allows to analyse the effect of increasing atmospheric CO_2 on the inter-annual variability and decadal changes of DMS emissions from 1951 to 2007 in the SBNS

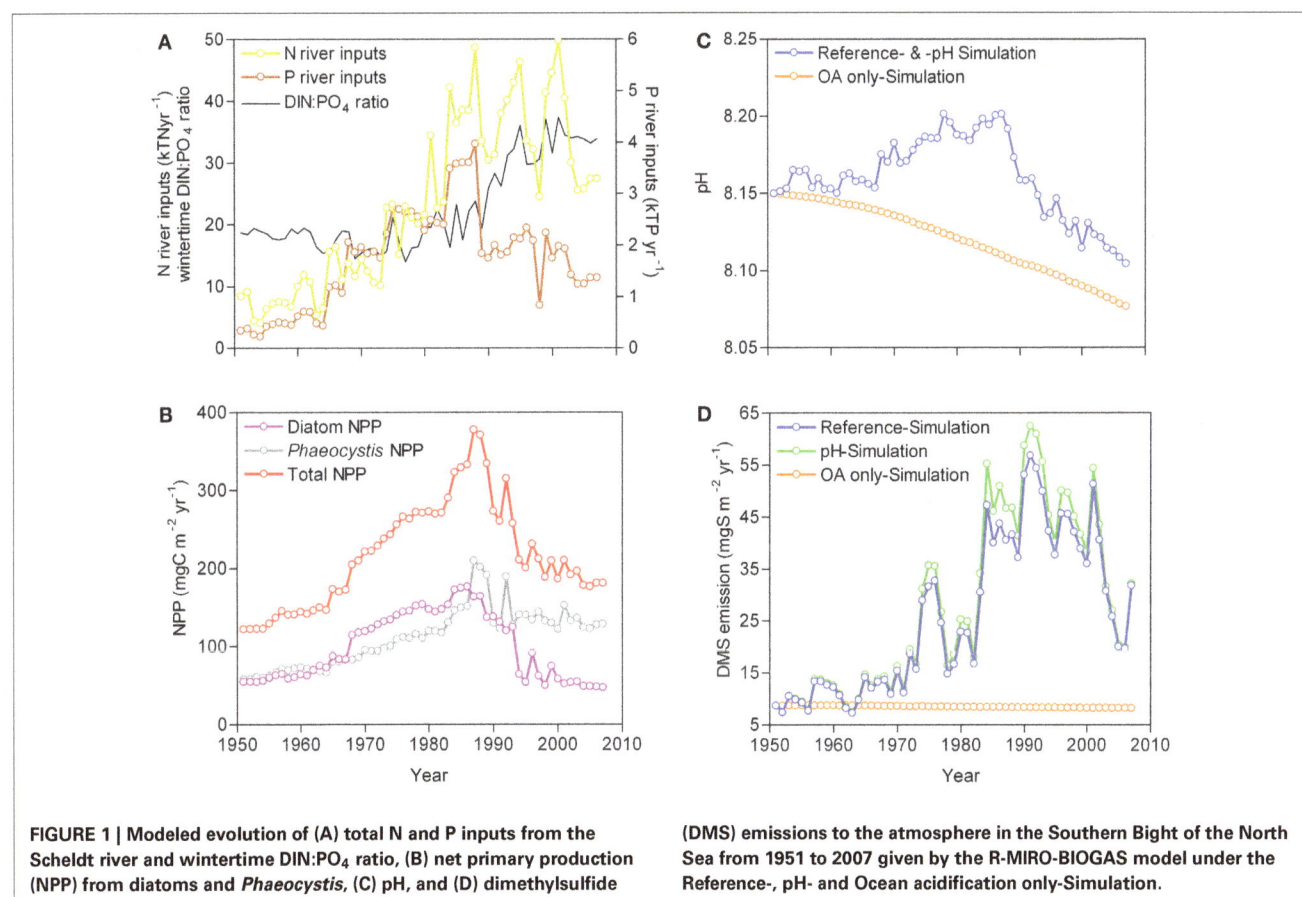

FIGURE 1 | Modeled evolution of (A) total N and P inputs from the Scheldt river and wintertime DIN:PO$_4$ ratio, (B) net primary production (NPP) from diatoms and *Phaeocystis*, (C) pH, and (D) dimethylsulfide (DMS) emissions to the atmosphere in the Southern Bight of the North Sea from 1951 to 2007 given by the R-MIRO-BIOGAS model under the Reference-, pH- and Ocean acidification only-Simulation.

the winter-time DIN:PO$_4$ ratios remained close to phytoplankton requirements (Redfield ratio = 16:1; Redfield et al., 1963) leading to an increase of NPP and pH. From 1991 to 2007, the decrease of the total P river inputs (mainly due to removal of polyphosphates from washing powders) led to unbalanced winter-time DIN:PO$_4$ ratios (>16:1) and a P limitation of primary production (Lancelot et al., 2007) accompanied by a decline of NPP and pH. As previously described, from 1966 to 1990, eutrophication by increasing NPP countered the decline of pH due to atmospheric CO$_2$ increase (ocean acidification), while the decline of NPP between 1991 and 2007 led to a decline of pH faster than the one predicted from atmospheric CO$_2$ increase alone (Borges and Gypens, 2010).

The emissions of DMS to the atmosphere (**Figure 1D**) responded strongly to the changes of NPP and in particular of *Phaeocystis* that accounts for about 80% of annual DMS emissions in the study area (Gypens et al., 2014). From 1951 to 1990, *Phaeocystis* accounted for about 50% of total annual NPP, and on average 64% from 1991 to 2007. From 1951 to 1965, the DMS emissions remained stable, but showed inter-annual variations that correspond to ±22% of the overall annual mean. From 1966 to 1990, the DMS emissions strongly increased at a rate of 14.4 mgS m^{-2} year^{-1} decade^{-1} ($r^2 = 0.69$, $p < 0.0001$), while from 1991 to 2007 the DMS emissions strongly decreased at a rate of -18.2 mgS m^{-2} year^{-1} decade^{-1} ($r^2 = 0.69$, $p < 0.0001$). The highest DMS emission of 56.8 mgS m^{-2} year^{-1} simulated in 1991 was 6.5 times higher than the one of 1951 (8.8 mgS m^{-2} $year^{-1}$). The annual DMS emission for the last year of the simulation period, 2007, (31.8 mgS m^{-2} $year^{-1}$) remained 3.6 times higher than the one of 1951. This shows that eutrophication strongly enhanced the DMS emissions to the atmosphere and also countered the effect of ocean acidification on DMS emissions. Indeed, a small (-0.11 mgS m^{-2} $year^{-1}$ $decade^{-1}$) but highly significant ($r^2 = 0.98$, $p < 0.0001$) decrease of DMS emissions was simulated between 1951 and 2007 when the system only reacted to the increase of atmospheric CO_2 (Ocean acidification only-Simulation). Since pH actually increased from 1966 to 1990 (**Figure 1B**), the inclusion in the model of the parameterization of DMS production as function of pH given by Six et al. (2013) led to DMS emissions (pH-Simulation) on average ~9% higher compared to the Reference-Simulation (**Figure 1D**).

DMS EMISSION IN THE GLOBAL COASTAL OCEAN

The most recent estimate of global marine emission of DMS is 28.1 TgS $year^{-1}$ (Lana et al., 2011), although the respective part of emissions from continental shelves are not provided. We roughly evaluate DMS emissions from continental shelves assuming that DMS emissions are proportional to primary production (Bell et al., 2010; Miles et al., 2012; Kameyama et al., 2013). We used the average primary production value for coastal waters of 230 mgC m^{-2} $year^{-1}$, and for open ocean of 100 mgC m^{-2} $year^{-1}$ (Wollast, 1998). Using the global surface area of 27×10^6 km^2 for continental shelves and of $335 \times 10^6 km^2$ for the open ocean, globally integrated primary production in coastal waters is 6.2 PgC $year^{-1}$ vs. 33.5 PgC $year^{-1}$ in the open ocean. This yields a global emission of DMS from continental shelves of 4.4 TgS $year^{-1}$ vs. 23.7 TgS $year^{-1}$ from the open ocean. We computed the primary production of coastal waters influenced by rivers based on the relative present day inputs of N from rivers of 43 TgN $year^{-1}$ (Seitzinger et al., 2010) and the input of N by mixing and upwelling at continental margins of 380 TgN $year^{-1}$ (Wollast, 1998) that sustains primary production in the remaining continental shelves. This implies that coastal waters influenced by rivers have a DMS emission of 0.4 TgS $year^{-1}$ corresponding to 2% of total oceanic DMS emissions, and 10% of DMS emissions from continental shelves. Using a value of 18 TgN $year^{-1}$ for pristine N river inputs (Meybeck, 1982), we can roughly evaluate that eutrophication has led to a 58% increase of DMS emissions from coastal waters influenced by rivers globally. This corresponds to the average increase globally, but regionally it is much higher, such as in the SBNS (510% from 1951 to 1990). In the SBNS, the delivery of nutrients has been declining since the early 1990's, however, it is expected that globally N river inputs will continue to increase in future (Galloway et al., 2004). By 2050, the total N inputs by rivers could range between 63 TgN $year^{-1}$ (Galloway et al., 2004) and 77 TgN $year^{-1}$ (Mackenzie et al., 2011). Using the average value of 70 TgN $year^{-1}$, DMS emissions from coastal waters influenced by rivers could increase by 2050 by 63% (0.7 TgS $year^{-1}$) that would correspond to about 3% of the present day total oceanic DMS emissions, for a region corresponding to only ~0.5% of the total surface of the ocean.

CONCLUSIONS

We acknowledge that the change of marine DMS emissions in response to ocean acidification involves numerous and complex changes in DMS(P) sources and sinks. In particular, some studies point to an increase of DMSP production in parallel to either an increase of primary production, as both are usually correlated (e.g., Bell et al., 2010), or a change in phytoplankton dominance as a response to ocean acidification, although generally associated to a decrease of DMS emissions (Archer et al., 2013) or dissolved DMSP concentration (Lee et al., 2009). This indicates major modifications of DMS(P) cycling in particular by bacteria and zooplankton. However, the majority of manipulative experiments (micro- and mesocosms) converge to show a decrease of DMS emissions as a consequence of ocean acidification (Hopkins et al., 2010; Avgoustidi et al., 2012; Spielmeyer and Pohnert, 2012; Archer et al., 2013; Arnold et al., 2013), although some studies show the opposite response (Kim et al., 2010; Hopkins and Archer, 2014) for undetermined reasons. DMSP production by phytoplankton can also be modified by nutrient availability, not taken into account in this application. It has been suggested that Fe- and N-limitation could increase DMSP cell quota and/or DMS emission (Sunda et al., 2002, 2007; Harada et al., 2009), but primary production in the studied zone is P-limited in general and Si-limited for diatoms (Lancelot et al., 2005, 2007). There are no or little changes in DMSP cell quota under Si- or P-limitation of primary production (Bucciarelli and Sunda, 2003).

Coastal environments are hotspots of DMS emissions and are subject to eutrophication. We showed that this major anthropogenic disturbance of marine ecosystems can lead to a stronger response of DMS emissions that actually counters the effect of ocean acidification in our particular case study in the SBNS. The response of DMS emissions to eutrophication at global scale cannot be currently modeled because the resolution of general ocean circulation models or earth system models is insufficient to represent coastal areas influenced by rivers. At present, these responses can only be investigated with regional models, hence, necessarily at local scales. We nevertheless attempted a first-order estimate of the global change of DMS emissions in response to eutrophication, and we evaluated an increase of about 63% of DMS emissions from coastal waters from present to 2050 as a result of increased NPP.

The parameterization of DMS concentration as function of pH used by Six et al. (2013) covers the range of relative changes of DMS concentrations observed so far in manipulative experiments. Hence, it is unlikely that a better representation of DMS(P) responses to ocean acidification would change the overall conclusion of the present work regarding to the relative effect of ocean acidification and eutrophication on DMS emissions in eutrophied coastal environments such as the SBNS.

AUTHOR CONTRIBUTIONS

Both authors contributed equally to this work Nathalie Gypens and Alberto V. Borges jointly conceived the study, designed the model simulations, and prepared the manuscript. Nathalie Gypens run the model and Alberto V. Borges conducted the global estimates of DMS emissions.

ACKNOWLEDGMENTS

We are grateful for the comments from three anonymous reviewers. This contributes to project 2.4580.11 (1882638) of the Fonds National de la Recherce Scientifique (FNRS) and to the EMOSEM (Ecosystem Models as Support to Eutrophication Management in the North Atlantic Ocean) EU FP7 ERA-NET Seas-era project funded by the Belgian Science Policy (BELSPO). Alberto V. Borges is a senior research associate at the FNRS, and co-first author.

REFERENCES

Archer, S. D., Kimmance, S. A., Stephens, J. A., Hopkins, F. E., Bellerby, R. G. J., Schulz, K. G., et al. (2013). Contrasting responses of DMS and DMSP to ocean acidification in Arctic waters. *Biogeosciences* 10, 1893–1908. doi: 10.5194/bg-10-1893-2013

Arnold, H. E., Kerrison, P., and Steinke, M. (2013). Interacting effects of ocean acidification and warming on growth and DMS-production in the haptophyte coccolithophore Emiliania huxleyi. *Glob. Chang. Biol.* 19, 1007–1016. doi: 10.1111/gcb.12105

Avgoustidi, V., Nightingale, P. D., Joint, I., Steinke, M., Turner, S. M., Hopkins, F. E., et al. (2012). Decreased marine dimethyl sulphide production under elevated CO_2. *Environ. Chem.* 9, 399–404. doi: 10.1071/EN11125

Bell, T. G., Poulton, A. J., and Malin, G. (2010). Strong linkages between dimethyl-sulphoniopropionate (DMSP) and phytoplankton community physiology in a large subtropical and tropical Atlantic Ocean data set. *Global Biogeochem. Cy.* 24, GB3009. doi: 10.1029/2009GB003617

Billen, G., Garnier, J., Ficht, A., and Cun, C. (2001). Modeling the response of water quality in the Seine river estuary to human activity in its watershed over the last 50 years. *Estuaries* 24, 977–993. doi: 10.2307/1353011

Billen, G., Garnier, J., and Rousseau, V. (2005). Nutrient fluxes and water quality in the drainage network of the Scheldt basin over the last 50 years. *Hydrobiologia* 540, 47–67. doi: 10.1007/s10750-004-7103-1

Blanchard, J. L., Jennings, S., Holmes, R., Harle, J., Merino, G., Allen, J. I., et al. (2012). Potential consequences of climate change for primary production and fish production in large marine ecosystems. *Philos. Trans. R. Soc. Lond. B Biol. Sci.* 367, 2979–2989. doi: 10.1098/rstb.2012.0231

Bopp, L., Aumont, O., Belviso, S., and Monfray, P. (2003). Potential impact of climate change on marine dimethyl sulfide emissions. *Tellus* 55B, 11–22. doi: 10.1034/j.1600-0889.2003.042.x

Borges, A. V., and Gypens, N. (2010). Carbonate chemistry in the coastal zone responds more strongly to eutrophication than to ocean acidification. *Limnol. Oceanogr.* 55, 346–353. doi: 10.4319/lo.2010.55.1.0346

Bucciarelli, E., and Sunda, W. (2003). Influence of CO_2, nitrate, phosphate, and silicate limitation on intracellular dimethylsulfoniopropionate in batch cultures of the coastal diatom Thalassiosira pseudonana. *Limnol. Oceanogr.* 48, 2256–2265. doi: 10.4319/lo.2003.48.6.2256

Cameron-Smith, P., Elliott, S., Maltrud, M., Erickson, D., and Wingenter, O. (2011). Changes in dimethyl sulfide oceanic distribution due to climate change. *Geophys. Res. Lett.* 38, L07704. doi: 10.1029/2011GL047069

Carslaw, K. S., Boucher, O., Spracklen, D. V., Mann, G. W., Rae, J. G. L., Woodward, S., et al. (2010). A review of natural aerosol interactions and feedbacks within the Earth system. *Atmos. Chem. Phys.* 10, 1701–1737 doi: 10.5194/acp-10-1701-2010

Charlson, R. J., Lovelock, J. E., Andreae, M. O., and Warren, S. G. (1987). Oceanic phytoplankton, atmospheric sulphur, cloud albedo and climate. *Nature* 326, 655–661. doi: 10.1038/326655a0

Cloern, J. E. (2001). Our evolving conceptual model of the coastal eutrophication problem. *Mar. Ecol. Prog. Ser.* 210, 223–253. doi: 10.3354/meps210223

Doney, S. C., Fabry, V. J., Feely, R. A., and Kleypas, J. A. (2009). Ocean acidification: the other CO_2 problem. *Annu. Rev. Mar. Sci.* 1, 169–192. doi: 10.1146/annurev.marine.010908.163834

Flynn, K. J., Blackford, J. C., Baird, M. E., Raven, J. A., Clark, D. R., Beardall, J., et al. (2012). Changes in pH at the exterior surface of plankton with ocean acidification. *Nat. Clim. Change* 2, 510–513. doi: 10.1038/nclimate1696

Gabric, A. J., Simó, R., Cropp, R. A., Hirst, A. C., and Dachs, J. (2004). Modeling estimates of the global emission of dimethylsulfide under enhanced greenhouse conditions. *Global Biogeochem. Cy.* 18, GB2014. doi: 10.1029/2003GB002183

Galloway, J. N., Dentener, F. J., Capone, D. G., Boyer, E. W., Howarth, R. W., Seitzinger, S. P., et al. (2004). Nitrogen cycles: past, present, and future. *Biogeochemistry* 70, 153–226. doi: 10.1007/s10533-004-0370-0

Garnier, J., Billen, G., Hannon, E., Fonbonne, S., Videnina, Y., and Soulie, M. (2002). Modeling transfer and retention of nutrients in the drainage network of the Danube River. *Estuar. Coast. Shelf Sci.* 54, 285–308. doi: 10.1006/ecss.2000.0648

Gypens, N., Borges, A. V., and Lancelot, C. (2009). Effect of eutrophication on air-sea CO_2 fluxes in the coastal Southern North Sea: a model study of the past 50 years. *Glob. Change Biol.* 15, 1040–1056. doi: 10.1111/j.1365-2486.2008.01773.x

Gypens, N., Borges, A. V., Speeckaert, G., and Lancelot, C. (2014). The dimethyl-sulfide cycle in the eutrophied Southern North Sea: a model study integrating phytoplankton and bacterial processes. *PLoS ONE* 9:e85862. doi: 10.1371/journal.pone.0085862

Gypens, N., Lacroix, G., Lancelot, C., and Borges, A. V. (2011). Seasonal and inter-annual variability of air-sea CO2 fluxes and seawater carbonate chemistry in the Southern North Sea. *Prog. Oceanogr.* 88, 59–77. doi: 10.1016/j.pocean.2010.11.004

Gypens, N., Lancelot, C., and Borges, A. V. (2004). Carbon dynamics and CO_2 air-sea exchanges in the eutrophied coastal waters of the Southern Bight of the North Sea: a modelling study. *Biogeosciences* 1, 147–157. doi: 10.5194/bg-1-147-2004

Harada, H., Vila-Costa, M., Cebrian, J., and Kiene, R. P. (2009). Effects of UV radiation and nitrate limitation on the production of biogenic sulfur compounds by marine phytoplankton. *Aquat. Bot.* 90, 37–42. doi: 10.1016/j.aquabot.2008.05.004

Hopkins, F. E., and Archer, S. D. (2014). Consistent increase in dimethyl sulphide (DMS) in response to high CO_2 in five shipboard bioassays from contrasting NW European waters. *Biogeosci. Discuss* 11, 2267–2303. doi: 10.5194/bgd-11-2267-2014

Hopkins, F. E., Turner, S. M., Nightingale, P. D., Steinke, M., Bakker, D., and Liss, P. S. (2010). Ocean acidification and marine trace gas emissions. *Proc. Natl. Acad. Sci. U.S.A.* 107, 760–765. doi: 10.1073/pnas.0907163107

Kameyama, S., Tanimoto, H., Inomata, S., Yoshikawa-Inoue, H., Tsunogai, U., Tsuda, A., et al. (2013). Strong relationship between dimethyl sulfide and net community production in the western subarctic Pacific. *Geophys. Res. Lett.* 40, 3986–3990. doi: 10.1002/grl.50654

Keller, M. D., Bellows, W. K. K., and Guillard, R. R. (1989). "Dimethyl sulfide production in marine phytoplankton," in *Biogenic Sulfur in the Environment*, eds E. S. Saltzman and W. J. Cooper (Washington, DC: American Chemical Society), 167–182. doi: 10.1021/bk-1989-0393.ch011

Kim, J.-M., Lee, K., Yang, E. J., Shin, K., Noh, J. H., Park, K.-T., et al. (2010). Enhanced production of oceanic dimethylsulfide resulting from CO_2-induced grazing activity in a high CO_2 world. *Environ. Sci. Technol.* 44, 8140–8143. doi: 10.1021/es102028k

Kloster, S., Six, K. D., Feichter, J., Maier–Reimer, E., Roeckner, E., Wetzel, P., et al. (2007). Response of dimethylsulfide (DMS) in the ocean and atmosphere to global warming. *J. Geophys. Res.* 112, G03005. doi: 10.1029/2006JG000224

Lana, A., Bell, T. G., Simó, R., Vallina, S. M., Ballabrera-Poy, J., Kettle, A. J., et al. (2011). An updated climatology of surface dimethlysulfide concentrations and emission fluxes in the global ocean. *Global Biogeochem. Cy.* 25, 1–17. doi: 10.1029/2010GB003850

Lancelot, C., Gypens, N., Billen, G., Garnier, J., and Roubeix, V. (2007). Testing an integrated river–ocean mathematical tool for linking marine eutrophication to land use: the Phaeocystis-dominated Belgian coastal zone (Southern North Sea) over the past 50 years. *J. Mar. Syst.* 64, 216–228. doi: 10.1016/j.jmarsys.2006.03.010

Lancelot, C., Spitz, Y., Gypens, N., Ruddick, K., Becquevort, S., Rousseau, V., et al. (2005). Modelling diatom and Phaeocystis blooms and nutrient cycles in the Southern Bight of the North Sea: the MIRO model. *Mar. Ecol. Prog. Ser.* 289, 63–78. doi: 10.3354/meps289063

Lee, P. A., Rudisill, J. R., Neeley, A. R., Maucher, J. M., Hutchins, D. A., Feng, Y., et al. (2009). Effects of increased pCO2 and temperature on the North Atlantic Spring Bloom. III. Dimethylsulfoniopropionate. *Mar. Ecol. Prog. Ser.* 388, 41–49. doi: 10.3354/meps08135

Mackenzie, F. T., De Carlo, E. H., and Lerman, A. (2011). "Coupled C, N, P, and O biogeochemical cycling at the Land–Ocean interface," in *Treatise on Estuarine and Coastal Science, Vol. 5*, eds E. Wolanski and D. S. McLusky (Waltham: Academic Press), 317–342.

Meybeck, M. (1982). Carbon, nitrogen and phosphorus transport by world rivers. *Am. J. Sci.* 282, 401–450. doi: 10.2475/ajs.282.4.401

Miles, C. J., Bell, T. G., and Suntharalingam, P. (2012). Investigating the inter-relationships between water attenuated irradiance, primary production and DMS(P). *Biogeochemistry* 110, 201–213. doi: 10.1007/s10533-011-9697-5

Passy, P., Gypens, N., Billen, G., Garnier, J., Thieu, V., Rousseau, V., et al. (2013). A model reconstruction of riverine nutrient fluxes and eutrophication in the Belgian Coastal Zone since 1984. *J. Mar. Syst.* 128, 106–122. doi: 10.1016/j.jmarsys.2013.05.005

Quinn, P. K., and Bates, T. S. (2011). The case against climate regulation via oceanic phytoplankton sulphur emissions. *Nature* 480, 51–56. doi: 10.1038/nature10580

Redfield, A. C., Ketchum, B. A., and Richards, F. A. (1963). "The influence of organisms on the composition of sea-water," in *The Sea*, ed M. N. Hill (New York, NY: Wiley), 26–77.

Richier, S., Achterberg, E. P., Dumousseaud, C., Poulton, A. J., Suggett, D. J., Tyrrell, T., et al. (2014). Carbon cycling and phytoplankton responses within highly-replicated shipboard carbonate chemistry manipulation experiments conducted around Northwest European Shelf Seas. *Biogeosci. Discuss* 11, 3489–3534. doi: 10.5194/bgd-11-3489-2014

Seitzinger, S. P., Mayorga, E., Bouwman, A. F., Kroeze, C., Beusen, A. H. W., Billen, G., et al. (2010). Global river nutrient export: a scenario analysis of past and future trends. *Global Biogeochem. Cy.* 24, GB0A08. doi: 10.1029/2009GB003587

Six, K. D., Kloster, S., Ilyina, T., Archer, S. D., Zhang, K., and Maier-Reimer, E.(2013). Global warming amplified by reduced sulphur fluxes as a result of ocean acidification. *Nat. Clim. Change* 3, 975–978. doi: 10.1038/nclimate1981

Spielmeyer, A., and Pohnert, G. (2012). Influence of temperature and elevated carbon dioxide on the production of dimethylsulfoniopropionate and glycine betaine by marine phytoplankton. *Mar. Environ. Res.* 73, 62–69. doi: 10.1016/j.marenvres.2011.11.002

Stefels, J., Steinke, M., Turner, S. M., Malin, G., and Belviso, S. (2007). Environmental constraints on the production and removal of the climatically active gas dimethylsulphide (DMS) and implications for ecosystem modelling. *Biogeochemistry* 83, 245–275. doi: 10.1007/s10533-007-9091-5

Sunda, W., Hardison, R., Kiene, R. P., Bucciarelli, E., and Harada, H. (2007). The effect of nitrogen limitation on cellular DMSP and DMS release in marine phytoplankton: climate feedback implications. *Aquat. Sci.* 69, 341–351. doi: 10.1007/s00027-007-0887-0

Sunda, W., Kieber, D. J., Kiene, R. P., and Huntsman, S. (2002). An antioxidant function for DMSP and DMS in marine algae. *Nature* 418, 317–320. doi: 10.1038/nature00851

Uher, G. (2006) Distribution and air-sea exchange of reduced sulphur gases in European coastal waters. *Estuar. Coast. Shelf Sci.* 70, 338–360. doi: 10.1016/j.ecss.2006.05.050

Vallina, S. M., Simó, R., and Manizza, M. (2007). Weak response of oceanic dimethylsulfide to upper mixing shoaling induced by global warming. *Proc. Natl. Acad. Sci. U.S.A.* 104, 16004–16009. doi: 10.1073/pnas.0700843104

Vogt, M., Steinke, M., Turner, S., Paulino, A., Meyerhöfer, M., Riebesell, U., et al. (2008). Dynamics of dimethylsulphoniopropionate and dimethylsulphide under different CO_2 concentrations during a mesocosm experiment. *Biogeosciences* 5, 407–419. doi: 10.5194/bg-5-407-2008

Wollast, R. (1998). "Evaluation and comparison of the global carbon cycle in the coastal zone and in the open ocean," in *The Global Coastal Ocean*, eds K. H. Brink and A. R. Robinson (New York, NY: John Wiley & Sons), 213–252.

Conflict of Interest Statement: The authors declare that the research was conducted in the absence of any commercial or financial relationships that could be construed as a potential conflict of interest.

2

Mapping ecosystem services provided by benthic habitats in the European North Atlantic Ocean

Ibon Galparsoro *, Angel Borja and María C. Uyarra

Marine Research Division, AZTI-Tecnalia, Pasaia, Spain

Edited by:
Michael Arthur St. John, Danish Technical University, Denmark

Reviewed by:
Steve Whalan, Southern Cross University, Australia
Jose M. Riascos, Universidad de Antofagasta, Chile
Maria Salomidi, HCMR, Greece

*Correspondence:
Ibon Galparsoro, Marine Research Division, AZTI-Tecnalia, Herrera kaia portualdea z/g, Pasaia 20110, Spain
e-mail: igalparsoro@azti.es

The mapping and assessment of the ecosystem services provided by benthic habitats is a highly valuable source of information for understanding their current and potential benefits to society. The main objective of this research is to assess and map the ecosystem services provided by benthic habitats in the European North Atlantic Ocean, in the context of the "Mapping and Assessment of Ecosystems and their Services" (MAES) programme, the European Biodiversity Strategy and the implementation of the Marine Strategy Framework Directive (MSFD). In total, 62 habitats have been analyzed in relation to 12 ecosystem services over 1.7 million km^2. Results indicated that more than 90% of the mapped area provides biodiversity maintenance and food provision services; meanwhile, grounds providing reproduction and nursery services are limited to half of the mapped area. Benthic habitats generally provide more services closer to shore—rather than offshore—and in shallower waters. This gradient is likely to be explained by difficult access (i.e., distance and depth) and lack of scientific knowledge for most of the services provided by distant benthic habitats. This research has provided a first assessment of the benthic ecosystem services on the Atlantic-European scale, with the provision of ecosystem services maps and their general spatial distribution patterns. Regarding the objectives of this research, conclusions are: (i) benthic habitats provide a diverse set of ecosystem services, being the food provision, with biodiversity maintenance services more extensively represented. In addition, other regulating and cultural services are provided in a more limited area; and (ii) the ecosystem services assessment categories are significantly related to the distance to the coast and to depth (higher near the coast and in shallow waters).

Keywords: ecosystem service, benthic habitat, Regional Seas, Marine Strategy Framework Directive, habitat classification

INTRODUCTION

Functioning ecosystems are essential for maintaining the oceans in a healthy state (Tett et al., 2013). While being healthy, they provide numerous and diverse goods and services that contribute "for free" to the general well-being and health of humans (Van Den Belt and Costanza, 2012). The "ecosystem goods and services" term integrates two concepts: (i) the ecosystem goods, which represent marketable material products that are obtained from natural systems for human use, such as food and raw materials (De Groot et al., 2002); and (ii) ecosystem services, which refers to all "the conditions and processes through which natural ecosystems, and the species that make them up, sustain and fulfill human life" (Daily, 1997). The latter are not directly marketable services, and include nutrient recycling, biodiversity maintenance, climate regulation or cultural and esthetic services (Costanza et al., 1997). Ecosystem services occur at multiple spatial scales; from the global, such as climate regulation, primary production, and carbon sequestration, to a more regional or local scale, such as coastal protection and leisure.

Previous studies show that coastal ecosystem services provide an important portion of the total contribution of ecosystem services to human welfare (Pimm, 1997; Pearce, 1998). Costanza et al. (1997) showed that, while the coastal zone only covers 8% of the world's surface, the services that this zone provides are responsible for approximately 43% of the estimated total value of global ecosystem services. Despite our dependence on biodiversity and ecosystem services, population expansion and economic growth are leading to increasing anthropogenic pressures on coastal areas (Wilson et al., 2013) and consequently, to a decreasing supply of ecosystem services worldwide (Costanza et al., 2014). Recognizing that human pressures directly impact on ecosystem services and that in turn, ecosystem services directly benefit human well-being, they have sparked interest amongst coastal planners and have led to the integration of ecosystem services in conservation management measures (Cimon-Morin et al., 2013).

Due to the above-mentioned reasons, ecologists, social scientists, economists and environmental managers are increasingly interested in assessing the economic values associated with the ecosystem services of coastal and marine ecosystems (Bingham et al., 1995; Costanza et al., 1997; Daily, 1997; Farber et al., 2002; Liquete et al., 2013a). Different approaches and frameworks have been proposed to identify, define, classify and quantify services provided by marine biodiversity (MEA, 2003; Ten Brink et al., 2009; Cices, 2013; Liquete et al., 2013a). Neither of these approaches being a straight forward one; the accurate estimation of the values of services, and in particular their temporal and spatial variation, is relatively new and has not been extensively researched (Schägner et al., 2013).

Indeed, the complexity of the processes and functioning of marine ecosystems, and their highly dynamic nature, translates into the absence or low resolution of spatially explicit information. Furthermore, the deep sea, and in particular benthic habitats, is mostly lacking in ecosystem services assessments (Armstrong et al., 2012; Thurber et al., 2013). Due to these limiting factors, there are few published studies, and they mainly focus on food production, such as fisheries, with other services receiving minor attention (Murillas-Maza et al., 2011; Liquete et al., 2013a; Seitz et al., 2014). Mapping and assessing ecosystem services may help to overcome such hindrances. Maps not only enable the characterization of current benefits that services provide to society, but also the adoption management measures that guarantee their future provision and contribution to human welfare (Egoh et al., 2012).

To date, several habitat mapping efforts have been carried out at different spatial and temporal resolutions (Liquete et al., 2013a). Within Europe, Mapping and Assessment of Ecosystems and their Services (MAES) is one of the keystones of the EU Biodiversity Strategy to 2020 (Maes et al., 2013). This strategy demands Member States to map and assess the state of ecosystems and their services in their national territory (including their marine waters) with the assistance of the European Commission. The results of this mapping and assessment should support the maintenance and restoration of ecosystems and the services they provide (Maes et al., 2013). It will also contribute to the assessment of the economic value of ecosystem services, and promote the integration of these values into accounting and reporting systems at EU and national level by 2020. The results are expected to be used to inform policy decision makers and policy implementation in many fields, such as nature and biodiversity, territorial cohesion, agriculture, forestry, and fisheries. Outputs can also inform policy development and implementation in other domains, such as transport and energy (Maes et al., 2013). For example, the Marine Strategy Framework Directive (MSFD, 2008/56/EC) requires the availability of ecosystem services valuation for the assessment of the environmental status and to define the measures that make sustainable human activities at sea (Cardoso et al., 2010). Hence, according to the MSFD, the assessment of the environmental status should be undertaken for the Exclusive Economic Zone (EEZ) of the Member States within the four European Regional Seas: North Eastern Atlantic, Baltic, Mediterranean, and Black Seas.

In this context, the objectives of this research were: (i) the qualitative assessment and mapping of the ecosystem services provided by benthic habitats within the European North Atlantic Ocean; and (ii) to determine if ecosystem services assessment categories are related to the habitat distance to the coast and depth. The analysis was based on available cartographic information and ecosystem services assessment, focusing on the benefits that

they provide in the Regional Seas and sub-regions defined by the MSFD.

MATERIALS AND METHODS

The implementation of ecosystem services valuation involves two dimensions: (i) a biophysical assessment of services supply; and (ii) a socio-economic assessment of the value per unit of services (Schägner et al., 2013). Within this investigation, we focused only on the first approach of trying to map and assess the ecosystem services provided by benthic habitats at the European North Atlantic Ocean scale. This is because the economic value of the services is still poorly known, needing comprehensive data supply, which the results from this investigation can provide.

GEOGRAPHIC AREA

For this investigation, the North Eastern Atlantic was selected. According to MSFD, the North Eastern Atlantic Ocean is divided into four sub-regions: Greater North Sea, Celtic Seas, Bay of Biscay and Iberian coasts, and Macaronesia (**Figure 1**). It should be noted that at the time of this investigation, no official geographical delimitations of the sub-regions were adopted, and therefore, they were defined according to the EEZs. The total area of the European North Atlantic Ocean covered by the MSFD is 4,540,025 km^2, which corresponds to the EEZ of 10 European Member States and part of Norway (**Figure 1**).

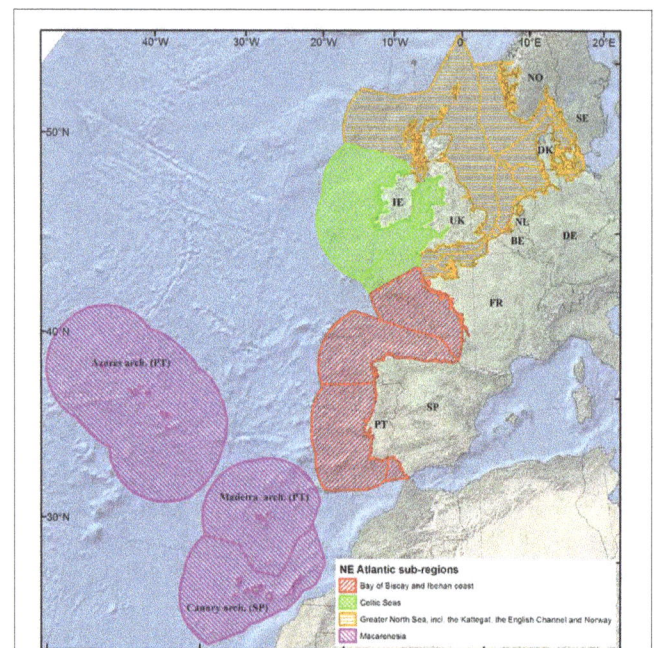

FIGURE 1 | European North Atlantic Ocean sub-regions. Spatial limits are based on the Marine Strategy Framework Directive and Exclusive Economic Zone of the countries located in each sub-region. BE, Belgium; DK, Denmark; FR, France; DE, Germany; IE, Ireland; NL, Netherlands; NO, Norway; PT, Portugal (including Azores archipelago and Madeira archipelago); SP, Spain (including Canary archipelago); SE, Sweden; and UK, United Kingdom.

BACKGROUND INFORMATION USED IN THE ANALYSIS

In order to proceed with the mapping of ecosystem services, main bathymetric and habitat data were obtained from the following sources:

- EMODnet—European Marine Observation and Data Network [http://www.emodnet-hydrography.eu/; European Commission; Directorate-General for Maritime Affairs and Fisheries (DG MARE)]. EMODnet-Hydrography portal provides hydrographic data collated for a number of sea regions in Europe. Bathymetric information was available as Digital Elevation Model at 500 m (c.a. 0.0042°) grid resolution.
- EUSeaMap—Mapping European seabed habitats (http://jncc.defra.gov.uk/page-6266). EUSeaMap is a broad-scale modeled habitat map built in the framework of MESH (Mapping European Seabed Habitats) and BALANCE (Baltic Sea Management—Nature Conservation and Sustainable Development of the Ecosystem through Spatial Planning) INTERREG IIIB-funded projects. EUSeaMap covers over 2 million km^2 of European seabed (Cameron and Askew, 2011). This information layer was available in polygon format.
- MeshAtlantic project (www.meshatlantic.eu; Atlantic Area Transnational Cooperation Programme 2007–2013 of the European Regional Development Fund). It covers over 356,000 km^2 of seabed habitats of the European North Atlantic Ocean produced 250 m (c.a. 0.0027°) grid resolution. This information layer was available in polygon format (Vasquez et al., in press).

DIGITAL ELEVATION MODEL

To produce the digital elevation model information layer, bathymetric information from MeshAtlantic and EMODnet was mosaicked. The information on this layer enabled the investigation of the depth distribution of benthic habitats in the sub-regions of the mapped areas.

BENTHIC HABITATS INFORMATION

For practical purposes of mapping and assessment (i.e., data availability) this investigation focused on "benthic habitats," as a means to assess the provision of ecosystem goods and services.

Habitats were classified according to EUNIS (European Union Nature Information System) habitat classes (Davies et al., 2004). The EUNIS habitat classification aims to provide a common European reference set of habitat types to allow the reporting of habitat data in a comparable manner for use in nature conservation (e.g., inventories, monitoring, and assessments) (Davies and Moss, 2002; Davies et al., 2004; Galparsoro et al., 2012). The classification is organized into hierarchical levels (EUNIS habitat type hierarchical view is available at http://eunis.eea.europa.eu/habitats-code-browser.jsp). The present version of the classification starts at level 1, where "Marine habitats" are defined, up to level 6, by using different abiotic and biological criteria at each level of the classification. For seabed habitats for which EUNIS classes were not defined, underwater features defined under EUSeaMap (e.g., infralittoral seabed) were retained.

Habitat maps were transformed into raster format and mosaicked to obtain a total broad-scale habitat map. In overlapping cells, MeshAtlantic habitat classes were kept, according to the criteria that this represents the most recent information. The mapped area outside EEZ of Ireland was excluded from the later analysis, in order to make results comparable among different countries, in which only EEZ areas were included.

Finally, to analyse the spatial distribution of benthic habitats (in terms of their distance to shore) and therefore, that of the ecosystem services that they provide, the distance of each cell, assigned to each habitat type, to the nearest coastline point was estimated using Euclidean distance algorithm, in a Geographic Information System (GIS).

ECOSYSTEM SERVICES ASSESSMENT

In total, twelve ecosystem services were considered in this investigation: (i) Food provision; (ii) Raw materials (biological) (incl. biochemical, medicinal, and ornamental); (iii) Air quality and climate regulation; (iv) Disturbance and natural hazard prevention; (v) Photosynthesis, chemosynthesis, and primary production; (vi) Nutrient cycling; (vii) Reproduction and nursery; (viii) Maintenance of biodiversity; (ix) Water quality regulation and bioremediation of waste; (x) Cognitive value; (xi) Leisure, recreation and cultural inspiration; and (xii) Feel good or warm glow.

Ecosystem services were classified into: (i) Provisioning services (i.e., 1 and 2 from the above list); (ii) Regulating services (i.e., 3–9); and (iii) Cultural services (i.e., 10–12). The qualitative ecosystem services categories offered by each habitat were based on Table 1 from Salomidi et al. (2012), which, in turn, classified them based on an adaptation of the categories proposed by the Millennium Ecosystem Assessment (MEA, 2003) and Beaumont et al. (2007). Rather than using absolute metrics to classify services of each habitat, the assessment was based on the expert judgment of Salomidi et al. (2012), collated in the aforementioned **Table 1** of that manuscript, and the following guidelines: (i) when the provision of a specific service is well documented in the scientific literature and is widely accepted as important for the specific benthic habitat analyzed, it was considered as providing a "High" value for such ecosystem service (e.g., the role of seagrass beds in sediment retention and prevention of coastal erosion); (ii) when a service was or could be provided by a habitat but to a substantially lower magnitude than by other habitats and without being vital for the persistence of an important human activity, a "Low" value was assigned; and (iii) in all other cases, ecosystem services were classified as "Negligible/Irrelevant/Unknown." For the purpose of the present investigation, ecosystem services categories were rated into the following numerical values for further analysis: "High = 3," "Low = 1," "Negligible/Irrelevant/Unknown = 0." A similar classification and scores were successfully used in smaller areas (Potts et al., 2014) (see **Figures 3, 4** in that manuscript).

The ecosystem services provisioning categories of each habitat type, was linked to the final habitat map. For those habitat classes that were included in the map, but not listed in Salomidi et al. (2012), the categories were assigned according to the knowledge of the authors, in a similar way to that of Potts et al. (2014).

To analyse the spatial distribution pattern of ecosystem services provisioning levels, the total area and its percentage cover of the total mapped area, mean depth, and mean distance to the coastline were calculated. The values of all cells encompassed within a polygon representing the extent of a habitat, were averaged to assign a unique value to each polygon for each variable (i.e., mean depth value within a polygon) To assess whether the distance to the coastline and depth had an effect on the categories at which the different ecosystem services are provided (i.e., high, low, and negligible values), Kruskal-Wallis non-parametric tests were applied using Statgraphics v.5.0. Then, differences in ecosystem services categories within the subregions were tested using Chi-Square tests. Finally, Friedman test, followed by *post-hoc* Wilcoxon tests, was undertaken to explore statistical differences between ecosystem services typologies (i.e., provision, regulation, and cultural).

RESULTS

The European North Atlantic Ocean (EEZ only) covers more than 4.5 million km^2 (**Table 1**), of which 26% corresponds to continental shelf (up to 200 m depth) and 74% to deeper areas (**Figure 2**). To date, 88% of the continental shelf and 18% of the deeper areas have been mapped, accounting for 38.9 % of the total EEZ area of the European North Atlantic Ocean.

The Macaronesia accounts for the highest proportion of the European North Atlantic EEZ, followed by the Extended North Sea (**Table 1**). However, differences in the amount of mapped area can be found among sub-regions. Whereas countries located in the Celtic Sea and North Sea have already mapped almost all their EEZ seabed surface (i.e., 98 and 93%, respectively), countries located in Macaronesia, Bay of Biscay, and Iberian coasts (i.e., France, Portugal, and Spain) have still more than 80% of the seabed area without cartographic information (**Table 1** and

Table 1 | Total spatial contribution of each sub-region to the Exclusive Economic Zone (EEZ) of the European North Atlantic Ocean, and their mapped area, represented in total and relative (%) terms.

Subregion	EEZ of the European North Atlantic Ocean		Mapped area of the EEZ of the European Atlantic Ocean	
	Total area (km^2)	Total area (%)	Total mapped area (km^2)	Total mapped area (%)
Macaronesia	2,119,095	47	88,150	4
Bay of Biscay and Iberian peninsula	818,491	18	154,472	19
Celtic Sea	550,606	12	541,042	98
Extended North Sea	1,051,611	23	981,633	93
TOTAL	4,539,803	100	1,765,297	39

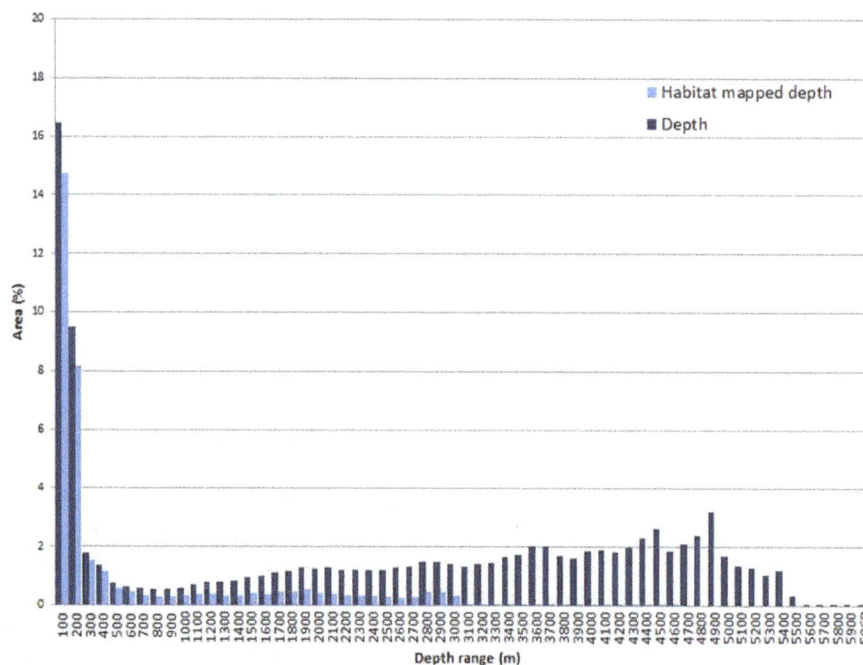

FIGURE 2 | Depth distribution of the Exclusive Economic Zone of the European North Atlantic Ocean (dark blue) and depth distribution of habitat-mapped areas (light blue).

Figure S1). Indeed, habitat maps for the Canary and Madeira Archipelagos, in Macaronesia, are not available. It should be highlighted that these countries have some of the most extensive and deepest EEZs areas of the European North Atlantic Ocean.

The 1.7 million km^2 covered by the integrated broad-scale habitat map encompassed 62 different benthic habitats and seabed seascape features (**Figure 3**). The North Sea and the Celtic Sea encompassed 58 and 55 habitats respectively, while the Bay of Biscay and Macaronesia only covered 42 and 20 habitats, respectively. Furthermore, very few habitats accounted for a large section of the mapped area (**Figure 4**). Ten habitats covered more than 75% of the total mapped area, of which deep sea mud (18.3%), deep circalittoral sand (16.2%), circalittoral fine sands, or circalittoral muddy sand (9.7%) were the most dominant ones. Opposite, a large number of habitats (i.e., 33) covered less than 10,000 km^2 or 0.5% of the mapped seabed. The least dominant habitats in the European North Atlantic Ocean were the low energy infralittoral mixed hard sediments, Atlantic and Mediterranean low energy infralittoral rock and sponge communities on deep circalittoral rock, all of which cover less than 100 km^2.

Of the 62 habitats identified in European North Atlantic Ocean, none of them provides the 12 ecosystem services considered in this study at the highest value (**Table 2**). However, four of these habitats (i.e., Infralittoral rock and other hard substrata, Atlantic and Mediterranean high energy infralittoral rock, High energy infralittoral seabed, and High energy infralittoral mixed

hard sediments) provide high values for 11 services (excluding nutrient cycling). Another seven infralittoral habitats also provide high values for 10 of the services. On the other hand, 12 deep and bathyal habitats are considered as providing negligible values for 10 or more ecosystem services. The upper, mid, and lower bathyal seabed habitats provide the lowest number of ecosystem services and values.

Results also indicate that the highest provision of services is that of habitats located close to the coastline and in shallow waters ($p < 0.001$ for all services and in both cases—distance and depth; see **Tables 3, 4**). Thus, there is a gradient on the level of services provision, from high to lower or negligible values, seawards and toward deeper areas. For example, areas providing high food provision services are located close to the coast (16 ± 35 km) and in shallow areas (47 ± 50 m). Furthermore, it is also observed that the level of service provision significantly varies across subregions (Chi-Square test: p always < 0.001), with the North Sea being the region generally providing services at the highest levels.

Table 2 also suggests that none of the ecosystem services is provided by all the habitats. "Food," "biodiversity maintenance" and "nursery grounds" (i.e., "reproduction") are the ecosystem services most commonly provided by habitats (and to the highest level). Opposite, "photosynthesis," "disturbance prevention," "air quality" and "cultural services" are provided on a high level by a limited number of habitats. This pattern is also observed when considering not only the number of habitats providing specific ecosystem services, but also the area providing such

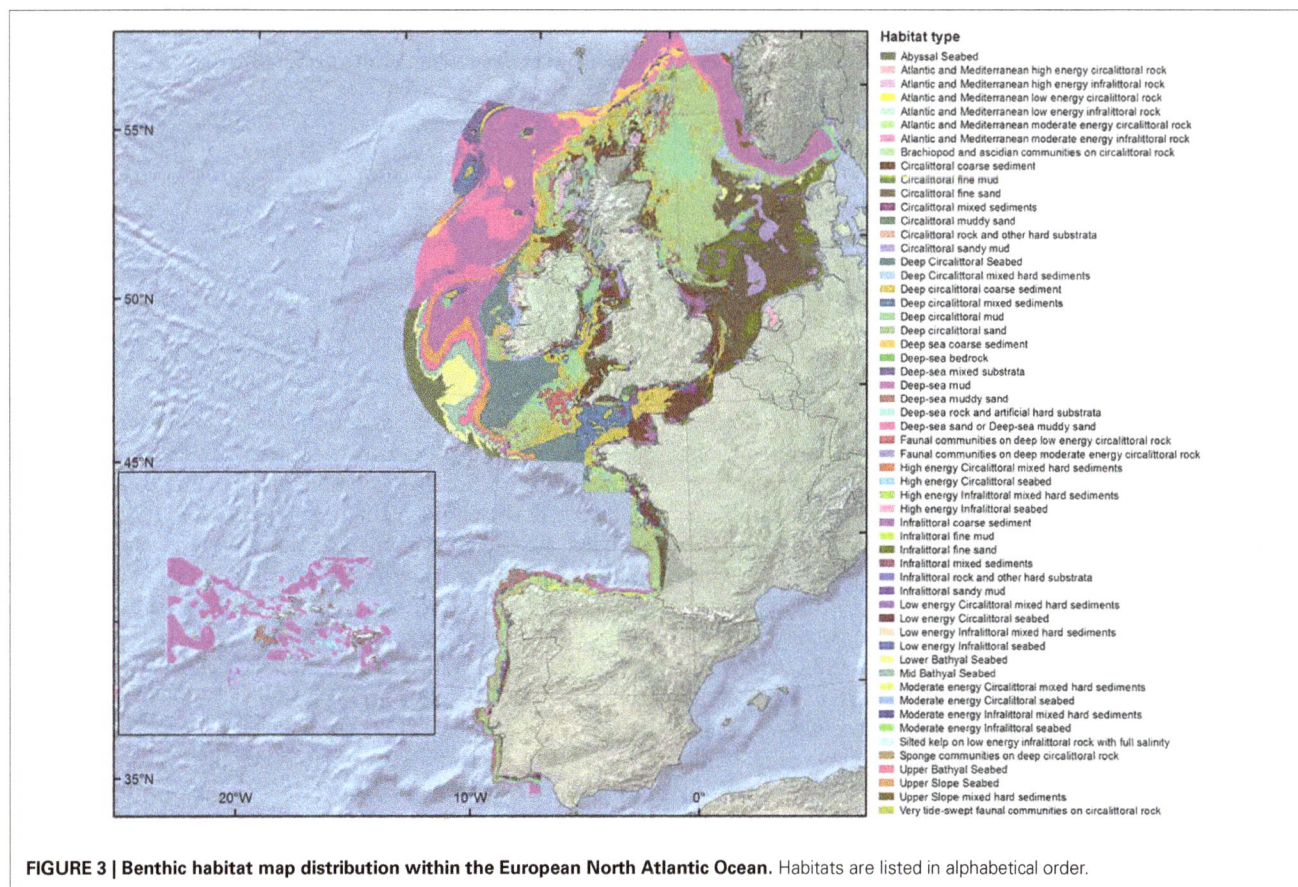

Habitat type
- Abyssal Seabed
- Atlantic and Mediterranean high energy circalittoral rock
- Atlantic and Mediterranean high energy infralittoral rock
- Atlantic and Mediterranean low energy circalittoral rock
- Atlantic and Mediterranean low energy infralittoral rock
- Atlantic and Mediterranean moderate energy circalittoral rock
- Atlantic and Mediterranean moderate energy infralittoral rock
- Brachiopod and ascidian communities on circalittoral rock
- Circalittoral coarse sediment
- Circalittoral fine mud
- Circalittoral fine sand
- Circalittoral mixed sediments
- Circalittoral muddy sand
- Circalittoral rock and other hard substrata
- Circalittoral sandy mud
- Deep Circalittoral Seabed
- Deep Circalittoral mixed hard sediments
- Deep circalittoral coarse sediment
- Deep circalittoral mixed sediments
- Deep circalittoral mud
- Deep circalittoral sand
- Deep sea coarse sediment
- Deep-sea bedrock
- Deep-sea mixed substrata
- Deep-sea mud
- Deep-sea muddy sand
- Deep-sea rock and artificial hard substrata
- Deep-sea sand or Deep-sea muddy sand
- Faunal communities on deep low energy circalittoral rock
- Faunal communities on deep moderate energy circalittoral rock
- High energy Circalittoral mixed hard sediments
- High energy Circalittoral seabed
- High energy Infralittoral mixed hard sediments
- High energy Infralittoral seabed
- Infralittoral coarse sediment
- Infralittoral fine mud
- Infralittoral fine sand
- Infralittoral mixed sediments
- Infralittoral rock and other hard substrata
- Infralittoral sandy mud
- Low energy Circalittoral mixed hard sediments
- Low energy Circalittoral seabed
- Low energy Infralittoral mixed hard sediments
- Low energy Infralittoral seabed
- Lower Bathyal Seabed
- Mid Bathyal Seabed
- Moderate energy Circalittoral mixed hard sediments
- Moderate energy Circalittoral seabed
- Moderate energy Infralittoral mixed hard sediments
- Moderate energy Infralittoral seabed
- Silted kelp on low energy infralittoral rock with full salinity
- Sponge communities on deep circalittoral rock
- Upper Bathyal Seabed
- Upper Slope Seabed
- Upper Slope mixed hard sediments
- Very tide-swept faunal communities on circalittoral rock

FIGURE 3 | Benthic habitat map distribution within the European North Atlantic Ocean. Habitats are listed in alphabetical order.

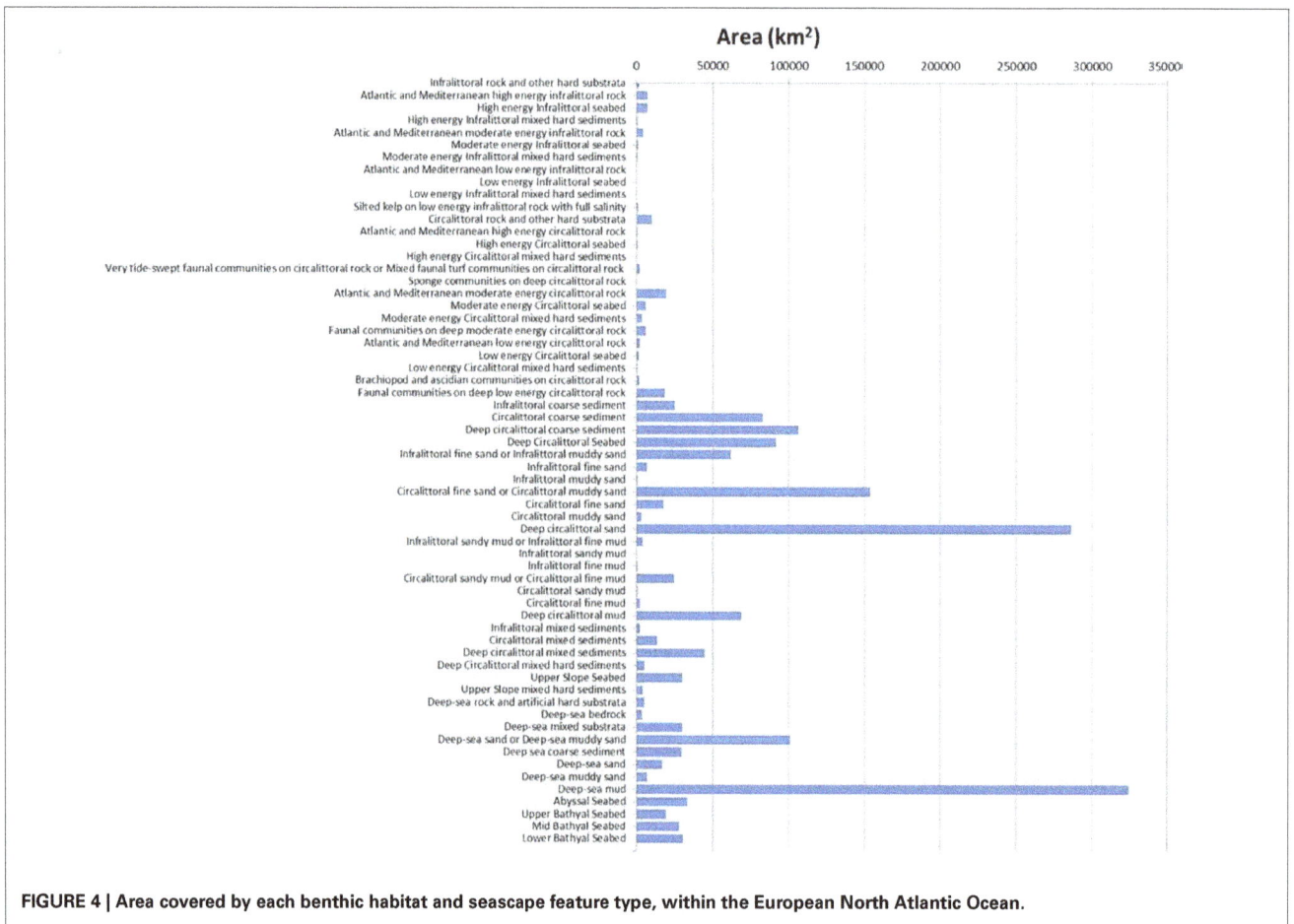

FIGURE 4 | Area covered by each benthic habitat and seascape feature type, within the European North Atlantic Ocean.

ecosystem services (**Table 3** and **Figures S2–S13**, in Supporting Information).

Indeed, 93% of the studied area provides food provision services, of which 62% corresponds with high food provision values. Similarly, a high proportion of the mapped area (99%) is considered as providing high (41%) and low (58%) biodiversity maintenance services.

The next ecosystem services, in terms of area coverage, are reproduction and nursery, which are provided by 53% of the mapped area. For the remaining ecosystem goods and services (i.e., air quality and climate regulation, water quality regulation and bioremediation, nutrient cycling, raw material provision, photosynthesis, chemosynthesis, and primary production), the area covered by habitats providing them at high values is much smaller. The disturbance and natural hazard prevention service has the smallest spatial coverage.

Finally, cultural services (i.e., cognitive value, leisure, recreation and cultural inspiration, and feel good and warm glow), showed similar patterns on their spatial distribution. The area covered by the habitats providing such type of services (both, at high and low levels) is very limited (around 11% of the total).

On the other hand, significant differences are observed in the spatial distribution of provision levels of aggregated ecosystem services (i.e., provisioning, regulating, and cultural), (Friedman test $\chi^2 = 47,858$; $p < 0.001$) (**Figure 5**). The provisioning services are supplied at significantly higher levels than both regulating (Wilcoxon *post-hoc* test $z = -154$, $p < 0.001$) and cultural services (Wilcoxon *post-hoc* test $z = -171$, $p < 0.001$); and in turn, regulating services are also provided at significantly higher levels than cultural services (Wilcoxon *post-hoc* test $z = -130$, $p < 0.001$).

DISCUSSION

Seafloor maps are an essential source of information for resource exploitation and management purposes (Rice, 2010). Nevertheless, in Europe, it is worth noting that countries such as Spain, Portugal and France, with large EEZ areas have less mapped areas. This is probably due to the steepness of the seafloor, with large bathyal and abyssal areas, and the technical and economic challenge associated with mapping areas with such characteristics. Among others, marine shallow water areas support most of the human activities associated with the use and benefit of the ecosystem services provided by benthic habitats (Ramirez-Llodra et al., 2011; Korpinen et al., 2013), but accurate estimation of the values of services and their spatial distribution is not available for extensive areas. Within this research, the assessment and mapping of the ecosystem services provided by benthic habitats of the European North Atlantic Ocean has been undertaken for the first time.

Table 2 | Ecosystem services assessment for each habitat and seabed feature type (H, high; L, low; and N, Negligible).

Habitat name	EUNIS code	Food	Raw material	Air quality	Disturbance	Photosynthesis	Nutrient	Reproduction	Biodiversity	Wate	Cognitive	Leisure	Feelgood
Infralittoral rock and other hard substrata	A3*	H	H	H	H	H	L	H	H	H	H	H	H
Atlantic and Mediterranean high energy infralittoral rock	A3.1*	H	H	H	H	H	L	H	H	H	H	H	H
High energy infralittoral seabed		H	H	H	H	H	L	H	H	H	H	H	H
High energy infralittoral mixed hard sediments		H	H	H	H	H	L	H	H	H	H	H	H
Atlantic and Mediterranean moderate energy infralittoral rock	A3.2*	H	H	H	L	H	H	H	H	H	H	H	L
Moderate energy infralittoral seabed		H	H	H	L	H	H	H	H	H	H	H	L
Moderate energy infralittoral mixed hard sediments		H	H	H	L	H	H	H	H	H	H	H	L
Atlantic and Mediterranean low energy infralittoral rock	A3.3*	H	H	H	L	H	H	H	H	H	H	H	L
Low energy infralittoral seabed		H	H	H	N	H	H	H	H	H	H	H	L
Low energy infralittoral mixed hard sediments		H	H	H	N	H	H	H	H	H	H	H	L
Silted kelp on low energy infralittoral rock with full salinity	A3.31	H	H	H	N	H	H	H	H	H	H	H	L
Circalittoral rock and other hard substrata	A4*	H	H	L	H	N	H	H	H	H	H	L	L
Atlantic and Mediterranean high energy circalittoral rock	A4.1*	H	H	L	H	N	H	H	H	H	H	L	L
High energy circalittoral seabed		H	H	L	H	N	H	H	H	H	H	L	L
High energy circalittoral mixed hard sediments		H	H	L	H	N	H	H	H	H	H	L	L
Very tide-swept faunal communities on circalittoral rock or mixed faunal turf communities on circalittoral rock	A4.11 or A4.13*	H	H	N	H	N	H	H	H	H	L	L	L
Sponge communities on deep circalittoral rock	A4.12	H	H	N	H	N	H	H	H	H	H	L	L
Atlantic and Mediterranean moderate energy circalittoral rock	A4.2*	L	L	L	N	N	H	H	H	H	H	L	L
Moderate energy circalittoral seabed		L	N	L	N	N	H	H	H	H	H	L	L
Moderate energy circalittoral mixed hard sediments		L	N	L	N	N	H	H	H	H	H	L	L
Faunal communities on deep moderate energy circalittoral rock	A4.27	L	L	L	N	L	H	H	H	H	H	L	L
Atlantic and Mediterranean low energy circalittoral rock	A4.3*	H	L	H	N	L	H	H	H	H	H	H	L
Low energy circalittoral seabed		H	L	L	N	N	H	H	H	H	H	H	L
Low energy circalittoral mixed hard sediments		H	L	L	N	N	H	H	H	H	H	H	L
Brachiopod and ascidian communities on circalittoral rock	A4.31	L	L	L	L	L	L	L	H	L	H	H	L
Faunal communities on deep low energy circalittoral rock	A4.33	H	L	H	N	L	H	H	H	H	H	H	H
Infralittoral coarse sediment	A5.13*	H	H	N	N	N	L	H	N	N	N	L	L
Circalittoral coarse sediment	A5.14*	H	H	N	N	N	L	L	L	N	N	N	N
Deep circalittoral coarse sediment	A5.15*	H	L	N	N	N	L	N	L	N	N	N	N
Deep circalittoral seabed		H	L	N	N	N	L	N	L	N	N	N	N
Infralittoral fine sand or infralittoral muddy sand	A5.23* or A5.24*	H	L	N	N	N	L	H	L	N	N	L	L
Infralittoral fine sand	A5.23*	H	L	N	N	N	L	H	L	N	N	L	L
Infralittoral muddy sand	A5.24*	H	L	N	N	N	L	H	L	N	N	L	L
Circalittoral fine sand or circalittoral muddy sand	A5.25* or A5.26*	H	L	N	N	N	L	H	L	N	N	N	N
Circalittoral fine sand	A5.25*	H	L	N	N	N	L	H	L	N	N	N	N
Circalittoral muddy sand	A5.26*	H	L	N	N	N	L	L	L	L	N	N	N
Deep circalittoral sand	A5.27	H	L	N	L	N	L	L	L	L	N	N	N

(Continued)

Table 2 | Continued

Habitat name	EUNIS code	Food	Raw material	Air quality	Disturbance	Photosynthesis	Nutrient	Reproduction	Biodiversity	Wate	Cognitive	Leisure	Feelgood
Infralittoral sandy mud or infralittoral fine mud	A5.33* or A5.34*	H	N	N	N	N	L	L	L	L	N	N	N
Infralittoral sandy mud	A5.33*	H	N	N	N	N	L	L	L	L	N	N	N
Infralittoral fine mud	A5.34*	L	N	N	N	N	L	N	L	L	N	N	N
Circalittoral sandy mud or circalittoral fine mud	A5.35* or A5.36*	H	N	N	N	N	L	L	L	L	N	N	N
Circalittoral sandy mud	A5.35*	H	N	N	N	N	L	L	L	L	N	N	N
Circalittoral fine mud	A5.36*	H	N	N	N	N	L	L	L	L	N	N	N
Deep circalittoral mud	A5.37*	H	N	N	N	N	L	L	L	L	N	N	N
Infralittoral mixed sediments	A5.43*	H	L	N	N	N	L	L	H	L	N	N	N
Circalittoral mixed sediments	A5.44*	H	L	N	N	N	L	L	H	L	N	N	N
Deep circalittoral mixed sediments	A5.45*	H	L	N	N	N	L	L	H	L	N	N	N
Deep circalittoral mixed hard sediments		H	N	N	N	N	N	H	H	N	N	N	N
Upper slope seabed		H	N	N	N	N	N	L	H	N	N	N	N
Upper slope mixed hard sediments		H	N	N	N	N	N	L	H	N	N	N	N
Deep-sea rock and artificial hard substrata	A6.1*	L	N	N	N	N	N	N	H	N	N	N	N
Deep-sea bedrock	A6.11	N	N	N	N	N	N	N	H	N	N	N	N
Deep-sea mixed substrata	A6.2	L	N	N	N	N	N	N	H	N	N	N	N
Deep-sea sand or deep-sea muddy sand	A6.3* or A6.4	L	N	N	N	N	N	N	H	N	N	N	N
Deep sea coarse sediment		L	N	N	N	N	N	N	H	N	N	N	N
Deep-sea sand	A6.3*	L	N	N	N	N	N	N	H	N	N	N	N
Deep-sea muddy sand	A6.4	L	N	N	N	N	N	N	H	N	N	N	N
Deep-sea mud	A6.5	L	N	N	N	N	N	N	H	N	N	N	N
Abyssal seabed		N	N	N	N	N	N	N	L	N	H	N	N
Upper bathyal seabed		N	N	N	N	N	N	N	L	N	L	N	N
Mid bathyal seabed		N	N	N	N	N	N	N	L	N	L	N	N
Lower bathyal seabed		N	N	N	N	N	N	N	L	N	L	N	N

*EUNIS habitat code is given for those habitats included in the classification; * indicates that the assessment was based upon Salomidi et al. (2012).*

In the studied area, a clear gradient has been identified for the provision of ecosystem services, with significantly higher provision levels for habitats located in shallow waters and close to the shore. This is coherent with the fact that habitats provide more ecosystem services as people have easier access to them. In fact, accessibility is a crucial factor and it is typically included in the monetization of some services, especially for cultural services (Milcu et al., 2013). In the case of benthic habitats, access depends on depth, and generally, on the distance from the coastline. Therefore, deep-sea habitats and habitats located further away from the coast generally provide fewer ecosystem services and at lower degree due to limited access and lack of scientific knowledge for most of them. However, as exploration of the deep-sea improves with recent technological advances, access to such habitats (Ramirez-Llodra et al., 2011) will become less difficult, increasing the ecosystem services that they provide in the near future (Thurber et al., 2013).

According to our estimations, between 93 and 99% (depending on the sub-regions) of the benthic habitats of the European North Atlantic Ocean deliver food provision and biodiversity maintenance services; meanwhile, reproduction and nursery services are provided by 53% of the area. We consider that the assessment of this last service could be underestimated due the fact that knowledge on life-cycles is mainly limited to commercially important species. But it should be taken into account that other non-commercial species, with unknown life cycles, also play an important role in food webs. Thus, the reproduction and nursery grounds are likely to cover a wider area than the one resulting from this investigation. In contrast, areas providing other services are smaller or have much more limited spatial distribution. For example, the area corresponding to habitats that supply raw materials is very limited, and the highest proportion of this area only provides low or negligible resources. To explain this pattern, it should be considered that few raw materials are exploited at present, and that their exploitation is regulated by national and international regulations as the impacts associated with such exploitation may be high. However, there may be high potential for habitats to provide higher provision of this service as new raw materials are discovered and exploited (i.e., pharmaceutical).

Another interesting pattern is that observed for the provision of coastal protection as an ecosystem service. Liquete et al. (2013b) propose the use of 14 biophysical and socio-economic

Table 3 | Depth, distance to the coast, and area covered by the ecosystem services assigned with different assessment categories (i.e., High, Low, and Negligible) and provided by benthic habitats, within the Atlantic Ocean, and for each of the sub-regions.

Ecosystem service	Categories	Macaronesia Area (km²)	Area (%)	Depth (m)	Distance (km)	Bay of Biscay Area (km²)	Area (%)	Depth (m)	Distance (km)	Celtic Sea Area (km²)	Area (%)	Depth (m)	Distance (km)	North Sea Area (km²)	Area (%)	Depth (m)	Distance (km)	Total Area (km²)	Area (%)	Mean depth (m) ± SD	Mean distance (km) ± SD
Food provision	High	1421	2	97±82	3±7	120811	78	37±42	7±12	278777	52	42±47	18±39	699171	71	37±48	15±35	1101365	62	47±50	16±35
	Low	86742	98	983±919	40±71	33450	22	191±268	17±18	153104	28	88±208	52±95	277165	28	91±203	15±44	550790	31	186±397	24±56
	Negligible	0	0	0	0	95	0	1091±254	247±2	109114	20	1116±799	230±112	4583	0	730±535	152±129	113787	6	917±579	193±122
Raw materials (biological) (incl. biochemical medicinal and ornamental)	High	662	1	78±82	2±5	13767	9	27±33	4±7	37213	7	26±27	6±20	95802	10	25±29	8±22	148244	8	33±33	8±22
	Low	759	1	111±78	3±8	100032	65	43±40	8±12	198619	37	53±42	25±43	541029	55	48±52	21±40	840706	48	57±50	21±39
	Negligible	86742	98	981±919	40±71	40556	26	195±273	18±25	305162	56	187±428	69±108	344088	35	145±308	27±69	776992	44	240±455	37±78
Air quality and climate regulation	High	267	<1	55±79	1±3	3939	3	24±32	4±8	13809	3	18±23	6±23	27256	3	29±37	4±17	45857	3	34±38	5±19
	Low	238	<1	107±94	3±6	10830	7	49±46	7±9	13329	2	43±34	16±39	26609	3	49±49	6±19	51251	3	59±47	9±25
	Negligible	87658	99	457±724	18±49	139586	90	68±142	9±16	513858	95	71±189	29±60	927054	95	63±164	26±50	1668835	95	91±237	26±51
Disturbance and natural hazard prevention	High	506	1	77±89	2±4	7655	5	34±42	4±7	9873	2	24±27	3±8	13447	1	18±24	2±5	32008	2	31±36	3±7
	Low	59	0	164±58	7±17	46154	30	47±46	8±11	44343	8	59±44	26±34	204484	21	40±39	11±26	295141	17	53±42	15±29
	Negligible	87599	99	474±741	19±50	100546	65	67±141	9±16	486779	90	69±192	29±63	762988	78	63±149	21±47	1438794	81	88±223	23±50
Photosynthesis, chemosynthesis and primary production	High	267	0	55±79	1±3	1710	1	11±9	2±3	5072	1	14±17	5±23	17273	2	25±32	3±9	24898	1	28±3	4±13
	Low	0	0	0	0	2229	1	76±37	13±14	10247	2	71±36	22±22	16548	2	67±47	15±37	29053	2	77±40	17±37
	Negligible	87896	100	415±690	16±46	150417	97	64±129	9±15	525676	97	65±169	26±56	947098	97	59±145	20±45	1711992	97	83±211	22±47
Nutrient cycling	High	238	<1	107±94	3±6	13537	9	47±45	7±10	23753	4	39±33	15±37	42393	4	44±47	5±20	80372	5	53±45	9±25
	Low	1183	1	95±79	3±7	111720	72	34±38	7±11	236946	44	42±41	18±36	668826	68	35±43	19±38	1019454	58	44±45	18±36
	Negligible	86742	98	983±919	40±71	29099	19	407±310	33±30	280296	52	627±691	224±105	269700	27	524±499	101±113	666116	38	642±633	94±110
Reproduction and nursery	High	795	1	73±80	2±6	27547	18	29±34	5±8	48651	9	28±29	11±31	291708	30	33±39	13±33	369716	21	39±39	13±33
	Low	566	1	121±75	3±8	78075	51	49±49	10±15	127503	24	59±60	26±41	364004	37	46±51	17±35	570291	32	59±55	19±36
	Negligible	86802	98	893±905	36±68	48733	32	229±283	22±23	364841	67	194±415	75±105	325207	33	230±372	45±82	825936	47	314±499	51±84
Maintenance of biodiversity	High	87641	99	456±733	18±49	58162	38	95±176	11±17	213522	39	50±114	24±60	355537	36	61±146	11±38	716000	41	95±236	16±45
	Low	475	1	103±73	4±10	95430	62	36±43	7±12	323645	60	68±191	23±46	605026	62	45±100	23±42	1024886	58	55±122	23±42
	Negligible	48	0	54±26	4±9	763	0	12±9	2±2	3828	1	18±16	6±17	20355	2	20±17	21±42	25057	1	21±16	17±38
Water quality regulation and bioremediation of waste	High	506	<1	77±89	2±4	14769	10	39±42	6±9	28149	5	30±31	11±32	53342	5	38±44	5±18	97600	6	46±45	7±22
	Low	458	<1	125±78	3±8	73241	47	51±48	10±13	78298	14	58±46	26±36	301596	31	43±44	16±35	453678	26	58±49	18±34
	Negligible	87199	99	618±835	25±58	66345	43	78±175	9±17	434548	80	82±242	32±72	625980	64	72±194	30±55	1214665	69	108±287	30±59
Cognitive value	High	506	1	77±89	2±4	14769	10	39±42	6±9	60097	11	36±131	12±35	53946	5	39±45	5±18	130136	7	48±72	7±23
	Low	0	0	0	0	95	0	445±603	99±135	77252	14	393±688	62±113	2371	0	130±356	15±44	79733	5	202±469	28±71
	Negligible	87658	99	457±724	18±49	139491	90	68±141	9±15	403646	75	58±101	28±56	924601	94	60±150	26±51	1556074	88	87±222	26±50
Leisure, recreation and cultural inspiration	High	267	<1	55±79	1±3	3939	3	24±32	4±8	14603	3	24±29	13±37	30729	3	39±47	4±17	50162	3	44±47	7±23
	Low	411	<1	80±81	2±5	13948	9	31±39	5±7	23697	4	29±29	7±23	106159	11	30±32	14±36	144690	8	37±35	14±33
	Negligible	87485	99	514±756	20±52	136468	83	83±157	11±17	502696	93	85±209	35±65	844031	86	82±194	27±53	1571091	89	114±270	29±54
Feel good or warm glow	High	267	<1	55±79	1±3	1233	1	10±9	1±2	12957	2	16±23	3±12	20931	2	26±38	5±22	35761	2	30±40	5±19
	Low	411	<1	80±81	2±5	16654	11	33±39	5±8	25343	5	31±30	12±33	115957	12	36±40	11±30	159090	9	42±41	11±30
	Negligible	87485	99	514±756	20±52	136468	83	83±157	11±17	502696	93	85±209	35±65	844031	86	82±194	27±53	1571091	89	114±270	29±52

Table 4 | Differences (Kruskal-Wallis test) between ecosystem services categories provided by benthic habitats, according to the distance to coastline, and depth ($N = 55, 023$).

Ecosystem service	Distance to coastline			Depth		
	Category	Kruskal-Wallis (H)	p	Category	Kruskal-Wallis (H)	p
Food provision	High[a] Low[b] Negligible[c]	1024.4	<0.001***	High[a] Low[b] Negligible[c]	4181.0	<0.001***
Raw materials (biological) (incl. Biochemical. medicinal and ornamental)	High[a] Low[b] Negligible[c]	4842.1	<0.001***	High[a] Low[b] Negligible[c]	5531.1	<0.001***
Air quality and climate regulation	High[a] Low[b] Negligible[c]	8416.0	<0.001***	High[a] Low[b] Negligible[c]	2676.8	<0.001***
Disturbance and natural hazard prevention	High[a] Low[b] Negligible[c]	5595.6	<0.001***	High[a] Low[b] Negligible[c]	2799.6	<0.001***
Photosynthesis, chemosynthesis and primary production	High[a] Low[b] Negligible[c]	6354.9	<0.001***	High[a] Low[b] Negligible[b]	4426.9	<0.001***
Nutrient cycling	High[a] Low[b] Negligible[c]	5288.0	<0.001***	High[a] Low[b] Negligible[c]	7653.9	<0.001***
Reproduction and nursery	High[a] Low[b] Negligible[c]	4543.1	<0.001***	High[a] Low[b] Negligible[c]	8444.5	<0.001***
Maintenance of biodiversity	High[a] Low[b] Negligible[a]	3786.5	<0.001***	High[a] Low[b] Negligible[b]	1617.1	<0.001***
Water quality regulation and bioremediation of waste	High[a] Low[b] Negligible[c]	8391.6	<0.001***	High[a] Low[b] Negligible[c]	548.9	<0.001***
Cognitive value	High[a] Low[b] Negligible[b]	8252.1	<0.001***	High[a] Low[b] Negligible[c]	202.0	<0.001***
Leisure, recreation and cultural inspiration	High[a] Low[b] Negligible[c]	8687.9	<0.001***	High[a] Low[b] Negligible[c]	4065.5	<0.001***
Feel good or warm glow	High[a] Low[b] Negligible[c]	8105.2	<0.001***	High[a] Low[b] Negligible[c]	4785.2	<0.001***

***Indicates significant results at 0.001 significance level. The superscripts within each service have been used to indicate significant (different superscripts) or non-significant (equal superscripts) differences on post-hoc tests between pairs of data, at 0.05 significance level.

variables, from both terrestrial and marine datasets, in assessing coastal protection. In this investigation, we have only used benthic habitats, which may explain the relatively small area providing this service in the European North Atlantic Ocean. Furthermore, it is the limited distribution of biogenic structures and seagrass species within this ocean, considered as the main producer of this service, which may explain the limited provision to shallow and habitats located close to the

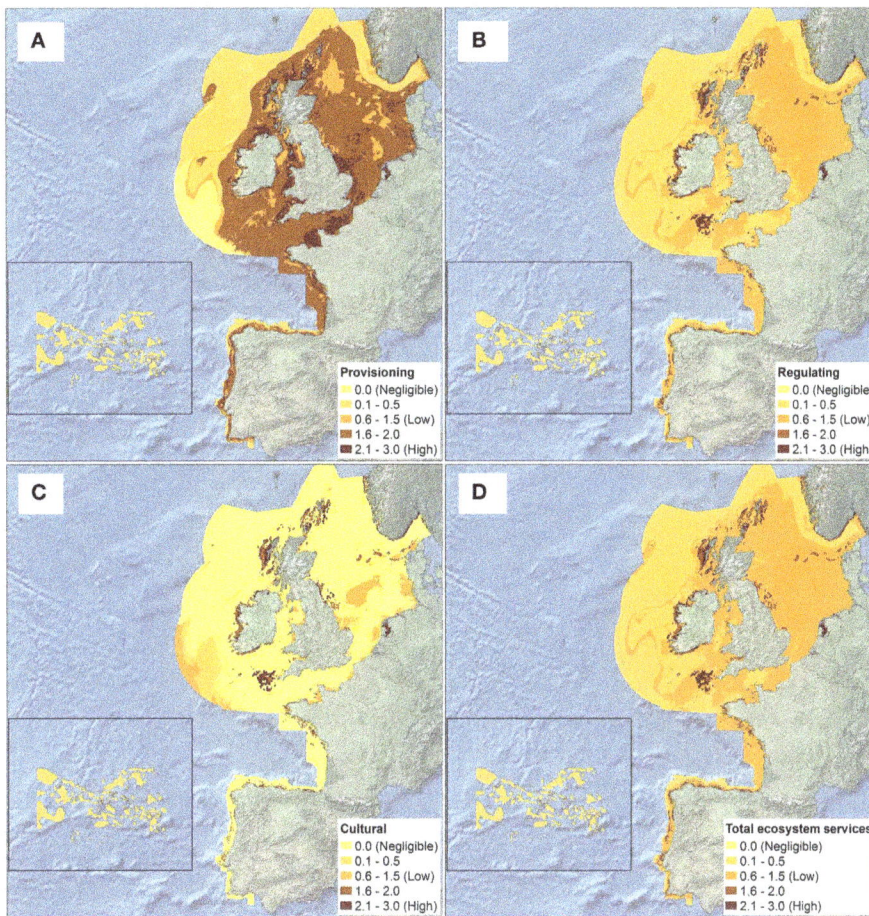

FIGURE 5 | Spatial distribution of the mean value of aggregated ecosystem: (A) Provisioning services; (B) Regulating services; (C) Cultural services; and (D) Total ecosystem services.

coast (Christianen et al., 2013; Cullen-Unsworth and Unsworth, 2013).

The remaining ecosystem services are provided in limited areas. This pattern is possibly explained by the fact that some of the services analyzed are provided by very specific, spatially limited benthic habitats (i.e., photic zones), or in a larger scale, by pelagic habitats, i.e., air quality and climate regulation, water quality regulation and bioremediation, nutrient cycling, photosynthesis, chemosynthesis, and primary production. For example, some of them, such as climate regulation or carbon sequestration, are very important in coastal margin habitats, rather than in subtidal habitats (Beaumont et al., 2014).

Very small areas (11%) have been identified as providing cultural services (i.e., cognitive, leisure, recreation and cultural inspiration, feel good, and warm glow). This result is likely to be a consequence of the dependence of these services on accessibility. Therefore, even if the current provision of these services is limited to few habitats and areas (which are probably heavily used), it is likely that over time, as access increases to certain areas, these services will increase their value and distribution (Ghermandi et al., 2012). The broad-scale spatial patterns of the ecosystem services

assessment resulting from this investigation could be considered consistent for different spatial scales of analysis if the approach is implemented elsewhere.

When considering the approach and results obtained through this research, authors would like to highlight that, rather than getting a valuation of the ecosystem services provided by the benthic habitats of the European North Atlantic Ocean, in our investigation a pragmatic approach for benthic services mapping is applied, based on the best available knowledge (De Groot et al., 2010). We recognize that the reliability of the results obtained in this investigation depend on, among other things, two major aspects: (i) the quality and reliability of benthic habitat maps used, which is an important but insufficiently assessed issue (Schägner et al., 2013); and (ii) the valuation of the ecosystem services carried out by scientific expert judgment (extracted from Salomidi et al., 2012), which could be biased toward the knowledge of the experts who published that research; meanwhile, social and economic aspects could be under-rated.

Some of the aforementioned weaknesses could be overcome: (i) enhancing the scientific knowledge of marine ecosystem functioning by finalizing detailed benthic habitat maps of the complete study area (especially, for the EEZ of France, Spain, and Portugal and deeper benthic habitats; Liquete et al., 2013a); and (ii) improving the assessment of services valuation, promoting the multidisciplinary discussions among environmental and social scientists and economists, to achieve consensus on benthic habitat services values.

A more adequate ecosystem services assessment and valuation could be carried out following the steps below:

(i) Definition of marine ecosystem services categories, based upon those already in use (see Liquete et al., 2013a). This definition should be carried out by experts from different scientific disciplines such as environmental, social (including stakeholders' participation) and economical sciences. In order to ensure consistency and allow for aggregation or comparison of results across the countries, there is a need for a common classification and to define which ecosystems and services will be considered as a priority by Member States (Maes et al., 2013).

(ii) Mapping services based on spatial distribution and patterns of different ecosystem components, processes and their relationships, including the need for future scenarios.

(iii) Biological and environmental valuation services by common procedures, undertaken by environmental, social, and economic scientists. Many ecosystem services cannot be directly quantified and thus, researchers must rely on indicators or proxy data for their quantification (Liquete et al., 2013a). Expert judgment may be a very important source of information, but the careful selection of a broad panel of experts may be required for ecosystem service assessment.

(iv) Economic valuation undertaken by economists and social scientists. No single ecological, social or economic methodology can capture the total value of these complex systems (Wilson et al., 2013). Assigning economic values to seascape features and habitat functions of marine ecosystems requires full understanding of the natural systems upon which they rely (Wilson et al., 2013). Probably, new economic valuation methods should be adopted (see Liquete et al., 2013a).

(v) Ecosystem services valuation assessment, which could assist in the determination of the ecological and environmental status under the Water Framework Directive (WFD) and MSFD, respectively (Katsanevakis et al., 2011; Vlachopoulou et al., 2014).

This process could result in the definition of proposals for management plans for different directives (e.g., MSFD, Habitats Directive) and instruments such as Marine Spatial Planning. Since oceans are facing an increasing number of human uses and threats, the inclusion of ecosystem services within management plans is growing in importance. In this context, the science of ecology must play a crucial role in bringing concepts like ecosystem goods and services to the forefront of the valuation debate (Bingham et al., 1995; Wilson and Carpenter, 1999; Liquete et al., 2013a).

The spatially explicit nature of the approach presented in this investigation is of special interest to support decision-making approaches and different aspects of the ecosystem-based marine spatial management *sensu* Katsanevakis et al. (2011). Among other things, the key to achieving a more comprehensive set of management mechanisms is, in the first instance, to know more about the ecosystem functions of benthic habitats (Martinez et al., 2011). In this way, there is a key goal of maintaining the delivery of ecosystem services, which must be based upon ecological principles that articulate the scientifically-recognized attributes of healthy functioning ecosystems (Foley et al., 2010), as required by the MSFD (Borja et al., 2013; Tett et al., 2013). This would require management measures for minimizing environmental impact and maximizing the socio-economic benefit of marine services (Salomidi et al., 2012); aspects that are basic to the Marine Spatial Planning.

This research has provided a first assessment of the benthic ecosystem services at Atlantic European scale, with the provision of ecosystem services maps and their general spatial distribution patterns. Related to the objectives of this research, the conclusions are: (i) benthic habitats provide a diverse set of ecosystem services, with the food provision and biodiversity maintenance services more extensively represented. In addition, other regulating and cultural services are provided in a more limited area; and (ii) the ecosystem services assessment categories are significantly related to the distance to the coast and with depth (higher near the coast and in shallow waters).

The results obtained in this investigation highlight the need for diverse, healthy and extensive benthic habitat areas to support the provision of important and valuable ecosystem services (i.e., food provisioning, disturbance prevention, nutrient cycling, etc.). Spatially explicit assessment and valuation of ecosystem services might be of crucial interest for future management measures adoption such as Marine Spatial Planning. The approach proposed here could be considered as a pragmatic way of getting a first snapshot of the distribution of ecosystem services based on the available information and we consider this as a promising starting point for further research and discussion on ecosystem services contribution of benthic habitats in Europe.

ACKNOWLEDGMENTS

This manuscript is a result of the projects MeshAtlantic (Atlantic Area Transnational Cooperation Programme 2007–2013 of the European Regional Development Fund) (www.meshatlantic.eu) and DEVOTES (DEVelopment Of innovative Tools for understanding marine biodiversity and assessing good Environmental Status) funded by the European Union under the 7th Framework Program "The Ocean of Tomorrow" Theme (grant agreement no. 308392) (www.devotes-project.eu), and also supported by the Basque Water Agency (URA), through a Convention with AZTI-Tecnalia. The funders had no role in study design, data collection and analysis, decision to publish, or preparation of the manuscript. We wish to thank Udane Martinez and Iñigo Muxika (AZTI-Tecnalia) for their significant contributions to the data analysis. This paper is contribution number 676 from AZTI-Tecnalia (Marine Research Division).

SUPPLEMENTARY MATERIAL

The Supplementary Material for this article can be found online at: http://www.frontiersin.org/journal/10.3389/fmars.2014.00023/abstract

Figure S1 | Depth distribution of the Exclusive Economic Zone (dark blue) and depth distribution of habitat-mapped areas (light blue), in the four subregions of the European North Atlantic Ocean; (A) Macaronesia; (B) Bay of Biscay and Iberian Coast; (C) Celtic Seas; and (D) Greater North Sea, including the Kattegat, the English Channel and Norway.

Figure S2 | Spatial distribution of food provision services.

Figure S3 | Spatial distribution of raw materials (biological, incl. biochemical, medicinal, and ornamental) services.

Figure S4 | Spatial distribution of air quality and climate regulation services.

Figure S5 | Spatial distribution of disturbance and natural hazard prevention services.

Figure S6 | Spatial distribution of photosynthesis, chemosynthesis, and primary production services.

Figure S7 | Spatial distribution of nutrient cycling services.

Figure S8 | Spatial distribution of reproduction and nursery services.

Figure S9 | Spatial distribution of maintenance of biodiversity services.

Figure S10 | Spatial distribution of water quality regulation and bioremediation of waste services.

Figure S11 | Spatial distribution of cognitive value services.

Figure S12 | Spatial distribution of leisure, recreation, and cultural inspiration services.

Figure S13 | Spatial distribution of feel good or warm glow services.

REFERENCES

Armstrong, C. W., Foley, N. S., Tinch, R., and Van Den Hove, S. (2012). Services from the deep: steps towards valuation of deep sea goods and services. *Ecosystem Services* 2, 2–13. doi: 10.1016/j.ecoser.2012.07.001

Beaumont, N. J., Austen, M. C., Atkins, J. P., Burdon, D., Degraer, S., Dentinho, T. P., et al. (2007). Identification, definition and quantification of goods and services provided by marine biodiversity: implications for the ecosystem approach. *Mar. Pollut. Bull.* 54, 253–265. doi: 10.1016/j.marpolbul.2006.12.003

Beaumont, N. J., Jones, L., Garbutt, A., Hansom, J. D., and Toberman, M. (2014). The value of carbon sequestration and storage in coastal habitats. *Estuar. Coast. Shelf Sci.* 137, 32–40. doi: 10.1016/j.ecss.2013.11.022

Bingham, G., Bishop, R., Brody, M., Bromley, D., Clark, E., Cooper, W., et al. (1995). Issues in ecosystem valuation: improving information for decision making. *Ecol. Econ.* 14, 73–90. doi: 10.1016/0921-8009(95)00021-Z

Borja, A., Elliott, M., Andersen, J. H., Cardoso, A. C., Carstensen, J., Ferreira, J. G., et al. (2013). Good Environmental Status of marine ecosystems: What is it and how do we know when we have attained it? *Mar. Pollut. Bull.* 76, 16–27. doi: 10.1016/j.marpolbul.2013.08.042.

Cameron, A., and Askew, N. (eds.). (2011). *EUSeaMap - Preparatory Action for Development and Assessment of a European Broad-Scale Seabed Habitat Map Final Report*, 226. Available online at: http://jncc.gov.uk/euseamap

Cardoso, A. C., Cochrane, S., Doemer, H., Ferreira, J. G., Galgani, F., Hagebro, C., et al. (2010). *Scientific support to the European Commission on the Marine Strategy Framework Directive. Management Group Report. EUR 24336 EN.* Luxembourg: Joint Research Centre, Office for Official Publications of the European Communities.

Christianen, M. J. A., Van Belzen, J., Herman, P. M. J., Van Katwijk, M. M., Lamers, L. P. M., Van Leent, P. J. M., et al. (2013). Low-canopy seagrass beds still provide important coastal protection services. *PLoS ONE* 8:e62413. doi: 10.1371/journal.pone.0062413

Cices. (2013). *Common International Classification of Ecosystem Services.* Available online at: http://cices.eu/

Cimon-Morin, J., Darveau, M., and Poulin, M. (2013). Fostering synergies between ecosystem services and biodiversity in conservation planning: a review. *Biol. Conserv.* 166, 144–154. doi: 10.1016/j.biocon.2013.06.023

Costanza, R., D'Arge, R., De Groot, R., Farber, S., Grasso, M., Hannon, B., et al. (1997). The value of the world's ecosystem services and natural capital. *Nature* 387, 253–260. doi: 10.1038/387253a0

Costanza, R., De Groot, R., Sutton, P., Van Der Ploeg, S., Anderson, S. J., Kubiszewski, I., et al. (2014). Changes in the global value of ecosystem services. *Global Environ. Change* 26, 152–158. doi: 10.1016/j.gloenvcha.2014.04.002

Cullen-Unsworth, L., and Unsworth, R. (2013). Seagrass meadows, ecosystem services, and sustainability. *Environment* 55, 14–28. doi: 10.1080/00139157.2013.785864

Daily, G. C. (1997). *Nature's Services: Societal Dependence on Natural Ecosystems.* Washington, DC: Island Press. ISBN: 1-55963-475-8 (hbk), 1 55963 476 6 (soft cover).

Davies, C. E., and Moss, D. (2002). "EUNIS habitat classification," in *Final Report to the European Topic Centre on Nature Protection and Biodiversity, European Environment Agency*, 125.

Davies, C. E., Moss, D., and Hill, M. O. (2004). "EUNIS habitat classification revised 2004," in *Report to European Environmental Agency and European Topic Centre on Nature Protection and Biodiversity*, 307.

De Groot, R. S., Alkemade, R., Braat, L., Hein, L., and Willemen, L. (2010). Challenges in integrating the concept of ecosystem services and values in landscape planning, management and decision making. *Ecol. Complexity* 7, 260–272. doi: 10.1016/j.ecocom.2009.10.006

De Groot, R. S., Wilson, M. A., and Boumans, R. M. J. (2002). A typology for the classification, description and valuation of ecosystem functions, goods and services. *Ecol. Econ.* 41, 393–408. doi: 10.1016/S0921-8009(02)00089-7

Egoh, B., Drakou, E. G., Dunbar, M. B., Maes, J., and Willemen, L. (2012). *Indicators for Mapping Ecosystem Services: a Review JRC Scientific and Policy Reports.* Luxembourg. doi: 10.2788/41823

Farber, S. C., Costanza, R., and Wilson, M. A. (2002). Economic and ecological concepts for valuing ecosystem services. *Ecol. Econ.* 41, 375–392. doi: 10.1016/S0921-8009(02)00088-5

Foley, M. M., Halpern, B. S., Micheli, F., Armsby, M. H., Caldwell, M. R., Crain, C. M., et al. (2010). Guiding ecological principles for marine spatial planning. *Marine Policy* 34, 955–966. doi: 10.1016/j.marpol.2010.02.001

Galparsoro, I., Connor, D. W., Borja, A., Aish, A., Amorim, P., Bajjouk, T., et al. (2012). Using EUNIS habitat classification for benthic mapping in European seas: Present concerns and future needs. *Mar. Pollut. Bull.* 64, 2630–2638. doi: 10.1016/j.marpolbul.2012.10.010

Ghermandi, A., Nunes, P. A. L. D., Portela, R., Rao, N., and Teelucksingh, S.S. (2012). "Recreational, cultural and aesthetic services from estuarine and coastal ecosystems," in *Ecological Economics of Estuaries and Coasts,* Vol. 12, Chapter 11, eds M. van den Belt and R. Costanza, Treatise on estuarine and coastal science, Series eds E. Wolanski and D. McLusky (Waltham, MA: Academic Press), 217–237.

Katsanevakis, S., Stelzenmüller, V., South, A., Sorensen, T. K., Jones, P. J. S., Kerr, S., et al. (2011). Ecosystem-based marine spatial management: review of concepts, policies, tools, and critical issues. *Ocean Coast. Manag.* 54, 807–820. doi: 10.1016/j.ocecoaman.2011.09.002

Korpinen, S., Meidinger, M., and Laamanen, M. (2013). Cumulative impacts on seabed habitats: an indicator for assessments of good environmental status. *Mar. Pollut. Bull.* 74, 311–319. doi: 10.1016/j.marpolbul.2013.06.036

Liquete, C., Piroddi, C., Drakou, E. G., Gurney, L., Katsanevakis, S., Charef, A., et al. (2013a). Current status and future prospects for the assessment of marine and coastal ecosystem services: a systematic review. *PLoS ONE* 8:e67737. doi: 10.1371/journal.pone.0067737

Liquete, C., Zulian, G., Delgado, I., Stips, A., and Maes, J. (2013b). Assessment of coastal protection as an ecosystem service in Europe. *Ecol. Indic.* 30, 205–217. doi: 10.1016/j.ecolind.2013.02.013

Maes, J., Teller, A., Erhard, M., Liquete, C., Braat, L., Berry, P., et al. (2013). *Mapping and Assessment of Ecosystems and their Services. An Analytical Framework for Ecosystem Assessments under action 5 of the EU Biodiversity Strategy to 2020.* Brussels: Publications office of the European Union. doi: 10.2779/12398

Martinez, M. L., Costanza, R., and Pérez-Maqueo, O. (2011). "12.07 - ecosystem services provided by estuarine and coastal ecosystems: storm protection as a service from estuarine and coastal ecosystems," in *Treatise on Estuarine and Coastal Science,* eds W. Eric and M. Donald (Editors-in-Chief) (Waltham, MA: Academic Press), 129–146.

MEA. (2003). "Ecosystems and their services," in *Ecosystems and Human Wellbeing: a Framework for Assessment,* Chapter 2, Millenium Ecosystem Assessment. Available online at: http://www.millenniumassessment.org/

Milcu, A. I., Hanspach, J., Abson, D., and Fischer, J. (2013). Cultural ecosystem services: a literature review and prospects for future research. *Ecol. Soc.* 18. doi: 10.5751/ES-05790-180344

Murillas-Maza, A., Virto, J., Gallastegui, M. C., González, P., and Fernández-Macho, J. (2011). The value of open ocean ecosystems: a case study for the Spanish exclusive economic zone. *Nat. Resour. Forum* 35, 122–133. doi: 10.1111/j.1477-8947.2011.01383.x

Pearce, D. (1998). Auditing the earth: the value of the world's ecosystem services and natural capital. *Environment* 40, 23–28. doi: 10.1080/00139159809605092

Pimm, S. L. (1997). The value of everything. *Nature* 387, 231–232. doi: 10.1038/387231a0

Potts, T., Burdon, D., Jackson, E., Atkins, J., Saunders, J., Hastings, E., et al. (2014). Do marine protected areas deliver flows of ecosystem services to support human welfare? *Marine Policy* 44, 139–148. doi: 10.1016/j.marpol.2013.08.011

Ramirez-Llodra, E., Tyler, P. A., Baker, M. C., Bergstad, O. A., Clark, M. R., Escobar, E., et al. (2011). Man and the last great wilderness: human impact on the deep sea. *PLoS ONE* 6:e22588. doi: 10.1371/journal.pone.0022588

Rice, J. (2010). "Science dimensions of an Ecosystem Approach to Management of Biotic Ocean Resources (SEAMBOR)," *Marine Board-ESF Position Paper 14.* (Ostend), 92.

Salomidi, M., Katsanevakis, S., Borja, A., Braeckman, U., Damalas, D., Galparsoro, I., et al. (2012). Assessment of goods and services, vulnerability, and conservation status of European seabed biotopes: a stepping stone towards ecosystem-based marine spatial management. *Mediterr. Mar. Sci.* 13, 49–88. doi: 10.12681/mms.23

Schägner, J. P., Brander, L., Maes, J., and Hartje, V. (2013). Mapping ecosystem services' values: current practice and future prospects. *Ecosystem Services* 4, 33–46. doi: 10.1016/j.ecoser.2013.02.003

Seitz, R. D., Wennhage, H., Bergström, U., Lipcius, R. N., and Ysebaert, T. (2014). Ecological value of coastal habitats for commercially and ecologically important species. *ICES J. Mar. Sci.* 71, 648–665. doi: 10.1093/icesjms/fst152

Ten Brink, P., Berghöfer, A., Schröter-Schlaack, C., Sukhdev, P., Vakrou, A., White, S., et al. (2009). "TEEB – The economics of ecosystems and biodiversity for national and international policy makers 2009," in *United Nations Environment Programme and the European Commission,* 1–48.

Tett, P., Gowen, R. J., Painting, S. J., Elliott, M., Forster, R., Mills, D. K., et al. (2013). Framework for understanding marine ecosystem health. *Mar. Ecol. Prog. Ser.* 494, 1–27. doi: 10.3354/meps10539

Thurber, A. R., Sweetman, A. K., Narayanaswamy, B. E., Jones, D. O. B., Ingels, J., and Hansman, R. L. (2013). Ecosystem function and services provided by the deep sea. *Biogeosci. Discuss.* 10, 18193–18240. doi: 10.5194/bgd-10-18193-2013

Van Den Belt, M., and Costanza, R. (2012). "Ecological economics of estuaries and coasts," in *Treatise on Estuarine and Coastal Science,* Vol. 12, eds E. Wolanski and D. S. McLusky (Waltham, MA: Academic Press), 525.

Vasquez, M., Chacón, D. M., Tempera, F., O'keeffe, E., Galparsoro, I., Alonso, J. L. S., et al. (in press). Mapping at broad scale seabed habitats of the North-East Atlantic using environmental data. *J. Sea Res.*

Vlachopoulou, M., Coughlin, D., Forrow, D., Kirk, S., Logan, P., and Voulvoulis, N. (2014). The potential of using the ecosystem approach in the implementation of the eu water framework directive. *Sci. Tot. Environ.* 470–471, 684–694. doi: 10.1016/j.scitotenv.2013.09.072

Wilson, M. A., and Carpenter, S. R. (1999). Economic valuation of freshwater ecosystem services in the United States: 1971-1997. *Ecol. Appl.* 9, 772–783. doi: 10.1890/1051-0761(1999)009[0772:EVOFES]2.0.CO;2

Wilson, M., Costanza, R., Boumans, R., and Liu, S. (2013). "Integrated assessment and valuation of ecosystem goods and services provided by coastal systems," in *The Intertidal Ecosystem: The Value of Ireland's Shores,* ed J. G. Wilson (Dublin: Royal Irish Academy), 1–24.

Conflict of Interest Statement: The authors declare that the research was conducted in the absence of any commercial or financial relationships that could be construed as a potential conflict of interest.

3

Interpreting environmental change in coastal Alaska using traditional and scientific ecological knowledge

William G. Ambrose Jr.[1,2]*, *Lisa M. Clough*[3], *Jeffrey C. Johnson*[4,5], *Michael Greenacre*[2,6], *David C. Griffith*[7], *Michael L. Carroll*[2] *and Alex Whiting*[8]

[1] Department of Biology, Bates College, Lewiston, ME, USA
[2] Akvaplan-niva, FRAM- High North Research Centre for Climate and the Environment, Tromsø, Norway
[3] Division of Polar Programs, National Science Foundation, Arlington, VA, USA
[4] Department of Sociology and Institute for Coastal Science and Policy, East Carolina University, Greenville, NC, USA
[5] Department of Anthropology, University of Florida, Gainesville, FL, USA
[6] Department of Economics and Business, Universitat Pompeu Fabra, and Barcelona Graduate School of Economics, Barcelona, Spain
[7] Department of Anthropology and Institute for Coastal Science and Policy, East Carolina University, Greenville, NC, USA
[8] Native Village of Kotzebue, Kotzebue, AK, USA

Edited by:
Dorte Krause-Jensen, Aarhus University, Denmark

Reviewed by:
Brezo Martínez, Universidad Rey Juan Carlos, Spain
Mikael K. Sejr, Aarhus University, Denmark

*Correspondence:
William G. Ambrose Jr., Department of Biology, Bates College, 44 Campus Ave., Lewiston, Maine 04240, USA
e-mail: wambrose@bates.edu

Humans who interact directly with local ecosystems possess traditional ecological knowledge that enables them to detect and predict ecosystem changes. Humans who use scientific ecological methods can use species such as mollusks that lay down annual growth rings to detect past environmental variation and use statistical models to make predictions about future change. We used traditional ecological knowledge shared by local Iñupiaq, combined with growth histories of two species of mollusks, at different trophic levels, to study local change in the coastal ecosystems of Kotzebue, Alaska, an area in the Arctic without continuous scientific monitoring. For the mollusks, a combination of the Arctic Oscillation and total Arctic ice coverage, and summer air temperature and summer precipitation explained 79–80% of the interannual variability in growth of the suspension feeding Greenland cockle (*Serripes groenlandicus*) and the predatory whelk (*Neptunea hero*), respectively, indicating these mollusks seem to be impacted by local and regional environmental parameters, and should be good biomonitors for change in coastal Alaska. The change experts within the Kotzebue community were the elders and the fishers, and they observed changes in species abundance and behaviors, including benthic species, and infer that a fundamental change in the climate has taken place within the area. We conclude combining traditional and scientific ecological knowledge provides greater insight than either approach alone, and offers a powerful way to document change in an area that otherwise lacks widespread quantitative monitoring.

Keywords: environmental change, scientific ecological knowledge, traditional ecological knowledge, sclerochronology, knowledge networks, Arctic

INTRODUCTION

The average atmospheric temperature in the Arctic has increased twice as fast as the average temperature for the rest of the world over the past 50 years, and is predicted to continue to increase rapidly over the next 100 years (Arctic Climate Impact Assessment, 2005). The marine and terrestrial ecosystem changes accompanying these rising temperatures have especially strong impacts on the humans who depend on these ecosystems for their survival and quality of life (Morison et al., 2000; Huntington et al., 2012). We assert that some humans who interact with ecosystems are more attuned to observe changes than others, and are therefore able to report more accurately on such changes (Davis and Wagner, 2003). Similarly, not all species within the ecosystem will be affected equally; rather some species are more sensitive to rapid climate change than others (Wassmann et al., 2011). While the emphasis on sensitivity is typically focused on an organism's ability to withstand change, there is also a component of sensitivity related to the ability to detect or record change. Whereas many studies of climate change only use key indicators that are physical in nature (e.g., atmospheric concentrations of carbon dioxide, sea surface temperature), we advocate here for combining knowledge from key ecosystem components and key human observers in an integrated approach to monitoring and assessing environmental change, especially in areas without continual monitoring of physical variables.

Traditional ecological knowledge (TEK) accumulates in individuals who regularly interact with the natural environment, often via a subsistence lifestyle, making it possible to discern changes occurring over several human generations. TEK can provide information on time scales of 100 years or more (Davis and Wagner, 2003; Shackeroff et al., 2011), and people with high levels of TEK often successfully predict the behaviors of fish, mammals, and other higher trophic level organisms by monitoring how natural resources respond to natural and anthropogenic conditions, and tracking environmental change over time (Griffith, 2006; Menzies and Butler, 2006) (**Figure 1**). TEK derived from subsistence and commercial resource extraction activities is especially sensitive to environmental changes, as success in obtaining resources is tied to an ability to predict and respond to changing conditions (Huntington, 2000). Further, TEK can be studied using structured methodologies that afford the systematic documentation of cultural beliefs about species, climate, food webs, and other dimensions of natural environments (Boster and Johnson, 1989; García Quijano, 2007). This approach allows a comparison with, and integration into, scientific ecological knowledge (SEK) models. Incorporating TEK into more traditional SEK studies can improve the breadth of research findings while also providing legitimacy to scientific findings for local communities, broadening the knowledge set that local communities can draw on as they develop effective responses to changing ecosystems.

We used both TEK and SEK to understand coastal environmental change in Kotzebue Sound, Alaska, an area without continuous scientific monitoring from, for example, oceanographic moorings. Our scientific knowledge of the history of change within the system was based on the growth patterns of two common mollusks. There is a close relationship between benthic and water column processes in the Arctic (Grebmeier et al., 1988; Ambrose and Renaud, 1995; Dunton et al., 2005), making long-lived, sessile, benthic organisms particularly good biomonitors of environmental change on Arctic shelves (Wanamaker et al., 2011; Kortsch et al., 2012; Mann et al., 2013). A wide range of marine climate conditions can be reconstructed from the growth and shell chemistry of mollusks (Richardson, 2001; Wanamaker et al., 2011). In Kotzebue Sound, Alaska, we used the shell growth of the suspension feeding Greenland cockle *Serripes groenlandicus* and the predatory whelk *Neptunea hero* to provide a temporally

consistent and uninterrupted record of change over decadal time scales (**Figure 1**).

Ducklow et al. (2009) maintain that attribution of longer-term changes in marine ecosystems is best assessed with a minimum 50-year dataset, and they assert that documenting a regime shift takes at least a decade of data. Mollusks shells can provide high-resolution seasonal records and while at least one species is known to live in excess of 500 years (Butler et al., 2013), most live on the order of 50 years or less (Gröcke and Gillikin, 2008). The mollusks we examined lived 15–20 years, so use of TEK was necessary to distinguish climate change from climate oscillation in the mollusk growth data. The objectives of our study were to: (1) determine the derived annual growth patterns of mollusks; (2) determine if the patterns of change in mollusk growth were related to environmental parameters for which records are available (with 1 and 2 providing us with information about annual change on decadal time scales); and then (3) determine if there was a shared consensus about ecosystem change within the Kotzebue community, and (4) if within that community there were change experts (with 3 and 4 providing us with information about broader change, over longer temporal scales). We then combine these two forms of knowledge, which we feel results in a much more comprehensive assessment of local environmental change than either type of knowledge could alone.

MATERIALS AND METHODS
STUDY SITE AND ORGANISMS

Kotzebue, Alaska (67° 00′ N, 163° 00′ W), is a town of approximately 3500 people, around 85% of whom are Iñupiaq, with the majority enrolled in the Native Village of Kotzebue (NVK), the federally recognized tribal government. Tribal organizations and corporations, federal, state, tribal, and city government, a hospital, school, and service and transportation industries provide the bulk of the employment in the community. Many Kotzebue households are still dependent on caribou, seal, salmon, sheefish, berries, and other flora and fauna for food, clothing, crafts manufacture, and cultural wellness. The town's population fluctuates seasonally, with many families residing in hunting and fishing camps at various times of the year. Because subsistence is so much a part of the local economy, most jobs provide paid subsistence leave.

Kotzebue Sound is a shallow (average water depth 10–18 m) embayment in the southeast Chukchi Sea. The area is characterized by long, severely cold winters and short, cool summers. Sea ice is typically present from October to June, leaving the sound ice-free for a maximum duration of 4 months. Sediments, poorly to very poorly sorted, are primarily muddy sand to sandy mud, with a minor portion of gravel (Feder et al., 1991). The predominant currents are counterclockwise, with clockwise circulation occasionally occurring at shallower depths (Kinder et al., 1977). During the period of sea ice formation, cold, high salinity bottom water flows out of the Sound via the deepest (28 m) channel.

The shallow depth of the Sound and the influence of two major rivers (the Noatak and Kobuk), result in large annual changes in both the temperature and salinity of the Sound's near shore waters (annual water temperature range of −0.8–15°C and a

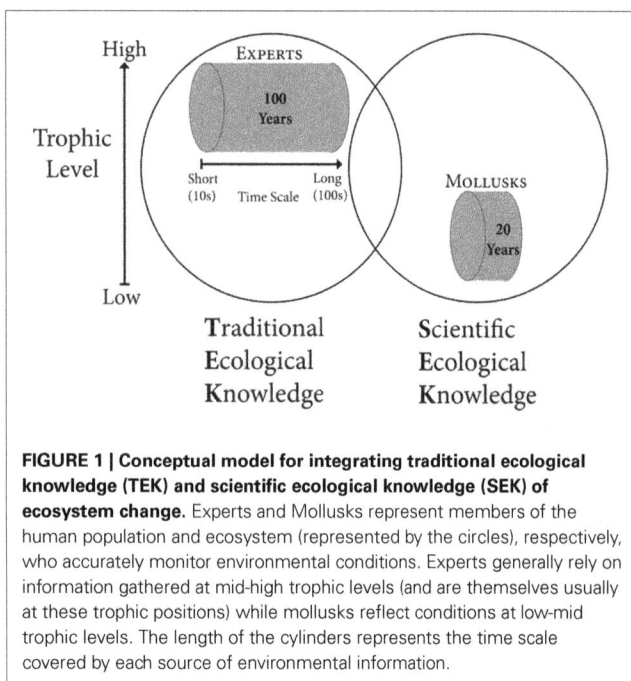

FIGURE 1 | Conceptual model for integrating traditional ecological knowledge (TEK) and scientific ecological knowledge (SEK) of ecosystem change. Experts and Mollusks represent members of the human population and ecosystem (represented by the circles), respectively, who accurately monitor environmental conditions. Experts generally rely on information gathered at mid-high trophic levels (and are themselves usually at these trophic positions) while mollusks reflect conditions at low-mid trophic levels. The length of the cylinders represents the time scale covered by each source of environmental information.

salinity range of 0.1–35.09 ppt. in 3 m water). The mollusks used in this study were collected at ca. 10 m water depth near Cape Krusenstern (67° 04.9′ N, 163° 41.5′ W) and Cape Blossom (66° 45.1′ N, 162° 39.2′ W). We measured water column properties at collection sites at the time of sampling with a CTD. The mean bottom water temperature during sampling for the years 2002, 2003, and 2004 was 5.5°C (range 4.8–7.3°C) at Cape Krusenstern and 10.2°C (range 6.6–13.6°C) at Cape Blossom. Bottom salinity averaged 28.94 ppt. (range 27.34–29.92 ppt.) at Cape Krusenstern and 21.99 ppt. (range 18.15–27.9 ppt.) at Cape Blossom.

S. groenlandicus (hereafter *Serripes*) is a large suspension feeding cockle that has a circumpolar distribution (Kafanov, 1980; Koszteyn et al., 1990; Coan et al., 2000). Throughout the Arctic it is an important food for walruses (Fisher and Stewart, 1997; Born et al., 2003), bearded seals (Lowry et al., 1980; Finley and Evans, 1983), and bottom-feeding birds (Merkel et al., 2007). *N. heros* (hereafter *Neptunea*), is a large predatory whelk that is common in the Beaufort, Chukchi, and Bering Seas (Wagner, 1977; Feder et al., 1991, 2005). It preys on mollusks, including *Serripes* and other infaunal taxa (Ambrose, personal observation 2002). Although whelks are sometimes found in the stomachs of marine mammals (Finley and Evans, 1983) they are not common prey for marine mammals.

STANDARD GROWTH INDEX FOR *SERRIPES* AND *NEPTUNEA*

Serripes were collected during July in 2002 ($N = 16$ individuals), 2003 ($N = 6$), and 2004 ($N = 7$) using an otter trawl (2 m mouth) fitted with a tickler chain from approximately 10 m water depth off Cape Krusenstern (2002 and 2003) and off Cape Blossom (2004). The trawl was towed at 2 knots parallel to shore for 20–25 min. Only live cockles with undamaged shells were used for analyses. *Neptunea* were collected by two means: (1) using the same trawl used to collect *Serripes*, and (2) provided to us by local fishers from their crab traps. In 2003, *Neptunea* ($N = 27$) were trawled from the same areas off Cape Krusenstern and Cape Blossom as *Serripes*. In 2006, all individuals ($N = 80$) were collected from crab traps placed 2–6 nautical miles off Cape Krusenstern. A total of 29 *Serripes* and 103 *Neptunea* were used in our subsequent analyses.

We only used the hard parts of the mollusks in this study. *Serripes* deposit annual lines that are visible on the external shell surface, and appear as thin dark lines deposited during the slow growth periods in the winter, separated by thicker light bands representing the faster summer growth. These lines have been verified as annual in *Serripes* in both the Chukchi Sea (Khim et al., 2003) and in two fjords on Svalbard (Ambrose et al., 2012). The distances between the ventral edges of successive growth lines along the line of maximum growth (shell height) were measured with a digital caliper to the nearest 0.01 mm. We excluded growth beyond the last growth line in analyses because this represents an incomplete growth year. The articulated growth steps, or striate, on the internal face of the opercula of *Neptunea* species and other genera of Buccinidae also correspond to the summer growth season, and the interrupting depressions to winter quiescence (Richardson et al., 2005). While we did not validate the annual deposition of striate in *Neptunea heros* for Kotzebue Sound, we presume the same pattern that occurs in other *Neptunea* species

(Richardson et al., 2005). Annual growth of each *Neptunea* was thus measured as the distance between each annulus as described above for *Serripes*.

Annual mollusk growth declines with age, so it is necessary to standardize growth increments within an individual and among individuals of different ages before growth can be compared among years. Each growth increment can be assigned to a calendar year because we collected all individuals live and lines are deposited annually. The same methods were applied to *Serripes* and *Neptunea* data. We followed established methods (Jones et al., 1989; Ambrose et al., 2006) to remove the ontogentic change in growth rate. Briefly, we used the von Bertalanffy growth function and its first derivative with respect to time to derive a predicted change in shell height for each age based on all individuals in the population. Then we calculated the expected increase in shell height for each individual for each calendar year of its life. Finally, we divided the measured shell growth for each calendar year by the expected growth for that year to generate a standardized growth index (SGI). This standardization process removes the ontogenetic changes in growth and equalizes the variance for the entire series (Fritts, 1976). Once annual changes in shell or operculum growth were standardized, we calculated the mean SGI for each calendar year from all individuals, although individual growth (rather than means) was used for the linear mixed models. The result is an annually resolved growth record, reflecting relatively better and poorer growth years compared to the expected von Bertalanffy fit of the data. An SGI greater than 1 indicates a better than average year for growth, while a value less than 1 reflects a worse than average growth year.

CLIMATOLOGICAL AND METEOROLOGICAL DATA

We obtained climatological and meteorological data from published sources. We examined two Arctic climate indices with potential influence on the region: the Arctic Climate Regime Index (ACRI), and the Arctic Oscillation (AO). Recently, Proshutinsky has refined and re-evaluated the ACRI, resulting in some slight changes to the originally published ACRI values (Johnson et al., 1999). Data for the AO were obtained from http://www.cpc.ncep.noaa.gov/products/precip/CWlink/daily_ao_index/ao_index.html.

We also related growth patterns to three regional indices that influence North Pacific ecosystems: the North Pacific Index NPI-Aleutian Low, reflecting the intensity of the mean winter (November through March) Aleutian low pressure cell; the Siberian/Alaskan Index, the difference between mean winter (December through March) pressure anomalies in eastern Siberia and the Yukon (Alaska); and the Pacific Decadal Oscillation (PDO), a recurring pattern of variability in climate with pan-Pacific effects on marine ecosystems (Overland and Wang, 2005; Overland et al., 2008). For the PDO, we examined the impact of the annual PDO and the summer (June through September) and winter (December through March) indices. The North Pacific Index-Aleutian Low, Siberian/Alaskan Index, and the PDO values were obtained from http://www.beringclimate.noaa.gov/data/.

Meteorological data for the Kotzebue Airport were obtained from the Western Regional Climate Center (http://www.wrcc.dri.edu). We used these data to calculate mean winter (December through March) and summer (June through September) values for wind and mean summer, winter, and annual values for air temperature and precipitation.

Local ice conditions were estimated from the data point nearest Kotzebue Sound collected by the Nimbus-7 SMMR and DMSP SSM/I passive microwave satellite and obtained from the National Snow and Ice Data Center (http://nsidc.org/data/seaice/pm.html). The spatial resolution of the satellite imagery is 25×25 km, and the cell used for the ice analysis was located at $67°\ 5.4'$ N, $163°\ 41.5'$ W. The temporal resolution is daily from 1990 to 2005. For our analysis we used the Julian date of freeze-up and break-up defined as the day when ice concentration first falls above or below, respectively, 50%. The threshold used to define ice-free days was ice cover <25%.

We obtained data on total Arctic-wide spatial extent (million km^2) of pack ice from the US National Snow and Ice Data Center (https://nsidc.org/data/seaice_index/). We used these data to determine several measures of regional ice conditions: annual average extent of total ice in the Arctic, maximum ice extent (typically in March), autumn ice (average October through December ice coverage), ice coverage the previous 6 months (January through June of the year preceding growth), and an ice anomaly index (percent difference between the average annual total ice coverage and the average Arctic ice cover from 1980 to 2006).

MOLLUSK STATISTICAL ANALYSIS

To verify if averaging the growth rates of the individuals sampled was warranted, we computed the Cronbach α measure of reliability (Bland and Altman, 1997) for the available growth data on a common set of years (1990–2003). The reliability coefficient measures the homogeneity of the mollusk growth rates. Cronbach α was 0.81 for *Serripes*; values of α above 0.7 are considered fairly reliable. The growth of individual *Neptunea* was remarkably homogeneous with a very high Cronbach α measure of reliability (0.94).

We used linear mixed modeling (for example, Pinheiro and Bates, 2000) on individual growth rates, as implemented in the R package nlme, to identify significant relationships between, in this case, mollusk growth rates and the environmental variables. We also investigated the time-dependence between data in consecutive years, leading us to incorporate two data transformations: 2-year running means of environmental data to reduce the magnitude of interannual variability of environmental data, and a 1-year lag in both the original variables and in their running means to account for the time it may take for physical processes to be reflected in shell growth. In these models the mollusks define the random effect, and an autoregressive lag-1 correlation is incorporated into the modeling at the mollusk level across the years.

All subsets of two predictor variables were then investigated (over 6000 models) as well as all subsets of three predictors (over 200,000 models) in an attempt to detect optimum combinations of variables to explain mollusk growth. For the best subsets, interaction effects were also investigated. Models were selected based on the AIC criterion, which penalizes the number of parameters in the model. In order to measure the success of the model in recovering mean growth rates, the predicted growth rates from the models were correlated with mean observed growth rates and then squared to give an R^2 measure similar to that obtained in regression.

ASSESSING TRADITIONAL ECOLOGICAL KNOWLEDGE

There were two phases of sampling for the TEK portion of the study (Johnson and Weller, 2002). In Phase I, in-depth interviews were conducted with a non-probability sample of Iñupiaq hunters and fishers in the region who were identified as being knowledgeable about the Kotzebue Sound ecosystem, including hunters and fishers from the villages of Kotzebue and Noatak. In Phase II, the sample consisted of the top 79 hunters and fishers as determined from hunting and fishing records provided by the Native Village of Kotzebue (NVK). The interview protocol for the study was approved by both the East Carolina University Institutional Review Board and the NVK and written consent was obtained from all interviewees.

The Phase I open-ended interviews focused on individuals' uses of local natural resources and the behavior of marine organisms, including their views about how natural resources functioned and changed over time, yielding 25 ecological narratives. During these interviews, respondents routinely spoke of the ways that various features of the natural environment had either changed or not changed over their lifetimes. The narratives were thematically coded and common themes and associated propositions were compiled in an agree/disagree format (Johnson and Weller, 2002). We developed a list of 102 propositions, and subsequently asked the 79 Iñupiaq hunter-fishers (Phase II sample) whether or not they agreed or disagreed with each of the propositions (e.g., see **Table 1**).

Lists of agree-disagree propositions are central to eliciting what is called "cultural consensus," or consensus among informants regarding specific domains of knowledge. It is formally called the Cultural Consensus Model (CCM), and is a way to understand culture as a matter of belief and knowledge agreement (Romney et al., 1986). The CCM allows for an assessment of the extent to which individuals within a culture have a shared understanding of a set of beliefs and allows for an assessment of intracultural variation within a shared understanding. Here we used the formal model to assess cultural consensus (Romney et al., 1986). For respondents' dichotomous responses (agree/disagree) to fit the model, the rule of thumb is the ratio of the first to second eigenvalue in a minimum residual factor analysis of the respondent's responses should be greater than 3, there should be no negative factor scores on the first factor, and the mean of the factor scores should be >0.5. **Table 1** provides the propositions for the marine ecosystem domain (a subset of the 102 propositions) and the culturally correct answers derived from a Bayesian weighting method described by Romney et al. (1987). Also included for the first five propositions are the responses reflecting the change index described below. Supplemental Table 1 lists the 37 propositions (out of the 102) used in the analysis of the bearded seal (*Erignathus barbatus*, known as ugruk in the Iñupiaq language) knowledge domain for comparison.

We further refined the marine ecosystem domain propositions, and derived a change index while investigating the

Table 1 | (A) The five propositions comprising the Climate Change Knowledge Index (CCKI) as derived from a factor analysis of all 35 propositions, the classification of the statement as either change related or system related, the statement topic, the culturally correct answer (as determined by the cultural consensus model using Bayesian modeling), and the answer for the index. (B) The remaining thirty agree/disagree propositions, the classification of the statement as either change related or system related, the statement topic, and the culturally correct answer (as determined by the cultural consensus model using Bayesian modeling).

(A)

Statement	Change/System	Topic	Culturally correct answer	Change index answer
The temperature of the water is a lot warmer than 10 years ago.	Change	Climate	Agree	Agree
The first salmon are arriving earlier than they used to.	Change	Fish/Invertebrates	Agree	Agree
The increase in water temperature in the Sounds is bringing in more crabs to the area.	Change	Fish/Invertebrates	Disagree	Agree
People are getting more flounders in their nets today than in the past.	Change	Fish/Invertebrates	Disagree	Agree
The trout are going out earlier than usual.	Change	Fish/Invertebrates	Disagree	Agree

(B)

Statement	Change/System	Topic	Culturally correct answer
Over the past few years, freeze-up has been longer and break-up a little bit earlier.	Change	Climate	Agree
The west winds in the summer are not coming as much as they used to.	Change	Climate	Agree
The temperatures on the whole are warmer throughout the year.	Change	Climate	Agree
People are beginning to get more pink salmon in the Sounds.	Change	Fish/Invertebrates	Disagree
The last 3 or 4 years there have been less trout.	Change	Fish/Invertebrates	Disagree
There has been an increase in dirty ice.	Change	Ice	Agree
The ice has been staying longer in the spring than it used to.	Change	Ice	Disagree
There is very little difference in ice conditions from 1 year to the next.	Change	Ice	Disagree
The ugruks and the seals aren't any skinnier or fatter, but are about the same as always.	Change	Marine mammals	Agree
Over the last 15 years the ugruk population in the Sounds has stayed about the same.	Change	Marine mammals	Agree
Some years there's so many boats out there that the ugruk won't stay up on the ice.	Change	Marine mammals	Agree
There is less beluga today because of all the outboard noise and exhaust.	Change	Marine mammals	Agree
More porpoise have been showing up in recent years.	Change	Marine mammals	Agree
Beaver moving into this country are blocking the ability of the whitefish to spawn.	Change	Terrestrial and Fish/Invertebrates	Disagree
People have begun taking more animals than they can use.	Change	Terrestrial and Marine mammals	Agree
Mussels and clams come up along the beach whenever you get a good west wind or storm.	System	Fish/Invertebrates	Agree
Herring come in right when the Kobuk ice starts breaking up.	System	Fish/Invertebrates	Agree
Break-up is a good time to get sheefish.	System	Fish/Invertebrates	Disagree

(Continued)

Table 1 | Continued

(B)

Statement	Change/System	Topic	Culturally correct answer
Sheefish are the first fish that come out from the rivers under the ice.	System	Fish/Invertebrates	Agree
Sheefish very seldom are taken in the oceanfront along Sisaulik.	System	Fish/Invertebrates	Agree
The adult tomcods come out with the freshwater flush of the Noatak in springtime.	System	Fish/Invertebrates	Disagree
The tomcods lay their eggs in the waters just in front of Kotzebue in December and January.	System	Fish/Invertebrates	Disagree
When ice fishing in front of Kotzebue for tomcod the best time is when the tide is going out.	System	Fish/Invertebrates	Agree
As the ice first breaks up trout migrate right along the coast of Krusenstern and Sisaulik.	System	Fish/Invertebrates	Agree
The less snow covering in the winter, the thicker the ice.	System	Ice	Agree
When there are a lot of heavy east winds in the spring the ice leaves the Sounds quickly.	System	Ice	Agree
It is difficult to read the ice after a fresh snow.	System	Ice	Agree
When there is less rain, there are fewer berries.	System	Terrestrial	Agree
The wind affects our tidal changes here in the Kotzebue Sounds more than anything else.	System	Weather	Agree
A lot of east wind in the winter can lead to thinner ice in the spring.	System	Weather	Agree

Table 2 | Factor loadings from a minimum residual factor analysis for the top five propositions used in the Climate Change Knowledge Index (CCKI).

1st factor loadings	Propositions
0.601	The temperature of the water is a lot warmer than 10 years ago.
0.540	The first salmon are arriving earlier than in the past.
0.508	The trout are going out earlier than usual.
0.481	People are getting more flounders in their nets today than in the past.
0.463	The increase in water temperature in the Sound is bringing in more crabs to the area.

relationship between expertise and normative cultural ecosystem beliefs. The 35 change propositions (**Table 1**) were intercorrelated and subjected to minimum residual factor analysis. The first factor contained five propositions with high factor scores (**Table 2** and the first five propositions of **Table 1**). These five propositions were related to increases in water temperature, earlier salmon returns, increases in flounder catches, trout leaving earlier, and increases in the Sound crab populations. Responses to the 5 change propositions (1, 0) were summed across respondents to produce the Climate Change Knowledge Index (CCKI).

During Phase II interviews, we also elicited information that allowed us to further define the fishers' and hunters' knowledge networks. The fishers' knowledge network was developed by asking the 79 respondents to name the five individuals they thought were most knowledgeable about fish and fishing in the Kotzebue Sound (**Figure 2A**). The hunters' knowledge network was derived similarly (**Figure 2B**). This resulted in two n × m matrices of respondents (rows) reports of whom they perceived as knowledgeable about fishing/hunting (columns). These two-mode networks were transformed into bipartite graphs and symmetrized. This yielded two n × n matrices where the *i, j*th entry is the presence or absence of a knowledge relation between two respondents. Betweenness centrality (Freeman, 1977) was used to determine knowledge experts in the network. The definition of betweenness centrality is:

$$b_j = \sum_{i,k} \frac{g_{ijk}}{g_{ik}}$$

where b_j is the betweenness centrality of node j and g_{ijk} is the number of geodesic paths (shortest paths) connecting i and k through j and g_{ik} is the total number of geodesic paths connecting i and k (Borgatti et al., 2013). The measure, as used here, is normalized by dividing b_j by the maximum possible betweenness thereby expressing the measure as a percentage. The more an individual respondent connects respondents who are themselves not connected, the higher their betweenness centrality.

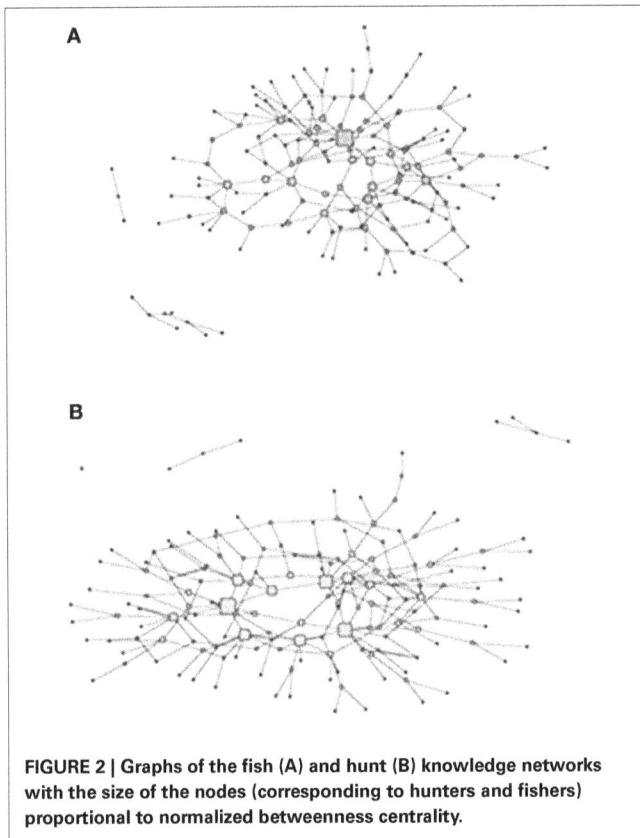

FIGURE 2 | Graphs of the fish (A) and hunt (B) knowledge networks with the size of the nodes (corresponding to hunters and fishers) proportional to normalized betweenness centrality.

Table 3 | Multiple regression for Climate Change Knowledge Index (CCKI) as dependent variable with models using demographic and non-network independent variables (Model 1; $R^2 = 0.279$) and a model including network independent variables (Model 2; $R^2 = 0.421$).

Effect	t	t
Constant	2.816	2.631
Age	**2.593****	**2.909****
Store bought (%)	−1.016	−1.436
Kotzebue resident (dummy)	−0.187	−0.016
Education	−1.662	**−2.074***
Wage labor (dummy)	**−2.212***	**−2.401****
Marine subsistence (%)	0.253	0.095
Hunt know expert		**−2.627****
Fish know expert		**2.480****

[Age, age of respondent; Store Bought (%), percentage of diet from store food; Kotzebue Resident (dummy variable: 1, 0); Education, years of education; Wage Labor (dummy variable: 1, 0); Marine Subsistence (%), percentage of food from marine systems; Hunt Know Expert and Fish Know Expert, the normalized betweenness centralities in the knowledge networks of hunter and fishers, respectively]. Significant relationships are in bold. The levels of significance are: *p < 0.05, **p < 0.01.

This generally reflects expertise and brokerage abilities in knowledge and communication networks (Maiolo et al., 1992). The two independent expertise variables using betweenness centrality are "Fish Know Expert" and "Hunt Know Expert."

We used a general linear model to investigate the relationship between perceived change (CCKI) and expertise while controlling for a number of demographic independent variables including: Age, Education (number of years of formal schooling), Store Bought (percent of food purchased from store), Kotzebue resident (dummy variable, where 1 = Kotzebue residence, 0 otherwise), Wage Labor (dummy variable, where 1 = engaged in wage labor, 0 otherwise), and Percent Marine Food (percent of food that is marine subsistence including mammals) (**Table 3**). Intercorrelations among the independent variables were conducted in order to limit any potential problems with multicollinearity. Finally, the two expertise variables were simply the normalized betweenness centrality measures in both the hunting and fishing knowledge networks as calculated in UCINET (Borgatti et al., 2002).

RESULTS

SCIENTIFIC ECOLOGICAL KNOWLEDGE

As mentioned in the methods, the growth patterns of individual cockles and of individual whelks were homogeneous within a year, meaning that individuals of the same species were responding similarly to environmental conditions and individual SGIs could be reliably averaged. The mean standard growth index (SGI) for each calendar year for both species of mollusks varied considerably over the 22 years of the data set (**Figure 3**), and

patterns for both species exhibit two distinct phases. Before 1996 (for *Serripes*) and 1995 (for *Neptunea*) SGIs are consistently below 1.0, representing relatively slow growth, and with little interannual variability. Subsequently, SGIs are near or above 1.0 for both species, with a high degree of interannual variability for *Serripes*. The growth of both species declines after 2001 (*Neptunea*) and 2002 (*Serripes*), with this decline continuing to the end of the chronology in 2005 for *Neptunea*.

In general, the relationship between the SGI's and potential environmental predictors was improved with the 2-year running means and 1-year lags of the environmental variables, as compared to the original variables. Two environmental examples, and the corresponding lagged predictors, are shown in **Figure 3**, illustrating the improved relationship, although we caution that the effects of single predictors are not comparable to the multiple predictor models reported next, which additionally include interaction effects. The amount of variability in *Serripes* growth explained by AO summer increased by more than a factor of 2 from 9 to 21% when the 2-year running mean of AO summer was used compared to the yearly AO summer (**Figure 3**). Total Arctic Ice the previous winter explains 11.6% of *Neptunea* growth when the predictor is lagged a year compared to only 3.7% unlagged, though this relationship is not significant as a single predictor of *Neptunea* growth.

Based on the linear mixed modeling, the best model for *Serripes* includes two regional parameters and their interaction:

$$\text{SGI} = 1.1314 - 0.9832 \, \text{AOsummer}_{m2} \qquad p = 0.004$$

$$- 0.8607 \, \text{TotalArcticIce}_{-1} \qquad p < 0.0001$$

$$+ 3.4427 \, \text{AOsummer}_{m2} \times \text{TotalArcticIce}_{-1} \qquad p = 0.005$$

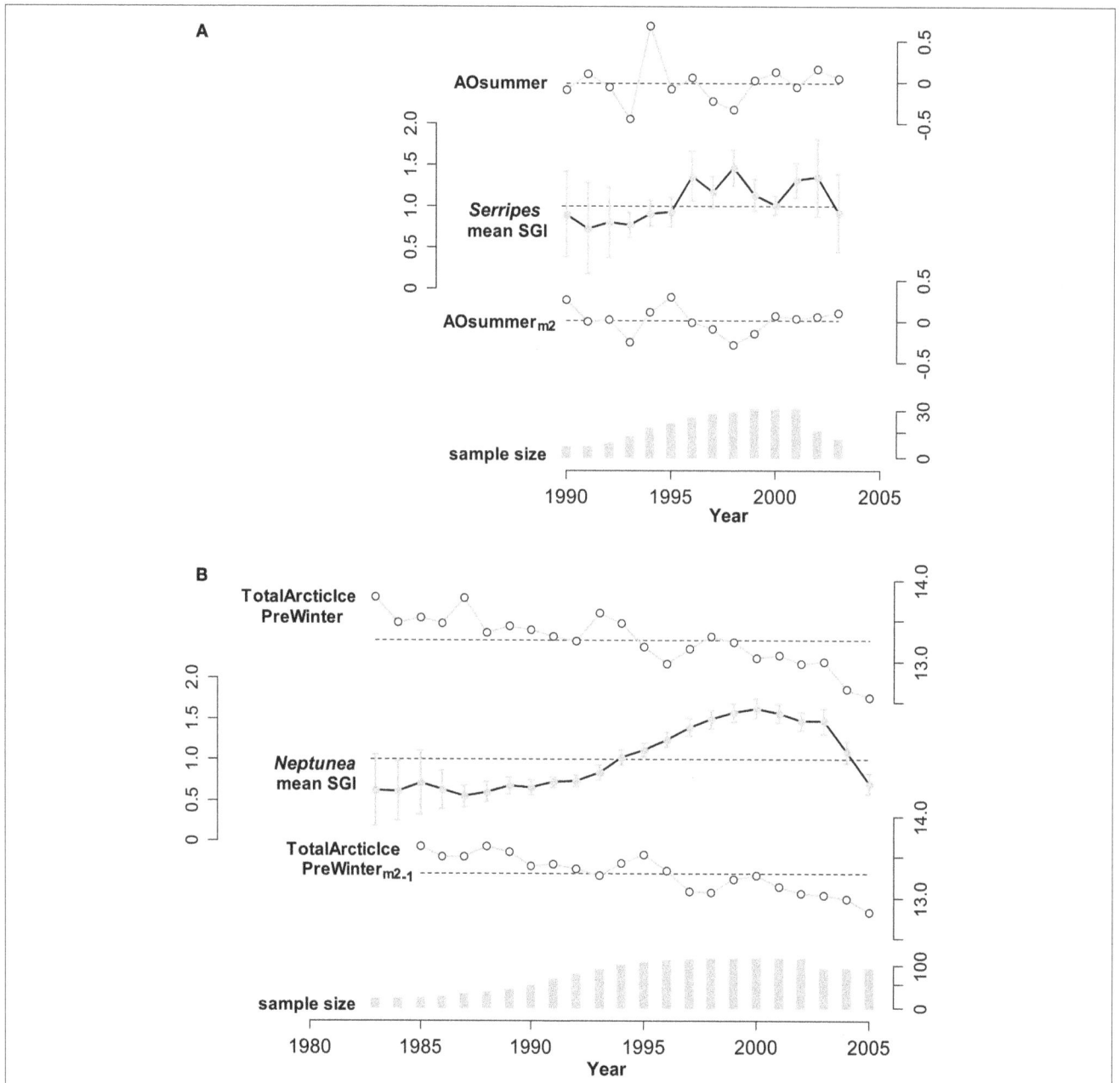

FIGURE 3 | Illustrations of annual means for an environmental predictor unmanipulated, and modified as suggested by the linear mixed modeling results (described in the text), and the SGI (Standard Growth Index; units are relative to expected growth) for: (A) *Serripes,* **using predictor AOSummer (Arctic Oscillation; relative index), and (B)** *Neptunea,* **using predictor TotalArcticIcePreWinter (Previous Winter; units are millions of square kilometers).** The scales of the variables are shown alongside the corresponding series. For each predictor, the mean across all years is shown as a dashed line; for the SGI the dashed line shows a value of 1 (expected growth = observed growth). Vertical bars are 95% confidence intervals for the SGI means, using the *t*-distribution appropriate for each annual sample. For **(A)**, the unmanipulated variable (AOsummer) has an effect size of −0.124 on *Serripes* SGI (estimated from mixed model, variance explained in mean *Serripes* SGI is $R^2 = 0.092$, $p = 0.32$), and the 2-year running mean has a stronger effect size of −0.690 ($R^2 = 0.212$, $p = 0.002$). For **(B)**, the unmanipulated variable has an effect size of 0.052 ($R^2 = 0.037$, $p = 0.46$) on *Neptunea* SGI, whereas the 2-year running mean, lagged by 1 year, has an effect size of −0.117 ($R^2 = 0.116$, $p = 0.23$). In this latter case, variance explained does increase, from 3.7 to 11.6%, but the variable is not significant as a single predictor and becomes important in combination with the other two predictors in the final *Neptunea* model.

where the subscript m2 represents the 2-year running mean of the variable, and −1 a 1-year lag. In order to make the model coefficients (estimated effect sizes) more meaningful, the predictors have been centered with respect to their means, so that AOsummer$_{m2}$ is actually [AOsummer$_{m2}$ − (− 0.02065)] and TotalArcticIce$_{-1}$ is [TotalArcticIce$_{-1}$ − 11.87]. The constant 1.1314 is then the predicted value of SGI at the mean values of the two predictors.

This model explains 80.3% of the variability in mean growth. The interaction term means that the negative effect of Total Arctic Ice (lagged 1 year) is less as the 2-year running mean of the AO in summer increases. For example, if AOsummer$_{m2}$ is increased by 0.1 over its mean value, then the effect of TotalArcticIce$_{-1}$ is increased by $0.1 \times 3.4427 = 0.3443$, becoming less negative. The modeled relationship between these variables and the effect of the interaction term can be clearly seen when the terms are plotted in three dimensions (see the Video in Supplemental Figure 1).

The best model for *Neptunea* includes a mixture of regional and local parameters and is more complicated than for *Serripes* because it contains three terms and an interaction term:

$$SGI = 1.2278 - 0.2316\ \text{AirTempSummer}_{m2} \qquad p < 0.0001$$
$$- 1.2047\ \text{TotalArcticIcePreWinter}_{m2_{-1}} \qquad p < 0.0001$$
$$+ 0.004756\ \text{PrecipSummer}_{m2_{-1}} \qquad p < 0.0001$$
$$+ 0.5027\ \text{AirTempSummer}_{m2}$$
$$\times\ \text{TotalArcticIcePreWinter}_{m2_{-1}} \qquad p < 0.0001$$

Subscripts are as above for *Serripes* and m2$_{-1}$ is a 1-year lag of the 2-year running mean. Once again, the environmental variables have been centered with respect to their means: 9.197°C for AirTempSummer$_{m2}$, 13.22 million square kms for TotalArcticIcePreWinter$_{m2_{-1}}$ and 154.9 mm for PrecipSummer$_{m2_{-1}}$.

This model explains 79.3% of the variability in *Neptunea* growth and uses the 2-year running mean for all parameters, with ice and precipitation also lagged a year. The interaction between summer air temperature and total Arctic ice the previous winter (2-year running mean lagged 1 year) means that the negative effect on SGI of the previous winter total Arctic ice extent is lessened with increasing summer air temperature.

TRADITIONAL ECOLOGICAL KNOWLEDGE

TEK, integrating over longer time scales than the mollusks we examined, can further elucidate the apparent shift in the coastal ecosystem that began in the mid-1990s. Of the 35 separate agree/disagree propositions concerning coastal ecosystems, 20 addressed change/variability, and 16 specifically addressed marine fish or invertebrates (with 7 of the 16 fish/invertebrate propositions related to change) (**Table 1**). These propositions derived from in-depth interviews with Iñupiaq hunter-fishers allowed for the systematic modeling of cultural ecosystem beliefs and their variation across individuals and groups. In the factor analysis of the inter-correlations among respondents' answers, (i.e., factor analysis of the people rather than the propositions), the ratio of the first to second eigenvalue for coastal ecosystem knowledge was 3.19 and all the factor scores were positive, indicating cultural consensus among the respondents for propositions in this domain (Romney et al., 1987). It is not, however, nearly as strong as the consensus found for bearded seal (ugruk) knowledge (ratio of 7.78, **Table 4**, with questions presented in Supplemental Table 1). In fact, the coastal ecosystem knowledge is approaching the classification of "proto-cultural" (Caulkins, 2004) (knowledge just entering the cultural system) due to

Table 4 | Comparison of the cultural consensus tests between two domains.

Cultural domain	1st to 2nd eigenvalue ratio	Mean competence	Range
Change/Fish	3.192	0.409	0–0.75
Ugruk (bearded seal)	7.778	0.632	0.14–0.89

Both fit the model but the ugruk (bearded seal) domain shows higher consensus in comparison to the change/fish domain.

higher levels of intra-cultural variability within the shared understanding. A comparison among the various cultural and TEK knowledge domains assessed indicates that cultural knowledge competency in one domain does not necessarily translate to such competency in others for the Kotzebue community (**Table 5**). It is important to note that our overall study addressed several topics, and included both hunting and fishing experts.

Further breakdown of the 35 coastal ecosystem propositions resulted in the previously mentioned CCKI a subset of 5 statements that dealt explicitly with change. **Table 1A** compares the culturally correct answers for the 5 statements as determined by the CCM with the responses for the change index. There is general agreement between the two that the water is getting warmer and that the salmon are returning earlier. However, the index reflects more change, particularly with respect to changes in the behavior of some fish and in increasing numbers of some species being observed in the Sound. The CCKI showed that respondents who perceived more change were often at odds with the overall normative ecological beliefs, particularly for the fish/climate domain (**Table 5**, $r = -0.526$, $p = 0.0001$). We subsequently used the knowledge network information to further characterize the hunters/fishers who believe the ecosystem is changing.

The size of the nodes in the networks for both the hunters and the fishers (**Figure 2**) is proportional to their betweenness centrality and is used as index of expertise in each of the domains as described in the Section Materials and Methods. The two networks are similar in structure, but vary slightly in terms of the distribution of centrality values. Whereas both have a core periphery structure, the fish knowledge network core is more dominated by a single fisher in the core (including a number of fishers with moderate centrality), while the hunt knowledge network has a more uniform distribution of centrality among core members. In both cases, however, the hunters and fishers with higher centrality are in the core of the network linking to other actors in the periphery of the network. The extent to which hunters and fishers are central, in terms of betweenness centrality, in the two networks reflects domain expertise in that they receive knowledge nominations from a broader range of hunters and fishers who are themselves not connected in the network. In addition, hunters and fishers in these central positions in the knowledge network would have access to a wider range of shared ecological knowledge as well as more novel ecological information.

We compared expertise, as determined by normalized betweenness centrality in the knowledge networks, and the change index (CCKI) while controlling for a set of other possible

Table 5 | Relationships among cultural ecological knowledge domains.

	Ugruk (bearded seal) knowledge	Food web knowledge	Change knowledge Fish/Climate	Fall seal knowledge	Change index
Ugruk knowledge	1.000				
Food web knowledge	−0.141	1.000			
Change knowledge	−0.013	0.180	1.000		
Fall seal knowledge	−0.040	0.061	**0.460***	1.000	
Change index	**0.379***	−0.196	**−0.526***	**−0.327***	1.000

*Pearson correlation coefficients relating various knowledge domains and the Change Index. Significant correlations are in bold. The levels of significance are: *p < 0.05, ***p < 0.001.*

independent variables including age, education, residence, engagement in wage labor, percent of total food purchased at stores, and percent of food from marine subsistence. Interestingly, the two expert subsets of the population (hunters and fishers) had different perspectives on coastal ecosystem change. In a regression of factors that may be driving the increased perceptions of change, both age and fish expertise (Fish Know Expert) were positively related to the change index (CCKI), while involvement in wage labor, education, age, and hunting expertise (Hunt Know Expert) were negatively related (**Table 3**). This suggests that the fish experts were more likely than others, particularly the hunting experts, to perceive increased coastal change that was outside the range of the normative ecological beliefs as reflected in the consensus analysis of the change statements. Hunters and fishers are aware of the natural variability in the climate over time and as such these normative ecological beliefs already incorporate the normal range of variation that might result from such things as the Arctic Oscillation. For example, in response to the statement on variation in ice conditions from 1 year to the next (**Table 1B**), the culturally correct answer clearly pointed to recognition of annual variability.

DISCUSSION

Local knowledge of ecosystems has become increasingly valued and used in ecosystem and resource management over the past three decades (Johannes, 1981, 1984; Berkes et al., 2000; Le Fur et al., 2010). When combined with SEK, this often yields a more holistic view of ecosystems than either knowledge base alone (Huntington et al., 2011; Ferguson et al., 2012). Combining TEK and SEK could be especially useful in the Arctic where long-term historical data are lacking (Wassmann et al., 2011), and indigenous peoples have accumulated environmental information for many generations (Huntington et al., 2011). Despite the acknowledgement that TEK can inform SEK and lead to an enhanced environmental understanding in the Arctic (Huntington et al., 2004; Nicholas et al., 2004; Laidler, 2006), few studies have successfully combined the two ways of knowing in Arctic systems (Mahoney et al., 2009; Weatherhead et al., 2010; Carter and Nielson, 2011; Huntington et al., 2011; Riseth et al., 2011). We demonstrate that Iñupiaq fishers are especially attuned to perceiving changes in coastal climate and they can provide the longer time frame needed to interpret the high-resolution changes we see in the growth rate of mollusks. This provides a better understanding of climate change in Kotzebue than if we

had relied on either TEK or mollusk growth alone as a climate proxy.

Annual growth patterns of both *Neptunea* and *Serripes* reflect variations in large-scale climate drivers in the Kotzebue Sound/Chukchi Sea system and local manifestations of these drivers. The relationship we document between *Serripes* growth and the Arctic Oscillation (AO), which is closely related to the PDO (Sun and Wang, 2006), is consistent with studies of *Serripes* growth in the European Arctic where the relationship between *Serripes* growth and regional climate oscillations is well established (Ambrose et al., 2006; Carroll et al., 2009, 2011a). Even though the relationship was not strong enough to enter the *Neptunea* growth model, *Neptunea* growth was also negatively related to the annual AO (1 year lag of the 2-year running mean; effect size $= -0.3943$, $R^2 = 0.332$, $p < 0.001$). No study has linked growth of a buccinid to a regional climate index or environmental conditions in the field, but the deposition of annual lines in the operculum and statolith and the longevity of some taxa (Richardson et al., 2005) make members of this genus a good candidate for future climate change studies.

There is not always a direct relationship between a climate index and local conditions (Stenseth et al., 2003), but ultimately the growth of organisms is determined by the manifestation of climate oscillations on the local environment (Ambrose et al., 2006). *Serripes* growth is best explained by large-scale patterns of ice extent. Interestingly, it is the Arctic-wide ice pattern (total Arctic ice) that is a better predictor of growth than the local ice conditions (freeze up, ice free days) in the mixed-effects model. The local conditions are based on conditions in a 25 × 25 km area, which may not be as robust at predicting larger-scale conditions in Kotzebue Sound as Arctic-wide metrics. Annual phytoplankton production in the Arctic is directly proportional to the length of the open water period (Rysgaard et al., 1999) and this relationship may be driving the effect of ice cover on growth of *Serripes*, as has been proposed for another suspension feeding cockle (Sjer et al., 2009).

Both regional and local factors enter the mixed model explaining *Neptunea* growth. As with *Serripes*, *Neptunea* growth is also negatively affected by a large-scale measure of ice cover, total Arctic ice the previous winter. Unlike *Serripes*, however, local parameters are also important predictors of *Neptunea* growth. Whelks feed primarily on bivalves (including *Serripes*, Ambrose, personal observation 2002) and polychaetes (Shimek, 1984) and benthic biomass on Arctic shelves is inversely related to ice cover

(Ambrose and Renaud, 1995), so ice cover the preceding years might affect the abundance of *Neptunea* prey and the predator's growth. The mechanisms whereby summer temperature and precipitation affect *Neptunea* growth are not clear and complicated by an interaction between these two parameters; any explanations without experimentation would be speculative.

Even similar species in the same production regime can have opposite responses to a regime shift (Benson and Trites, 2002), so it is remarkable that species at two trophic levels exhibited relatively simultaneous and significant shifts in annual growth patterns. Only one other study has documented an effect of the AO on two trophic levels simultaneously (Aanes et al., 2002). The AO index shifted from a strong positive to a negative or neutral phase after 1995 (**Figure 3**; Thompson and Wallace, 1998; Overland and Wang, 2005) concurrent with the PDO switching from a warm to a cool phase (Matua and Hare, 2002) and an increase in the growth of *Serripes* and *Neptunea* (**Figure 3**). A major restructuring of the ecosystem in the northern Bering Sea has been hypothesized to have occurred around 1996 and is attributed to a reduction in the strong positive phase of the AO resulting in stronger southerly winds, less ice, and warmer temperatures over the northern Bering Sea and eastern Siberia (Grebmeier et al., 2006). Our growth data clearly support the documented changes in the regional climate in the mid-1990s.

Growth of the mollusks was best explained by local and regional parameters when parameters were lagged a year relative to growth or when a 2 year running mean of the parameter was used to incorporate the previous year's conditions, or both (**Figure 3**). Lagged response to climate oscillations are common in marine systems (Overland et al., 2010) and can typically span many trophic levels (Post, 2004) from benthic infauna (Tunberg and Nelson, 1998), including bivalves (Witbaard et al., 2003; Ambrose et al., 2006; Carroll et al., 2011b), to zooplankton (Pershing et al., 2005), fish (Ottersen et al., 2005), and birds (Thompson and Ollason, 2001; Hovinen et al., 2014). This lagged response is well explained by the double integration hypothesis where atmospheric forcing affects large-scale environmental factors (e.g., sea surface temperature, ocean circulation), which in turn affect population dynamics (Bestelmeyer et al., 2011; Di Lorenzo and Ohman, 2013; Doney and Sailley, 2013).

Mollusks are frequently touted as excellent biomonitors for reconstructing environmental conditions (Wanamaker et al., 2011), especially in the Arctic (Mann et al., 2013; Carroll et al., 2014), based on their close relationship between growth and environmental conditions and also because the chemical and mineral content of their shells can be a valuable proxy for environmental conditions. While we only make use of variation in growth in our study, the close correspondence between growth rates and environmental conditions we documented indicates that *Serripes* and *Neptunea* growth rates are good proxies for environmental conditions. Without a much longer dataset, though, it is unclear from SEK alone whether the shifts we see in growth are a result of a decadal oscillation, as the relationships between growth and regional climate indices would suggest, or, in contrast, is related to a more sustained climatic change.

To further explore longer term change, we draw on our TEK results that indicate the Kotzebue Sound ecosystem has been undergoing changes on a broader time scale than would be evident from natural oscillations alone. Also in support of this assertion, Moerlein and Carothers (2012) collected TEK via ethnographic methods in the Northwest Alaska communities of Noatak and Selawik, very near Kotzebue, and concluded that the changes in Northwest Alaska over the last 20–30 years are "without precedent and outside of the normal range of variation." Many of the more observable changes are occurring within the Sound ecosystem as reflected in the clam growth analysis. It is therefore not surprising to find that the older, more experienced fish experts were the first to observe such changes, which includes reported changes in the behavior and increased presence of several marine species. Furthermore, the fact that experts are seeing these changes before other marine mammal hunters and fishers points to the beginnings of the diffusion of new cultural ecological knowledge and understandings. Over time this incipient knowledge should gain broader cultural consensus, eventually representing a new ecological normative understanding, possibly at the same shared level of understanding as exhibited with the bearded seal (ugruk) behavior.

One reasonable hypothesis to make from our results is that it is the experts who would be the most likely to see major ecosystem changes before anyone else. After all, they are the ones who are more experienced and knowledgeable about the ecological and environmental factors, such as air and water temperature, essential for being successful as a hunter or fisher. In addition, they are more committed to the subsistence way of life in that they are older, have less formal Western education, and do not tend to engage in forms of wage labor. They have spent most of their lives on the water and ice fishing. Therefore, it is the fish experts who have intimate knowledge of spawning behaviors and marine species assemblages that are seeing unprecedented increases in some species, particularly benthic species such as crabs. Although we interviewed hunters and fishers at a single point in time, we argue here that the difference in beliefs between fish experts and the traditional cultural beliefs reflects the beginnings of the diffusion of new cultural knowledge.

This is not to say that it is only the fish experts that perceive change in Kotzebue. There was clearly agreement that the ice is breaking up earlier and freezing later, the west summer winds are becoming less frequent, and the air temperatures are getting warmer across the seasons (**Table 1B**). Further, as is evident from a comparison of responses to the change statements in **Table 1A**, there is overall cultural agreement among hunters and fishers that the water in the Sound is getting warmer and the salmon are returning earlier. It is the fish experts, however, more than any other group, who recognize the connection between the warming of the water and the increasing numbers of crabs entering the Sound (**Table 1A**). In addition, they are also observing increases in flounder numbers and changes in trout behavior, changes that have yet to be noticed by others, in particular, the hunting experts. The hunting experts tend to concentrate more on marine mammals, such as bearded seals (ugruk), who interface with hunters on the surface of the water or on the ice. In the consensus analysis there was general agreement that the bearded seal population numbers have been relatively stable over the last 15 years and that the fat content of the seals has stayed relatively the same

(**Table 1B**). What this suggests is that ecosystem changes due to a shift in climate may be more readily observed in the marine ecosystem, particularly the benthos. It also seems to be the case that the shifting climate has had less effect on marine mammal populations and behavior, at least at the time of our study. Recent scientific research has suggested that the pace of a shifting climate may be more pronounced in the ocean than on land at similar latitudes (Burrows et al., 2011). If this is the case, then the fact that the fish experts are noticing ecosystem change before others certainly follows.

Although there is not a direct mollusk-to-mollusk comparison for the two types of data, the fish experts observed changes and increases in the abundance of some benthic species that clearly suggests ongoing changes in the environmental conditions in the Kotzebue Sound, conditions that would also impact the mollusks. The recent decreases in ice cover will increase the growth rates of both *Serripes* and *Neptunea*. These changes are likely to have profound impacts on the structure of the marine community, especially the benthos, and on subsistence hunting. There will be a longer open-water fishing season due to less ice, more crabs to fish, and faster growing clams (*Serripes*) will provide more food for bearded seals (ugruk). On the other hand, the reported changes in ice conditions do not bode well for traditional, ice-based seal hunting and new immigrant marine species may reduce the abundances of clams. While speculative, we feel our predictions, based on a combination of TEK and SEK, are robust, and should be useful for future local (town), regional (borough), and statewide planning, as well as for scientific modeling of ecosystem responses to climate change.

It is challenging to determine whether a deviation in environmental conditions at a given time is due to a shift to a new climate regime or to natural cycling. The 15–20 years of mollusk growth data clearly indicate a change in growth conditions in the middle of the 1990s (**Figure 3**). The SEK data set alone cannot discern if this change is part of a decadal climate oscillation, a fundamental change in climate affecting the near-shore ecosystem, or a combination of the two. The TEK shared by the Iñupiaq ecomonitors provide insight into ecosystem change not revealed by mollusks. The experts' knowledge of the ecosystem is typically very local, integrative, and is longer in duration than the time frame provided by the mollusks we studied. The perceived changes in climate and subsequently in the coastal ecosystem are outside the range of natural variation and are best understood by the older, more experienced fishers, who are less involved in wage labor, and these change perceptions appear to represent the beginnings of the diffusion of new cultural beliefs related to climate and ecosystem change. The TEK data indicate the perceived change is so different that it is "protocultural," a recent shift in shared understanding that is making its way through the population.

Together, SEK and TEK provide more insight than either would alone; the mollusks indicate precisely when change occurred and the Iñupiaq tell us the change is not only a decadal oscillation. Recently the United States Arctic Research Commission recommended the incorporation of TEK into long-term monitoring of Arctic climate change (United States Arctic Research Commission, 2013). We have demonstrated that such a combination of scientific and TEK provides a much more holistic view of local climate change in one Arctic location than by relying solely on either approach. Application of this method across the Arctic would provide an assessment of the extent to which local ecosystems are affected by the changing Arctic climate even in the absence of continuous environmental monitoring with scientific instruments.

ACKNOWLEDGMENTS

This work was funded by a grant from the US National Science Foundation (NSF) Office of Polar Programs (OPP-0222423) to William G. Ambrose Jr., Lisa M. Clough, David C. Griffith and Jeffrey C. Johnson, the Norwegian Research Council to Michael L. Carroll, the BBVA Foundation and Spanish Ministry of Education and Competitiveness grant MTM2012-37195 to Michael Greenacre, and with funds from the Howard Hughes Medical Institute through Bates College. Lisa M. Clough is now an employee of the US NSF, however any opinion, finding, and conclusions or recommendations expressed in this material are those of Lisa M. Clough and her coauthors, and do not necessarily reflect the views of the US NSF. We thank John Goodwin for piloting the boat for sample collection and for allowing us to collect snails from his crab traps and Gerald Goodwin, Stephen Jewett, Janice Lewis, Kate Meltzer, Melinda Reynolds and Terry Reynolds for field help and Greg Henkes, Jessica Edgerly and William Locke for laboratory help. William Locke drafted **Figure 1**. Comments from R. Bernard, E. Jones, J. Muratori, P. Renaud, and G. Scheldeman and the two Frontiers in Marine Science reviewers significantly improved the manuscript.

REFERENCES

Aanes, R., Saether, B.-E., Smith, F. M., Cooper, E. J., Wookery, P. A., and Øritsland, N. A. (2002). the Arctic Oscillation predicts effects of climate change in two trophic levels in a high-Arctic ecosystem. *Ecol. Lett.* 5, 445–453. doi: 10.1046/j.1461-0248.2002.00340.x

Ambrose, W. G. Jr., Carroll, M. L., Greenacre, M., Thorrold, S. R., and McMahon, K. (2006). Variation in bivalve growth in a Norwegian high-Arctic fjord: evidence for local- and large-scale climatic forcing. *Global Chang. Biol.* 12, 1595–1607. doi: 10.1111/j.1365-2486.2006.01181.x

Ambrose, W. G. Jr., and Renaud, P. E. (1995). Benthic response to water column productivity: evidence for benthic pelagic coupling in the Northeast Water Polynya. *J. Geophys. Res.* 100, 4411–4421. doi: 10.1029/94JC01982

Ambrose, W. G. Jr., Renaud, P. E., Locke, V. W. L., Cottier, F. R., Berge, J. R., Carroll, M. L., et al. (2012). Growth line deposition and variability in growth of two circumpolar bivalves (*Serripes groenlandicus* and *Clinocardium ciliatum*). *Polar Biol.* 35, 345–354. doi: 10.1007/s00300-011-1080-4

Arctic Climate Impact Assessment. (2005). *Impacts of a Warming Arctic: Arctic Climate Impact Assessment.* Cambridge, UK: Cambridge Univ. Press.

Benson, A. J., and Trites, A. W. (2002). Ecological effects of regime shifts in the Bering Sea and eastern Pacific Ocean. *Fish Fisher.* 3, 95–113. doi: 10.1046/j.1467-2979.2002.00078.x

Berkes, F., Colding, J., and Folke, C. (2000). Rediscovery of traditional ecological knowledge as adaptive management. *Ecol. Appl.* 10, 1251–1261. doi: 10.1890/1051-0761(2000)010[1251:ROTEKA]2.0.CO;2

Bestelmeyer, B. T., Ellison, A. M., Fraser, W. R., Gorman, K. B., Holbrook, S. J., Laney, C. M., et al. (2011). Analysis of abrupt transitions in ecological systems. *Ecosphere* 2, 129. doi: 10.1890/ES11-00216.1

Bland, J. M., and Altman, D. G. (1997). Cronbach's alpha. *Br. Med. J.* 314, 572. doi: 10.1136/bmj.314.7080.572

Borgatti, S., Everett, M., and Johnson, J. (2013). *Analyzing Social Networks.* London: Sage.

Borgatti, S. P., Everett, M. G., and Freeman, L. C. (2002). *Ucinet 6 for Windows: Software for Social Network Analysis.* Harvard, MA; Analytic Technologies.

Born, E. W., Rysgaard, S., Ehlme, G., Sejr, M., Acquarone, M., and Levermann, N. (2003). Underwater observations of foraging free-living Atlantic walruses (*Odobenus rosmarus rosmarus*) and estimates of their food consumption. *Polar Biol.* 26, 348–357. doi: 10.1007/s00300-003-0486-z

Boster, J. S., and Johnson, J. (1989). Form or function: a comparison of expert and novice judgments of similarity among fish. *Am. Anthropol.* 91, 866–899. doi: 10.1525/aa.1989.91.4.02a00040

Burrows, M. T., Schoeman, D. S., Buckley, L. B., Moore, P., Poloczanska, E. S., Brander, K. M., et al. (2011). The pace of shifting climate in marine and terrestrial ecosystems. *Science* 334, 652–655. doi: 10.1126/science.1210288

Butler, P. G., Wanamaker, A. D. Jr., Scourse, J. D., Richardson, C. A., and Reynolds, D. J. (2013). Variability of marine climate on the North Icelandic Shelf in a 1357-year crossdated *Arctica islandica* chronology. *Palaeoceanogr. Palaeoclimatol. Palaeoecol.* 373, 141–151. doi: 10.1016/j.palaeo.2012.01.016

Carroll, M. L., Ambrose, W. G. Jr., Levin, B. S., Locke, W. L., Henkes, G. A., Hop, H., et al. (2011a). Pan-Svalbard growth rate variability and environmental regulation in the Arctic bivalve *Serripes groenlandicus. J. Mar. Syst.* 88, 239–251. doi: 10.1016/j.jmarsys.2011.04.010

Carroll, M. L., Ambrose, W. G. Jr., Levin, B., Ratner, A., Ryan, S., and Henkes, G. A. (2011b). Climatic regulation of *Clinocardium ciliatum* (bivalvia) growth in the northwestern Barents Sea. *Palaeoceanogr. Palaeoclimatol. Palaeoecol.* 302, 10–20. doi: 10.1016/j.palaeo.2010.06.001

Carroll, M. L., Ambrose, W. G. Jr., Locke, W. L., Ryan, S. K., and Johnson, B. J. (2014). Bivalve growth rate and isotopic variability across the Barents Sea Polar Front. *J. Mar. Syst.* 130, 167–180. doi: 10.1016/j.jmarsys.2013.10.006

Carroll, M. L., Johnson, B., Henkes, G. A., McMahon, K. W., Voronkov, A., Ambrose, W. G. Jr., et al. (2009). Bivalves as indicators of environmental variation and potential anthropogenic impacts in the southern Barents Sea. *Mar. Poll. Bull.* 59, 193–206. doi: 10.1016/j.marpolbul.2009.02.022

Carter, B. T. G., and Nielson, E. A. (2011). Exploring ecological changes in Cook Inlet beluga whale habitat though traditional and local ecological knowledge of contributing factors for population decline. *Mar. Policy* 35, 299–308. doi: 10.1016/J.marpol.2010.10.009

Caulkins, D. (2004). Identifying culture as a threshold of shared knowledge: a consensus analysis method. *Int. J. Cross Cult. Manage.* 4, 317–333. doi: 10.1177/1470595804047813

Coan, E. V., Valentich, P., and Bernard, F. R. (2000). *Bivalve seashells of western North America.* Santa Barbara, CA: Santa Barbara Museum of Natural History. Monograph 2, Studies in Biodiversity 2.

Davis, A., and Wagner, J. R. (2003). Who knows? On the importance of identifying "experts" when researching local ecological knowledge. *Hum. Ecol.* 31, 463–489. doi: 10.1023/A:1025075923297

Di Lorenzo, E., and Ohman, M. D. (2013). A double-integration hypothesis to explain ocean ecosystem response to climate forcing. *Proc. Natl. Acad. Sci. U.S.A.* 110, 2496–2499. doi: 10.1073/pnas.1218022110

Doney, S. C., and Sailley, S. F. (2013). When an ecological regime shift is really just stochastic noise. *Proc. NatL. Acad. Sci. U.S.A.* 110, 2438–2439. doi: 10.1073/pnas.1222736110

Ducklow, H. W., Doney, S. C., and Steinberg, D. K. (2009). Contributions of long-term research and time-series observations to marine ecology and biogeochemistry. *Ann. Rev. Mar. Sci.* 1, 279–302. doi: 10.1146/annurev.marine.010908.163801

Dunton, K. H., Goodall, J. L., Schonberg, S. V., Grebmeier, J. M., and Maidment, D. R. (2005). Multi-decadal synthesis of benthic pelagic coupling in the western Arctic: role of cross-shelf advective processes. *Deep-Sea Res. II* 52, 3462–3477. doi: 10.1016/j.dsr2.2005.09.007

Feder, H. M., Jewett, S. C., and Blanchard, A. (2005). Southeastern Chukchi Sea (Alaska) epibenthos. *Polar Biol.* 28, 402–421. doi: 10.1007/s00300-004-0683-4

Feder, H. M., Naidu, A. S., Baskaran, M., Frost, K., Hameedi, M. J., Jewett, S. C., et al. (1991). *Bering Strait-Hope Basin: Habitat Utilization and Ecological Characterization.* Institute of Marine Science technical report 92-2. Fairbanks: University of Alaska.

Ferguson, S. H., Higdon, J. W., and Westdal, K. H. (2012). Prey items and predation behavior of killer whales (*Orcinus orca*) in Nunavut, Canada based on Inuit hunter interviews. *Aquatic Biosyst.* 8:3. doi: 10.1186/2046-9063-8-3

Finley, K. J., and Evans, C. R. (1983). Summer diet of the bearded seal (*Erignathus barbatus*) in the Canadian high Arctic. *Arctic* 36, 82–89. doi: 10.14430/arctic2246

Fisher, K. I., and Stewart, R. E. A. (1997). Summer foods of atlantic walrus, odobenus *rosmarus rosmarus*, in northern Foxe Basin, North West Territories. *Can. J. Zool.* 75, 1166–1175. doi: 10.1139/z97-139

Freeman, L. C. (1977). A set of measures of centrality based on betweenness. *Sociometry* 40, 35–41. doi: 10.2307/3033543

Fritts, H. C. (1976). *Tree Rings and Climate.* New York, NY: Academic Press.

García Quijano, C. (2007). Fishers' knowledge of marine species assemblages: bridging scientific and local ecological knowledge in southeastern Puerto Rico. *Am. Anthropol.* 109, 529–536. doi: 10.1525/aa.2007.109.3.529

Grebmeier, J. M., McRoy, C. P., and Feder, H. M. (1988). Pelagic-benthic coupling on the shelf of the northern Bering and Chukchi Seas. I. Food supply source and benthic biomass. *Mar. Ecol. Prog. Ser.* 48, 57–67. doi: 10.3354/meps048057

Grebmeier, J. M., Overland, J. E., Moore, S. E., Farley, E. V., Carmack, E. C., Cooper, L. W., et al. (2006). A major ecosystem shift in the Northern Bering Sea. *Science* 311, 1461–1464. doi: 10.1126/science.1121365

Griffith, D. C. (2006). "Local knowledge, multiple livelihoods, and the use of natural and social resources in coastal North Carolina," in *Traditional Ecological Knowledge and Natural Resource Management*, ed C. Menzies (Lincoln, NE: University of Nebraska Press), 153–174.

Gröcke, D. R., and Gillikin, D. P. (2008). Advances in mollusk sclerochronology and sclerochemistry: tools for understanding climate and environment. *Geo-Marine Lett.* 28, 265–268. doi: 10.1007/s00367-008-0108-4

Hovinen, J. E. H., Welcker, J., Descamps, S., Strøm, H., Jerstad, K., Berge, J., et al. (2014). Climate warming decreases the survival of the little auk (*Alle alle*), a high Arctic avian predator. *Ecol. Evol.* 4, 3127–3138. doi: 10.1002/ece3.1160

Huntington, H. P. (2000). Using traditional ecological knowledge in science: methods and applications. *Ecol. Appl.* 10, 1270–1274. doi: 10.1890/1051-0761(2000)010[1270:UTEKIS]2.0.CO;2

Huntington, H. P., Callaghan, T. V., Fox Gearheard, S., and Krupnik, I. (2004). Matching traditional and scientific observations to detect environmental change: a discussion on Arctic Terrestrial Ecosystems. *Ambio* 33, 18–23.

Huntington, H. P., Gearhead, S., Mahoney, A. R., and Salomon, A. E. (2011). Integrating traditional and scientific knowledge through collaborative natural science field research: identifying elements for success. *Arctic* 64, 437–445. doi: 10.14430/arctic4143

Huntington, H. P., Goodstein, E., and Duskirechen, E. (2012). Towards a tipping point in responding to climate change. *Ambio* 41, 66–74. doi: 10.1007/s13280-011-0226-5

Johannes, R. E. (1981). Working with fishermen to improve coastaltropical fisheries and resource management. *Bull. Mar. Sci.* 31, 673–680.

Johannes, R. E. (1984). Marine conservation in relation to traditional lifestyles of tropical artisanal fishermen. *Environmentalist* 4, 30–35. doi: 10.1007/BF01907290

Johnson, J. C., and Weller, S. (2002). "Elicitation techniques for interviewing" in: *Handbook of Interview Research*, eds J. F. Gubrium and J. A. Holstein (Newbury Park: Sage), 491–514.

Johnson, M. A., Proshutinsky, A. Y., and Polyakov, I. V. (1999). Atmospheric patterns forcing two regimes of Arctic circulation: a return to anticyclonic conditions? *Geophys. Res. Lett.* 26, 1621–1624. doi: 10.1029/1999GL900288

Jones, D. S., Arthus, M. A., and Allard, D. J. (1989). Sclerochronological records of temperature and growth from shells of *Mercenaria mercenaria* from Narragansett Bay, Rhode Island. *Mar. Biol.* 102, 225–234. doi: 10.1007/BF00428284

Kafanov, A. (1980). Systematics of the subfamily Clinocardiinae Kafanov, 1975 (Bivalvia, Cardiidae). *Malacologia* 19, 297–328.

Khim, B.-K., Kranz, D. E., Cooper, L. W., and Grebmeier, J. M. (2003). Seasonal discharge to the western Chukchi Sea shelf identified in stable isotope profiles of mollusk shells. *J. Geophys. Res.* 108, 3300–3309. doi: 10.1029/2003JC 001816

Kinder, T. H., Schumacher, J. D., Tripp, T. B., and Pashinski, D. (1977). *The Physical Oceanography of Kotzebue Sound, Alaska, During Late Summer, 1976.* Technical report M77-99. Seattle, WA: University of Washington.

Kortsch, S., Primicerio, R., Beuchel, F., Renaud, P. E., Rodrigues, J., Lønne, O. J., et al. (2012). Climate-driven shifts in Arctic marine benthos. *Proc. Natl. Acad. Sci. U.S.A.* 109, 14052–14057. doi: 10.1073/pnas.12075 09109

Koszteyn, J., Kwasniewski, S., Róż'ycki, O., and Węslawski, J. M. (1990). *Atlas of the Marine Fauna of Southern Spitzbergen.* Poland: Institute of Oceanology.

Laidler, G. J. (2006). Inuit and scientific perspectives on the relationship between sea ice and climate change: the ideal complement? *Clim. Chang.* 78, 407–444. doi: 10.1007/s10584-006-9064-z

Le Fur, J., Guilavogui, A., and Teitelbaum, A. (2010). Contribution of local fishermen to improving knowledge of the marine ecosystem and resources in the Republic of Guinea, West Africa. *Can. J. Fish. Aquat. Sci.* 68, 1454–1469. doi: 10.1139/f2011-061

Lowry, L. F., Frost, K. J., and Burns, A. J. (1980). Feeding of bearded seals in the Bering and Chukchi Seas and tropic interactions with Pacific walruses. *Arctic* 33, 330–342. doi: 10.14430/arctic2566

Mahoney, A., Gearheard, S., Oshima, T., and Qillaq, T. (2009). Sea ice thickness measurement from a community-based observing network. *Am. Bull. Meteoro. Sci.* 90, 371–377. doi: 10.1175/2008BAMS2696.1

Maiolo, J., Johnson, J. C., and Griffith, D. C. (1992). Applications of social science theory to fisheries management: three examples. *Soc. Nat. Res.* 5, 391–407. doi: 10.1080/08941929209380801

Mann, R., Munroe, D. M., Powell, E. N., Hofmann, E. E., and Klinck, J. M. (2013). "Bivalve mollusks: barometers of climate change in Arctic marine systems," in *Responses of Arctic Marine Ecosystems to Climate Change*, eds F. J. Mueter, D. M. S. Dickson, H. P. Huntington, J. R. Irvine, E. A. Logerwell, S. A. MacLean, et al. (Alaska Sea Grant, University of Alaska Fairbanks), 61–82.

Matua, N. J., and Hare, S. R. (2002). The pacific decadal oscillation. *J. Ocean.* 58, 35–44. doi: 10.1023/A:1015820616384

Menzies, C., and Butler, C. (2006). "Understanding ecological knowledge," in *Traditional Ecological Knowledge and Natural Resource Management*, ed C. Menzies (Lincoln, NE: University Nebraska Press), 1–17.

Merkel, F. R., Jamieson, S. E., Falk, K., and Mosbech, A. (2007). The diet of common eiders winter in Nuuk, Southwest Greenland. *Polar Biol.* 390, 227–234. doi: 10.1007/s00300-006-0176-8

Moerlein, K. J., and Carothers, C. (2012). Total environment of change: impacts of climate change and social transitions on subsistence fisheries in Northwest Alaska. *Ecol. Soc.* 17, 10–19. doi: 10.5751/ES-04543-170110

Morison, J., Aagaard, K., and Steele, M. (2000). Recent environmental changes in the Arctic: a review. *Arctic* 53, 359–371. doi: 10.14430/ arctic867

Nicholas, T., Berkes, F., Jolly, D., Snow, N. B., and The Community of Sachs Harbour. (2004). Climate change and sea ice: local observations from the Canadian Western Arctic. *Arctic* 57, 68–79. doi: 10.14430/arctic484

Ottersen, G., Alheit, A., Drinkwater, K., Friedland, K., Hagen, E., and Stenseth, N. C. (2005). "The response of fish populations to ocean climate fluctuations" in: *Marine Ecosystems and Climate Variation The North Atlantic A Comparative Perspective*, eds N. C. Stenseth, G. Ottersen, J. W. Hurrell, and A. Belgrano (Oxford: Oxford University Press), 73–94.

Overland, J., Rodionov, S., Minobe, S., and Bond, N. (2008). North Pacific regime shifts: definitions, issues and recent transitions. *Prog. Ocean.* 77, 92–102. doi: 10.1016/j.pocean.2008.03.016

Overland, J., and Wang, M. (2005). The Arctic climate paradox: the recent decrease of the Arctic Oscillation. *Geophys. Res. Lett.* 32, 1–5. doi: 10.1029/2004GL 021752

Overland, J. E., Alheit, J., Bakun, A., Hurrell, J. W., Mackas, D. L., and Miller, A. J. (2010). Climate controls on marine ecosystems and fish populations. *J. Mar. Syst.* 79, 305–315. doi: 10.1016/j.jmarsys.2008.12.009

Pershing, A. J., Greene, C. H., Planque, B., and Fromentin, J.-M. (2005). "The influence of climate variability on North Atlantic zooplankton populations" in *Marine Ecosystems and Climate Variation: The North*

Atlantic A Comparative Perspective, eds N. C. Stenseth, G. Ottersen, J. W. Hurrell, and A. Belgrano (Oxford: Oxford University Press), 59–94. doi: 10.1093/acprof:oso/9780198507499.003.0005

Pinheiro, J. C., and Bates, D. M. (2000). *Mixed-Effects Models in S and S-PLUS.* New York, NY: Springer. doi: 10.1007/978-1-4419-0318-1

Post, E. (2004). "Time lags in terrestrial and marine environments," in *Marine Ecosystems and Climate Variation The North Atlantic A Comparative Perspective*, eds N. C. Stenseth and G. Ottersen (Oxford: Oxford University Press), 165–167.

Richardson, C. A. (2001). Molluscs as archives of environmental change. *Ocean. Mar. Biol. Ann. Rev.* 39, 103–164.

Richardson, C. A., Saurel, C., Barroso, C. M., and Thain, J. (2005). Evaluation of the age of the red welk *Neptunea* antique using statoliths, opercula and element ratios in the shell. *J. Exper. Mar. Biol. Ecol.* 325, 55–64. doi: 10.1016/j.jembe.2005.04.024

Riseth, J. Å., Tømmervik, H., Helander-Renvall, E., Labba, N., Johansson, C., Malnes, E., et al. (2011). Sámi traditional ecological knowledge as a guide to science: snow, ice and reindeer pasture facing climate change. *Polar Rec.* 47, 202–217. doi: 10.1017/S0032247410000434

Romney, A. K., Batchelder, W., and Weller, S. (1987). Recent applications of cultural consensus theory. *Am. Behav. Sci.* 31, 163–177. doi: 10.1177/000276487031 002003

Romney, A. K., Weller, S., and Batchelder, W. (1986). Culture as consensus: a theory of culture and informant accuracy. *Am. Anthropol.* 88, 13–338. doi: 10.1525/aa.1986.88.2.02a00020

Rysgaard, S., Nielsen, T., and Hansen, B. W. (1999). Seasonal variation innutrients, pelagic primary production and grazing in a high-Arctic marine ecosystem, Young Sound, Northeast Greenland. *Mar. Ecol. Prog. Ser.* 179, 13–25. doi: 10.3354/meps179013

Shackeroff, J. M., Campbell, L. M., and Crowder, L. B. (2011). Social-ecological guilds: putting people into marine historical ecology. *Ecol. Soc.* 16, 1–20.

Shimek, R. L. (1984). The diet of Alaskan *Neptunea*. *Veliger* 26, 274–281.

Sjer, M., Blicher, M. E., and Rysgaard, S. (2009). Sea ice affects inter-annual and geographic variation in the growth of the Arctic cockle *Clinocardium cilitium* (Bivalvia) in Greenland. *Mar. Ecol. Prog. Ser.* 389, 149–158. doi: 10.3354/meps08200

Stenseth, N. C., Ottersen, G., Hurrell, J. W., Mysterud, A., Lima, M., Chan, K-S., et al. (2003). Studying climate effects on ecology through the use of climate indices: the North Atlantic Oscillation, El Niño Southern Oscillation, and beyond. *Proc. R. Soc. Lond. B* 270, 2087–2096. doi: 10.1098/rspb. 2003.2415

Sun, J., and Wang, H. (2006). Relationship between Arctic oscillation and pacific decadal oscillation on decadal timescale. *Chinese Sci. Bull.* 51, 75–79. doi: 10.1007/s11434-004-0221-3

Thompson, D. W. J., and Wallace, J. M. (1998). The Arctic Oscillation signature in the wintertime geopotential height and temperature fields. *Geophys. Res. Lett.* 25, 1297–1300. doi: 10.1029/98GL00950

Thompson, P. M., and Ollason, J. (2001). Lagged effects of ocean climate change on fulmar population dynamics. *Nature* 413, 417–420. doi: 10.1038/350 96558

Tunberg, B. G., and Nelson, W. G. (1998). Do climate oscillations influence cyclic patterns of soft bottom macrobenthic communities on the Swedish west coast? *Mar. Ecol. Prog. Ser.* 170, 85–94. doi: 10.3354/meps170085

United States Arctic Research Commission. (2013). *Report on the Goals and Objective for Arctic Research 2013-2014 for the U. S. Arctic Research Program Plan.* Washington, DC: U. S. Arctic Research Commission.

Wagner, F. J. E. (1977). Recent mollusk distribution patterns and palaeobathymetry, southeastern Beaufort Sea. *Can. J. Earth Sci.* 14, 2013–2028. doi: 10.1139/ e77-173

Wanamaker, A. D. Jr., Hetzinger, S., and Halfar, J. (2011). Reconstructing mid- to high-latitude marine climate and ocean variability using bivalves, coralline algae, and marine sediment cores from the Northern Hemisphere. *Paleogeogr. Paleoclimatol. Paleoecol.* 302, 1–9. doi: 10.1016/j.palaeo.2010.12.024

Wassmann, P., Duarte, C. M., Agustí, S., and Sejr, M. K. (2011). Footprints of climate change in the Arctic marine ecosystem. *Global Chang. Biol.* 17, 1235–1249. doi: 10.1111/j.1365-2486.2010.02311.x

Weatherhead, E., Gearheard, S., and Barry, R. G. (2010). Changes in weather persistence: insight from Inuit knowledge. *Global Environ. Chang.* 20, 523–528. doi: 10.1016/j.gloenvcha.2010.02.002

Witbaard, R., Jansma, E., Sass Klaassen, U. (2003). Copepods link quahog growth to climate. *J. Sea Res.* 50, 77–83. doi: 10.1016/S1385-1101(03)00040-6

Conflict of Interest Statement: The authors declare that the research was conducted in the absence of any commercial or financial relationships that could be construed as a potential conflict of interest.

4

A glimpse into the future composition of marine phytoplankton communities

Esteban Acevedo-Trejos[1,2]*, Gunnar Brandt[1], Marco Steinacher[3,4] and Agostino Merico[1,2]

[1] Systems Ecology Group, Leibniz Center for Tropical Marine Ecology, Bremen, Germany
[2] School of Engineering and Science, Jacobs University Bremen, Bremen, Germany
[3] Climate and Environmental Physics, Physics Institute, University of Bern, Bern, Switzerland
[4] Oeschger Centre for Climate Change Research, University of Bern, Bern, Switzerland

Edited by:
Susana Agusti, The University of Western Australia, Australia

Reviewed by:
Aleksandra M. Lewandowska, German Centre for Integrative Biodiversity Research (iDiv) Halle-Jena-Leipzig, Germany
Angel Lopez-Urrutia, Instituto Español de Oceanografía, Spain

*Correspondence:
Esteban Acevedo-Trejos, Systems Ecology Group, Leibniz Center for Tropical Marine Ecology, Fahrenheitstrasse 6, 28359 Bremen, Germany
e-mail: esteban.acevedo@zmt-bremen.de

It is expected that climate change will have significant impacts on ecosystems. Most model projections agree that the ocean will experience stronger stratification and less nutrient supply from deep waters. These changes will likely affect marine phytoplankton communities and will thus impact on the higher trophic levels of the oceanic food web. The potential consequences of future climate change on marine microbial communities can be investigated and predicted only with the help of mathematical models. Here we present the application of a model that describes aggregate properties of marine phytoplankton communities and captures the effects of a changing environment on their composition and adaptive capacity. Specifically, the model describes the phytoplankton community in terms of total biomass, mean cell size, and functional diversity. The model is applied to two contrasting regions of the Atlantic Ocean (tropical and temperate) and is tested under two emission scenarios: SRES A2 or "business as usual" and SRES B1 or "local utopia." We find that all three macroecological properties will decline during the next century in both regions, although this effect will be more pronounced in the temperate region. Being consistent with previous model predictions, our results show that a simple trait-based modeling framework represents a valuable tool for investigating how phytoplankton communities may reorganize under a changing climate.

Keywords: climate change, phytoplankton cell size, trait, trade-offs, adaptive dynamics

1. INTRODUCTION

Scientists agree that Earth's climate is changing at an unprecedented rate (Pachauri and Reisinger, 2007; Wolff et al., 2014). This is expected to have an impact on all living organisms (Parmesan, 2006; Mooney et al., 2009). For example, some studies have observed a declining trend in global phytoplankton biomass during the last century (Falkowski and Wilson, 1992; Boyce et al., 2010; Mackas, 2011; Rykaczewski and Dunne, 2011), while others have proposed an increase in the abundance of marine photoautotrophs (McQuatters-Gollop et al., 2011). Irrespective of the direction of the change, it appears indisputable that any shift in the abundance of marine primary producers will have an impact on higher trophic levels (Parmesan, 2006). Scientists therefore face the challenge of better comprehending the factors and the mechanisms that will shape the marine phytoplankton community in the decades to come.

Natural ecological communities are typically studied by gathering information about, for example, physiological and morphological properties on a predefined taxon. This generally treats all species as identical and neglects property variations within the taxon (Violle et al., 2012). An alternative approach is to gather information on the distribution of such properties within the entire community (McGill et al., 2006). This is the so-called trait-based approach (McGill et al., 2006), which provides a conceptual framework for understanding how communities are organized and adapted to given environmental conditions and how they will reorganize under a changing climate. Fundamental constraints, such as energy allocation, prevent organisms to invest equally in all traits and lead to the emergence of trade-offs, which in turn give rise to functionally diverse communities (Tilman, 2001).

By identifying key traits and trade-offs, the trait-based approach has advanced our understanding of phytoplankton community structure and functioning (Litchman et al., 2007; Litchman and Klausmeier, 2008) and provided the basis for constructing mechanistic links between community composition and environmental conditions (Follows and Dutkiewicz, 2011; Smith et al., 2011). In addition, the distribution of traits reflects the functional diversity of the community and hence its adaptive capacity (Abrams et al., 1993). Cell size, for example, has been proposed as the most characterizing morphological trait of phytoplankton organisms (Litchman and Klausmeier, 2008; Finkel et al., 2009). It spans over several orders of magnitude only within phytoplankton and influences many ecological and physiological processes, such as nutrient uptake (Aksnes and Egge, 1991; Tang, 1995; Litchman et al., 2007; Marañón et al., 2013), light harvesting (Ciotti et al., 2002; Finkel et al., 2004), respiration (Laws, 1975; López-Urrutia et al., 2006), sinking (Kiørboe, 1993), and grazing (Kiørboe, 1993; Fuchs and Franks, 2010; Wirtz, 2012a,b).

Therefore, investigating the dynamics of phytoplankton cell size helps to better understand the adaptive capacity of this group of organisms under a changing environment.

Size-based models of phytoplankton communities in combination with climate projections provide a useful tool for these investigations. Models following the species-by-species approach (e.g., Baird and Suthers, 2007; Banas, 2011) typically comprise many state variables and parameters. While valuable, such models have been criticized for the high number of free parameters required to adequately describe the observed macroecological properties of planktonic communities (Anderson, 2005; Ward et al., 2010). Alternatively, many scientists (e.g., Litchman and Klausmeier, 2008; Follows and Dutkiewicz, 2011; Smith et al., 2011; Edwards et al., 2013) have called for the use of trait-based modeling approaches to investigate how phytoplankton community structure and diversity may reorganize under a changing climate. No study to date has yet attempted to investigate the effects of future climate on the combined dynamics of phytoplankton biomass and trait distributions.

Here we use a novel size-based model of aggregate group properties that mechanistically describes phytoplankton community structure and functional diversity in two contrasting regions of the Atlantic Ocean (i.e., temperate and tropical). The model focuses on three macroecological properties, namely: total phytoplankton biomass, mean cell size, and size variance, the latter reflecting the functional diversity of the community and therefore its adaptive capacity. We run the model into the future using two distinct projections of the IPCC emission scenarios (Nakićenović et al., 2000), which are routinely used for studying the transitions of ecosystems. These are: the "a divided world" (SRES A2) scenario (also known as "business as usual"), which assumes emphasis on national identities, and the more optimistic "local utopia" (SRES B1) scenario, which assumes a moderate reduction in material demand and a moderate use of clean and efficient technologies but still with a pronounced regionalization of the economies. With this model we investigate how the structure and adaptive capacity of phytoplankton communities will respond to changing environmental conditions.

2. METHODS
2.1. SIZE-BASED MODEL

We developed a size-based model of the upper mixed-layer to study the future composition of phytoplankton communities under different climate change scenarios. The model focuses on the description of three macroecological properties: total phytoplankton biomass (P), phytoplankton cell size ($\overline{S}$, expressed as Equivalent Spherical Diameter (ESD)), and size variance (V). In this type of models the trait variance reflects the functional diversity of the community (Wirtz and Eckhardt, 1996; Norberg et al., 2001; Merico et al., 2009) and quantifies the speed of the adaptive process, i.e., the speed with which the mean trait changes (Abrams et al., 1993). Given that adaptive responses are considered to be more robust for more diverse communities (Visser, 2008), the size variance captured by our model is also a representation of the adaptive capacity of the phytoplankton community. This means that the higher the trait variance, the higher the functional diversity and the adaptive capacity of the system, and vice versa.

Instead of resolving the full spectrum of species or functional types, this modeling approach centers on aggregate properties of the entire planktonic community thus reducing model complexity (Merico et al., 2009) and mechanistically linking community structure with environmental conditions (Wirtz and Eckhardt, 1996; Norberg et al., 2001). In the model, the phytoplankton communities self-assemble based on a trade-off emerging from relationships between phytoplankton mean size and: (1) phytoplankton nutrient uptake, (2) zooplankton grazing, and (3) phytoplankton sinking. Equations for nutrient (N), zooplankton (Z), and detritus (D) complete the model system (see Appendix for a detailed description of all model equations).

2.2. STUDIED REGIONS

The model is applied to two contrasting regions of the Atlantic Ocean: temperate (45.5–49.5 °N, 10.5–20.5 °W) and tropical (4.5–14.5 °N, 19.5–24.5 °W). The temperate region is characterized by seasonal changes of mixed layer depth (MLD), sea surface temperature (SST), photosynthetic active radiation (PAR), and concentration of nutrients below the mixed layer (N_0), whereas environmental conditions in the tropical region are relatively constant. Note that the model parameterization is identical for the two regions and that different results emerge only from the contrasting environmental forcing (see **Table 1**). Hereafter, we will

Table 1 | Model parameters (identical for both regions).

Name	Symbol (Units)	Value	Source
P growth rate	μ_P (d^{-1})	1.4	Edwards and Brindley, 1996
P mortality rate	m_P (d^{-1})	0.05	Fasham et al., 1990
P Immigration rate	δ_I (d^{-1})	0.008	This study
Z growth rate	μ_Z (d^{-1})	0.8	Edwards and Brindley, 1996
Z mortality rate	m_Z (d^{-1})	0.3	Edwards and Brindley, 1996
P assimilation coefficient	δ_Z (-)	0.3	Edwards and Brindley, 1996
P half-saturation	K_P (mmol N m^{-3} μm^{-1} ESD)	0.08	This study
Cross-thermocline mixing	κ (m·d^{-1})	0.01	Fasham, 1993
Mineralization rate	δ_D (d^{-1})	0.2	This study
Light attenuation constant	k_w (m^{-1})	0.1	Edwards and Brindley, 1996
Optimum irradiance	I_s (W m^{-2})	100	This study
Intercept of the K_N allometric function	β_U	0.14257	Litchman et al., 2007
Slope of the K_N allometric function	α_U	0.81	Litchman et al., 2007
Intercept of the ν allometric function	β_ν	0.01989	Kiørboe, 1993
Slope of the ν allometric function	α_ν	1.17	Kiørboe, 1993

The parameters with source "this study" were considered as free parameters and allowed to vary in order to obtain a better model to data fit.

use the terminology "environmental variables," to refer to all variables that force the model (i.e., MLD, SST, PAR, N_0).

2.3. ENVIRONMENTAL FORCING DATA

The environmental variables used to force the ecosystem model are derived from simulations with the NCAR CSM1.4-carbon model (Fung et al., 2005; Doney et al., 2006). The NCAR model is one of the global Earth system models that are used to study past and future impacts of anthropogenic emissions on the environment (e.g., Steinacher et al., 2010) and contributed to previous assessment reports of the IPCC. The NCAR model was driven by historical CO_2 emissions over the industrial

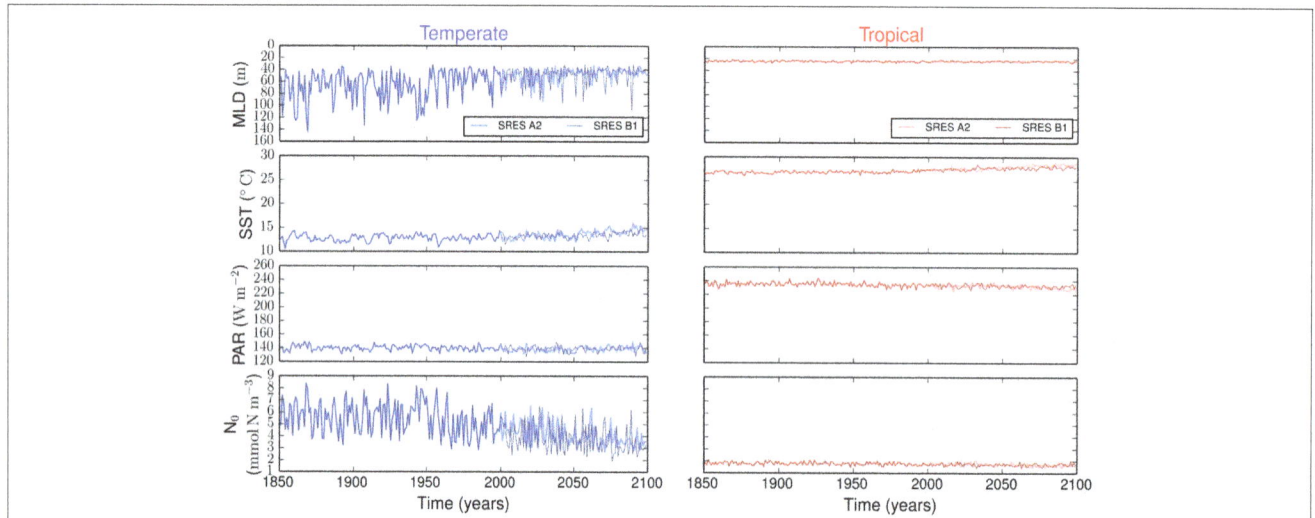

FIGURE 1 | Environmental forcing for temperate and tropical regions under the SRES A2 and the SRES B1 emission scenarios. MLD is the mixed layer depth, PAR is the photosynthetically active radiation, SST is the sea surface temperature, and N_0 is the concentration of nutrients below the upper mixed-layer.

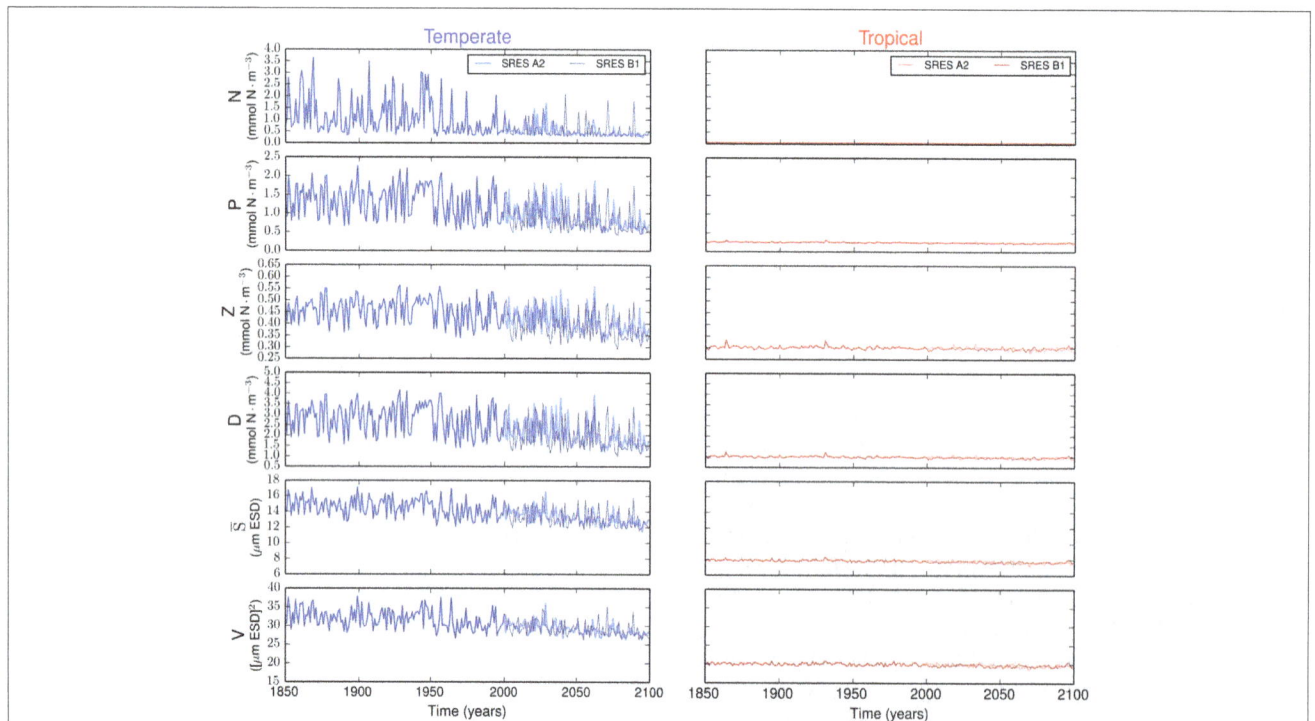

FIGURE 2 | Time series of the state variables for the temperate and tropical regions under the SRES A2 and the SRES B1 emission scenarios. N is the nutrient concentration, P is the phytoplankton concentration, Z is the zooplankton concentration, D is the detritus concentration, $\bar{S}$ is the phytoplankton mean size, and V is the size variance or functional diversity of phytoplankton.

period, followed by two IPCC scenarios for the 21st century: SRES A2, a "business-as-usual" scenario in which atmospheric CO_2 reaches 840 ppm by 2100 AD, and SRES B1, a convergent world scenario in which CO_2 concentrations stabilize at about 540 ppm by 2100 AD (Nakićenović et al., 2000). A more detailed description of the simulations is given by Steinacher et al. (2009).

From the comprehensive NCAR model output, we use MLD, SST, PAR, and N_0, which are available at a monthly resolution on a grid with a resolution of 3.6° in longitude and 0.8–1.8° in latitude. The gridded variables are then spatially averaged within each region to obtain time series over the period of the simulation for both scenarios. PAR (in $W \cdot m^{-2}$) is calculated from the solar heat flux at the ocean surface (in $K \cdot cm \cdot s^{-1}$) by applying a conversion factor of $4.1 \cdot 10^4 \, J \cdot m^{-2} \cdot cm^{-1} \cdot K^{-1}$. Since the biogeochemistry module of the NCAR model is defined in units of phosphorus, we consider the Redfield N:P ratio of 16:1 whenever a conversion to units of nitrogen is needed.

2.4. ANALYSIS OF ANNUAL TRENDS OF P, $\overline{S}$, AND V

We apply a Principal Component Analysis, PCA (Mantua, 2004) to conduct an exploratory statistical analysis with annually averaged variables of the size-based model and the environmental forcing. With the PCA we identify the general patterns of variability from a number of time series by means of reducing the dimensionality of the dataset into a small number of uncorrelated principal components (von Storch and Zwiers, 2001). Typically, the first principal component has the highest eigenvalue and, as new orthogonal components are added, the eigenvalues of new components decrease. These eigenvalues are interpreted as the amount of variability explained by each principal component (Supplementary Figure 1). Another important output of the PCA is represented by the eigenvectors or loadings, which reflect the relative importance of each variable on each principal component. PCA has been widely used to explore multivariate time series in particular with respect to the detection of regime shifts (e.g., Hare and Mantua, 2000; Mantua, 2004; Weijerman et al., 2005; Schlüter et al., 2012).

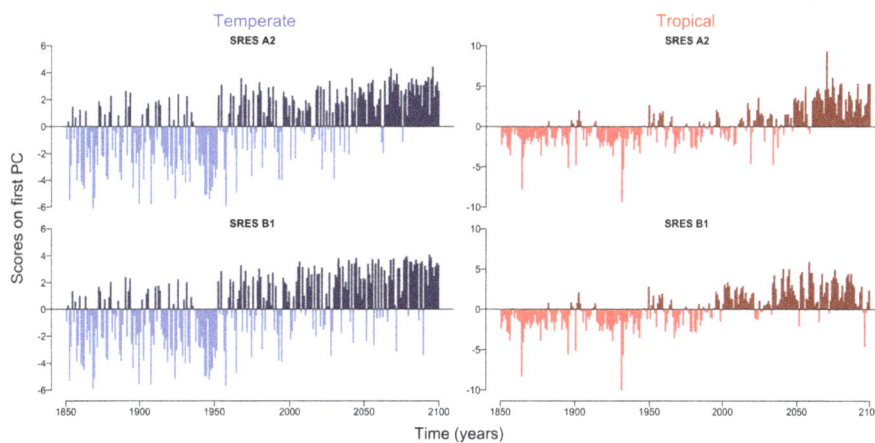

FIGURE 3 | Scores on the first principal component for temperate and tropical regions under the SRES A2 and SRES B1 emission scenarios.

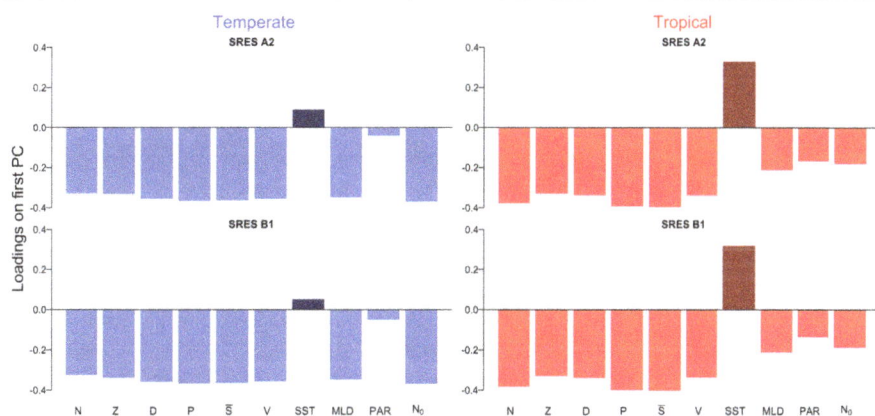

FIGURE 4 | Loadings on the first principal component for temperate and tropical regions under the SRES A2 and SRES B1 emission scenarios.

In addition, we use probabilistic models based on the generalized least squares method to analyse the long-term trends of the state variables that describe structure and diversity of the two phytoplankton communities (i.e., total biomass, mean trait, and trait variance). We adopted this statistical approach, because it allows us to correct the data for temporal correlations and for unequal variances (Zuur et al., 2009). Each fit is evaluated by a visual inspection of the residuals. To better understand the long-term trends of phytoplankton community structure and diversity under the two emission scenarios we subdivide the time series into three main periods: (1) past, which corresponds to the period from 1850 to 2000 and is the same for both scenarios, (2) future-A2, which corresponds to the period from 2000 to 2100 of the business as usual scenario, and (3) future-B1, which corresponds to the period from 2000 to 2100 of the convergent world scenario. First, the probabilistic models are fitted to each of the three periods and two regions using the variables P, $\overline{S}$, and V as response variables and using time as the explanatory variable. Second, we perform an analysis of variance (ANOVA) with the slopes of each probabilistic model as response variables, and periods (past, future-A2, and future-B1) and regions (temperate and tropical) as explanatory variables.

3. RESULTS

Figure 1 summarizes all the environmental variables used to force our size-based model. The emission scenarios are identical in the period from 1850 to 2000, but differ from 2000 to 2100. The strongest difference between the two scenarios is in SST, whereas the other three environmental variables show similar variability (**Figure 1**).

The resulting simulations reveal no major differences in phytoplankton community composition between emission scenarios (Supplementary Figure 1). Strong differences are observed between regions. Our results show that the variability is stronger in the temperate region than in the tropical region (**Figure 2**). In the temperate region, the projected increase in SST leads to a more stratified ocean, which is expressed by the shoaling of the MLD (**Figure 1**). The amplitude of the fluctuations in nutrient concentration reduces after the year 2000. This long-term decline in nutrient concentration reflects the change exhibited by the MLD and drives the decreases in P, $\overline{S}$, and V. The changes in total phytoplankton biomass appear to cause a baseline shift toward smaller concentrations of Z and D. In contrast, the tropical region shows weaker or no trends in all state variables (**Figure 2**).

The first principal component explains over 55 % of the variation in the datasets (Supplementary Figure 1). The scores on the first principal component of the PCA reveal the existence of two regimes, which are linked by a rather smooth transition (**Figure 3**). While the first regime from 1850 to approximately 2000 is dominated by negative scores on the first principal component, the second regime from 2000 to 2100 is characterized by mostly positive scores on the first principal component (**Figure 3**). This analysis also reveals that the environmental drivers of the major mode of variability represented by the first principal component are temperature (SST) and nutrient availability, the latter determined by MLD and N_0 (**Figure 4**). These variables, in fact, have the highest positive and negative loadings on the first principal component for both, the two scenarios and the two regions (**Figure 4**).

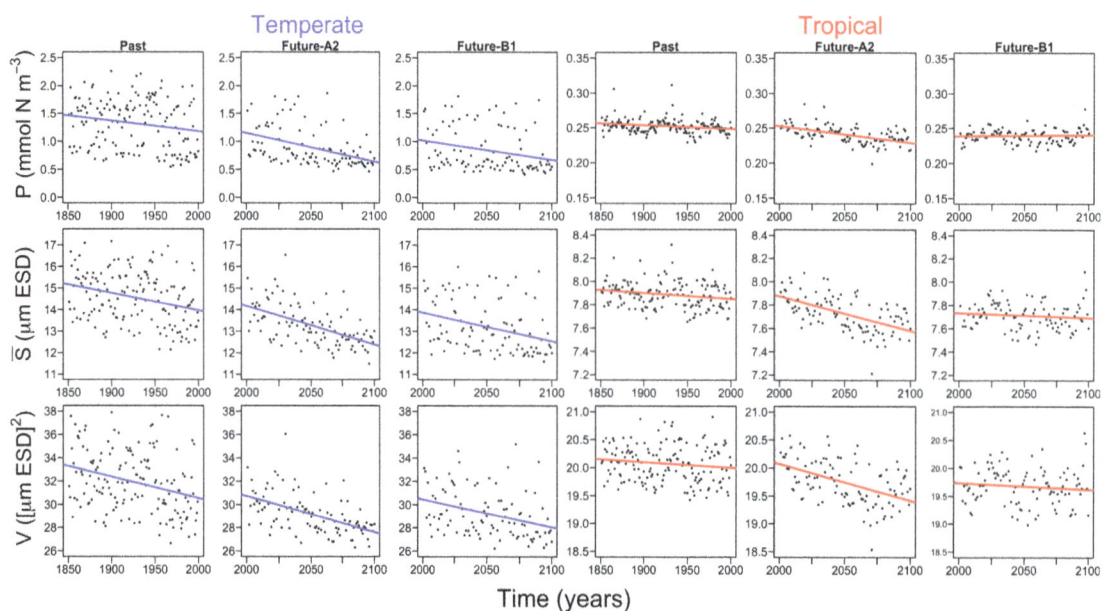

FIGURE 5 | Probabilistic model fits of the annual trends of the phytoplankton state variables for temperate and tropical regions and for the three different time periods (i.e., past, future-A2, and future-B1). P is the phytoplankton concentration, $\overline{S}$ is the phytoplankton mean size, and V is the size variance or functional diversity of phytoplankton. The color lines represent the generalized least square fits for temperate and tropical regions.

All three periods (i.e., past, future-A2, and future-B1), show declining trends in both regions (**Figure 5**). However, there are no differences in the slopes of the trends (Supplementary Figure 2), between periods (ANOVA, df: 2, F-value: 1.9542, p-value: 0.1842) or between regions by periods (ANOVA, df: 2, F-value: 0.2749, p-value: 0.7643). The only significant difference in the slopes is between regions (ANOVA, df: 1, F-value: 10.9999, p-value: 0.0061). The declining trends are steepest in the future-A2 period (Supplementary Figure 2). More specifically, the temperate Atlantic appears to be more sensitive since it will experience the largest decline among all scenarios in phytoplankton mean size ($\overline{S}$) and functional diversity (V), with approximately -1.8 (μm ESD century^{-1}) and -3.1 ($[\mu$m ESD$]^2$ century^{-1}) (Supplementary Figure 2). Our results also suggest that the tropical region will be less affected than the temperate region with less than 0.3 decline in P (mmol N m^{-3}century^{-1}), $\overline{S}$ (μm ESD century^{-1}), and V ($[\mu$m ESD$]^2$ century^{-1}) (Supplementary Figure 2).

In addition, we performed a sensitivity analysis in which we varied all the model parameters by $\pm 25\%$ and quantified the effect of such change on the annual trends of the three periods (**Figures 6–8**). The greatest effects arise in the parameters that control phytoplankton grazing, i.e., μ_Z. A decrease in μ_Z by 25% produces up to an 1.5-fold decrease in the declining trend of P

and also an increase in the slopes of $\overline{S}$ and V of, respectively, 6-fold and 3-fold (**Figures 6–8**). As expected, increasing μ_Z by 25% causes opposite effects on the slopes of P, $\overline{S}$, and V, however, with a lower magnitude compared to the equivalent decrease of μ_Z (**Figures 6–8**). Another sensitive parameter is the density-dependent immigration rate (δ_I), which mainly affects V and $\overline{S}$ for the tropical region under scenario B1 (**Figures 6–8**). Changes in the light attenuation coefficient k$_w$ showed an impact only in the temperate region under past conditions (**Figure 6**). This probably reflects an overestimation of the effects of light attenuation in the temperate region (see Appendix) due to the weaker stratification in the past as compared to the future scenarios.

4. DISCUSSION

We used a size-based model to explore the composition of two different phytoplankton communities (tropical and temperate) under past environmental conditions and future climate change scenarios. The model was previously calibrated and tested against present day observations of nutrient concentrations and phytoplankton biomass (see Supplementary Figure 3).

Most projections of the future climate suggest that the world oceans will become warmer and more stratified than present (Pachauri and Reisinger, 2007; Steinacher et al., 2010; Reusch and Boyd, 2013; Wolff et al., 2014). Such warming will likely

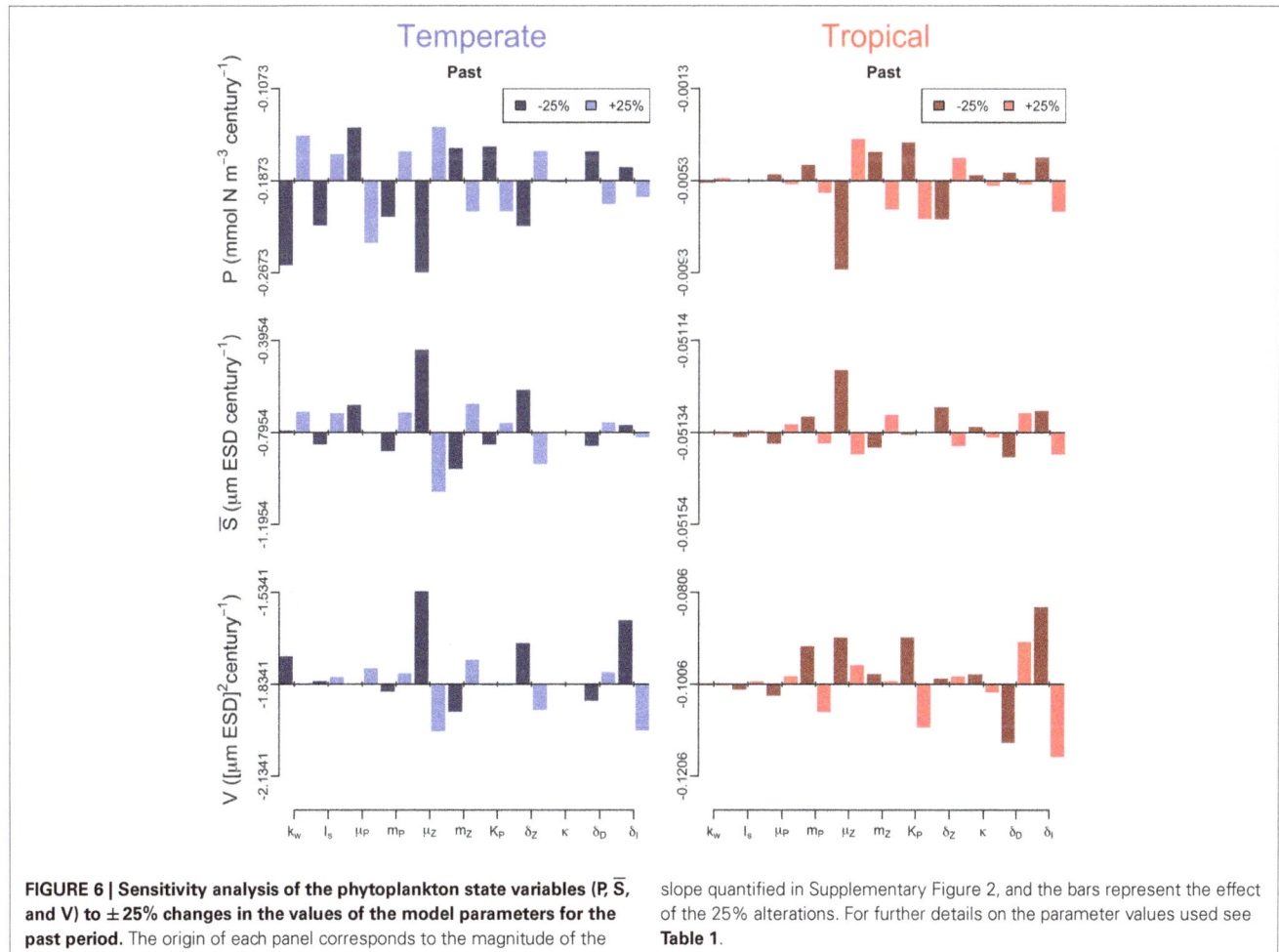

FIGURE 6 | Sensitivity analysis of the phytoplankton state variables (P, $\overline{S}$, and V) to $\pm 25\%$ changes in the values of the model parameters for the past period. The origin of each panel corresponds to the magnitude of the slope quantified in Supplementary Figure 2, and the bars represent the effect of the 25% alterations. For further details on the parameter values used see **Table 1**.

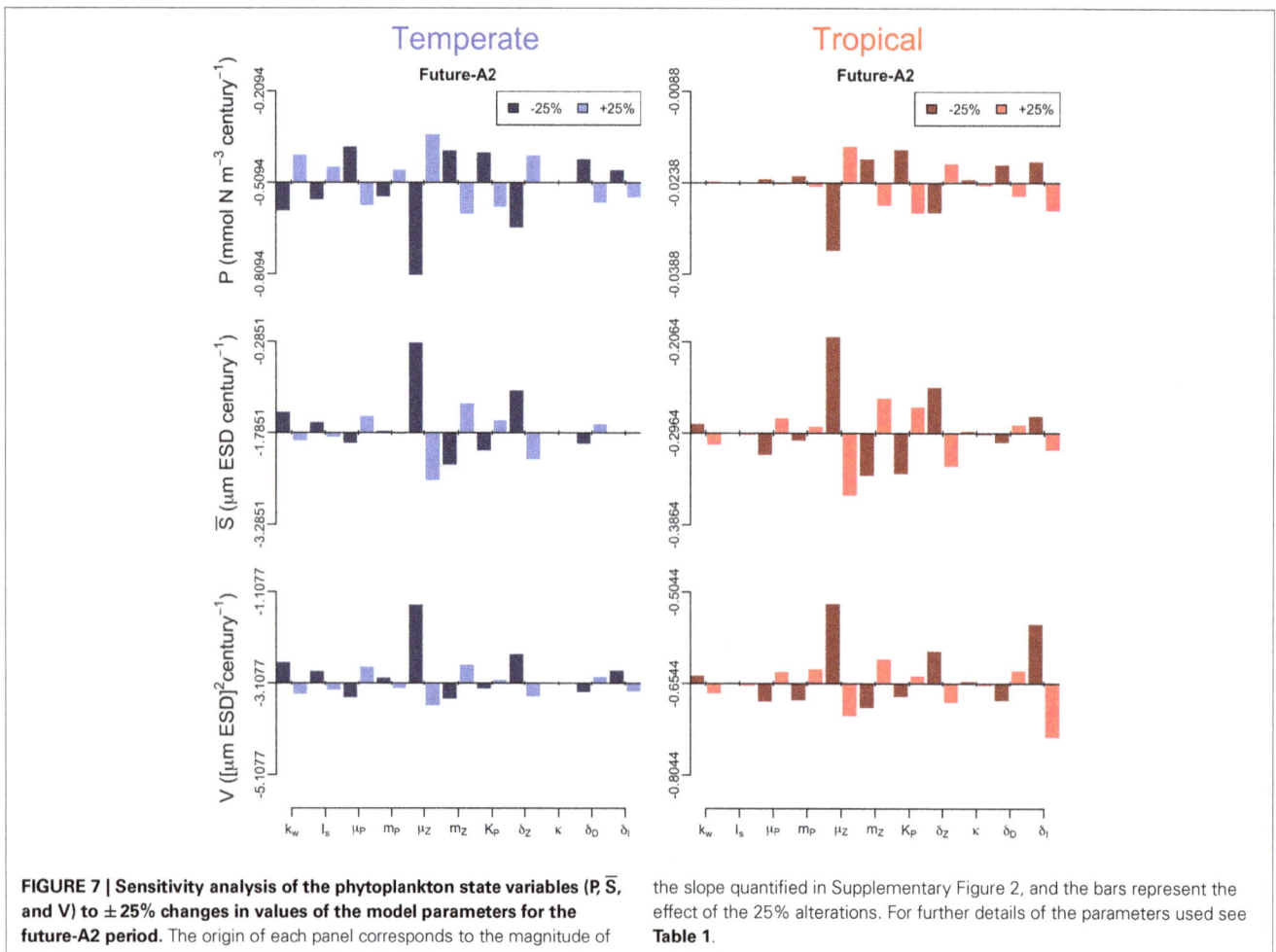

FIGURE 7 | Sensitivity analysis of the phytoplankton state variables (P, $\overline{S}$, and V) to ± 25% changes in values of the model parameters for the future-A2 period. The origin of each panel corresponds to the magnitude of the slope quantified in Supplementary Figure 2, and the bars represent the effect of the 25% alterations. For further details of the parameters used see **Table 1**.

lead to a shallower upper mixed-layer and thus to a reduced nutrient supply from the deep ocean (Steinacher et al., 2010; Reusch and Boyd, 2013). Our model experiments generally confirm this view, but further show that the expected changes will differ considerably by ocean region. In contrast, the different future scenarios produce minor changes in the model results (**Figure 1**). In our simulations, the temperate region will experience strongest changes in MLD, SST, and N_0 by the year 2100 (**Figure 1**). These changes will lead to a decline in average phytoplankton biomass, mean cell size, and functional diversity (i.e., adaptive capacity), see **Figure 2**. The tropical region will experience comparatively less environmental changes than the temperate region (**Figure 1**) and, consequently, less alterations in phytoplankton community composition (**Figure 2**). Contrasting differences in phytoplankton community structure between tropical and temperate regions are well known biogeographical features of the modern ocean (Marañón et al., 2001) and our results suggest that these differences will be maintained also in the future. Analogously to the results presented here, a previous multi-species and multi-nutrient modeling study showed that a surface ocean with less mixing and hence less nutrients will favor smaller phytoplankton species (Litchman et al., 2006).

Accumulating evidence suggests that a changing climate has and will have dramatic consequences on the Earth's biota (Parmesan, 2006). In planktonic ecosystems, for example, changes in the environment have entailed shifts in the phenology of phytoplankton (Schlüter et al., 2012) and zooplankton (Schlüter et al., 2010) causing mismatches between trophic levels and functional groups (Edwards and Richardson, 2004). The quantitative analysis of the temporal patterns of the two planktonic ecosystems in the Atlantic Ocean show two apparently distinct regimes, the first from 1850 to 2000 and the second from 2000 to 2100 (**Figure 3**). The two regimes are, however, not separated by an abrupt shift as observed, for example, in the Pacific Ocean (Hare and Mantua, 2000), in the North Sea (Beaugrand, 2004; Weijerman et al., 2005) and in the German Bight (Schlüter et al., 2008), but are rather linked by a very smooth transition (**Figure 3**). Note that the low signal to noise ratio typical of phytoplankton time-series may hinder the detection of patterns or key drivers of change (Winder and Cloern, 2010). To minimize the effect of potential noise on our results and to avoid the over-interpretation of short-term fluctuations, we focused our analyses on long-term trends (i.e., decades to centuries) of temporally and spatially averaged variables.

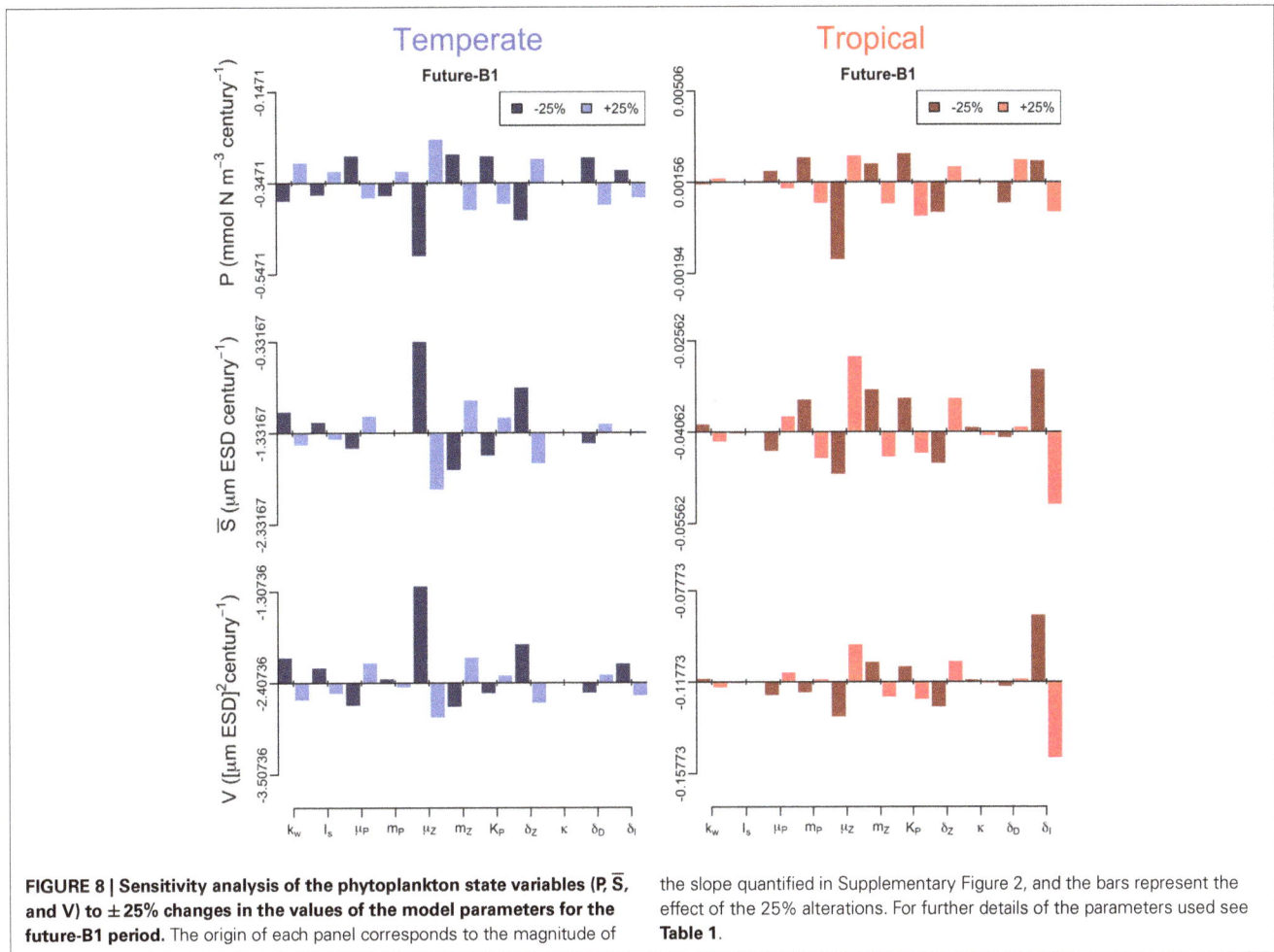

FIGURE 8 | Sensitivity analysis of the phytoplankton state variables (P, $\overline{S}$, and V) to ±25% changes in the values of the model parameters for the future-B1 period. The origin of each panel corresponds to the magnitude of the slope quantified in Supplementary Figure 2, and the bars represent the effect of the 25% alterations. For further details of the parameters used see **Table 1**.

The quantitative analyses of the long-term trends in the three macroecological properties show a decline in P, $\overline{S}$, and V (**Figure 5**), with the major differences arising between regions. The temperate Atlantic will be the region to experience the highest rates of decline (Supplementary Figure 2), in terms of phytoplankton cell sizes and functional diversity. It has been suggested that a changing climate, specifically an increase in SST, will lead to a dominance of small phytoplankton cells in the oceans (Daufresne, 2009; Morán et al., 2010). However, other studies have questioned this conclusion providing empirical evidence for a stronger effect of nutrients rather than of SST on the size structure of phytoplankton communities (Marañón et al., 2012). Our simulations suggest a combined effect of SST and nutrient availability, with the latter being governed by changes in MLD and N_0 (**Figure 4**). This finding is also consistent with a recent empirical biogeographical classification of the phytoplankton size distribution in the global ocean (Acevedo-Trejos et al., 2013) suggesting that the most parsimonious model of size structure can be defined by nutrient availability and temperature. Nevertheless, none of the mentioned theoretical and empirical studies has linked in a single framework, as we do here, the size composition and the adaptive capacity of the community to future changes of the environment.

Our modeling approach focuses on phytoplankton community structure and functional diversity and treats zooplankton as an assemblage of many identical individuals. We thus disregard the variety of different feeding mechanisms in zooplankton and their potential impact on the phytoplankton community (see Gentleman et al., 2003, for a review on functional responses of zooplankton). However, to assess the sensitivity of our results to this model simplification, we performed an in-depth sensitivity analysis on the role of zooplankton grazing in structuring the two phytoplankton communities (**Figures 6–8**). Specifically, we found that an increase in grazing pressure would exacerbate the declining trends in $\overline{S}$, and V. Analogously, a decrease in grazing pressure would attenuate the declining trends in the two variables. While this finding highlights the fundamental role of grazing in defining phytoplankton size distributions, a more detailed description of zooplankton, e.g., including different zooplankton size classes, life stages, feeding preferences, etc. (e.g., Banas, 2011; Prowe et al., 2012; Wirtz, 2012a; Mariani et al., 2013), would probably improve the quality of our projections (Wirtz, 2012a; Smith et al., 2014). However, an adequate description of the adaptive responses of grazers is still a topic of ongoing research (Litchman et al., 2013; Smith et al., 2014) and a consistent and well accepted formulation that takes into account all relevant zooplankton-phytoplankton interaction processes is still lacking.

In summary, our study suggests that under a warmer and more stratified ocean phytoplankton communities will shift toward smaller mean size and will lose adaptive capacity in the process. These changes will be more pronounced in regions with strong seasonal variations and will be additionally modulated by potential modifications in grazing pressure.

AUTHOR CONTRIBUTIONS

Agostino Merico and Gunnar Brandt conceived the idea of the study; Esteban Acevedo-Trejos coded the model and performed the analyses; Marco Steinacher provided the environmental forcing data from GCM simulations; all authors contributed to interpret the results and to write the manuscript.

REFERENCES

Abrams, P., Matsuda, H., and Harada, Y. (1993). Evolutionarily unstable fitness maxima and stable fitness minima of continuous traits. *Evol. Ecol.* 7, 465–487. doi: 10.1007/BF01237642

Acevedo-Trejos, E., Brandt, G., Merico, A., and Smith, S. L. (2013). Biogeographical patterns of phytoplankton community size structure in the oceans. *Glob. Ecol. Biogeogr.* 22, 1060–1070. doi: 10.1111/geb.12071

Aksnes, D. L., and Egge, J. K. (1991). A theoretical model for nutrient uptake in phytoplankton. *Mar. Ecol. Prog. Ser.* 70, 65–72. doi: 10.3354/meps070065

Anderson, T. R. (2005). Plankton functional type modelling: running before we can walk? *J. Plankton Res.* 27, 1073–1081. doi: 10.1093/plankt/fbi076

Baird, M. E., and Suthers, I. M. (2007). A size-resolved pelagic ecosystem model. *Ecol. Modelling* 203, 185–203. doi: 10.1016/j.ecolmodel.2006.11.025

Banas, N. S. (2011). Adding complex trophic interactions to a size-spectral plankton model: emergent diversity patterns and limits on predictability. *Ecol. Modelling* 222, 2663–2675. doi: 10.1016/j.ecolmodel.2011.05.018

Beaugrand, G. (2004). The North Sea regime shift: evidence, causes, mechanisms and consequences. *Prog. Oceanogr.* 60, 245–262. doi: 10.1016/j.pocean.2004.02.018

Boyce, D. G., Lewis, M. R., and Worm, B. (2010). Global phytoplankton decline over the past century. *Nature* 466, 591–596. doi: 10.1038/nature09268

Ciotti, A. M., Lewis, M. R., and Cullen, J. J. (2002). Assessment of the relationships between dominant cell size in natural phytoplankton communities and the spectral shape of the absorption coefficient. *Limnol. Oceanogr.* 47, 404–417. doi: 10.4319/lo.2002.47.2.0404

Daufresne, M. (2009). Global warming benefits the small in aquatic ecosystems. *Proc. Natl. Acad. Sci. U.S.A.* 106, 12788–12793. doi: 10.1073/pnas.0902080106

Doney, S., Lindsay, K., Fung, I., and John, J. (2006). Natural variability in a stable, 1000-yr global coupled climate–carbon cycle. *J. Clim.* 19, 3033–3054. doi: 10.1175/JCLI3783.1

Edwards, A., and Brindley, J. (1996). Oscillatory behaviour in a three-component plankton population model. *Dyn. Stabil. Syst.* 11, 347–370. doi: 10.1080/02681119608806231

Edwards, K. F., Litchman, E., and Klausmeier, C. A. (2013). Functional traits explain phytoplankton community structure and seasonal dynamics in a marine ecosystem. *Ecol. Lett.* 16, 56–63. doi: 10.1111/ele.12012

Edwards, M., and Richardson, A. J. (2004). Impact of climate change on marine pelagic phenology and trophic mismatch. *Nature* 430, 881–884. doi: 10.1038/nature02808

Falkowski, P. G., and Wilson, C. (1992). Phytoplankton productivity in the North Pacific ocean since 1900 and implications for absorption of anthropogenic CO_2. *Nature* 358, 741–743. doi: 10.1038/358741a0

Fasham, M. (1993). "Modelling the marine biota," in *The Global Carbon Cycle*, ed M. Heimann (Heidelberg: Springer-Verlag), 457–504.

Fasham, M., Ducklow, H., and Mckelvie, S. M. (1990). A nitrogen-based model of plankton dynamics in the oceanic mixed layer. *J. Mar. Res.* 48, 591–639. doi: 10.1357/002224090784984678

Finkel, Z. V., Beardall, J., Flynn, K. J., Quigg, A., Rees, T. A. V., and Raven, J. A. (2009). Phytoplankton in a changing world: cell size and elemental stoichiometry. *J. Plankton Res.* 32, 119–137. doi: 10.1093/plankt/fbp098

Finkel, Z. V., Irwin, A. J., and Schofield, O. (2004). Resource limitation alters the 3/4 size scaling of metabolic rates in phytoplankton. *Mar. Ecol. Prog. Ser.* 273, 269–279. doi: 10.3354/meps273269

Follows, M. J., and Dutkiewicz, S. (2011). Modeling diverse communities of marine microbes. *Ann. Rev. Mar. Sci.* 3, 427–451. doi: 10.1146/annurev-marine-120709-142848

Fuchs, H. L., and Franks, P. J. (2010). Plankton community properties determined by nutrients and size-selective feeding. *Mar. Ecol. Prog. Ser.* 413, 1–15. doi: 10.3354/meps08716

Fung, I. Y., Doney, S. C., Lindsay, K., and John, J. (2005). Evolution of carbon sinks in a changing climate. *Proc. Natl. Acad. Sci. U.S.A.* 102, 11201–11206. doi: 10.1073/pnas.0504949102

Gentleman, W., Leising, A., Frost, B., Strom, S., and Murray, J. (2003). Functional responses for zooplankton feeding on multiple resources: a review of assumptions and biological dynamics. *Deep Sea Res. II Top. Stud. Oceanogr.* 50, 2847–2875. doi: 10.1016/j.dsr2.2003.07.001

Hare, S., and Mantua, N. (2000). Empirical evidence for North Pacific regime shifts in 1977 and 1989. *Prog. Oceanogr.* 47, 103–145. doi: 10.1016/S0079-6611(00)00033-1

Kiørboe, T. (1993). Turbulence, phytoplankton cell size, and the structure of pelagic food webs. *Adv. Mar. Biol.* 29, 1–72. doi: 10.1016/S0065-2881(08)60129-7

Laws, E. (1975). The importance of respiration losses in controlling the size distribution of marine phytoplankton. *Ecology* 56, 419–426. doi: 10.2307/1934972

Litchman, E., and Klausmeier, C. A. (2008). Trait-based community ecology of phytoplankton. *Ann. Rev. Ecol. Evol. Syst.* 39, 615–639. doi: 10.1146/annurev.ecolsys.39.110707.173549

Litchman, E., Klausmeier, C. A., Miller, J. R., Schofield, O., and Falkowski, P. G. (2006). Multi-nutrient, multi-group model of present and future oceanic phytoplankton communities. *Biogeosciences* 3, 585–606. doi: 10.5194/bg-3-585-2006

Litchman, E., Klausmeier, C. A., Schofield, O., and Falkowski, P. G. (2007). The role of functional traits and trade-offs in structuring phytoplankton communities: scaling from cellular to ecosystem level. *Ecol. Lett.* 10, 1170–1181. doi: 10.1111/j.1461-0248.2007.01117.x

Litchman, E., Ohman, M. D., and Kiorboe, T. (2013). Trait-based approaches to zooplankton communities. *J. Plankton Res.* 35, 473–484. doi: 10.1093/plankt/fbt019

López-Urrutia, A., San Martin, E., Harris, R. P., and Irigoien, X. (2006). Scaling the metabolic balance of the oceans. *Proc. Natl. Acad. Sci. U.S.A.* 103, 8739–8744. doi: 10.1073/pnas.0601137103

Mackas, D. L. (2011). Does blending of chlorophyll data bias temporal trend? *Nature* 472, E4–E5. doi: 10.1038/nature09951

Mantua, N. (2004). Methods for detecting regime shifts in large marine ecosystems: a review with approaches applied to North Pacific data. *Prog. Oceanogr.* 60, 165–182. doi: 10.1016/j.pocean.2004.02.016

Marañón, E., Cermeño, P., Latasa, M., and Tadonléké, R. D. (2012). Temperature, resources, and phytoplankton size structure in the ocean. *Limnol. Oceanogr.* 57, 1266–1278. doi: 10.4319/lo.2012.57.5.1266

Marañón, E., Cermeño, P., López-Sandoval, D. C., Rodríguez-Ramos, T., Sobrino, C., Huete-Ortega, M., et al. (2013). Unimodal size scaling of phytoplankton growth and the size dependence of nutrient uptake and use. *Ecol. Lett.* 16, 371–379. doi: 10.1111/ele.12052

Marañón, E., Holligan, P., Barciela, R., González, N., Mouriño, B., Pazó, M., et al. (2001). Patterns of phytoplankton size structure and productivity in contrasting open-ocean environments. *Mar. Ecol. Prog. Ser.* 216, 43–56. doi: 10.3354/meps216043

Mariani, P., Andersen, K. H., Visser, A. W., Barton, A. D., and Kiørboe, T. (2013). Control of plankton seasonal succession by adaptive grazing. *Limnol. Oceanogr.* 58, 173–184. doi: 10.4319/lo.2013.58.1.0173

McGill, B. J., Enquist, B. J., Weiher, E., and Westoby, M. (2006). Rebuilding community ecology from functional traits. *Trends Ecol. Evol.* 21, 178–185. doi: 10.1016/j.tree.2006.02.002

McQuatters-Gollop, A., Reid, P., Edwards, M., Burkill, P., Castellani, C., Batten, S., et al. (2011). Is there a decline in marine phytoplankton? *Nature* 472, E6–E7. doi: 10.1038/nature09950

Merico, A., Bruggeman, J., and Wirtz, K. (2009). A trait-based approach for downscaling complexity in plankton ecosystem models. *Ecol. Modelling* 220, 3001–3010. doi: 10.1016/j.ecolmodel.2009.05.005

Merico, A., Tyrrell, T., Lessard, E. J., Oguz, T., Stabeno, P. J., Zeeman, S. I., et al. (2004). Modelling phytoplankton succession on the Bering Sea shelf: role of climate influences and trophic interactions in generating Emiliania huxleyi blooms 19972000. *Deep Sea Res. I* 51, 1803–1826. doi: 10.1016/j.dsr.2004.07.003

Mooney, H., Larigauderie, A., Cesario, M., Elmquist, T., Hoegh-Guldberg, O., Lavorel, S., et al. (2009). Biodiversity, climate change, and ecosystem services. *Curr. Opin. Environ. Sustain.* 1, 46–54. doi: 10.1016/j.cosust.2009.07.006

Morán, X. A. G., López-Urrutia, A., Calvo-Díaz, A., and Li, W. K. W. (2010). Increasing importance of small phytoplankton in a warmer ocean. *Glob. Change Biol.* 16, 1137–1144. doi: 10.1111/j.1365-2486.2009.01960.x

Nakićenović, N., Alcamo, J., Davis, G., Grübler, A., Kram, T., Lebre La Rovere, E., et al. (2000). *Special Report on Emissions Scenarios*. New York, NY: Intergovernmental Panel on Climate Change, Cambridge University Press.

Norberg, J., Swaney, D. P., Dushoff, J., Lin, J., Casagrandi, R., and Levin, S. A. (2001). Phenotypic diversity and ecosystem functioning in changing environments: a theoretical framework. *Proc. Natl. Acad. Sci. U.S.A.* 98, 11376–11381. doi: 10.1073/pnas.171315998

Pachauri, R. K., and Reisinger, A. (eds.). (2007). *Climate Change 2007: Synthesis Report*. Geneva: IPPC.

Parmesan, C. (2006). Ecological and evolutionary responses to recent climate change. *Ann. Rev. Ecol. Evol. Syst.* 37, 637–669. doi: 10.1146/annurev.ecolsys.37.091305.110100

Prowe, A. F., Pahlow, M., Dutkiewicz, S., Follows, M. J., and Oschlies, A. (2012). Top-down control of marine phytoplankton diversity in a global ecosystem model. *Prog. Oceanogr.* 101, 1–13. doi: 10.1016/j.pocean.2011.11.016

Reusch, T. B. H., and Boyd, P. W. (2013). Experimental evolution meets marine phytoplankton. *Evolution* 67, 1849–1859. doi: 10.1111/evo.12035

Rykaczewski, R. R., and Dunne, J. P. (2011). A measured look at ocean chlorophyll trends. *Nature* 472, E5–E6. doi: 10.1038/nature09952

Schlüter, M. H., Kraberg, A., and Wiltshire, K. H. (2012). Long-term changes in the seasonality of selected diatoms related to grazers and environmental conditions. *J. Sea Res.* 67, 91–97. doi: 10.1016/j.seares.2011.11.001

Schlüter, M. H., Merico, A., Reginatto, M., Boersma, M., Wiltshire, K. H., and Greve, W. (2010). Phenological shifts of three interacting zooplankton groups in relation to climate change. *Glob. Change Biol.* 16, 3144–3153. doi: 10.1111/j.1365-2486.2010.02246.x

Schlüter, M. H., Merico, A., Wiltshire, K. H., Greve, W., and Storch, H. (2008). A statistical analysis of climate variability and ecosystem response in the German Bight. *Ocean Dyn.* 58, 169–186. doi: 10.1007/s10236-008-0146-5

Smith, S. L., Merico, A., Wirtz, K. W., and Pahlow, M. (2014). Leaving misleading legacies behind in plankton ecosystem modelling. *J. Plankton Res.* 36, 613–620. doi: 10.1093/plankt/fbu011

Smith, S. L., Pahlow, M., Merico, A., and Wirtz, K. W. (2011). Optimality-based modeling of planktonic organisms. *Limnol. Oceanogr.* 56, 2080–2094. doi: 10.4319/lo.2011.56.6.2080

Steinacher, M., Joos, F., Frölicher, T., Bopp, L., Cadule, P., Cocco, V., et al., (2010). Projected 21st century decrease in marine productivity: a multi-model analysis. *Biogeosciences* 7, 979–1005. doi: 10.5194/bg-7-979-2010

Steinacher, M., Joos, F., Frölicher, T. L., Plattner, G.-K., and Doney, S. C. (2009). Imminent ocean acidification in the Arctic projected with the NCAR global coupled carbon cycle-climate model. *Biogeosciences* 6, 515–533. doi: 10.5194/bg-6-515-2009

Tang, E. P. (1995). The allometry of algal growth rates. *J. Plankton Res.* 17, 1325–1335. doi: 10.1093/plankt/17.6.1325

Tilman, D. (2001). An evolutionary approach to ecosystem functioning. *Proc. Natl. Acad. Sci. U.S.A.* 98, 10979–10980. doi: 10.1073/pnas.211430798

Violle, C., Enquist, B. J., McGill, B. J., Jiang, L., Albert, C. H., Hulshof, C., et al. (2012). The return of the variance: intraspecific variability in community ecology. *Trends Ecol. Evol.* 27, 244–252. doi: 10.1016/j.tree.2011.11.014

Visser, M. E. (2008). Keeping up with a warming world; assessing the rate of adaptation to climate change. *Proc. R. Soc. B* 275, 649–659. doi: 10.1098/rspb.2007.0997

von Storch, H., and Zwiers, F. (2001). *Statistical Analysis in Climate Research*. Cambridge: Cambridge University Press.

Ward, B. A., Friedrichs, A. M. M., Anderson, T. R., and Oschlies, A. (2010). Parameter optimisation techniques and the problem of underdetermination in marine biogeochemical models. *J. Marine Syst.* 81, 34–43. doi: 10.1016/j.jmarsys.2009.12.005

Weijerman, M., Lindeboom, H., and Zuur, A. (2005). Regime shifts in marine ecosystems of the North Sea and Wadden Sea. *Mar. Ecol. Prog. Ser.* 298, 21–39. doi: 10.3354/meps298021

Winder, M., and Cloern, J. E. (2010). The annual cycles of phytoplankton biomass. *Philos. Trans. R. Soc. Lond. B* 365, 3215–3226. doi: 10.1098/rstb.2010.0125

Wirtz, K. (2012a). Who is eating whom? Morphology and feeding type determine the size relation between planktonic predators and their ideal prey. *Mar. Ecol. Prog. Ser.* 445, 1–12. doi: 10.3354/meps09502

Wirtz, K. W. (2012b). How fast can plankton feed? Maximum ingestion rate scales with digestive surface area. *J. Plankton Res.* 35, 33–48. doi: 10.1093/plankt/fbs075

Wirtz, K. W., and Eckhardt, B. (1996). Effective variables in ecosystem models with an application to phytoplankton succession. *Ecol. Modelling* 92, 33–53. doi: 10.1016/0304-3800(95)00196-4

Wolff, E., Fung, I., Hoskins, B., Mitchell, J., Palmer, T., Santer, B., et al. (2014). "Climate change evidence and causes," in *Technical Report, The National Academy of Science, and The Royal Society* (Washington, DC: The National Academy Press).

Zuur, A., Ieno, E., Walker, N., Saveliev, A., and Smith, G. (2009). *Mixed Effects Models and Extensions in Ecology With R*. New York, NY: Springer. doi: 10.1007/978-0-387-87458-6

Conflict of Interest Statement: The authors declare that the research was conducted in the absence of any commercial or financial relationships that could be construed as a potential conflict of interest.

APPENDIX

Following the classical approach of Fasham et al. (1990), the effect of the physical forcing (and therefore the connection with climate) on the ecosystem is modeled implicitly by the seasonal dynamics of the upper mixed layer depth M(t), which is applied as a forcing to the model. h(t) = dM(t)/dt is used to calculate the time rate of change of the upper mixed layer depth. Material exchange between the upper mixed layer and the bottom layer are typically modeled as two processes (Fasham et al., 1990; Merico et al., 2004): (1) vertical turbulent diffusion and (2) entrainment or detrainment caused by deepening or shallowing of the upper mixed layer. Following Fasham et al. (1990) and Merico et al. (2004), we use the variable $h^+(t) = max[h(t), 0]$ in order to account for the effects of entrainment and detrainment. Zooplankton organisms are considered capable of maintaining themselves within the upper mixed layer and thus the simple function h(t) is used in that case. Diffusive mixing across the thermocline, κ, is parameterized by means of a constant factor. The whole diffusion term is written as

$$ K = \frac{\kappa + h^+}{M(t)}. \tag{A1} $$

The model is based on adaptive dynamics (Wirtz and Eckhardt, 1996; Norberg et al., 2001; Merico et al., 2009). This modeling approach focuses on a characteristic trait of the phytoplankton community to reduce the complexity of multispecies models (Merico et al., 2009). A moment closure technique is then adopted to approximate the whole community dynamics with three macroecological properties: (1) total biomass, (2) mean trait, and (3) trait variance, the latter reflecting the functional diversity of the community (Merico et al., 2009). The considered trait is phytoplankton cell size, S, which is expressed as Equivalent Spherical Diameter, ESD, in units of μm. The changes in total community biomass (P) over time t depend on the mean cell size $\bar{S}$ and are described by

$$ \frac{dP}{dt} = [r(\bar{S}) + \epsilon]P, \tag{A2} $$

where $\epsilon = \frac{1}{2} V \frac{\partial^2 r(\bar{S})}{\partial^2 \bar{S}}$ denotes higher order moments resulting from the moment closure technique (Merico et al., 2009), with V indicating the variance of the distribution of cell sizes (see Equations A9, A10 below). The term $r(\bar{S})$ is the net growth rate of the total phytoplankton biomass, i.e., gains minus losses in P, which is given by:

$$ r(\bar{S}) = \mu_P \cdot f(T) \cdot \Psi(I) \cdot U(\bar{S}, N) + \delta_I - m_P $$
$$ - \mu_Z \cdot G(\bar{S}, P) \cdot Z - \nu(\bar{S}) + K, \tag{A3} $$

where μ_P is the maximum specific growth rate at temperature $T = 0°C$ and $f(T) = e^{0.063 \cdot T}$ represents Eppley's formulation of temperature-dependent growth. The light limitation term, $\Psi(I)$, integrates the photosynthetically active radiation (PAR) I through the mixed layer by using Steele's formulation:

$$ \Psi(I) = \frac{1}{M(t)} \int_0^M \left[\frac{I(z)}{I_s} \cdot e^{\left(1 - \frac{I(z)}{I_s}\right)} \right] dz, \tag{A4} $$

where I_s is the light level at which photosynthesis saturates and $I(z)$ is the PAR at depth z. The exponential decay of light with depth is computed according to the Beer-Lambert law with a generic extinction coefficient k_w

$$ I(z) = I_0 \cdot e^{-k_w \cdot z}, \tag{A5} $$

with I_0 representing light at the surface of the ocean (i.e., $z = 0$). Light attenuation is simulated with a constant attenuation coefficient and it is assumed to include the effects of different substances in the water, such as chlorophyll and other suspended particles. This means that the effect of chlorophyll concentration on light attenuation is uniform throughout the year in both locations. While this may not impinge on the accuracy of our results in the tropics, because of the relative low and constant biomass there throughout the year, it may lead to an overestimation of the biomass during high productivity periods in the temperate region. However, the value of the extinction coefficient that we consider ($k_w = 0.1$ per meter, see **Table 1**) is relatively higher, albeit within the range suggested in the literature (Edwards and Brindley, 1996), than the one used in the classical approach of Fasham et al. (1990), $k_w = 0.04$ per meter, and this can partly compensate for the absence of an explicit self-shading mechanism. Despite the simplicity of the light limitation term, our model is able to reasonably capture the observed nutrient and phytoplankton seasonal cycles in both regions (see Supplementary Figure 3).

The nutrient-limited uptake term $U(\bar{S}, N)$ depends on the nutrient concentration and scales with phytoplankton cell size:

$$ U = \frac{N}{N + K_N} = \frac{N}{N + (\beta_U \cdot \bar{S}^{\alpha_U})}, \tag{A6} $$

where β_{K_N} and α_{K_N} are, respectively, intercept and slope of the K_N allometric function. This empirical relationship is based on observations of different phytoplankton groups (Litchman et al., 2007), with the regression parameters rescaled from cell volume to Equivalent Spherical Diameter (ESD).

The term $G(\bar{S}, P)$ denotes zooplankton grazing, which is a function of phytoplankton cell size:

$$ G = \frac{\bar{S}^{-1}}{\frac{P}{\bar{S}} + K_P}, \tag{A7} $$

where K_P is the half saturation constant for grazing.

The term $\nu(\bar{S})$ represents the size-dependent sinking

$$ \nu = \frac{\beta_\nu \cdot \bar{S}^{\alpha_\nu}}{M(t)}, \tag{A8} $$

where the constants α_ν and β_ν are the parameters of the allometric function proposed by Kiørboe (1993), which transformed here to obtain units in meters per day.

Finally, the term δ_I accounts for the dispersal rate of phytoplankton (i.e., immigration) into the considered community (Norberg et al., 2001), m_P accounts for all possible phytoplankton losses other than grazing and mixing.

The temporal changes in mean cell size are described by the adaptive dynamics equation

$$\frac{d\overline{S}}{dt} = V \frac{\partial r(\overline{S})}{\partial \overline{S}}, \qquad (A9)$$

where V is the size variance or functional diversity of the community. The size variance determines the adaptive capacity of the phytoplankton community and its temporal evolution is given by

$$\frac{dV}{dt} = -V^2 \frac{\partial^2 r(\overline{S})}{\partial^2 \overline{S}} + \left[\frac{\delta_I \cdot P}{P}(V_0 - V) \right], \qquad (A10)$$

where V_0 is a source of size variance from an immigrating community outside the modeled region.

Differential equations for nutrients (N), zooplankton (Z), and detritus (D) complete the model system:

$$\frac{dN}{dt} = -\mu_P \cdot f(T) \cdot \Psi(I) \cdot U(\overline{S}, N) \cdot P + \delta_D \cdot D \qquad (A11)$$

$$+ K \cdot (N_0 - N) + \epsilon_N,$$

$$\frac{dZ}{dt} = \delta_Z \cdot \mu_Z \cdot G(\overline{S}, P) \cdot P \cdot Z - m_Z \cdot Z^2 \qquad (A12)$$

$$- \frac{h(t)}{M(t)} \cdot Z + \epsilon_Z,$$

$$\frac{dD}{dt} = (1 - \delta_Z) \cdot \mu_Z \cdot G(\overline{S}, P) \cdot P \cdot Z + m_P \cdot P + m_Z \cdot Z^2$$

$$- \delta_D \cdot D - K \cdot D + \epsilon_D, \qquad (A13)$$

where $\epsilon_N = \frac{1}{2} V \frac{\partial^2 \mu_P \cdot f(T) \cdot \Psi(I) \cdot U(\overline{S}, N)}{\partial^2 \overline{S}} P$, $\epsilon_Z = \frac{1}{2} V \frac{\partial^2 \delta_Z \cdot \mu_Z G(\overline{S}, P) \cdot Z}{\partial^2 \overline{S}} P$, $\epsilon_D = \frac{1}{2} V \frac{\partial^2 (1 - \delta_Z) \cdot \mu_Z G(\overline{S}, P) \cdot Z}{\partial^2 \overline{S}} P$ account for higher order moments resulting from the moment closure technique (Merico et al., 2009) and N_0 is the concentration of nutrients below the mixed-layer. This variable is a forcing obtained from NCAR model simulations and changes, therefore, with mixed-layer depth and time.

Note that the only differences between the model applications to the two regions are represented by the environmental forcing. Model parameters and model structure (including the formulation of the size-dependent processes and therefore of the emerging trade-off) are identical in both cases.

For a detailed description of all the parameters used refer to **Table 1**.

5

High salinity tolerance of the Red Sea coral *Fungia granulosa* under desalination concentrate discharge conditions: an *in situ* photophysiology experiment

Riaan van der Merwe[1]*[†], Till Röthig[2][†], Christian R. Voolstra[2], Michael A. Ochsenkühn[3], Sabine Lattemann[1] and Gary L. Amy[1]

[1] Water Desalination and Reuse Center, Biological and Environmental Sciences and Engineering Division, King Abdullah University of Science and Technology, Thuwal, Saudi Arabia

[2] Red Sea Research Center, Biological and Environmental Sciences and Engineering Division, King Abdullah University of Science and Technology, Thuwal, Saudi Arabia

[3] Biological and Organometallic Catalysis Laboratories, Physical Sciences and Engineering Division, King Abdullah University of Science and Technology, Thuwal, Saudi Arabia

Edited by:
Hans Uwe Dahms, Kaohsiung Medical University, Taiwan

Reviewed by:
Xiaoshou Liu, Ocean University of China, China
Rajesh Kumar Ranjan, Central University of Bihar, India

*Correspondence:
Riaan van der Merwe, Red Sea Research Center, Marine Monitoring and Environmental Management, King Abdullah University of Science and Technology, Building 2, Level 2, Thuwal 23955-6900, Saudi Arabia
e-mail: riaan.vandermerwe@kaust.edu.sa

[†] These authors have contributed equally to this work.

Seawater reverse osmosis desalination concentrate may have chronic and/or acute impacts on the marine ecosystems in the near-field area of the discharge. Environmental impact of the desalination plant discharge is supposedly site- and volumetric- specific, and also depends on the salinity tolerance of the organisms inhabiting the water column in and around a discharge environment. Scientific studies that aim to understand possible impacts of elevated salinity levels are important to assess detrimental effects to organisms, especially for species with no mechanism of osmoregulation, e.g., presumably corals. Previous studies on corals indicate sensitivity toward hypo- and hyper-saline environments with small changes in salinity already affecting coral physiology. In order to evaluate sensitivity of Red Sea corals to increased salinity levels, we conducted a long-term (29 days) *in situ* salinity tolerance transect study at an offshore seawater reverse osmosis (SWRO) discharge on the coral *Fungia granulosa*. While we measured a pronounced increase in salinity and temperature at the direct outlet of the discharge structure, effects were indistinguishable from the surrounding environment at a distance of 5 m. Interestingly, corals were not affected by varying salinity levels as indicated by measurements of the photosynthetic efficiency. Similarly, cultured coral symbionts of the genus *Symbiodinium* displayed remarkable tolerance levels in regard to hypo- and hypersaline treatments. Our data suggest that increased salinity and temperature levels from discharge outlets wear off quickly in the surrounding environment. Furthermore, *F. granulosa* seem to tolerate levels of salinity that are distinctively higher than reported for other corals previously. It remains to be determined whether Red Sea corals in general display increased salinity tolerance, and whether this is related to prevailing levels of high(er) salinity in the Red Sea in comparison to other oceans.

Keywords: desalination, salinity tolerance, *Fungia granulosa*, *Symbiodinium*, coral reef, Red Sea, marine monitoring, environmental impact assessment

INTRODUCTION

A growing demand of freshwater in semi-arid and arid regions (e.g., Arabian Peninsula) leads to the construction of an increasing number of seawater desalination plants, especially the low energy consuming seawater reverse osmosis (SWRO) desalination plants (Fritzmann et al., 2007). As a result, more hypersaline concentrate discharge (brine) reaches the marine environments (Lattemann and Höpner, 2008). Environmentally safe disposal of this brine is one of the key factors determining the environmental impacts of a desalination plant. The highest salinity that marine

organisms can cope with in a desalination discharge area is defined as a salinity tolerance threshold. It depends on the species and the exposure time to elevated salinity levels (Voutchkov, 2009).

Euryhaline marine organisms can commonly tolerate changes in salinity (in contrast to stenohaline species) and series of small increments are generally better tolerated than direct exposure to high salinities (Voutchkov, 2009). Effects of concentrate discharges depend on exposure intensities, frequencies, the environment the brine is released into, and the brine temperature (Roberts et al., 2010). Accordingly, the effects of discharged brine can range from no significant impacts on microbial abundance or plankton communities, to widespread alterations in community structures of seagrass, invertebrates, soft-sediment infauna, and corals

---

Abbreviations: $\Delta F/Fm'$, effective quantum yield; Brine, hypersaline concentrate discharge; DO, dissolved oxygen; FACS, fluorescence activated cell-sorting; PAM, pulse-amplitude modulation (fluorometry); PAR, photosynthetically active radiation; PBS, *phosphate buffered saline*; PSII, photosystem II; PSU, practical salinity units; SWRO, seawater reverse osmosis (desalination plant).

(Roberts et al., 2010 and references therein; van der Merwe et al., 2014).

Hermatypic corals are the key stone species of coral reefs, which are among the most diverse and productive ecosystems on this planet (Moberg and Folke, 1999). Coral health and survival fundamentally depend on the interaction between coral host and photosynthetic algae (zooxanthellae, genus *Symbiodinium*) that can be found in the endodermal tissues of reef-building corals. Unfortunately, research on the effects of brine discharge on corals is scarce. Mabrook (1994) reported corals disappearing from coastal areas in the Red Sea (Egypt) as a result of desalination plants discharge, but no reproducible data are presented. Corals are generally considered stenohaline osmoconformers and very sensitive to the effects of desalination plant discharge (Ferrier-Pages et al., 1999; Manzello and Lirman, 2003; Elimelech and Phillip, 2011).

Besides a desalination discharge context, more data on coral salinity tolerance are available; especially for decreased salinities. Generally, changes in salinity may affect metabolism and/or photophysiology of the coral animal and/or the corals' algal symbionts due to salinity stress (Muscatine, 1967; Chartrand et al., 2009). Moberg et al. (1997) suggested that photosynthetic rates are lowered in proportion to the reductions in salinity (10–20 PSU), whereas respiration rates were either slightly decreased or unaltered for two hermatypic corals upon hyposaline treatments. Hoegh-Guldberg and Smith (1989) concluded moderately reduced salinity (30 PSU) exposure for 4–10 days at several temperatures does not induce bleaching in *Stylophora pistillata* and *Seriatopora hystrix*. At lower salinities (i.e., at 23 PSU) the authors reported death within 48 h. Only a limited number of studies have included the impacts of hypersalinity on corals (and the possible effect it might have on dinoflagellate symbiont functionality within the host). Generally, the effects of hyper- and hyposaline treatments are similar (Muthiga and Szmant, 1987; Lirman and Manzello, 2009). Severity of effects observed commonly coincides with salinity concentration, exposure time, coral species, and the speed of salinity change.

In this study, our objective was to determine long-term effects (4 weeks) of a strong salinity increase on the coral *Fungia granulosa*. To do this, we transplanted corals along a 25 m transect from a SWRO facility discharge structure, and determined salinity, temperature, oxygen, and light levels regularly. At the same time, we measured the (photo)physiological state of the algal symbiont via PAM fluorometry and checked for signs of visual bleaching. PAM fluorometry quantifies the photosynthetic efficiency with which light energy is converted into chemical energy at photosystem II (PSII) level. Evaluating the chlorophyll fluorescence can indicate an organism's photosynthetic efficiency under changing or stressful conditions, e.g., varying salinity regimes (Chartrand et al., 2009), and therefore, serves as a stress indicator. Additionally, we assessed the salinity tolerance of cultured *Symbiodinium* to understand the contribution of the algal symbiont to salinity tolerance of the coral holobiont.

MATERIALS AND METHODS
EXPERIMENT OVERVIEW

The existing SWRO facility (i.e., the submerged discharge location) at the King Abdullah University of Science and Technology

(KAUST) was selected for this case study. The plant is located on the KAUST campus and is designed to provide all potable water needs. Under current operational conditions, the raw water intake is about $2825 \, m^3 \, h^{-1}$ with a recovery rate of 39%, resulting in an average brine flow of $1723.25 \, m^3 \, h^{-1}$ ($41,358 \, m^3 \, d^{-1}$) that is discharged to the Red Sea. The submerged outfall (discharge structure) is located at a water depth of 18 m, approximately 2.8 km from the pump station (22° 17.780N, 39° 04.444E). The concentrate is pumped through a 1.2 m diameter pipeline to the offshore structure where the concentrate is pushed up in a concrete riser and discharged horizontally through four discharge screens ($1800 \times 1000 \, mm$) approximately 6 m above the seafloor.

EXPERIMENTAL SETUP

We conducted an *in situ* salinity stress experiment on the coral *F. granulosa* collected from Fsar reef (22° 13.945N, 39° 01.783E, approximately 9 km from the study site) over 29 days (15.01.14–13.2.14). At Fsar reef salinity, light conditions, and effective quantum yields for *F. granulosa* were measured for reference purposes. Corals were handled with latex gloves and 18 specimens were collected from 16 to 19 m (similar depth to experimental study area) into separate zip lock bags. Corals were transported to the study site in shaded opac plastic boxes filled with ambient sea water. All specimens were placed on the roof of the discharge structure (**Figure 1**), labeled with nylon fishing line and under-water paper tags, and left for acclimation for 20 h. Three specimens were then randomly selected and placed at each of the 6 stations at the discharge screen (station 1) and along a 25 m transect (stations 2–6) (**Figures 2A,B**).

DATA COLLECTION

We collected data on temperature, dissolved oxygen (DO), and salinity. For each sampling time point (T0–T6) we sampled between 11:00 and 12:00 h. We measured the effective quantum yield ($\Delta F / Fm'$) and visually assessed all specimens for signs of bleaching. Temperature was logged continuously in 10 min intervals with HOBO Pendant® temperature data loggers at each station over the entire experiment. For DO and salinity measurements, water samples were collected during each dive from each station. Water samples for salinity measurements were collected using 50 mL Falcon Conical Centrifuge Tubes and 1 L low-density polyethylene (LDPE) cubitainer were used for DO samples. Salinity and DO were measured for all stations immediately after each dive. DO measurements were conducted using a WTW (Multi) 3500i Multi-Parameter Water Quality Meter with a CellOx® 325 DO electrode and salinity with a WTW Cond 3310 m with TetraCon® 325. Salinity and DO were analyzed for significant differences via One-Way ANOVA using Statistica 10 (StatSoft Inc. 2011, version 10). A diving PAM fluorometer (DIVING-PAM, Walz, Germany) was used to measure photosynthetically active radiation (PAR) at each station and to evaluate the effective quantum yields ($\Delta F / Fm'$) of the symbiotic algae of each coral specimen. The effective quantum yield ($\Delta F / Fm'$) of photochemical energy conversion in PSII for each measurement was calculated based on F and Fm' according to Genty et al. (1989):

$$\Phi \, PSII \; = \left(Fm' - F \right) / Fm' = \Delta F / Fm'.$$

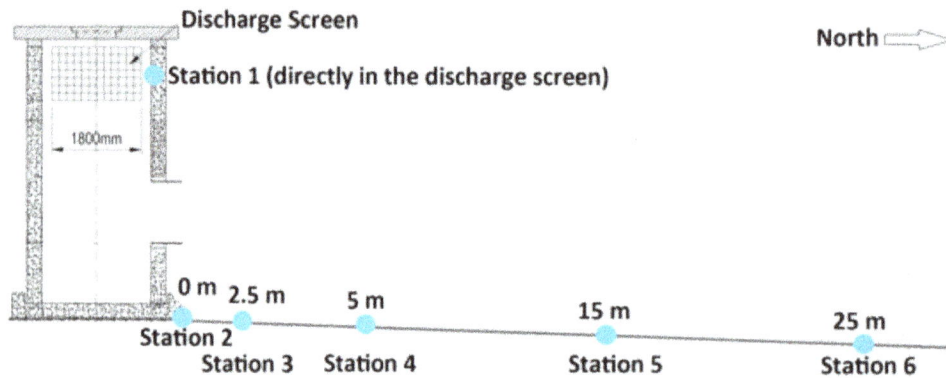

FIGURE 1 | Discharge structure and stations assayed in this experiment.

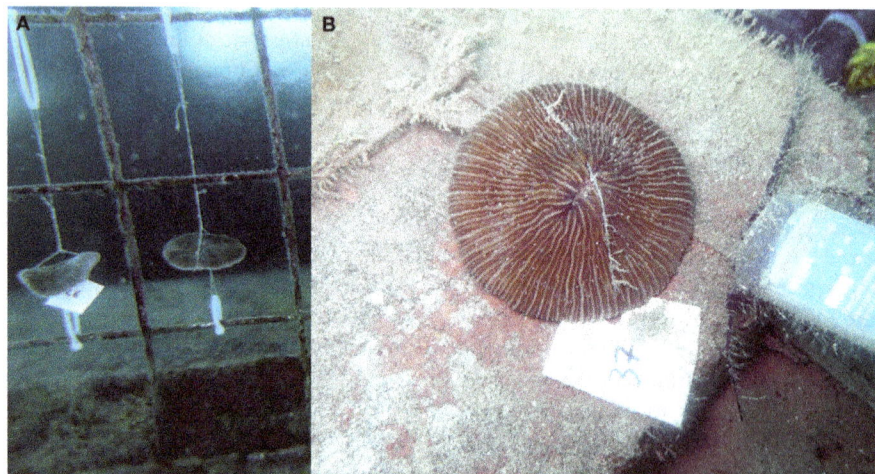

FIGURE 2 | (A) *Fungia granulosa* attached to the discharge screen. **(B)** Tagged *Fungia granulosa* specimen; blue temp logger visible on right hand side (station 6).

F and Fm' measurements for each specimen were conducted between 11:00 and 12:00 h to ensure comparable daytime conditions (in regard to physiology and prevailing light regime). All $\Delta F/Fm'$ measurements were taken in triplicate for each coral and specimens were only collected after the last sampling event (29 days).

SYMBIODINIUM CULTURING AND SALT GRADIENT STRESS EXPERIMENT

Symbiodinium microadriaticum CCMP2467 (clade A1 https://ncma.bigelow.org) was cultured in Guillard and Ryther F/2 media suspension without silicium in an incubator at a temperature of 26°C and with a light intensity set at 4 PAR (as measured by diving-PAM) (Guillard and Ryther, 1962). The F/2 media was newly prepared from sea water (obtained from 100 m depth in the Red Sea) and completed with 0.5 ml NaNO$_3$, NaH$_2$PO$_4$, vitamins and trace metals following Guillard and Ryther (1962). The cells were kept in the exponential growth phase with a cell density between 10^5 and 10^6 cells mL^{-1} for 2 weeks prior to experiments. For the comparability of experiments, the cells were then directly

transferred into a freshly prepared salt adjusted F/2 media with a cell density of $\sim 1 \times 10^5$ cells mL^{-1}. A salt gradient experiment was conducted in concentrations ranging from 25 to 55 PSU (in increments of 5 PSU). The F/2 media were prepared by adding appropriate amounts of double-distilled water (ddH$_2$O) for a diluted salinity range of 25–35 PSU. In order to obtain elevated salinity levels on the order of 45–55 PSU, the media was spiked with NaCl (Sigma). Cells were then transferred into sterilized plastic culture flasks in equal volumes (400 mL) of adjusted F/2 media to reach a cell density of $\sim 1 \times 10^5$ cells mL^{-1}. Cells were sampled over a 7 day period at 7 different time points (0, 1, 2, 3, 5, and 7 d) at a temperature of 28°C and under a 9 PAR light intensity. The 0 d sample was used as a control and only withdrawn from cells at the 40 PSU (ambient) condition. Sample volumes for fluorescence activated cell- sorting (FACS) analysis were 1 mL, which were directly harvested by centrifugation (5430 R centrifuge, Eppendorf) at 10,000× g for 10 min at 4°C. For cell fixation, 700 μL media were withdrawn, adding 100 μL of 40% formaldehyde in ddH$_2$0 to each of the samples in order to reach a final concentration of 10%. Cells were thoroughly resuspended by vortexing and kept at 4°C until further analysis. Following cell fixation, *Symbiodinium* cells were again harvested by centrifugation, supernatant discarded, and washed once with 500 μL buffer solution phosphate buffered saline (PBS). PBS was removed completely and 500 μL SYBR Green DNA (2× conc., Life Technologies) staining in PBS was added. Cells were resuspended by vortexing and stained for 1 h at room temperature at 400 rpm in a ThermoMixer® (Eppendorf). After staining, cells were again pelleted by centrifugation, washed once with PBS and finally resuspended in 1 mL PBS. For FACS measurements 200 μL of each sample were transferred into a 96 well flat bottom well plate and measured. FACS measurements were conducted on a BD LSRFortessa™ cell analyzer (BD Bioscience, US) using the 405 nm violet laser and QDot655 filters for chlorophyll fluorescence. SYBR Green fluorescence was excited via the 488 nm blue laser and emission detected via Alexa Fluor® 488 filters for total nucleotide detection. FACS data was analyzed by FlowJo 7.5 flow cytometry analysis software.

RESULTS

ECOLOGICAL CONDITIONS

Water temperature along the transect showed a range between 24.26 and 28.46°C (**Table 1**). Water temperatures recorded at the discharge screen (26.3 ± 0.78°C) were on average higher than at the other stations (25.93–26.08°C). The average temperatures were also slightly decreasing with increasing distance from the discharge structure. DO varied between 5.75 and 6.37 mg L^{-1}. Average DO levels were 6.07 ± 0.17 mg L^{-1}. We found no significant differences in DO between the stations ($P_{ANOVA} \geq 0.05$). In contrast to DO, salinity data showed significant differences (**Figure 3**). At the discharge screen (station 1) salinity differed significantly from the other stations ($P_{ANOVA} \leq 0.05$). We found no significant difference between all other stations ($P_{ANOVA} \geq 0.05$) with an average of 41.3 ± 0.7 PSU. However, salinity decreased slightly with increasing distance from the discharge (**Figure 3**). Control measurements at Fsar reef (site of coral collection) showed a salinity of 39 PSU.

CORAL SALINITY TOLERANCE

The effective quantum yield ($\Delta F/Fm'$) was measured for all 18 coral colonies in triplicates for all time points (i.e., T0–T6) (**Figure 4**). Control measurements from corals in their natural reef environment showed average $\Delta F/Fm'$ of 0.700 ± 0.010 (at 30–50 PAR; triplicate measurements on three specimens).

$\Delta F/Fm'$ at all stations were constant and in the same range as the controls. Average reads were 0.705 ± 0.009 (station 1), 0.694 ± 0.008 (station 2), 0.687 ± 0.007 (station 3), 0.687 ± 0.019 (station 4), 0.693 ± 0.017 (station 5), and 0.690 ± 0.02 (station 6). We found a drop in $\Delta F/Fm'$ at T1 for stations 4–6 which corresponds with elevated light conditions on this day compared to the other stations and sampling events (65–82 PAR; **Figures 4D–F**). PAR levels ranged between 4 and 82, with average reads of 14 ± 4 (station 1), 17 ± 5 (station 2), 33 ± 13 (station 3), 45 ± 27 (station 4), 34 ± 27 (station 5), and 28 ± 21 (station 6). By trend, PAR levels at station 1 and 2 and to a smaller extent at station 3 were lower and more stable compared to the other stations (**Figure 4**). This pattern is caused by a shading effect of the discharge structure which kept light levels in its direct surrounding lower and more stable. No PAR was measured at time point T5. During the study we did not observe bleaching characteristics on the measured coral specimens.

CORAL SYMBIONT SALINITY TOLERANCE

To confirm our *in vivo* observations, i.e., the absence of measurable detrimental effects of high salinity on the (photo)physiology of *F. granulosa*, we investigated the response of cultured coral symbionts, i.e., *Symbiodinium microadriaticum*, to a range of salinity levels. Chlorophyll a levels and growth rate of *S. microadriaticum* were investigated at salinities between 25 and 55 PSU and samples were taken at 7 time points (0, 1, 2, 3, 5, and 7 d). FACS cell counts and chlorophyll a level measurements showed that *Symbiodinium* cells reached the exponential growth phase after 4 days of incubation at salinities between 30 and 50 PSU (**Figure 5**). The quickest cell proliferation was observed at 35 PSU. For concentration levels of 25 and 55 PSU, respectively, the cell growth rates were inhibited. Measuring chlorophyll levels by FACS showed no differences (data not shown).

DISCUSSION

In this study we (1) characterized physicochemical conditions along a 25 m transect at a SWRO facility discharge structure in

Table 1 | Temperature data (HOBO Pendant® Temperature Data Loggers) from all transect stations.

	Station 1 (discharge)	Station 2 (0 m)	Station 3 (2.5 m)	Station 4 (5 m)	Station 5 (15 m)	Station 6 (25 m)
AVG	26.30	26.08	26.02	25.99	25.84	25.83
±SD	±0.78	±0.63	±0.62	±0.64	±0.73	±0.95
Min	24.74	24.64	24.55	24.45	24.35	24.26
Max	28.46	27.76	27.67	27.57	27.57	27.57

Logging interval 10 min; temperature [°C] mean ± SD; minimal and maximal measured temperature.

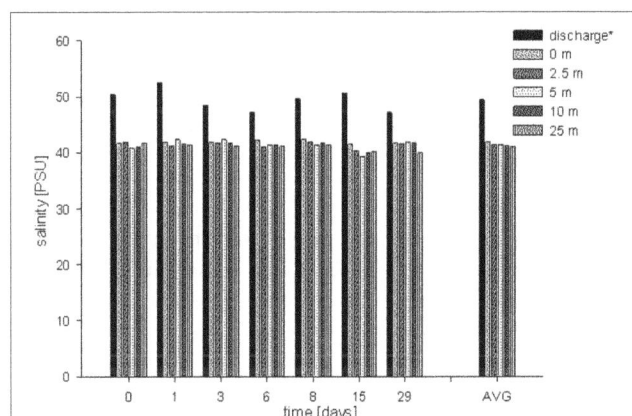

FIGURE 3 | Salinity concentrations measured for each sampling event at each station (6 time points, 6 stations). Average salinities: 49.4 ± 2.0 PSU (discharge screen), 41.8 ± 0.3 PSU (0 m), 41.3 ± 0.6 PSU (2.5 m), 41.3 ± 1.0 PSU (5 m), 41.2 ± 0.6 PSU (15 m) and 41.0 ± 0.7 PSU (25 m), respectively. AVG shows the average salinity at each station.

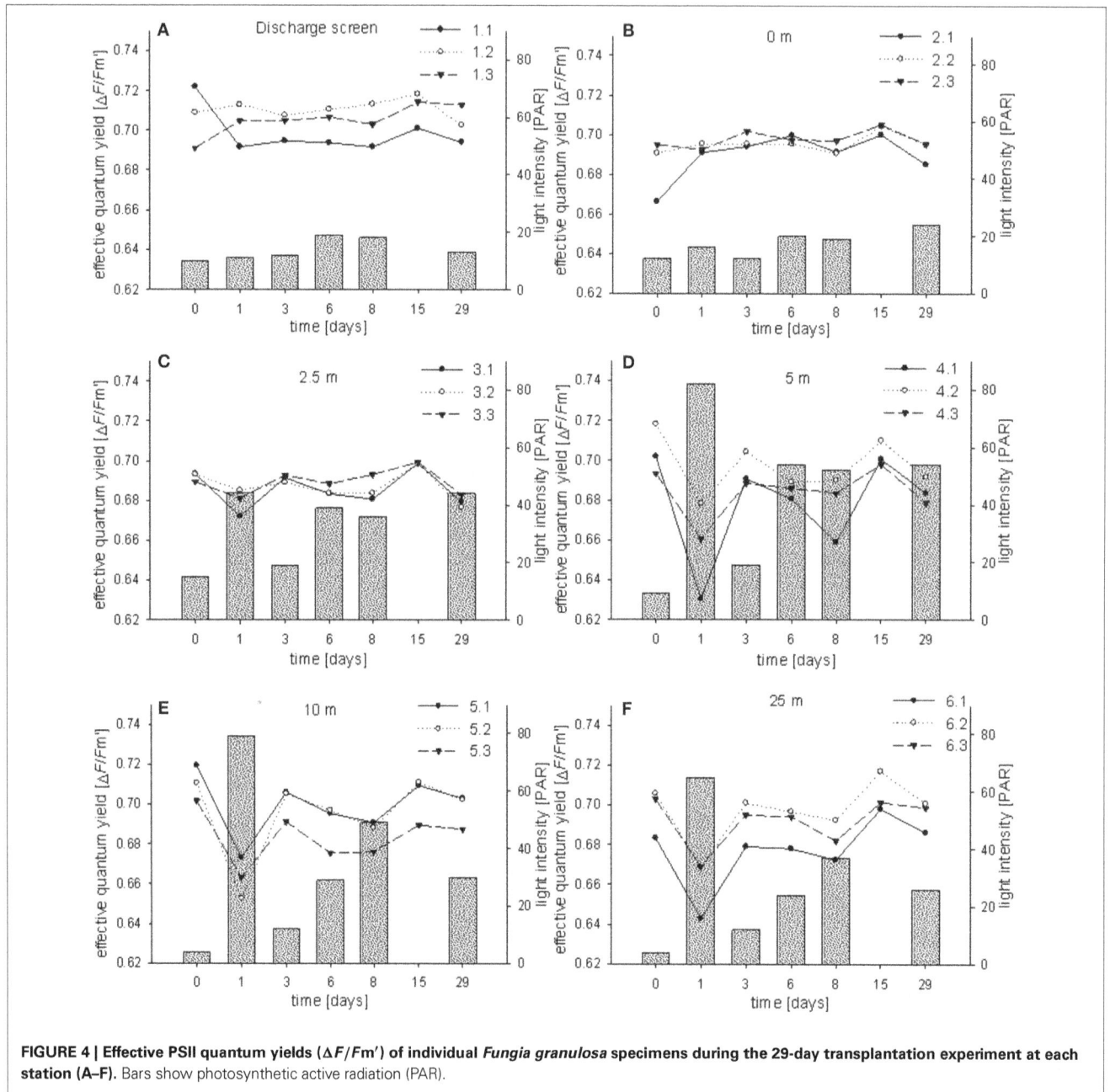

FIGURE 4 | Effective PSII quantum yields (ΔF/Fm′) of individual *Fungia granulosa* specimens during the 29-day transplantation experiment at each station (A–F). Bars show photosynthetic active radiation (PAR).

the Red Sea. We measured (2) photosynthetic characteristics of *F. granulosa* as a response to a sudden and strong increase and continued elevated levels in salinity. We also checked for (3) indications of bleaching for the coral specimens exposed to the highest salinity levels directly at the discharge screen. Last, we (4) exposed *Symbiodinium* cultures to low and high levels of salinity to test salinity tolerance of the coral symbiont.

With regard to the physicochemical conditions, we measured a decrease in salinity and temperature with increasing distance from the discharge structure. Of note, we already measured distinct lower values at stations 2 and 3 (seafloor, 0 and 2.5 m) compared to station 1 (discharge screen). This indicates a quick natural mixing of the brine in the study area to the point of

showing little to no discernible salinity abnormalities (within short distances). These findings are in line with the literature (Roberts et al., 2010). Light levels were dependent on water clarity and measured light levels were in a similar range to what we measured at the collection site (Fsar reef, 30–50 PAR). Stations 1–3 were shaded by the discharge structure and showed lower and more stable PAR levels. These light patterns were also reflected in our ΔF/Fm′ measurements, i.e., the photosynthetic characteristics of *F. granulosa* in response to changes in salinity (**Figure 6**). Measurements especially at station 1 and 2 (but also station 3) were more stable; in contrast stations 4–6 displayed stronger variations with a drop after 1 day (T1). This drop corresponds to noticeable clearer water and accordingly higher PAR values for

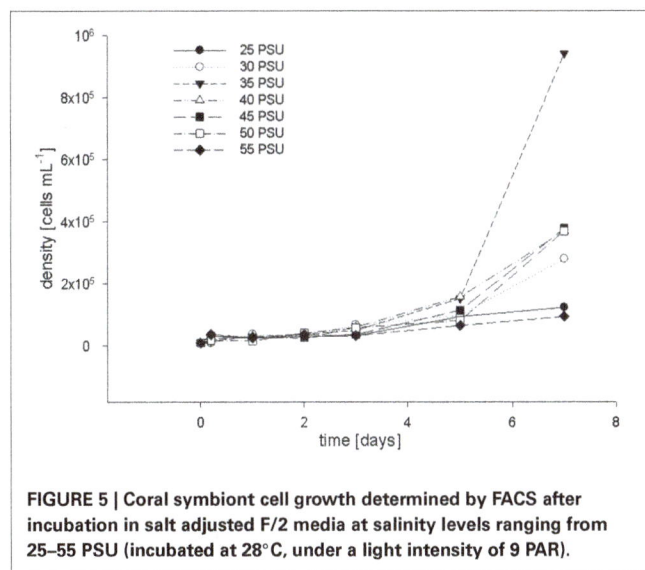

FIGURE 5 | Coral symbiont cell growth determined by FACS after incubation in salt adjusted F/2 media at salinity levels ranging from 25–55 PSU (incubated at 28°C, under a light intensity of 9 PAR).

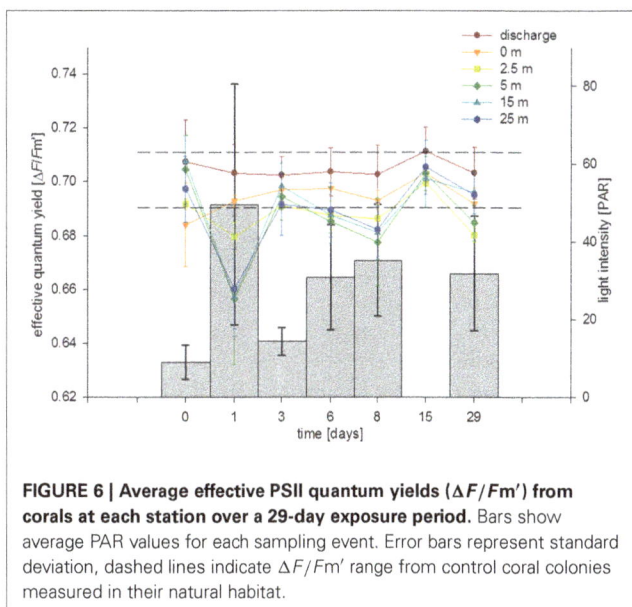

FIGURE 6 | Average effective PSII quantum yields ($\Delta F / Fm'$) from corals at each station over a 29-day exposure period. Bars show average PAR values for each sampling event. Error bars represent standard deviation, dashed lines indicate $\Delta F / Fm'$ range from control coral colonies measured in their natural habitat.

the unshaded stations 4–6. All results are on the order of the "baseline" yields measured under natural conditions considering variations in light intensity and suggest no discernable effect on PSII level of the dinoflagellate symbionts of the corals. Average $\Delta F/Fm'$ measures recorded at station 1 were higher than at the other stations (0.705 compared to 0.687–0.694). This again corresponds with lower average PAR values (14 compared to 17–45), but also underlines that the significantly higher salinity at station 1 does not negatively impact the photosynthetic efficiency of the coral's symbionts. However, effective quantum yield measurements, reflecting (chronic) photoinhibition, may have limitations. Measurements from bleached corals may result in apparent healthy PAM yields, rather depending on the physiological state of associated *Symbiodinium* and less on the symbiont density or total number of cells (Fitt et al., 2001). Bleaching may exhibit a loss of zooxanthellae and/or pigments and can indicate a breakdown in the essential symbiotic relationship between coral host and algal symbiont (Brown, 1997). Taking this into account, we visually inspected the corals during each sampling dive. We could not detect any apparent colony changes or loss in coloration (i.e., pigment loss) and thus exclude bleaching as a consequence of increased salinity during the duration of the experiment.

Our results demonstrate a high salinity tolerance of the Red Sea coral *F. granulosa*. In contrast, other studies observed substantial loss of pigmentation and/or symbionts at considerably smaller salinity changes. For example, Ferrier-Pages et al. (1999) used a Red Sea clone of *Stylophora pistillata* that has been maintained in aquaria for several months. At moderately increased salinity levels (+2 PSU), coral colonies showed significant effects on photosynthesis, respiration, and protein content. Interestingly, the increased salinity of 40 PSU corresponds with the salinity in *S. pistillata*'s natural habitat. In contrast, Lirman and Manzello (2009) found notable tolerance to salinity in *Siderastrea radians*, collected from Biscayne Bay, Florida where the authors measured highly variable levels of salinity. The authors also reported on *Porites furcata* (collected from the same bay) to be highly salinity tolerant (Manzello and Lirman, 2003). In a study from

Chartrand et al. (2009), *S. radians* was collected from several different sites with different levels of natural salinity variations. The authors found a correlative trend of local salinity regimes and hyposaline stress tolerance levels. Colonies originating from less stable environments showed higher photosynthetic efficiency in hyposaline treatments than corals from more stable surroundings. These observations are in line with a recently conducted study, where Barshis et al. (2013) found "front-loading" of genes to confer higher temperature tolerance levels in corals exposed to temperature-variable environments in comparison to cooler, stable environments.

So far, literature mostly assumed corals to be stenohaline osmoconformers (Hoegh-Guldberg and Smith, 1989; Ferrier-Pages et al., 1999; Kerswell and Jones, 2003). In contrast, Chartrand et al. (2009) stated that a threshold response is indicative of the coral maintaining and successfully regulating its internal osmotic balance, which would contradict corals to be stenohaline osmoconformers. This is supported by a broad range of salinity tolerance levels found in different coral species depending on their original environment, which also indicates differently effective osmotic regulation. Furthermore, Mayfield and Gates (2007) discuss potential mechanisms involved in corals maintaining their osmotic balance. Accordingly, osmoregulatory processes presumably play a role in the performance of *F. granulosa* in this experiment. Osmoregulatory processes might be reflected by an initial reaction to the sudden salinity increase. This period presumably happened in our experiment between T0 and T1 and could be addressed by short-term studies. Since ambient salinity levels in the Red Sea are higher than in most other oceans (Douabul and Haddad, 1970) salinity tolerance in Red Sea corals might generally be higher than for corals in other oceans.

Photophysiological resilience of *F. granulosa* toward high salinity is supported by our *Symbiodinium* culture study. Inhibited cell growth at extreme changes in salinity (i.e., 25 and 55 PSU)

might be caused by limits in the cellular salinity regulation capabilities of *Symbiodinium* cells, namely Na^+ pumps (Na^+-ATPase) (Goiran et al., 1997). Similar to plants, high NaCl exposure levels could lead to degradation of chlorophyll a as observed in sunflower leaves (Santos, 2004). Continuously lowered levels of active chlorophyll might result in a reduced energy uptake via photosynthesis and therefore, also directly influence cell growth. However, salinity levels up to 50 PSU (at par to highest measured *in situ* salinity levels) did not seem to inhibit cell growth in coral symbiont cultures. This demonstrates a wide plasticity (30–50 PSU) of *Symbiodinium* in regard to salinity changes, which has been found for hyposaline treatments before (Chartrand et al., 2009). This resilience furthermore indicates that the algal symbionts may generally not be determining the acclimation potential of the coral holobiont toward salinity changes.

In conclusion, we found a quick mixing of discharged brine with surrounding waters based on salinity and temperature measurements. The photophysiology of the coral *F. granulosa* exposed to the discharge environment along a 25 m transect was not influenced by rapid and prolonged changes in salinity (but varied according to changes in light conditions). Our data characterize *F. granulosa* coral holobionts to be remarkably resilient toward increased salinity levels, which are potentially brought about by acclimation to increased salinity levels in the Red Sea environment. Additionally, we showed that cell cultures of *Symbiodinium* only displayed inhibited cell growth at very high and low salinity levels. Based on our data we suggest *F. granulosa* from the Red Sea to possess a great acclimation potential to salinity changes, also in regard to future ocean scenarios.

AUTHOR CONTRIBUTIONS

Till Röthig, Christian Robert Voolstra, Riaan van der Merwe, Michael A. Ochsenkühn designed and conceived the experiments. Riaan van der Merwe, Till Röthig, Michael A. Ochsenkühn generated data. Riaan van der Merwe, Till Röthig, Christian Robert Voolstra, Michael A. Ochsenkühn analyzed and interpreted data. Sabine Lattemann, Gary L. Amy contributed reagents/materials/analysis tools. Riaan van der Merwe, Till Röthig, Christian Robert Voolstra, Michael A. Ochsenkühn wrote the manuscript.

ACKNOWLEDGMENTS

We would like to thank CMOR for assistance and support in field operations. Research in this study was supported by King Abdullah University of Science and Technology (KAUST).

REFERENCES

Barshis, D. J., Ladner, J. T., Oliver, T. A., Seneca, F. O., Traylor-Knowles, N., and Palumbi, S. R. (2013). Genomic basis for coral resilience to climate change. *Proc. Natl. Acad. Sci. U.S.A.* 110, 1387–1392. doi: 10.1073/pnas.1210224110

Brown, B. E. (1997). Coral bleaching: causes and consequences. *Coral Reefs* 16, S129–S138. doi: 10.1007/s003380050249

Chartrand, K., Durako, M., and Blum, J. (2009). Effect of hyposalinity on the photophysiology of *Siderastrea radians*. *Mar. Biol.* 156, 1691–1702. doi: 10.1007/s00227-009-1204-3

Douabul, A., and Haddad, A. M. (1970). *The Red Sea and Yemen's Red Sea Environments*. Hassell and Assoc., AMSAT and UNOPS.

Elimelech, M., and Phillip, W. A. (2011). The future of seawater desalination: energy, technology, and the environment. *Science* 333, 712–717. doi: 10.1126/science.1200488

Ferrier-Pages, C., Gattuso, J.-P., and Jaubert, J. (1999). Effect of small variations in salinity on the rates of photosynthesis and respiration of the zooxanthellate coral *Stylophora pistillata*. *Mar. Ecol. Prog. Ser.* 181, 309–314. doi: 10.3354/meps181309

Fitt, W., Brown, B., Warner, M., and Dunne, R. (2001). Coral bleaching: interpretation of thermal tolerance limits and thermal thresholds in tropical corals. *Coral Reefs* 20, 51–65. doi: 10.1007/s003380100146

Fritzmann, C., Löwenberg, J., Wintgens, T., and Melin, T. (2007). State-of-the-art of reverse osmosis desalination. *Desalination* 216, 1–76. doi: 10.1016/j.desal.2006.12.009

Genty, B., Briantais, J.-M., and Baker, N. R. (1989). The relationship between the quantum yield of photosynthetic electron transport and quenching of chlorophyll fluorescence. *Biochim. et Biophys. Acta Gen. Subj.* 990, 87–92. doi: 10.1016/S0304-4165(89)80016-9

Goiran, C., Allemand, D., and Galgani, I. (1997). Transient Na^+ stress in symbiotic dinoflagellates after isolation from coral-host cells and subsequent immersion in seawater. *Mar. Biol.* 129, 581–589. doi: 10.1007/s002270050199

Guillard, R. R. L., and Ryther, J. H. (1962). Studies of marine planktonic diatoms: I. *Cyclotella nana* Hustedt, and *Detonula confervacea* (cleve) Gran. *Can. J. Microbiol.* 8, 229–239. doi: 10.1139/m62-029

Hoegh-Guldberg, O., and Smith, G. J. (1989). The effect of sudden changes in temperature, light and salinity on the population density and export of zooxanthellae from the reef corals *Stylophora pistillata* Esper and *Seriatopora hystrix* Dana. *J. Exp. Mar. Biol. Ecol.* 129, 279–303. doi: 10.1016/0022-0981(89)90109-3

Kerswell, A. P., and Jones, R. J. (2003). Effects of hypo-osmosis on the coral *Stylophora pistillata*: nature and cause of 'low-salinity bleaching'. *Mar. Ecol. Prog. Ser.* 253, 145–154. doi: 10.3354/meps253145

Lattemann, S., and Höpner, T. (2008). Environmental impact and impact assessment of seawater desalination. *Desalination* 220, 1–15. doi: 10.1016/j.desal.2007.03.009

Lirman, D., and Manzello, D. (2009). Patterns of resistance and resilience of the stress-tolerant coral *Siderastrea radians* (Pallas) to sub-optimal salinity and sediment burial. *J. Exp. Mar. Biol. Ecol.* 369, 72–77. doi: 10.1016/j.jembe.2008.10.024

Mabrook, B. (1994). Environmental impact of waste brine disposal of desalination plants, Red Sea, Egypt. *Desalination* 97, 453–465. doi: 10.1016/0011-9164(94)00108-1

Manzello, D., and Lirman, D. (2003). The photosynthetic resilience of *Porites furcata* to salinity disturbance. *Coral Reefs* 22, 537–540. doi: 10.1007/s00338-003-0327-0

Mayfield, A. B., and Gates, R. D. (2007). Osmoregulation in anthozoan–dinoflagellate symbiosis. *Comp. Biochem. Physiol. A: Mol. Integr. Physiol.* 147, 1–10. doi: 10.1016/j.cbpa.2006.12.042

Moberg, F., and Folke, C. (1999). Ecological goods and services of coral reef ecosystems. *Ecol. Econ.* 29, 215–233. doi: 10.1016/S0921-8009(99)00009-9

Moberg, F., Nyström, M., Kautsky, N., Tedengren, M., and Jarayabhand, P. (1997). Effects of reduced salinity on the rates of photosynthesis and respiration in the hermatypic corals *Porites lutea* and *Pocillopora damicornis*. *Mar. Ecol. Prog. Ser.* 157, 53–59. doi: 10.3354/meps157053

Muscatine, L. (1967). Glycerol excretion by symbiotic algae from corals and tridacna and its control by the host. *Science* 156, 516–519. doi: 10.1126/science.156.3774.516

Muthiga, N. A., and Szmant, A. M. (1987). The effects of salinity stress on the rates of aerobic respiration and photosynthesis in the hermatypic coral *Siderastrea siderea*. *Biol. Bull.* 173, 539–551. doi: 10.2307/1541699

Roberts, D. A., Johnston, E. L., and Knott, N. A. (2010). Impacts of desalination plant discharges on the marine environment: a critical review of published studies. *Water Res.* 44, 5117–5128. doi: 10.1016/j.watres.2010.04.036

Santos, C. V. (2004). Regulation of chlorophyll biosynthesis and degradation by salt stress in sunflower leaves. *Sci. Hortic.* 103, 93–99. doi: 10.1016/j.scienta.2004.04.009

van der Merwe, R., Hammes, F., Lattemann, S., and Amy, G. (2014). Flow cytometric assessment of microbial abundance in the near-field area of seawater reverse osmosis concentrate discharge. *Desalination* 343, 208–216. doi: 10.1016/j.desal.2014.01.017

Voutchkov, N. (2009). Salinity tolerance evaluation methodology for desalination plant discharge. *Desalination Water Treat.* 1, 68–74. doi: 10.5004/dwt.2009.126

Conflict of Interest Statement: The authors declare that the research was conducted in the absence of any commercial or financial relationships that could be construed as a potential conflict of interest.

6

Integrated assessment of marine biodiversity status using a prototype indicator-based assessment tool

Jesper H. Andersen[1,2]*, Karsten Dahl[3], Cordula Göke[3], Martin Hartvig[4,5], Ciarán Murray[3], Anna Rindorf[4], Henrik Skov[6], Morten Vinther[4] and Samuli Korpinen[2]

[1] NIVA Denmark Water Research, Copenhagen, Denmark
[2] Marine Research Center, Finnish Environment Institute (SYKE), Helsinki, Finland
[3] Department of Bioscience, Aarhus University, Roskilde, Denmark
[4] DTU Aqua, Section for Marine Ecosystem-Based Management, Technical University of Denmark, Charlottenlund, Denmark
[5] Centre for Macroecology, Evolution and Climate, University of Copenhagen, Copenhagen, Denmark
[6] DHI, Hørsholm, Denmark

Edited by:
Christos Dimitrios Arvanitidis,
Hellenic Centre for Marine
Research, Greece

Reviewed by:
Christos Dimitrios Arvanitidis,
Hellenic Centre for Marine
Research, Greece
Marco Sigovini, National Research
Council of Italy, Italy
Céline Labrune, Centre National de
la Recherche Scientifique, France

*Correspondence:
Jesper H. Andersen, NIVA Denmark
Water Research, Ørestads
Boulevard 73, 2300 Copenhagen S,
Denmark
e-mail: jha@niva-danmark.dk

Integrated assessment of the status of marine biodiversity is and has been problematic compared to, for example, assessments of eutrophication and contamination status, mostly as a consequence of the fact that monitoring of marine habitats, communities and species is expensive, often collected at an incorrect spatial scale and/or poorly integrated with existing marine environmental monitoring efforts. The objective of this Method Paper is to introduce and describe a simple tool for integrated assessment of biodiversity status based on the HELCOM Biodiversity Assessment Tool (BEAT), where interim biodiversity indicators are grouped by themes: broad-scale habitats, communities, and species as well as supporting non-biodiversity indicators. Further, we report the application of an initial indicator-based assessment of biodiversity status of Danish marine waters where we have tentatively classified the biodiversity status of Danish marine waters. The biodiversity status was in no areas classified as "unaffected by human activities." In all the 22 assessment areas, the status was classified as either "moderately affected by human activities" or "significantly affected by human activities." Spatial variations in the biodiversity status were in general related to the eutrophication status as well as fishing pressure.

Keywords: biodiversity, marine, integrated assessment, habitats, communities, species, Marine Strategy Framework Directive

INTRODUCTION

Assessments of biological diversity have the ambitious objective of describing the state of an entire ecosystem, often by using only a few selected indicators. The challenge of this objective is to select a representative set of indicators, which fulfill the needs of science and marine policy. The EU Marine Strategy Framework Directive (MSFD) sets 11 qualitative descriptors for "good environmental status" (Anon, 2008), laying a common framework for all European marine biodiversity assessments. In this new assessment regime, biodiversity is considered to include not only the species diversity and the state of populations and habitats, but also seafloor integrity and food webs. Despite the detailed guidance on the selection of indicators (Anon, 2010), the MSFD does not provide a methodology to assess the overall state of marine ecosystems with the proposed criteria and indicators. Instead the EC tasked ICES with the production of detailed reports on the next steps of the implementation of the MSFD descriptors (see Cardoso et al., 2010 and relevant background reports).

Biodiversity assessments generally need to take into account the fact that marine biodiversity is sensitive to and also structured by natural factors such as salinity, currents, temperature, etc. More specifically, marine biodiversity assessments have been limited by the lack of integrated monitoring networks, high-quality biodiversity indicators, and indicator-based assessment tools (Borja, 2014), partly a consequence of the vast nature of biodiversity. We hypothesize that all three deficiencies are related to two shortcomings in monitoring. Firstly, monitoring of marine biodiversity is often expensive compared to the monitoring of eutrophication and contamination and good proxies for biodiversity changes have not been developed. Secondly, for certain features of marine biodiversity, e.g., seabirds, monitoring is inadequately integrated with the existing marine environmental monitoring and, hence, resources are wasted in uncoordinated efforts.

Consequently, assessments of marine biodiversity are not as well-developed as other types of assessments, where multi-metric indicator-based assessment tools are commonly used (HELCOM, 2010; Andersen et al., 2011). The regional sea conventions in the Baltic Sea (HELCOM; www.helcom.fi) and North-East Atlantic (OSPAR; www.ospar.org) as well as EU Directives (Habitats Directive and MSFD) call for assessments of biodiversity, but only HELCOM has thus far made an attempt to develop an prototype indicator-based tool for an assessment of biodiversity (HELCOM, 2009b, 2010).

A few recent studies of marine biodiversity in Northern Europe are based on data addressing a wide range of biodiversity features (such as phytoplankton, benthic communities, fish, seabirds, marine mammals) and robust and transparent scientific methods, e.g., Certain et al. (2011), Ojaveer et al. (2010), and Ojaveer and Eero (2011). These studies do not, however, take into account numerical biodiversity targets, and this is a shortcoming in regard to assessment of biodiversity status in the context of the MSFD (Anon, 2008).

In this study, we introduce and describe a simple indicator-based methodology (i.e., tool) for assessing the status of marine biodiversity. The tool is tested in Danish marine waters using provisional indicators with associated numerical target values and the results presented and discussed should accordingly be regarded as tentative. The assessment of biodiversity is made despite the lack of a commonly accepted definition of "marine biodiversity." Both the tool and the assessment are anchored in a Baltic Sea-wide conceptual understanding of "good biodiversity status" (HELCOM, 2010), where the overall vision is a healthy Baltic Sea with a favorable biodiversity status, including (1) natural marine and coastal landscapes, (2) thriving and balanced communities of plants and animals, and (3) viable populations of species. Hence, our understanding of "marine biodiversity" is broad and includes other elements than just a count of the number of species.

METHODS

We have developed a methodology for classification of "biodiversity status," employing a tool named Biodiversity Assessment Tool (BEAT) 2.0, which is an improved version of the HELCOM Biodiversity Status Assessment Tool (BEAT 1.0). This multimetric indicator-based tool was initially developed for integrated assessment of the status of biodiversity in the Baltic Sea (HELCOM, 2009a, 2010), but its updated version differs from its predecessor by having an improved fit with the EU MSFD descriptors, three status classes, a balanced approach to confidence rating as well as a more user-friendly appearance, where information about the Biodiversity Quality Objective (BQO) as well as interim (per category) and integrated classification results are presented.

BEAT 2.0 is an indicator-based assessment tool. For an individual indicator, synoptic information is required regarding reference conditions (RefCon), acceptable deviation from reference conditions (AcDev), and observations of the present state of biological diversity (Obs). AcDev is defined as a fraction or percentage of the RefCon, and is set site-specifically per indicator.

In calculating the status, we considered two types of indicators: (1) indicators that show a positive (+ve) response to human pressure factors, i.e., whose value increases with greater degradation in biodiversity (e.g., primary production, which is positively correlated to nutrient enrichment), and (2) indicators with a negative (−ve) response, i.e., whose value decreases with greater degradation (e.g., depth distribution of submerged aquatic vegetation, which is negatively correlated to nutrient enrichment or population size of a fish species, which is negatively correlated to fishing pressure).

As a first step, a BQO, which defines the border between "biodiversity status unaffected by human activities" (UN) and "biodiversity status moderately affected by human activities" (MO), is calculated per indicator:

$$\text{BQO} = \text{RefCon} \times (1 + \text{AcDev}) \quad (+\text{ve response})$$
$$= \text{RefCon} \times (1 - \text{AcDev}) \quad (-\text{ve response}) \quad (1)$$

Step 2 is calculating the state value for each indicator through comparison with the BQO to determine indicator status. For example, for an indicator with +ve response, if the observed state (Obs) does not exceed the BQO, then the status "unaffected by human activities" is achieved. If the BQO is exceeded, the status is "moderately" (MO) or "significantly affected by human activities" (SI).

$$\text{Status} = \text{UN} \quad (+\text{ve response, Obs} \leq \text{BQO})$$
$$= \text{MO/SI} \quad (+\text{ve response, Obs} > \text{BQO})$$
$$= \text{UN} \quad (-\text{ve response, Obs} \geq \text{BQO})$$
$$= \text{MO/SI} \quad (-\text{ve response, Obs} < \text{BQO}) \quad (2)$$

Step 3 is to calculate a Biodiversity Quality Ratio (BQR), which in principle is comparable with the Ecological Quality Ratio principle *sensu* the WFD (Anon, 2000; Andersen et al., 2011). The BQR approach used in this assessment marks the ratio (0–1) between Obs and RefCon. For indicators with a positive response the BQR is given by RefCon/Obs. For those having a negative response the BQR is the inverse, i.e., Obs/RefCon.

$$\text{BQR} = \text{RefCon/Obs} \quad (+\text{ve response})$$
$$= \text{Obs/RefCon} \quad (-\text{ve response}) \quad (3)$$

This step represents a transformation of indicator-specific information regarding the state of biodiversity to a numerical value, where the BQR values for different indicators can be compared and combined.

As a step 4, indicators are combined within four categories: (I) broad-scale habitats, (II) communities, (III) species, and (IV) supporting indicators. The classifications are based on a weighted average of the BQO and BQR values within each category. Weights are established by expert judgment and used to balance indicators among different biodiversity components or correlated indicators (e.g., several fish indicators are down-weighted against single indicators for seabirds or mammals). If not specified otherwise, the weighting is kept neutral by giving each of the indicators equal weights. On the basis of the BQR and AcDev values, each category is given a quantitative assessment according to the principles described above for a single indicator. Individual indicators have only two "classes," i.e., "unaffected" and "impaired/affected." There are three category classes from "unaffected," to "moderately affected" and "significantly affected" by human activities. Whilst the boundary between "unaffected by human activities" (UN) and "moderately affected by human activities" (MO) is a simple weighted average derived from the indicator-specific BQOs, the boundary between "moderately" and "significantly affected by human activities" (SI) is a value of two times the criteria-specific BQO.

At step 5, the results of the four categories are combined by applying the so-called "One out—All out" principle *sensu* the Precautionary Principle (MSFD Preamble, section 27; Anon, 2008) to the Categories I–IV. This implies that the category most sensitive to human activities, i.e., scoring lowest, defines the overall status of biodiversity within an assessment sector.

In addition to the above-described classification of biodiversity status, we estimate the confidence of the data and of the resulting classification by applying a simple scoring system (see Andersen et al., 2010). This system was initially developed for estimation of the confidence in eutrophication classifications but can be directly transferred and applied, when assessing biodiversity status. The approach, which scores the data on RefCon, AcDev and Obs gives equal weight to each of these three factors. In order to balance BQOs and Obs, we have modified the weighting of the factors with 25% to RefCon and AcDev and 50% to Status. The final confidence of the assessment can range between 100 and 0% and is according to Andersen et al. (2010) grouped in three classes: High (100–75%), Acceptable (75–50%), and Low (<50%). A description of the confidence rating method is available online as Supporting Material (Annex S3).

All calculations and subsequent classifications are made within a spreadsheet (see the Supplementary Material). Tracking calculations per indicator and also the integrations made per category and integration made in order to arrive at a final classification of biodiversity status is transparent and straightforward.

The BEAT 2.0 tool was tested and demonstrated using data from Danish marine waters, which are located in two distinct marine regions, the saline North Sea and the brackish Baltic Sea (**Figure 1**). Comprehensive descriptions of the study area and environmental status can be found in HELCOM (2010) and OSPAR (2010). The test was made on the basis of 22 assessment sectors in the Danish marine waters (**Figure 1**). The assessment sectors were larger in the offshore waters where spatial variation of the biodiversity indicators was considered smaller than in the coastal waters.

The data used for testing of BEAT 2.0 were compiled from various sources. Data on submerged aquatic vegetation as well as plankton (chlorophyll-a), benthic invertebrate communities, and nutrient concentrations originate from the Danish National Aquatic Monitoring and Assessment Programme (DNAMAP; see Conley et al., 2000; Carstensen et al., 2006; Dahl and Carstensen, 2008; Hansen, 2013). Data originates from three sources which are specific to the following areas: (1) offshore parts North Sea, Skagerrak and Kattegat (assessment sectors 1, 2, 4, 5), (2) offshore part of the Arkona Basin and Bornholm Basin, which are parts of the Baltic Sea (sectors 21 and 22), and (3) Danish coastal waters (sectors 3 and 6–20).

The indicators in regard to offshore fish, seabirds and marine mammals, which should be regarded as provisional, were developed specifically for this study and were also used for an interim assessment of biodiversity status in the North Sea (HARMONY project; unpublished data). Indicators used in previous assessments of the state of the North Sea (OSPAR, 2010) and Baltic Sea (HELCOM, 2010) were used for benthic and pelagic habitats and communities as well as supporting indicators. Detailed

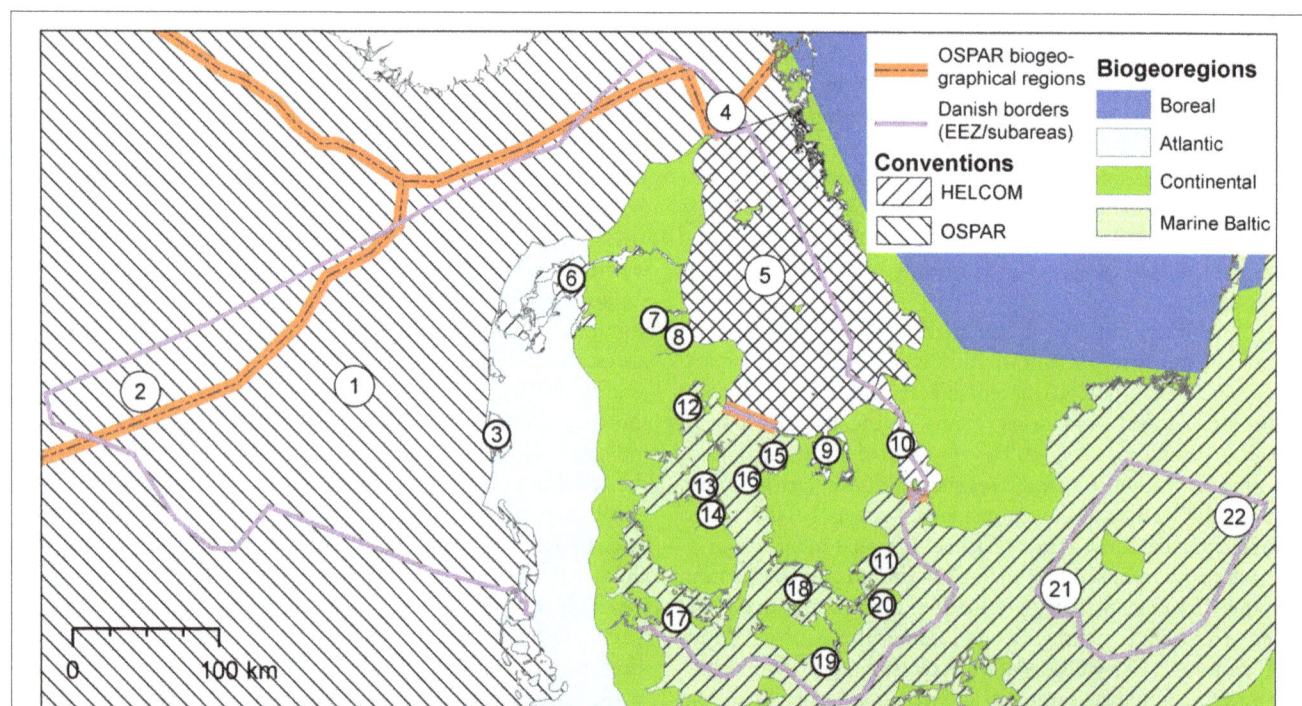

FIGURE 1 | Map of Danish marine waters. The borders indicated in the map represent the current MSFD boundary between the North Sea region and the Baltic Sea region, relevant OSPAR boundaries, relevant HELCOM boundaries as well as relevant Habitats Directive boundaries for biogeographical regions (BOR, Boreal; ATL, Atlantic; CON, Continental). Numbers indicates assessment sectors (see **Table 1** for names). Large circles indicate offshore assessment sectors, small circles coastal assessment sectors.

Table 1 | Assessment and classification of biodiversity status in Danish marine waters.

Assessment sector	Biodiversity Quality Ratio (BQR)				Integrated assessment
	C I	C II	C III	C IV	
1. NORTH SEA, eastern and southern parts	0.636*	0.907	0.656	–	SI
2. NORTH SEA, northern parts	0.700	0.904	0.619*	–	SI
3. Ringkøbing Fjord	–	0.377*	–	0.850	SI
4. SKAGERRAK, open parts	0.862	0.939	0.502*	–	SI
5. KATTEGAT, central parts	0.320*	0.749	0.482	0.733	SI
6. Limfjorden	–	0.351*	–	0.650	SI
7. Mariager Fjord	–	–	0.370*	0.519	SI
8. Randers Fjord	0.562	0.258*	0.485	0.369	SI
9. Isefjorden/Roskilde Fjord	–	0.613*	–	0.763	SI
10. The Sound, central parts	0.525*	0.823	–	0.560	MO
11. Fakse Bight/Stevns	0.843	0.704	–	0.336*	SI
12. Aarhus Bight	0.533*	0.671	–	0.548	MO
13. Marine waters north of Funen	0.353*	0.578	–	0.537	SI
14. Odense Fjord	0.294*	0.482	–	0.320	SI
15. Sejerø Bight	–	0.443*	–	–	SI
16. Kalundborg Fjord	–	0.357*	–	–	SI
17. Lillebælt, southern parts	0.230*	0.541	–	0.500	SI
18. Smålandsfarvandet	–	0.513*	–	–	SI
19. Rødsand	–	0.590*	–	–	SI
20. Hjelm Bight	0.838	0.702	–	0.533*	SI
21. ARKONA BASIN	0.534*	0.764	0.566	0.616	MO
22. BORNHOLM BASIN	0.553	0.239*	0.566	0.604	SI
Offshore assessment sectors (average)	0.601	0.750	0.565	0.651	–
Coastal assessment sectors (average)	0.522	0.534	0.428	0.540	–
All assessment sectors (average)	0.556	0.595	0.531	0.563	–

For each assessment sector, the weighted Biodiversity Quality Ratio (BQR) is presented. These values represent the perturbation in regard to the reference conditions. C I, marine landscapes (broad-scale marine habitats); C II, communities; C III, species; C IV, supporting indicators; MO, moderately affected by human activities; and SI, significantly affected by human activities. The category being decisive for the outcome of the integrated assessment and classification is marked with an asterisk. See Online Supporting material for details.

information about (1) the interim biodiversity indicators, (2) the sources for the monitoring data used as well as (3) the periods covered is available online as Supporting Material.

RESULTS

The average number of indicators per assessment sector was 10.2 ($n = 22$) ranging from 1 (no. 15 and 16) to 25 (no. 5). The average number of indicators in the four categories I–IV was 1.0, 4.0, 3.1, and 2.3, respectively. For the 6 offshore assessment sectors, the average number of indicators was 19.3 ranging from 8 (no. 22) to 25 (no. 2 and 5) and the average number of indicators in the four categories were 1.5, 5.8, 10.3, and 1.8 respectively. For the remaining 16 coastal assessment sectors, the average number of indicators was 6.8 ranging from 1 (no. 15 and 16) to 15 (no. 6) and the average number in the four categories were 0.9, 3.3, 0.3, and 2.4, respectively.

In the Danish marine waters, the average Biological Quality Ratio was 0.556, 0.595, 0.531, and 0.563 per category (**Table 1**). In category I, the BQR ranged from 0.230 to 0.862, in category II from 0.239 to 0.939, in category III from 0.370 to 0.656, and in category IV from 0.320 to 0.850.

For each assessment sector, a status classification was made per category and combined to a final integrated assessment of status per assessment sector (**Table 1**). The average of the lowest classified category was 0.433, ranging from 0.230 (sector 17: Southern Little Belt) to 0.639 (sector no. 1: North Sea, East+South). Areas with a BQR < 0.400 included Odense Fjord (sector 14), Little Belt (sector 17), and Bornholm Basin (sector 22), which all are significantly affected by eutrophication (HELCOM, 2010; Andersen et al., 2011). Areas with a BQR value above 0.600 were few and only found in the North Sea (sectors no. 1 and 2) and Isefjorden/Roskilde Fjord (sector 9). None of the assessment sectors were classified as unaffected by human activities. Three out of 22 assessment sectors were classified as moderately affected by human activities. The areas were Arkona Basin (no. 21), The Sound (no. 10) and Aarhus Bight (no. 12). The remaining 19 sectors were classified as significantly affected by human activities, and in 17 of these, the final classification was caused by categories I (broad-scale habitats), II (communities) or III (species). In two sectors, Hjelm Bight (no. 20) and Fakse Bight/Stevns (no. 11), the final classifications were a result of supporting indicators.

The confidence of the assessments was generally estimated to be above 50% and therefore considered acceptable (**Figure 2A**). However, two assessment sectors had a low confidence (no. 15 and 16: respectively, Sejerø Bay and Kalundborg Fjord) due to low number of indicators in the assessment in combination with challenges in regard to the setting of AcDev. Analysing the data per indicator revealed that monitoring data (State) and RefCon values on average had a higher confidence than the information on AcDev, which seemed to be slightly below the border between acceptable and low confidence (**Figure 2**). Scrutiny of the confidence per category revealed that all four categories on average had an acceptable confidence. All final classifications of the biodiversity status in the North Sea/Skagerrak area and the Kattegat had an acceptable confidence, while in the sub-division covering the Danish parts of the Baltic Sea, 2 out of 12 had an unacceptable confidence.

DISCUSSION

In this study we have presented a spreadsheet-based assessment tool for assessment of biodiversity, based on indicators, quantitative thresholds for good environmental status, and confidence rating. The assessment tool, tested by using both (i) existing and provisional indicators and (ii) recent data, showed that the marine biodiversity of Danish marine waters cannot be considered to be in good environmental status. The perturbations from reference conditions are indicative of human pressures in the assessment area (OSPAR, 2010; Korpinen et al., 2012).

Given the data and indicators available, we estimated the perturbations—understood as the deviation from reference conditions—represented by the lowest BQR values within an assessment sector. Parts of the North Sea and Skagerrak were less disturbed compared to the Kattegat and the Danish parts of the Baltic Sea (**Figure 3A**). The areas deviating most from reference conditions are all characterized by high nutrient inputs, high fishing pressure, and physical modification, sometimes caused by destructive fishing practices (HELCOM, 2010; Korpinen et al., 2012). Any measures to improve biodiversity status should as a priority address these key pressures.

An overview of the biodiversity status in the Danish marine waters revealed that a group of sectors being classified as moderately affected are interconnected (**Figure 3B**). The Sound is located downstream of Arkona Basin with a surface current from Arkona Basin to the west through Femernbelt between Denmark and Germany and to the north through the Sound. Hjelm Bight (sector no. 20) is located to the west and downstream of Arkona Basin. Fakse Bight/Stevns (sector no. 11) is located in between Arkona Basin and the Sound. The biodiversity status of the Arkona Basin and the Sound being classified as moderately affected by human activities is in line with the general understanding of the ecological status of these areas (HELCOM, 2010). Another sector having a slightly better status is Aarhus Bight (no. 12), where biodiversity status was classified as moderately affected by human activities in all the four categories. This, together with an estimated high confidence, does in our opinion confirm the classification. The reason for this slightly better status compared to adjacent sectors is most likely due to significant reductions in nutrient loads to Arhus Bight over past two decades (HELCOM, 2012).

Making an assessment without estimating the confidence of the result is a tendency, which in principle is unacceptable (**Figure 3C**). Estimating confidence is a statistical challenge, but the simple scoring system developed as a part of BEAT 2.0 overcomes this challenge in a non-statistical way and is able to cover confidence of threshold values, data and also the low number of indicators. This approach can be seen as temporary, leading to more sophisticated and data driven systems for assessment of confidence.

Many of the indicators in this assessment test have long traditions in previous assessments. Benthic communities and submerged aquatic vegetation have a long history in regard to assessments of eutrophication in the North Sea and Baltic Sea regions. Also indicators of fish communities have been used in previous assessments (Daan et al., 2005; Greenstreet et al., 2011), but reference levels had not yet been proposed for our study area, and for this analysis we used reference levels and acceptable deviations of 1 standard deviation based on the historic time series available.

Basin-wide biodiversity assessments have not hitherto included indicators for seabirds or marine mammals. The assessment in this respect can therefore be seen as a first attempt to use the trends in the population size of key species of seabirds or marine mammals as indicators of the status of the pelagic ecosystem in terms of habitat quality, food supply, and human-induced displacement. As the seabird data available for the assessment did not include data from the most recent period, the assessment used AcDev values of 50% and, hence, may give false positive impression of their status. Therefore, the reported changes in the abundance of fish-eating seabirds in the eastern parts of the North Sea, Skagerrak, and Kattegat should be regarded as strong indications of negative changes in the ecological status of these regions. Recent studies indicate that the regional reduction of fish-eating seabirds in the North Sea is mainly governed by changes in the large-scale abundance of herring (Fauchald et al., 2011). Reflecting the spatial caveats in the marine mammal data, the assessment used AcDev values of 50%. It is not known to what degree the impaired status of marine mammals in the eastern parts of the North Sea is a result of similar changes in the supply of pelagic fish which affected the abundance of seabirds in these regions. We did not include indicators for non-native species in this study. However, there is a growing understanding that, contrary to the normally negative perception of the ecological impact of non-native species, some species may provide significant ecosystem services in specific cases (Norkko et al., 2012).

In the current implementation process of the EU MSFD, there is a growing need to coordinate indicator development and agree on common sets of indicators, which allow coherent, trans-boundary assessments of the state of marine environment. By using existing indicators from the region, we noticed that several of the indicators were inherently correlated in nature (e.g., LFI and the slope of the size spectra, or chlorophyll a and Secchi depth) and using both as independent indicators in the present study may not be appropriate from a statistical point of view. In this study this correlation was accounted for by giving small weights to such indicators, but more stringent statistical

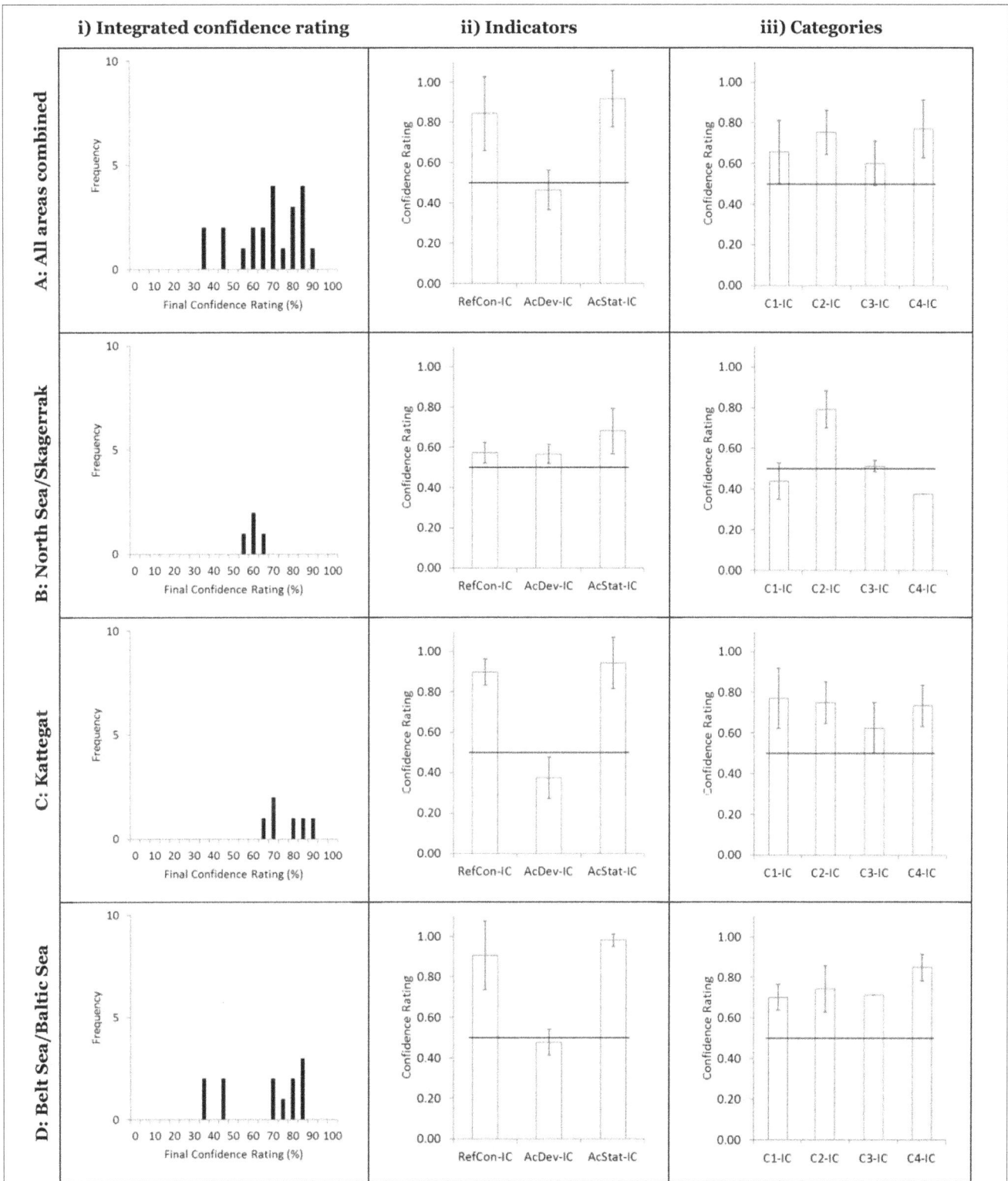

FIGURE 2 | (A) Confidence ratings made for (i) integrated assessments; (ii) information in regard to RefCon, AcDev, and AcStat of indicators, and (iii) categories I–IV. Values > 50% indicate an acceptable confidence (Andersen et al., 2010). **(B–D)** Sub-region-specific confidence assessments for the North Sea and Skagerrak, the Kattegat including the northern parts of the Sound and the Belt Sea and the western Baltic Sea. Please confer with Supplementary Material for details.

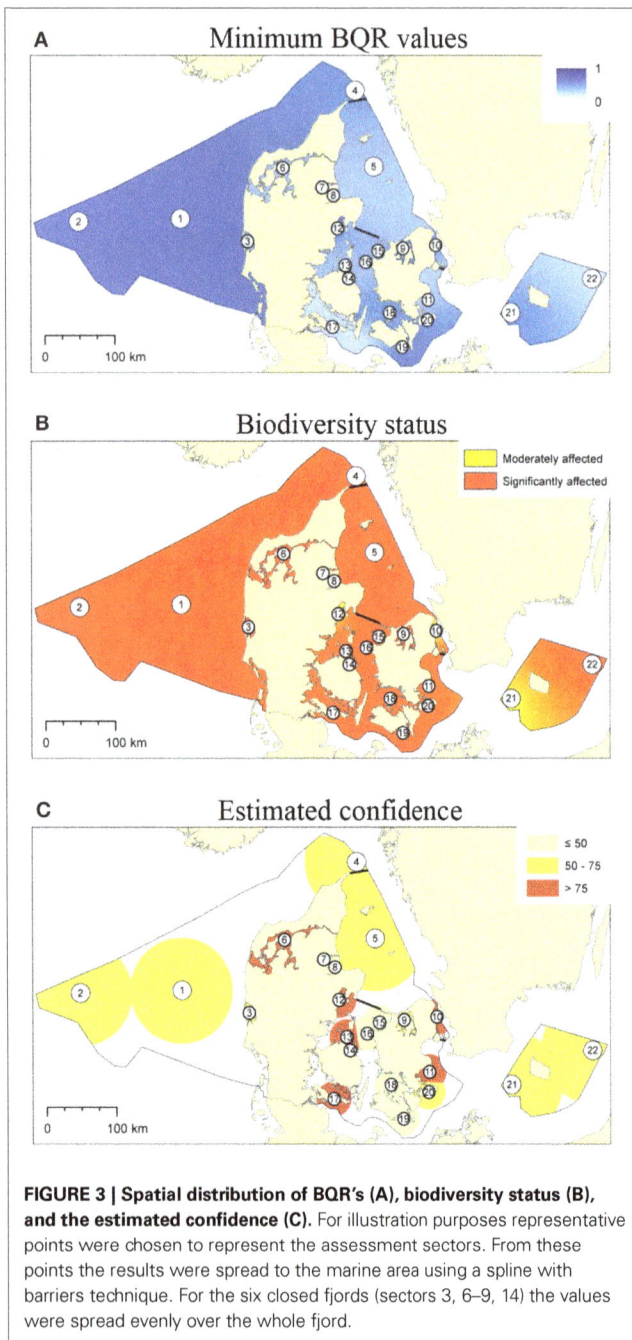

FIGURE 3 | Spatial distribution of BQR's (A), biodiversity status (B), and the estimated confidence (C). For illustration purposes representative points were chosen to represent the assessment sectors. From these points the results were spread to the marine area using a spline with barriers technique. For the six closed fjords (sectors 3, 6–9, 14) the values were spread evenly over the whole fjord.

in bottom waters (Conley et al., 2000; Nielsen et al., 2002a,b; Carstensen et al., 2004; Dahl and Carstensen, 2008). Thus, the water quality indicators can in a sense be called "true" indicators, as they can predict biological changes with simple methodology and relatively low costs. Nonetheless, in this study we considered them as "indirect" and prefer more direct measurements of biological parameters.

CONCLUDING REMARKS

Biological diversity in the Danish marine waters is significantly affected by human activities in most areas, but in a few sectors only moderately. None of the assessed sectors were classified as having a biodiversity status unaffected by human activities. The confidence of the assessments was estimated indirectly and generally regarded as acceptable, in a few cases even high. In two out of 22 sectors, the confidence was low indicating that monitoring of biodiversity in these sectors should be improved. The majority of the indicators were considered scientifically robust, but some indicators could, however, be further strengthened through production of peer reviewed scientific publications. Caution is also recommended in regard to the use of supporting indicators, especially in those few cases where they overrule biological indicators and thus determine the outcome of the integrated and final classification of biodiversity status. The BEAT 2.0 tool can support the EU Member States in the implementation of the MSFD, which specifically requires an overall assessment of the state of the marine environment as well as a specific assessment of biodiversity (Anon, 2008). The tool requires reliable indicators and quantitative thresholds for GES, but can function even with heterogeneous data availability. Assessments based on single indicators, though being simpler to link to human pressures, cannot reflect the variability and complexity of biodiversity responses required by the new assessments and therefore an integration of several indicators by an assessment tool is a prerequisite for the successful interface of science and environmental policy.

Finally, we would prudently like to remind the reader that there is no such thing as a perfect assessment tool. We do not promote the BEAT tool as such. We rather see this tool as a step for further development leading to better ecosystem-based tools for assessment, classification and adaptive management of marine biodiversity and human activities affecting marine life. The key challenges in regard to future integrated assessments of biodiversity status in marine waters are: (1) development of a wider range of biodiversity indicators representing different ecosystem components/food web categories, as well as (2) development of data driven methods for indicator integration and estimation of uncertainties.

ACKNOWLEDGMENTS

This article has been funded by the HARMONY project. The article has also been supported by the DEVOTES (DEVelopment Of innovative Tools for understanding marine biodiversity and assessing good Environmental Status) project funded by the European Union under the 7th Framework Programme, "The Ocean of Tomorrow" Theme (grant agreement no. 308392), www.devotes-project.eu. The authors would like to thank Johnny Reker and Joachim Raben as well as Stefan Heinänen, Alf B.

consideration should be given to the issue before the next regional MSFD assessments.

We used supporting indicators to reflect changes in water quality in the Danish waters, which are affected by eutrophication (Ærtebjerg et al., 2003; Andersen et al., 2011). The eutrophication indicators indirectly reflect the condition of pelagic and benthic habitats and can, thus, indicate an overall status for a range of species and communities. Significant relations have been identified between nutrient loads and concentrations, chlorophyll-a concentrations, Secchi depth, depth limit of eelgrass (*Zostera marina*), total cover of macroalgae, and oxygen concentration

Josefson, Alf Norkko, and Anna Villnäs. Martin Hartvig acknowledges the Danish National Research Foundation for support to the Center for Macroecology, Evolution and Climate. A prototype of the BEAT assessment tool was originally developed for HELCOM's integrated thematic assessment of biodiversity in the Baltic Sea and we would like to thank Hermanni Backer, Maria Laamanen, and Ulla Li Zweifel for constructive discussions of this prototype.

REFERENCES

Ærtebjerg, G., Andersen, J. H., and Hansen, O.S. (2003). *Nutrients and Eutrophication in Danish Marine Waters. A Challenge to Science and Management.* National Environmental Research Institute. Available online at: http://www2.dmu.dk/1_Viden/2_Publikationer/3_ovrige/rapporter/Nedmw 2003_alle.pdf

Andersen, J. H., Axe, P., Backer, H., Carstensen, J., Claussen, U., Fleming-Lehtinen, V., et al. (2011). Getting the measure of eutrophication in the Baltic Sea: towards improved assessment principles and methods. *Biogeochemistry* 106, 137–156. doi: 10.1007/s10533-010-9508-4

Andersen, J. H., Murray, C., Kaartokallio, H., Axe, P., and Molvær, J. (2010). A simple method for confidence rating of eutrophication status classifications. *Mar. Pollut. Bull.* 60, 919–924. doi: 10.1016/j.marpolbul.2010.03.020

Anon. (2000). Directive 200/60/EC of the European Parliament and of the Council of 23 October 2000 establishing a framework for Community action in the field of water policy. *Official J. Eur. Commun.* L 327, 1–72.

Anon. (2008). Directive 2008/56/EC of the European Parliament and the Council of 17 June 2008 establishing a framework for community action in the field of marine environmental policy (Marine Strategy Framework Directive). *Official J. Eur. Union Brussels L* 164, 19–40.

Anon. (2010). Commission decision of 1 September 2010 on criteria and methodological standards on good environmental status of marine waters. *Official J. Eur. Union L* 232, 14–24.

Borja, A. (2014). Grand challenges in marine ecosystems ecology. *Front. Mar. Sci.* 1:1. doi: 10.3389/fmars.2014.00001

Borja, A., Elliott, M., Andersen, J. H., Carstensen, J., Ferreira, J. G., Heiskanen, A.-C., et al. (2013). Good Environmental Status of marine ecosystems: What is it and how do we know when we have attained it? *Mar. Pollut. Bull.* 76, 16–27. doi: 10.1016/j.marpolbul.2013.08.042

Cardoso, A. C., Cochrane, S., Doerner, H., Ferreira, J. G., Galgani, F., Hagebro, C., et al. (2010). *Scientific Support to the European Commission on the Marine Strategy Framework Directive.* Luxemburg: Management Group Report. European Commission, Joint Research Centre.

Carstensen, J., Conley, D. J., Andersen, J. H., and Aertebjerg, G. (2006). Coastal eutrophication and trend reversal: a Danish case study. *Limnol. Oceanol.* 51, 398–408. doi: 10.4319/lo.2006.51.1_part_2.0398

Carstensen, J., Conley, D. J., and Henriksen, P. (2004). Frequency, composition, and causes of summer phytoplankton blooms in a shallow coastal ecosystem, the Kattegat. *Limnol. Oceanogr.* 49, 190–201. doi: 10.4319/lo.2004.49.1.0191

Certain, G., Skarpaas, O., Bjerke, J.-W., Framstad, E., Lindholm, M., Nilsen, J.-E., et al. (2011). The nature index: a general framework for synthesizing knowledge on the state of biodiversity. *PLoS ONE* 6:e18930. doi: 10.1371/journal.pone.0018930

Conley, D. J., Kaas, H., Møhlenberg, F., Rasmussen, B., and Windolf, J. (2000). Characteristics of Danish estuaries. *Estuaries* 23, 820–837. doi: 10.2307/1353000

Daan, N., Gislason, H., Pope, J. G., and Rice, J. C. (2005). Changes in the North Sea fish community: evidence of indirect effects of fishing? *ICES J. Mar. Sci.* 62, 177–188. doi: 10.1016/j.icesjms.2004.08.020

Dahl, K., and Carstensen, J. (2008). *Tools to Assess Conservation Status on Open Water Reefs in Nature-2000 Areas.* NERI Technical Report No. 663. National Environmental Research Institute, University of Aarhus. Available online at: http://www.dmu.dk/Pub/FR663.pdf

Fauchald, P., Skov, H., Skern-Mauritzen, M., Hausner, V. H., Johns, D., and Tveraa, T. (2011). Scale-dependent response diversity of seabirds to prey in the North Sea. *Ecology* 92, 228–239. doi: 10.1890/10-0818.1

Greenstreet, S., Rogers, S., Rice, J., Piet, G., Guirey, E., Fraser, H., et al. (2011). Development of the EcoQO for the North Sea fish community. *ICES J. Mar. Sci.* 68, 1–11. doi: 10.1093/icesjms/fsq156

Hansen, J. W. (2013). *Marine områder 2012. NOVANA. Tilstand og udvikling i miljø- og naturkvaliteten.* Aarhus Universitet, DCE - Nationalt Center for Miljø og Energi. 162 s. Videnskabelig rapport fra DCE - Nationalt Center for Miljø og Energi nr. 77. (In Danish).

HELCOM. (2009a). "Biodiversity in the Baltic Sea - An integrated thematic assessment on biodiversity and nature conservation in the Baltic Sea," in *Baltic Sea Environment Proceedings 116B* (Helsinki: Helsinki Commission), 188.

HELCOM. (2009b). "Eutrophication in the Baltic Sea. An integrated thematic assessment of eutrophication in the Baltic Sea region," in *Baltic Sea Environmental Proceedings No. 115B* (Helsinki: Helsinki Commission), 148.

HELCOM. (2010). "Ecosystem health of the Baltic Sea 2003–2007. HELCOM initial holistic assessment," in *Baltic Sea Environment Proceedings 120* (Helsinki), 63.

HELCOM. (2012). "Fifth Baltic Sea Pollution Load Compilation (PLC-5)" in *Baltic Sea Environment Proceedings 128* (Helsinki: Helsinki Commission), 217.

Korpinen, S., Meski, L., Andersen, J. H., and Laamanen, M. (2012). Human pressures and their potential impact on the Baltic Sea ecosystem. *Ecol. Indic.* 15, 105–114. doi: 10.1016/j.ecolind.2011.09.023

Nielsen, S. L., Sand-Jensen, K., Borum, J., and Geertz-Hansen, O. (2002a). Depth colonisation of eel-grass (*Zostera marina*) and macroalgae as determined by water transparency in Danish coastal waters. *Estuaries* 25, 1025–1032. doi: 10.1007/BF02691349

Nielsen, S. L., Sand-Jensen, K., Borum, J., and Geertz-Hansen, O. (2002b). Phytoplankton, nutrients and transparency in Danish coastal waters. *Estuaries* 25, 930-937. doi: 10.1007/BF02691341

Norkko, J., Reeds, D. C., Timmerman, K., Norkko, A., Gustafsson, B. G., Bonsdorff, E., et al. (2012). A welcome can of worms? Hypoxia mitigation by an invasive species. *Global Change Biol.* 18, 422–434. doi: 10.1111/j.1365-2486.2011.02513.x

Ojaveer, H., and Eero, M. (2011). Methodological challenges in assessing the environmental status of a marine ecosystem: case study of the baltic sea. *PLoS ONE* 6:e19231. doi: 10.1371/journal.pone.0019231

Ojaveer, H., Jaanus, A., MacKenzie, B. R., Martin, G., Olenin, S., Radziejewska, T., et al. (2010). Status of biodiversity in the Baltic Sea. *PLoS ONE* 5:e12467. doi: 10.1371/journal.pone.0012467

OSPAR. (2010). *Quality Status Report 2010.* London: OSPAR Commission.

Conflict of Interest Statement: The authors declare that the research was conducted in the absence of any commercial or financial relationships that could be construed as a potential conflict of interest.

7

Community structure and population genetics of Eastern Mediterranean polychaetes

Giorgos Chatzigeorgiou[1,2]*, Elena Sarropoulou[2], Katerina Vasileiadou[2], Christina Brown[3], Sarah Faulwetter[2], Giorgos Kotoulas[2] and Christos D. Arvanitidis[2]

[1] Biology Department, University of Crete, Heraklion, Greece
[2] Hellenic Centre for Marine Research, Institute of Marine Biology, Biotechnology and Aquaculture, Heraklion, Greece
[3] Department of Biology, Chemistry and Pharmacology, Institute of Biology, Free University of Berlin, Berlin, Germany

Edited by:
Alberto Basset, University of Salento, Italy

Reviewed by:
Guillem Chust, AZTI-Tecnalia, Spain
Katherine Dafforn, University of New South Wales, Australia

*Correspondence:
Giorgos Chatzigeorgiou, Hellenic Centre for Marine Research, Institute of Biology Biotechnology and Aquaculture, Former American Base at Gournes, 71500 Crete, Greece
e-mail: chatzigeorgiou@hcmr.gr

Species and genetic diversity are often found to co-vary since they are influenced by external factors in similar ways. In this paper, we analyse the genetic differences of the abundant polychaete *Hermodice carunculata* (Pallas, 1766) during two successive years at two locations in northern Crete (Aegean Sea) and compare them to other populations in the Mediterranean Sea and the Atlantic Ocean. The genetic analysis is combined with an analysis of ecological divergence of the total polychaete community structure (beta diversity) at these locations. The phylogenetic analysis of all included *H. carunculata* populations revealed two main clades, one exclusively found in the Mediterranean and a second occurring in both the Mediterranean and the Atlantic. Genetic diversity indices reveal unexpectedly high differences between the two Cretan populations, despite the absence of apparent oceanographic barriers. A similarly high divergence, represented by a high beta diversity index, was observed between the polychaete communities at the two locations. This comparatively high divergence of the genetic structure of a dominant species and the total polychaete community might be explained by the strong influence of local environmental factors as well as inter-specific interactions between the dominance of a single species and the members of the community.

Keywords: *Hermodice carunculata*, mtCOI, NaGISA, rocky shore, beta diversity, fixation index

INTRODUCTION

The marine environment provides many opportunities for dispersal of individuals within and among populations (Cowen and Sponaugle, 2009). Benthic invertebrate taxa usually have very limited dispersal potential as adults, whereas many species have pelagic larval stages that facilitate dispersal (Weersing and Toonen, 2009). The dispersal ability of a taxon is directed by a number of factors such as historical processes, environmental conditions, currents and life history traits such as the duration of the larval stage (Pringle and Wares, 2007; Jolly et al., 2009). These factors can lead to a variety of distribution pattens—from cosmopolitan distributions of species to the existence of populations with significant genetic differences even at close geographic distances (Hohenlohe, 2004; Hart and Marko, 2010; Derycke et al., 2013; Iacchei et al., 2013; Vergara-Chen et al., 2013).

The evolution of genetic diversity patterns depends mostly on the interplay of mutation, random genetic drift, gene flow and natural selection (Hartl and Clark, 2007). These processes can be studied through population genetics and phylogeographic approaches (Derycke et al., 2013). Over the last two decades, the development of molecular techniques has allowed the in-depth study of population structure and dynamics in benthic invertebrates (e.g., Duran et al., 2004). Mitochondrial genes and especially the cytochrome c oxidase subunit I (mtCOI) are widely used and have been proved useful for both DNA barcoding of

species (http://www.barcodeoflife.org/) and population genetic analysis (e.g., Avise, 1994; Galtier et al., 2009). For instance, mtDNA data have been widely used to assess the temporal and spatial fluctuations of haplotype diversity in natural populations. Many studies investigate cryptic species along environmental gradients (e.g., Jolly et al., 2005; Barroso et al., 2010), but few focus on the population structure within a specified area (e.g., Schulze et al., 2000; Craft et al., 2010; Chust et al., 2013).

At the community level, a series of mathematical approaches and indices exist to calculate quantitative estimates of divergence. Whittaker (1960) introduced the term "beta diversity" (often referred to as turnover diversity) to describe the diversification between assemblages within a certain geographic area. Since then, 24 different (dis-)similarity coefficients have been proposed to quantify beta diversity (Legendre and Cáceres, 2013).

There is a conceptual analogy between beta diversity of community structure and genetic divergence between populations of a species. Theory predicts that environmental and stochastic processes act at all levels of the biological organization and thus cause species diversity and genetic diversity to co-vary in time and space (Etienne and Olff, 2004; Vellend and Geber, 2005). A number of studies have identified the parallel influence of local processes on species and genetic diversity (e.g., Vellend, 2003; Papadopoulou et al., 2011), as well as the mutual shaping of community diversity and the genetic diversity of its members (e.g., Booth and Grime,

2003; Whitham et al., 2003; Hughes et al., 2008). If these theories hold true, beta diversity and population genetic diversity can be expected to change at similar rates between locations. Indeed, Baselga et al. (2013) have found such correlations in communities of water beetles across Europe, and Papadopoulou et al. (2011) revealed that beta diversity and haplotype diversity of tenebrionid beetle communities on the Aegean Islands declined at similar rates with increasing distances.

Furthermore, there are strong indications that the genetic diversity of a single species can have effects on the diversity of a whole community if that species has a dominant or keystone role in the ecosystem (Treseder and Vitousek, 2001; Whitham et al., 2003). If the genetic diversity of certain species is shown to change at similar rates to the species diversity of a community, these species could be used as a proxy for time-consuming and costly whole-community analyses (e.g., Féral et al., 2003). In this paper, we analyse the genetic differences of the abundant polychaete *Hermodice carunculata* (Pallas, 1766) during two successive years at two locations in northern Crete (Aegean Sea). The results are combined with an analysis of ecological divergence of the total polychaete community structure between these locations. This allows us to assess to what extent the genetic diversity of a single, dominant species reflects the overall community diversity, and provides a basis for the future exploration of such species as biological indicators.

MATERIALS AND METHODS
STUDY AREA

The analysis was based on samples collected previously in the framework of the NaGISA project [http://www.coml.org/projects/natural-geography-shore-areas-nagisa; datasets were published by Faulwetter et al. (2011) and Chatzigeorgiou et al. (2014)]. Samples were taken from two locations in northern Crete, Alykes and Elounda (35.41583, 24.98785; 35.25166, 25.75833; **Figure 1**). Both locations are characterized by a moderate wave exposure and a continuous hard bottom habitat, densely covered by algae, with *Cystoseira* spp. and *Sargassum* spp. being the most abundant in terms of surface coverage. The locations are 60 km away from each other and they seem to receive no detectable impact caused by human activities (Chatzigeorgiou et al., 2012).

SPECIMEN COLLECTION AND PROCESSING

Sampling was conducted following the NaGISA protocol (Iken and Konar, 2003); details on the sampling procedure can be found in Chatzigeorgiou et al. (2012). All samples were preserved in 98% ethanol. In total, more than 8000 individuals were sorted and identified as 182 different polychaetes species, using the most recent literature. The species composition varied in terms of sampling depth and sampling site. Very few species were found in all sampling depths and sampling sites. From these, 40 individuals of *Hermodice carunculata* (Pallas, 1766) were randomly selected from each sampling location (20 per sampling year) for genetic analysis (see also paragraph "Model species" below). Primary morphometric characters (wet weight, length, width and number of chaetigers) were measured on each individual. Subsequently, the internal parts of each individual were

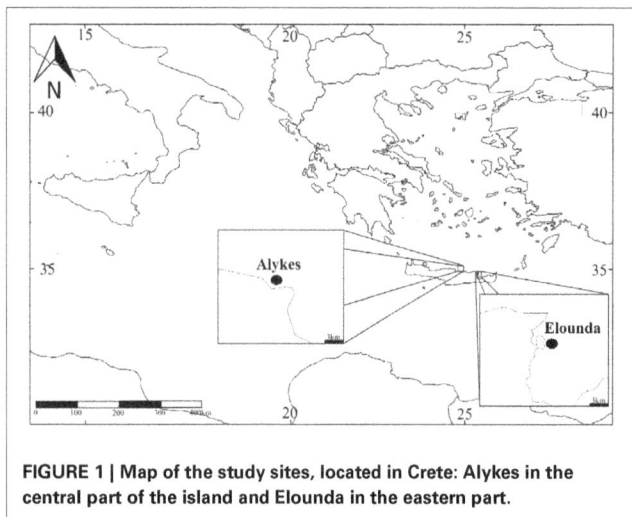

FIGURE 1 | Map of the study sites, located in Crete: Alykes in the central part of the island and Elounda in the eastern part.

removed and the remaining tissues were first cleaned and then processed for the analysis (Vasileiadou et al., 2012).

ECOLOGICAL ANALYSES

For the purpose of this study, the original dataset was modified by combining the abundances of each depth, and analyses of beta diversity were based on the total species pool in each location per year. Although many indices have been suggested, the b_w (Whittaker, 1960; Magurran, 1988; Southwood and Henderson, 2000) was selected as the most suitable for comparing geographic variation among populations, according to Wilson and Shmida (1984). The following equation was used:

$$b_w = [(a + b + c) / (2a + b + c) / 2] - 1,$$

where,

a is the total number of species occurring in both sampling locations, *b* the total number of species occurring only in the first sampling location and *c* the total number of species occurring only in second location. The beta diversity between two locations can assume values between zero (complete similarity) and one (complete dissimilarity) (Koleff et al., 2003).

A Mann-Whitney test (Mann and Whitney, 1947) was applied in order to test for differences between the measured morphological characters of the individuals belonging to the two populations.

MODEL SPECIES

Hermodice carunculata (Polychaeta, Amphinomidae) is an abundant and important predator and scavenger on hard substrates in circumtropical coastal waters. The larvae of amphinomid species are unique among polychaetes, in possessing two elongated feeding tentacles (Mileikovsky, 1961). This allows the larva to cover great distances by staying for a longer period in the water column and probably provides distant populations with a high connectivity potential. This hypothesis was confirmed by Ahrens et al. (2013), who found no substantial genetic differentiation in the *H. carunculata* populations he examined, from the Mediterranean Sea (Malta and Crete) to the Atlantic Ocean (Caribbean Sea, Gulf of Mexico and Gulf of Guinea), and consequently assigned them to a single species. The species was chosen for this analysis for four reasons: (a) its ecological function, as one of the major predators in rocky and reef environments (e.g., Vreeland and Lasker, 1989), (b) its high frequency and abundance in the above habitats, (c) stable COI protocols for the species exist and, (d) a number of sequence data sets are available from a large geographic scale (e.g., Ahrens et al., 2013).

DNA PURIFICATION, AMPLIFICATION AND SEQUENCING

A small tissue sample (maximally 2 mg) was removed from the abdominal part of the individuals. In cases where individuals were too small, the tissue sample was collected from the widest part of the body. DNA purification was accomplished using a Nucleospin Tisseuekit XS (MACHEREY-NAGEL GmbH and Co. KG, Dueren, Germany). In order to determine the average concentrations of extracted DNA and the contamination with protein and phenol in the samples, a spectrophotometer was used for nucleic acid measurement (Nanodrop -1000). The universal DNA primers COI- LCO1490 (5′-GGTCAACAAATCATAAAGATATTGG-3′) and HCO2198 (5′-TAAACTTCAGGGTGACCAAAAAATCA-3′) were used for the amplification of the COI gene (Folmer et al., 1994). The amplification reaction mix contained $1\,\mu l$ 10x buffer (with $1.5\,mM\ Mg^{2+}$), $0.6\,\mu l$ $MgCl_2$ ($25\,\mu M$), $0.2\,\mu l$ dNTPs ($100\,mM$), $0.25\,\mu l$ of each primer ($10\,mM$) and $0.1\,\mu l$ Taq DNA Polymerase ($5\,U/\mu l$) for a total volume of $10\,\mu l$ per reaction. The DNA template concentration was $\sim 50\,ng/\mu l$. The optimal PCR conditions were found by means of a temperature gradient PCR: $94°C$ for 5 min; 36 cycles with $94°C$ for 1 min, $44°C$ for 1 min, $72°C$ for 1 min; final extension at $72°C$ for 3 min. Amplifications were carried out in a BioRad Thermal Cycler T-100 PCR machine and PCR products were purified by using the Ethanol precipitation protocol. Cycle sequencing using the BigDye Terminator (Life Technologies) chemistry was conducted in $10\,\mu l$ volumes, contained $2\,\mu l$ BigDye Enzyme, $0.6\,\mu l$ of primer, $1\,\mu l$ of reaction buffer and $2\,\mu l$ of PCR product. The sequence reaction were performed with: an initial step at $96°C$ for 3 min, 35 cycles at $96°C$ for 10 s, $50°C$ for 15 s, $60°C$ for 4 min. Reactions were purified with EDTA (0,5 M, pH 8), NaAc (3 M, pH 4,3) and 98% Ethanol. Sequences were analyzed using an ABI 3730 Genetic Analyzer (Applied Biosystems, in the Laboratory of Genetics in HCMR), and resulting electrochromatograms were edited in BioEdit ver 7.2 by assembling the forward and reverse fragments from each specimen and deleting the primer regions. All sequences were submitted to GenBank with accession numbers KF878397-KF878476.

SEQUENCE ANALYSIS

Edited sequences were aligned in MEGA 5 (Tamura et al., 2011) via ClustalW using default settings (Larkin et al., 2007). The number of distinct haplotypes for COI was calculated using Arlequin 3.5.1.3 (Excoffier and Lischer, 2010). Phylogenetic trees were constructed using three different methods: (a) Bayesian inference (BI), (b) maximum likelihood (ML) and (c) neighbor-joining (NJ). The first method was done with the MrBayes 3.2.2 software (Huelsenbeck et al., 2002), using average branch lengths, while the other two were conducted with the Mega 5

software, using the Tamura-Nei model. For all the above methods a bootstrap option ($n = 1,000$) was used to calculate branch supporting values. Previously published data from Crete, Malta, Gulf of Mexico, Caribbean Sea and Gulf of Guinea were also included in the analyses (Ahrens et al., 2013, with accession numbers: Crete KC017526-8, 30; Malta KC017553-56; Caribbean Sea KC017521,24-25, KC017552; Gulf of Mexico KC017587-90 and Gulf of Guinea KC0 17480-2, KC017569). Taking into account the high rates (around 30%) of the genetic distances observed among other amphinomid genera (e.g., Borda et al., 2012) no out-group could be selected for the phylogenetic analyses. Therefore, the tree constriction midpoint rooted technique was selected, following Ahrens et al. (2013).

Additionally, the following indices were calculated, for each sampling year and for both years combined: haplotype diversity (h), nucleotide diversity (π) and fixation index (F_{ST}) using Arlequin 3.5.1.3. These analyses were run at the scale of (a) sampling location, which means that samples from the same location were grouped together (b) sites within Mediterranean sea and (c) sites between Mediterranean and Atlantic. For (b) and (c) data from Ahrens et al. (2013) were used. Phylogenetic networks were derived from the analyses using Network 4.6.1.1 software (Bandelt et al., 1999). The values of the fixation index range from zero to one, where zero implies complete panmixis, that is, that the two populations are interbreeding freely, and one implies that all genetic variation is explained by the population structure, and that the two populations do not share any genetic diversity at all.

Finally, a Mann-Whitney test was performed order to identify differences in the distribution of the haplotypes between the sampling sites (Alykes and Elounda) and between the sampling years (2007 and 2008).

RESULTS
MORPHOMETRIC DIVERSITY

The total length of the individuals ranged from 1.54 to 9.97 cm and the maximum width from 0.54 to 1.75 cm. The number of the chaetigers varied from 21 to 78 and the wet weight of the individuals ranged from 0.002 to 2.9 g. The Mann-Whitney test showed no significant differences between the populations sampled from the two locations (**Table 1**).

BETA DIVERSITY

Overall, a very high species diversity was observed in the polychaete assemblages sampled in the two Cretan locations with a total number of 182 species, 126 of which occur in both sites. Out of the total species number, 18 species were exclusive to Alykes

Table 1 | Results of the Mann-Whitney test, comparing the values of the primary morphometric characters measured on the individuals from the two sampling locations.

Morphometric characters	U	p-value
Body Length:	731.9	0.348
Body Width:	716.3	0.209
Wet weight:	787.4	0.423
No of chaetigers:	77	0.367

and 38 to Elounda. The beta diversity (b_w) value calculated from the above numbers is 0.181. In addition, b_w values were calculated for each sampling year separately. In the first sampling year, the b_w value was higher than in the second year (0.213 and 0.167, respectively).

PHYLOGENY

The COI sequencing analyses resulted in a joined alignment of 607 bp for the 80 individuals, with a total number of 23 mutations. A total of 15 haplotypes was found, of which seven were found exclusively in Elounda, seven exclusively in Alykes and a single one was shared between the two locations. Among the exclusive haplotypes, four were found only once (singletons), three of them in Alykes and one in Elounda. The shared haplotype, as expected, had a high frequency at both locations, but was the most common only in Alykes. In addition, the Mann-Whitney test performed on the frequencies of the above haplotypes at two different levels (sampling site and year), demonstrated that they were independent only in the case of the site (Year, $U = 35.5$, ($N^1 = 7, N^2 = 9$) $P = 0.072$; Sampling site, $U = 38.5$, ($N^1 = 8$, $N^2 = 8$) $P = 0.023$). The above haplotypes from both sampling locations are illustrated on the phylogenetic network of **Figure 2**.

The phylogenetic analyses of COI sequence data via BI resulted in the consensus tree depicted in **Figure 3**. Two main clades can be distinguished, each with several subclades. Clades I and II were separated with high branch support values (BI posterior probability = 1.0, NJ Bootstrap = 98% and ML bootstrap = 97%).

MOLECULAR DIVERSITY

Haplotype diversity (h) and nucleotide diversity (π) indices were calculated for each sampling location separately, for each sampling year and for both years combined (**Table 2**). Both indices showed higher values in the site of Elounda. The fixation index F_{ST} calculated from the two *H. carunculata* populations resulted in a value of 0.192 ($P < 0.001$) between two NaGISA sites (2007: 0.194, $P < 0.001$; 2008: 0.189, $P < 0.001$) while the F_{ST} value was 0.204 between the other Mediterranean sites and 0.224 between the Mediterranean and Atlantic sites.

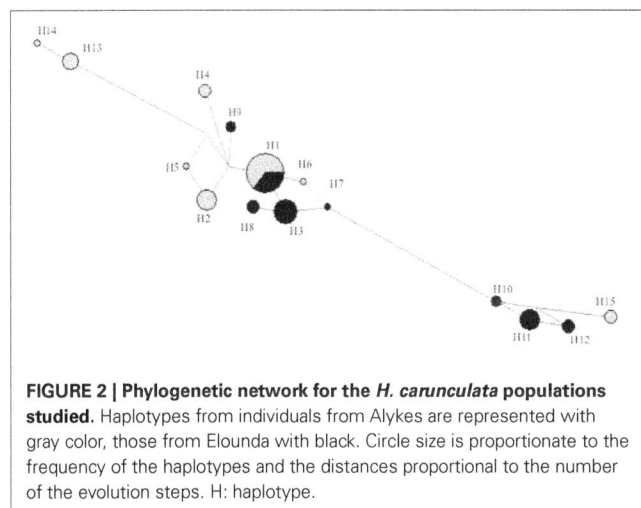

FIGURE 2 | Phylogenetic network for the *H. carunculata* populations studied. Haplotypes from individuals from Alykes are represented with gray color, those from Elounda with black. Circle size is proportionate to the frequency of the haplotypes and the distances proportional to the number of the evolution steps. H: haplotype.

DISCUSSION

Both the mitochondrial and nuclear markers such as COI, 16S and 18S have been extensively used for the assessment of the genetic structure of populations in many geographic areas (e.g., Reeb and Avise, 1990). The dispersion potential of the species, the presence or absence of clear barriers and the effective population size are important components affecting the genetic population structure (Foll and Gaggiotti, 2006). In the case of the annelids, a rather complex pattern seems to have emerged from these factors. Schulze (2006) found that despite the absence of planktonic larvae in eunicid polychaetes, long distance dispersal is possible in at least some lineages. Within the genera of the family Amphinomidae, both ends of the spectrum have been found: Strongly divergent populations have been detected in the genus *Eurythoe* (considered to be cryptic species, Barroso et al., 2010), and widely distributed populations representing single species, such as those of *Hermodice carunculata* (Ahrens et al., 2013). Although the existing references for the reproductive biology and larval stages of amphinomids are scarce (e.g., Allen, 1957; Kudenov, 1974; Giangrande, 1997), teleplanic and rostraria larval forms are typical of the family. Both forms are planktonic and especially that of the unique rostraria with elongated feeding tentacles supports the theory of a long distance larval dispersal potential. However, Glasby (2005), based on a study of local polychaete endemism, postulated that the presence of a pelagic larval form does not necessarily imply a gene flow capable of homogenizing distant populations or preventing speciation.

The phylogenetic analysis of *H. carunculata* populations has revealed two main clades. These clades are separated having strong branch support values. The sequences included in Clade I reflect the long isolation of Atlantic and Mediterranean populations, having led to divergence and lineage sorting, although selective factors cannot be excluded to have added to such divergence. Clade II includes subclades consisting of individuals from both Mediterranean and from Mediterranean populations. The latter is indicative of the gene flow mechanism through which high connection rates between distant populations can be recorded. Ahrens et al. (2013) proposed that this gene flow mechanism is largely facilitated by factors such as sea water currents and ship transportation (ballast water and hull fouling), and which are capable of establishing connections between the *H. carunculata* populations in the Mediterranean Sea and the Atlantic Ocean.

Despite the ability of the taxon to maintain gene flow over long distances, the two Cretan populations show an important degree of differentiation according to the F_{ST} index, when compared to the populations within the Mediterranean and between the Mediterranean—Atlantic sites. The small geographic distance between the two sampling sites in Crete is reflected by lineage sorting, which in turn leads to fairly distinct branches for each location on the haplotype network. Among the 15 found haplotypes, only one is shared between the two sampling sites. This pattern implies a very limited or no gene flow, which could be by explained by a small effective population size (Slatkin, 1987). Effective population size is a key component affecting genetic population structure, as exemplified by differences in the case of the genetic structure between two nematode species

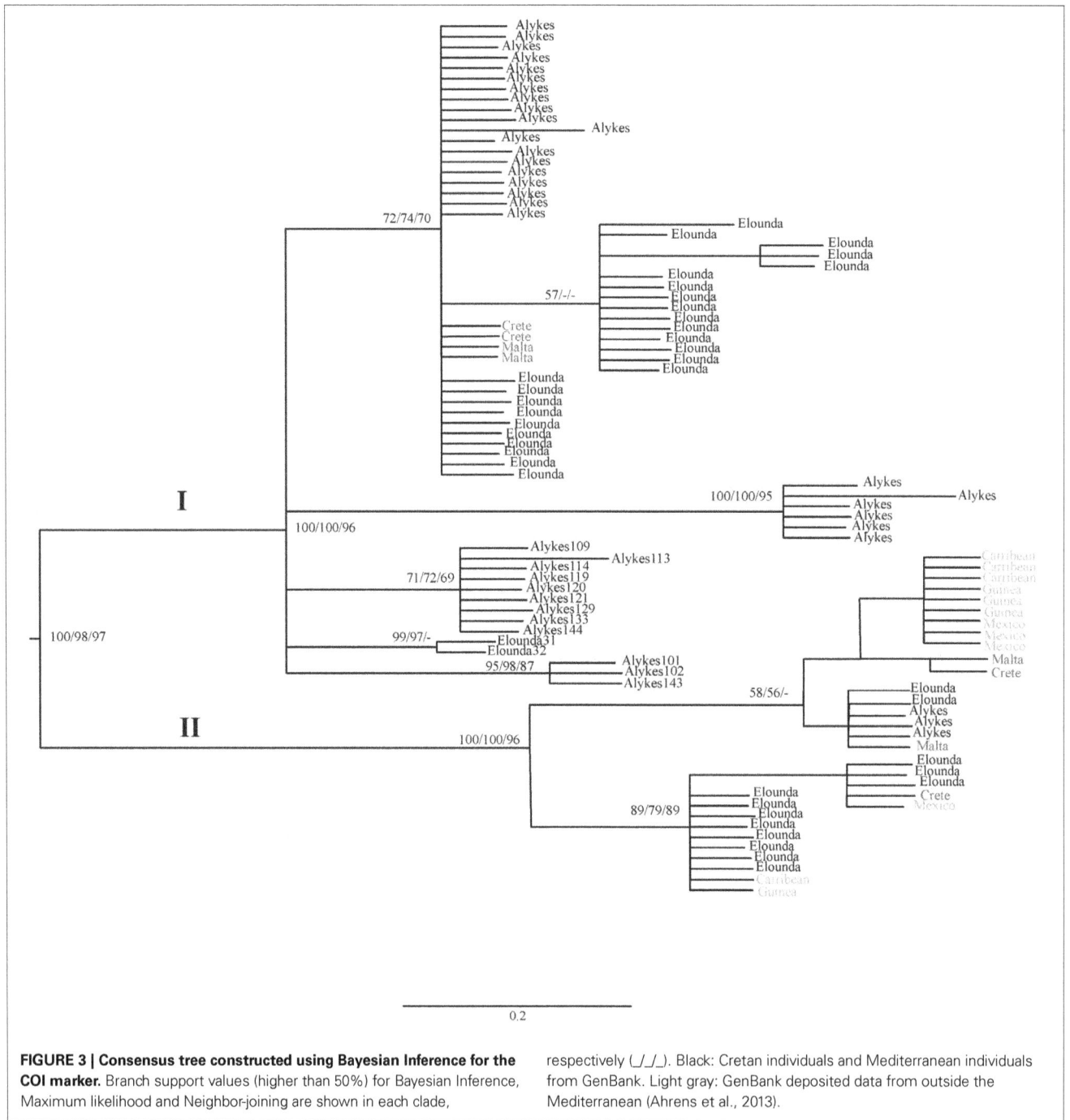

FIGURE 3 | Consensus tree constructed using Bayesian Inference for the COI marker. Branch support values (higher than 50%) for Bayesian Inference, Maximum likelihood and Neighbor-joining are shown in each clade, respectively (_/_/_). Black: Cretan individuals and Mediterranean individuals from GenBank. Light gray: GenBank deposited data from outside the Mediterranean (Ahrens et al., 2013).

Table 2 | Diversity indices values, calculated for the two sampling locations and two sampling years; ±SD: standard deviation.

	N	H	Haplotype diversity index h (±SD)	Nucleotide diversity index π (±SD)
Alykes total	40	8	0.7474 (±0.055)	0.0059 (±0.0034)
Elounda total	40	8	0.8256 (±0.0305)	0.0066 (±0.0037)
Alykes 2007	20	6	0.7278 (±0.0241)	0.0057 (±0.0031)
Elounda 2007	20	7	0.8136 (±0.0289)	0.0064 (±0.0041)
Alykes 2008	20	7	0.7631 (±0.0432)	0.0062 (±0.0036)
Elounda 2008	20	8	0.8301 (±0.0321)	0.0068 (±0.0034)

(*Caenorhabditis remanei* and *C. elegans*) with differing effective population sizes due to different breeding behavior (Cutter et al., 2006; Charlesworth and Charlesworth, 2010). The effective population size may also be the main factor that accounts for low divergence in benthic nematodes (Bik et al., 2010) which show huge population sizes and a the high number of cryptic species, and the high divergence usually found in polychaetes with much smaller effective population sizes (Ahrens et al., 2013 and references therein).

Considering such a relatively quick lineage sorting, it is difficult to understand the occurrence of Clade II with representative samples in both the Atlantic and the Mediterranean. On the other hand, in other invertebrate species, distinct oceanographic conditions, occurring at spatial scales of just a few kilometers, can be responsible for the detected genetic differentiation in populations (Palumbi, 2004). This has also been reported for many polychaete species (e.g., Jolly et al., 2005; Schulze, 2006; Rockman, 2012). In this study, haplotype and nucleotide diversity indices values present higher values in Elounda than in Alykes. This suggests that the populations of *H. carunculata* in Elounda may be larger and more stable than in Alykes (Lowe et al., 2004). The less stable a population is, the higher is the genetic drift, which in turn, may cause gene variants to disappear (Scaps, 2002). This is most probably the reason for the increased number of haplotype singletons found in Alykes.

The beta diversity between the two polychaete assemblages was found to be rather high, compared to cases in which larger geographic distances or contrasted ecosystems with geographically isolated populations were analyzed (e.g., Legendre and Cáceres, 2013). Such high values of beta diversity can be considered as the consequence of habitat diversification (Whittaker, 1972). Given the apparent homogeneity of the sublittoral habitats in the region, the absence of apparent oceanographic barriers, the shallow water currents along the coasts of Crete and the relatively small distance between the two locations, they cannot be considered as geographically isolated from each other. Therefore, the past and present environmental conditions have to be considered as the major factor responsible for the rather strong divergence of these assemblages.

The two indices (F_{ST} and b_w) represent different components of biodiversity in a certain area. Beta diversity reflects the differences between whole communities and thus represent species diversity, whereas F_{ST} is a measure of population differentiation due to genetic structure. In the Cretan locations, both the differences between the two populations of *H. carunculata* (based on molecular diversity) and the divergence of the polychaete taxo-communities (represented by b_w) are comparatively high, given the apparent absence of any barriers. This indicates that environmental processes in the two areas must act in a similar way on both the composition of the communities and on the genetic diversity (Vellend and Geber, 2005). Such positive correlations between species and genetic diversity have been reported before (e.g., Vellend, 2003; Papadopoulou et al., 2011; Struebig et al., 2011; Baselga et al., 2013), and our results are in line with these findings. The dominant role of *H. carunculata* in the studied ecosystem, being an abundant carnivore, might be an additional explanation for the corresponding patterns between species

and genetic divergence. Interactions between the species and the remaining community could therefore have led to a co-variation of the two levels of the biological organization (Whitham et al., 2003). If future analyses with additional locations show that these results are constant, the species could be a candidate for an indicator species, allowing rapid assessment of the divergence of two communities without the need for time-consuming identification of thousands of specimens.

Our research has highlighted a region and a taxonomic group worthy of further investigation, and is a promising approach for studies of community structuring and evolution, as well as for the determination of biological indicator species. In the present study, we have combined a standard community ecology analysis with population genetics and have successfully highlighted a scientific field of interest for follow up studies. Although the results of the present study are limited by the fact that only two locations are compared, the indices are rather constant over the two successive years, indicating that the populations and communities are not subjected to strong fluctuations over time. Furthermore, due to the high number of species in the community and the high number of individuals chosen for the genetic analysis the results are not statistically biased by a small sampling size, a common problem in similar analyses (Nazareno and Jump, 2012). Future studies should include additional locations as well as additional species and nuclear markers for the population genetics analysis. To complete this image, life history traits of the studied species as well as environmental factors such as oceanographic currents should be included to understand in detail the factors affecting the co-variation of species and genetic diversity.

ACKNOWLEDGMENTS

Authors thank Dr. P. Divanac and Mr. G. Skouradakis for assistance in the preparation of the sampling equipment. The two anonymous referees are thanked for their comments which significantly improved this paper. This study is an outcome of the EuroNaGISA Network and forms part of the biodiversity core project of the IMBBC.

REFERENCES

Ahrens, J. B., Borda, E., Barroso, R., Paiva, P. C., Campbell, A. M., Wolf, A., et al. (2013). The curious case of *Hermodice carunculata* (Annelida: Amphinomidae): evidence for genetic homogeneity throughout the Atlantic Ocean and adjacent basins. *Mol. Ecol.* 22, 2280–2291. doi: 10.1111/mec.12263

Allen, M. J. (1957). Histochemical studies on developmental stages of polychaetous annelids. *Anat. Rec.* 128, 515–516.

Avise, J. C. (1994). *Molecular Markers, Natural History and Evolution*. New York, NY;London:Chapman & Hall.

Bandelt, H. J., Forster, P., and Rohl, A. (1999). Median-joining networks for inferring intraspecific phylogenies. *Mol. Biol. Evol.* 16, 37–48. doi: 10.1093/oxford-journals.molbev.a026036

Barroso, R., Klautau, M., Solé-Cava, A. M., and Paiva, P. C. (2010). *Eurythoe complanata* (Polychaeta: Amphinomidae), the "cosmopolitan" fireworm, consists of at least three cryptic species. *Mar. Biol.* 157, 69–80. doi: 10.1007/s00227-009-1296-9

Baselga, A., Fujisawa, T., Crampton-Platt, A., Bergsten, J., Foster, P. G., Monaghan, M. T., et al. (2013). Whole-community DNA barcoding reveals a spatio-temporal continuum of biodiversity at species and genetic levels. *Nat. Commun.* 4:1892. doi: 10.1038/ncomms2881

Bik, H. M., Thomas, W. K., Lunt, D. H., and Lambshead, P. J. D. (2010). Low endemism, continued deep-shallow interchanges, and evidence for cosmopolitan distributions in free-living marine nematodes (order Enoplida). *Evol. Biol.* 10:389. doi: 10.1186/1471-2148-10-389

Booth, R. E., and Grime, J. P. (2003). Effects of genetic impoverishment on plant community diversity. *J. Ecol.* 91, 721–730. doi: 10.1046/j.1365-2745.2003.00804.x

Borda, E., Kudenov, J. D., Bienhold, C., and Rouse, G. W. (2012). Towards a revised Amphinomidae (Annelida, Amphinomida): description and affinities of a new genus and species from the Nile Deep-sea Fan, Mediterranean Sea. *Zool. Scri.* 41, 307–325. doi: 10.1111/j.1463-6409.2012.00529.x

Charlesworth, B., and Charlesworth, D. (2010). *Elements of Evolutionary Genetics.* Colorado, CO: Roberts and Company Publishers.

Chatzigeorgiou, G., Faulwetter, S., and Arvanitidis, C. (2014). *Polychaetes From Two Subtidal Rocky Shores of the North Coast of Crete, Collected for the NaGISA Project 2007-2008. 18300 Records.* Available online at: http://lifewww-00.her.hcmr.gr:8080/medobis/resource.do?r=nagisa_species_2007_2008

Chatzigeorgiou, G., Faulwetter, S., López, E., Sardá, R., and Arvanitidis, C. (2012). Can coastal biodiversity measured in four Mediterranean sites be representative of the region? a test for the robustness of the NaGISA protocol by using the hard substrate syllid (Annelida, Polychaeta) taxo-communities as a surrogate. *Hydrobiologia* 691, 147–156. doi: 10.1007/s10750-012-1065-5

Chust, G., Aitor Albaina, A., Aranburu, A., Borja, A., Diekmann, O. E., Estonba, A., et al. (2013). Connectivity, neutral theories and the assessment of species vulnerability to global change in temperate estuaries. *Estuar. Coast. Shelf Sci.* 131, 52–63. doi: 10.1016/j.ecss.2013.08.005

Cowen, R. K., and Sponaugle, S. (2009). Larval dispersal and marine population connectivity. *Ann. Rev. Mar. Sci.* 1, 443–466. doi: 10.1146/annurev.marine.01 0908.163757

Craft, K. J., Pauls, S. U., Darrow, K., Miller, S. E., Hebert, P. D. N., Helgen, L. E., et al. (2010). Population genetics of ecological communities with DNA barcodes: an example from New Guinea Lepidoptera. *Proc. Natl. Acad. Sci. U.S.A.* 107, 5041–5046. doi: 10.1073/pnas.0913084107

Cutter, A. D., Baird, S. E., and Charlesworth, D. (2006). High nucleotide polymorphism and the decay of linkage disequilibrium in wild populations of *Caenorhabditis remanei* and *C. elegans. Genetics* 174, 901–913. doi: 10.1534/genetics.106.061879

Derycke, S., Backeljau, T., and Moens, T. (2013). Dispersal and gene flow in free-living marine nematodes. *Front. Zool.* 10, 1–12. doi: 10.1186/1742-9994-10-1

Duran, S., Pascual, M., Estoup, A., and Turon, X. (2004). Strong population structure in the marine sponge *Crambe crambe* (Poecilosclerida) as revealed by microsatellite markers. *Mar. Ecol.* 13, 511–522. doi: 10.1046/j.1365-294X.2004.02080.x

Etienne, R. S., and Olff, H. (2004). A novel genealogical approach to neutral biodiversity theory. *Ecol. Lett.* 7, 170–175. doi: 10.1111/j.1461-0248.2004.00572.x

Excoffier, L., and Lischer, H. E. L. (2010). Arlequin suite ver 3.5: a new series of programs to perform population genetics analyses under Linux and Windows. *Mol. Ecol. Resour.* 10, 564–567. doi: 10.1111/j.1755-0998.2010.02847.x

Faulwetter, S., Chatzigeorgiou, G., Galil, B. S., and Arvanitidis, C. (2011). *Eastern Mediterranean Syllidae From Three Locations in Crete and Israel. 1123 Records.* Available online at: http://lifewww-00.her.hcmr.gr:8080/medobis/resource.do?r=easternmedsyllids

Féral, J. P., Fourt, M., Perez, T., Warwick, R. M., Emblow, C., Heip, C., et al. (2003). *European Marine Biodiversity Indicators.* Yerseke: NIOO-CEME.

Foll, M., and Gaggiotti, O. (2006). Identifying the environmental factors that determine the genetic structure populations. *Genetics* 174, 875–891. doi: 10.1534/genetics.106.059451

Folmer, O., Black, M., Hoeh, W., Lutz, R., and Vrijenhoek, R. (1994). DNA primers for amplification of mitochondrial cytochrome c oxidase subunit I from diverse metazoan invertebrates. *Mol. Mar. Biol. Biotech.* 3, 294–299.

Galtier, N., Nabholz, B., Glemin, S., and Hurst, G. D. D. (2009). Mitochondrial DNA as a marker of molecular diversity: a reappraisal. *Mol. Ecol.* 18, 4541–4550. doi: 10.1111/j.1365-294X.2009.04380.x

Giangrande, A. (1997). Polychaetes reproductive patterns, life cycle and life histories: an overview. *Oceanograph. Mar. Biol.* 35, 323–386.

Glasby, C. J. (2005). Polychaete distribution patterns revisited: an historical explanation. *Mar. Ecol.* 26, 235–245. doi: 10.1111/j.1439-0485.2005.00059.x

Hart, M. W., and Marko, P. B. (2010). It's about time: divergence, demography, and the evolution of developmental modes in marine invertebrates. *Integr. Comp. Biol.* 50, 643–661. doi: 10.1093/icb/icq068

Hartl, D., and Clark, A. (2007). *Principles of Population Genetics.* Sunderland, MA: Sinauer Associates.

Hohenlohe, P. A. (2004). Limits to gene flow in marine animals with planktonic larvae: models of *Littorina* species around Point Conception, California. *Biol. J. Lin. Soc.* 82, 169–187. doi: 10.1111/j.1095-8312.2004.00318.x

Huelsenbeck, J. P., Larget, B., Miller, R. E., and Ronquist, F. (2002). Potential applications and pitfalls of Bayesian inference of phylogeny. *Syst. Biol.* 51, 673–688. doi: 10.1080/10635150290102366

Hughes, A. R., Inouye, B. D., Johnson, M. T. J., Underwood, N., and Vellend, M. (2008). Ecological consequences of genetic diversity. *Ecol. Lett.* 11, 609–623. doi: 10.1111/j.1461-0248.2008.01179.x

Iacchei, M., Ben-Horin, T., Selkoe, K. A., Bird, C. E., García-Rodríguez, F. J., and Toonen, R. J. (2013). Combined analyses of kinship and FST suggest potential drivers of chaotic genetic patchiness in high gene-flow populations. *Mol. Ecol.* 22, 3476–3494. doi: 10.1111/mec.12341

Iken, K., and Konar, B. (2003). Natural Geography in Nearshore Areas (NaGISA): the Nearshore component of the Census of Marine Life. *Gayana* 67, 153–160. doi: 10.4067/S0717-65382003000200004

Jolly, M. T., Guyard, P., Ellien, C., Gentil, F., Viard, F., Thiébaut, E., et al. (2009). Population genetics and hydrodynamic modeling of larval dispersal dissociate contemporary patterns of connectivity from historical expansion into European shelf seas in the polychaete *Pectinaria koreni. Limnol. Oceanogr.* 54, 2089–2106. doi: 10.4319/lo.2009.54.6.2089

Jolly, M. T., Jollivet, D., Gentil, F., Thiébaut, E., and Viard, F. (2005). Sharp genetic break between Atlantic and English Channel populations of the polychaete *Pectinaria koreni,* along the North coast of France. *Heredity* 94, 23–32. doi: 10.1038/sj.hdy.6800543

Koleff, P., Gaston, K. J., and Lennon, J. J. (2003). Measuring beta diversity for presence-absence data. *J. Anim. Ecol.* 72, 367–382. doi: 10.1046/j.1365-2656.2003.00710.x

Kudenov, J. D. (1974). *The reproductive biology of Eurythoe complanata (Pallas, 1766) (Polychaeta, Amphinomidae).* Ph.D. dissertation

Larkin, M. A., Blackshields, G., Brown, N. P., Chenna, R., McGettigan, P. A., McWilliam, H., et al. (2007). ClustalW and ClustalX version 2. *Bioinformatics,* 23, 2497–2498. doi: 10.1093/bioinformatics/btm404

Legendre, P., and Cáceres, M. (2013). Beta diversity as the variance of community data: dissimilarity coefficients and partitioning. *Ecol. Lett.* 16, 951–963. doi: 10.1111/ele.12141

Lowe, A., Harris, S., and Ashton, P. (2004). *Ecological Genetics: Design, Analysis and Application.* Oxford: Blackwell.

Magurran, A. E. (1988). *Ecological Diversity and its Measurement.* London: Croom-Helm.

Mann, H. B., and Whitney, D. R. (1947). On a test of whether one of two random variables is stochastically larger than the other. *Ann. Math. Stat.* 18, 50–60. doi: 10.1214/aoms/1177730491

Mileikovsky, S. A. (1961). Assignment of two rostraria-type polychaete larvae from the plankton of the northwest Atlantic to species *Amphinome pallasi* Quatrefages, 1865 and *Chloenea atlantica* McIntosh, 1885 (Polychaeta, Errantia, Amphinomorpha). *Dokl. Biol. Sci.* 141, 1109–1112.

Nazareno, A. G., and Jump, A. S. (2012). Species-genetic diversity correlations in habitat fragmentation can be biased by small sample sizes. *Mol. Ecol.* 21, 2847–2849. doi: 10.1111/j.1365-294X.2012.05611.x

Pallas, P. S. (1766). *Miscellanea Zoologica Quibus Novae Imprimis Atque Obscurae Animalium Species Describuntur Et Observationibus Iconibusque Illustrantur.* Hagae Comitum: Apud Petrum van Cleef.

Palumbi, S. R. (2004). Marine reserves and ocean neighborhoods: the spatial scale of marine populations and their management. *Ann. Rev. Env. Resour.* 29, 31–68. doi: 10.1146/annurev.energy.29.062403.102254

Papadopoulou, A., Anastasiou, I., Spagopoulou, F., Stalimerou, M., Terzopoulou, S., Legakis, A., et al. (2011). Testing the species-genetic diversity correlation in the aegean archipelago: toward a haplotype-based macroecology? *Am. Nat.* 178, 241–255. doi: 10.1086/660828

Pringle, J. M., and Wares, J. P. (2007). Going against the flow: maintenance of alongshore variation in allele frequency in a coastal ocean. *Mar. Ecol. Prog. Ser.* 335, 69–84. doi: 10.3354/meps335069

Reeb, C. A., and Avise, J. C. (1990). A genetic discontinuity in a continu- ously distributed species: mitochondrial DNA in the American oyster, *Crassostrea virginica. Genetics* 124, 397–406.

Rockman, M. V. (2012). Patterns of nuclear genetic variation in the poecilogonous polychaete *Streblospio benedicti*. *Integr. Comp. Biol.* 52, 173–180. doi: 10.1093/icb/ics083

Scaps, P. (2002). A review of the biology, ecology and the potential use of the common ragworm *Hediste diversicolor* (O.F. Muller) (Annelida: Polychaeta). *Hydrobiologia* 470, 203–218. doi: 10.1023/A:1015681605656

Schulze, A. (2006). Phylogeny and genetic diversity of palolo worms (Palola, Eunicidae) from the Tropical North Pacific and the Caribbean. *Biol. Bull.* 210, 25–37. doi: 10.2307/4134534

Schulze, S. R., Rice, S. A., Simon, J. L., and Karl, S. A. (2000). Evolution of poecilogony and the biogeography of North American populations of the polychaete *Streblospio*. *Evolution* 54, 1247–1259. doi: 10.1111/j.0014-3820.2000.tb00558.x

Slatkin, M. (1987). Gene flow and geographic structure of natural populations. *Science* 236, 787–792. doi: 10.1126/science.3576198

Southwood, T. R. E., and Henderson, P. A. (2000). *Ecological Methods*. Oxford: Blackwell Science.

Struebig, M., Kingston, T., Petit, E. J., Zubaid, A., Mohd-Adnan, A., and Rossiteri, S. J. (2011). Parallel declines in species and genetic diversity in tropical forest fragments. *Ecol. Lett.* 14, 582–590. doi: 10.1111/j.1461-0248.2011.01623.x

Tamura, K., Peterson, D., and Peterson, N. (2011). MEGA5: molecular evolutionary genetics analysis using maximum likelihood, evolutionary distance, and maximum parsimony methods. *Mol. Biol. Evol.* 28, 2731–2739. doi: 10.1093/molbev/msr121

Treseder, K. K., and Vitousek, P. M. (2001). Potential ecosystem-level effects of genetic variation among populations of *Metrosideros polymorpha* from a soil fertility gradient in Hawaii. *Oecologia* 126, 266–275. doi: 10.1007/s004420000523

Vasileiadou, K., Sarropoulou, E., Tsigenopoulos, C., Reizopoulou, S., Nikolaidou, A., Orfanidis, S., et al. (2012). Genetic vs community diversity patterns of macrobenthic species : preliminary results from the lagoonal ecosystem. *Transitional Waters Bull.* 6, 20–33. doi: 10.1285/i1825229Xv6n2p20

Vellend, M. (2003). Island biogeography of genes and species. *Am. Nat.* 162, 358–365. doi: 10.1086/377189

Vellend, M., and Geber, M. A. (2005). Connections between species diversity and genetic diversity. *Ecol. Lett.* 8, 767–781. doi: 10.1111/j.1461-0248.2005.00775.x

Vergara-Chen, C., Gonzalez-Wanguemert, M., Marcos, C., and Perez-Ruzafa, A. (2013). Small-scale genetic structure of *Cerastoderma glaucum* in a lagoonal environment: potential significance of habitat discontinuity and unstable population dynamics. *J. Mollus. Stud.* 79, 230–240. doi: 10.1093/mollus/eyt015

Vreeland, H. V., and Lasker, H. R. (1989). Selective feeding of the polychaete *Hermodice carunculata* Pallas on Caribbean gorgonians. *J. Exp. Mar. Biol. Ecol.* 129, 265–277. doi: 10.1016/0022-0981(89)90108-1

Weersing, K., and Toonen, R. J. (2009). Population genetics, larval dispersal, and connectivity in marine systems. *Mar. Ecol. Prog. Ser.* 393, 1–12. doi: 10.3354/meps08287

Whitham, T. G., Young, W. P., Martinsen, G. D., Gehring, C. A., Schweitzer, J. A., Shuster, S. M., et al. (2003).Community and ecosystem genetics: a consequence of the extended phenotype. *Ecol.* 84, 559–573. doi: 10.1890/0012-9658(2003)084[0559:CAEGAC]2.0.CO;2

Whittaker, R. H. (1960). Vegetation of the Siskiyou Mountains, Oregon and California. *Ecol. Monogr.* 30, 279–338. doi: 10.2307/1943563

Whittaker, R. H. (1972). Evolution and measurement of species diversity. *Taxon* 21, 213–251. doi: 10.2307/1218190

Wilson, M. V., and Shmida, A. (1984). Measuring beta diversity with presence–absence data. *J. Ecol.* 72, 1055–1064. doi: 10.2307/2259551

Conflict of Interest Statement: The authors declare that the research was conducted in the absence of any commercial or financial relationships that could be construed as a potential conflict of interest.

8

Changes in the C, N, and P cycles by the predicted salps-krill shift in the southern ocean

Miquel Alcaraz[1]*, Rodrigo Almeda[2], Carlos M. Duarte[3,4], Burkhard Horstkotte[3], Sebastien Lasternas[3] and Susana Agusti[3,4]

[1] Institut de Ciències del Mar, CSIC, Barcelona, Spain
[2] Centre for Ocean Life, DTU Aqua, Technical University of Denmark, Charlottenlund, Denmark
[3] IMEDEA, CSIC-UIB, Esporles, Spain
[4] UWA Oceans Institute, Crawley, WA, Australia

Edited by:
Paul F. J. Wassmann, University of
Tromsø - Norway's Arctic University,
Norway

Reviewed by:
Joanna Carey, Marine Biological
Laboratory, USA
Maria Vernet, University of
California, San Diego, USA

*Correspondence:
Miquel Alcaraz, Institut de Ciènces
del Mar, CSIC, P. Marítim de la
Barceloneta 37-49,
08003 Barcelona, Catalonia, Spain
e-mail: miquel@icm.csic.es

The metabolic carbon requirements and excretion rates of three major zooplankton groups in the Southern Ocean were studied in February 2009. The research was conducted in the framework of the ATOS research project as part of the Spanish contribution to the International Polar Year. The objective was to ascertain the possible consequences of the predicted zooplankton shift from krill to salps in the Southern Ocean for the cycling of biogenic carbon and the concentration and stoichiometry of dissolved inorganic nutrients. The carbon respiratory demands and NH4-N and PO4-P excretion rates of <5 mm size copepods, krill and salps were estimated by incubation experiments. The carbon-specific metabolic rates and N:P metabolic quotients of salps were higher than those of krill (*furcilia* spp. and adults) and copepods, and as expected there was a significant negative relation between average individual zooplankton biomass and their metabolic rates, each metabolic process showing a particular response that lead to different metabolic N:P ratios. The predicted change from krill to salps in the Southern Ocean would encompass not only the substitution of a pivotal group for Antarctic food webs (krill) by one with an indifferent trophic role (salps). In a zooplankton community dominated by salps the respiratory carbon demand by zooplankton will significantly increase, and therefore the proportion of primary production that should be allocated to compensate for the global respiratory C-losses of zooplankton. At the same time, the higher production by salps of larger, faster sinking fecal pellets will increase the sequestration rate of biogenic carbon. Similarly, the higher N and P excretion rates of zooplankton and the changes in the N:P stoichiometry of the metabolic products will modify the concentration and proportion of N and P in the nutrient pool, inducing quantitative and qualitative changes on primary producers that will translate to the whole Southern Ocean ecosystem.

Keywords: Southern Ocean, zooplankton, community shifts, metabolism, carbon cycling, C:N:P stoichiometry

INTRODUCTION

The consequences of human-induced global perturbations in polar areas are predicted to include significant changes in structure and function of marine ecosystems (Smetacek and Nicol, 2005; Duarte, 2008; Wassmann et al., 2008), and although their specific nature is difficult to foretell, it is most likely that smooth environmental changes could result in non-linear and probably irreversible ecosystem shifts (Duarte et al., 2012).

Zooplankton play a fundamental role in the transfer and cycling of biogenic carbon in marine systems, controlling not only the fraction of primary production available to upper consumers, but the magnitude and fate of vertical carbon flux (either recycled or sequestered). Zooplankton can also modify the chemical environment of phytoplankton by increasing the "per cell" quota of nutrients (i.e., reduction of cell concentration by grazing), and change the N:P ratio of dissolved nutrients by excreting N and P at different rates (Sterner, 1986, 1990). In this sense, the inverse relation between the N:P quotient of the metabolic products of

Arctic zooplankton and temperature has been suggested as one of the tipping elements that could induce non-linear changes in the Arctic marine ecosystems by global warming (Alcaraz et al., 2013).

In the Southern Ocean the major mesozooplankton groups are copepods, krill, and salps. Krill constitute an essential node, directly transferring matter and energy from micro auto- and heterotrophs to upper consumers including birds, fish, seals and whales (Atkinson et al., 2004; Smetacek and Nicol, 2005), and recently being also the target of commercial fisheries (Omori, 1978; Constable et al., 2000; Atkinson et al., 2004). From the biogeochemical point of view, krill contributes decisively to the vertical flux of biogenic carbon (Pakhomov et al., 2002; Pakhomov, 2004; Tanimura et al., 2008; Ruiz-Halpern et al., 2011) and is an important source of recycled dissolved organic carbon and iron (Tovar-Sanchez et al., 2007).

Although the role of salps in the Antarctic food webs and biogeochemical cycles is less known, as a source of food for upper trophic levels seem to be of minor importance (but see Dubischar et al., 2012). However, their contribution to the vertical flux of biogenic carbon is higher than that of krill (Pakhomov et al., 2002; Pakhomov, 2004; Tanimura et al., 2008), with higher ingestion rates and the egestion of larger, faster sinking fecal pellets (Pakhomov et al., 2006; Ducklow et al., 2012).

Regarding the smaller size fractions of mesozooplankton (Copepods and furcilia) their role in Antarctic food webs is complex. Although their food include micro auto- and heterotrophs, copepods and furcilia show a clear preference for heterotrophic preys (Wickham and Berninger, 2007) and their contribution to the vertical flux is lower than that of salps or krill, as a large proportion of their fecal material is degraded while sinking (Dagg et al., 2003). However, their specific rates of carbon demand and nutrient cycling can be higher than those of krill and salps (Ikeda and Mitchell, 1982; Alcaraz et al., 1998).

During the last decades the Southern Ocean appears to be experiencing crucial structural and functional changes (Constable et al., 2014) that affect particularly the two main planktonic grazers, krill, and salps. All analyzed data on their relative abundance suggest, aside from a strong inter-annual variability, a sustained decreasing trend of krill (from 38 to 75% per decade, Atkinson et al., 2004) and their substitution by salps (Smetacek and Nicol, 2005; Murphy et al., 2007). In the zone west of the Antarctic Peninsula, for the decade 1993–2004 aside from the alternation of "salp years" (1994, 1997, 1999) with positive anomalies of krill biomass (1996, 1998), a constant decreasing tendency to negative biomass anomalies for krill, in opposition to positive anomalies for salps, has been also recorded (Ross et al., 2008). The reasons of this community shifts are not clear, but the changes of krill distribution appear to be related to chlorophyll concentration (Atkinson et al., 2004; Montes-Hugo et al., 2009) and their inter-annual variability to the changes in the extent of winter sea ice (Atkinson et al., 2004; Murphy et al., 2007). The decimation of baleen whales could also explain the present zooplankton shift by changes in the recycling characteristics of iron and nutrients in surface waters (Smetacek, 2008) that would have affected the structure and function of primary producers and of the whole Antarctic food web.

In order to ascertain the consequences of zooplankton shifts for the biogeochemical cycles in the Southern Ocean, we have analyzed the effects of community structure on the metabolic demand of biogenic carbon and on the stoichiometry of the recycled inorganic nutrients. The main objectives were (1) To determine how the predicted shift will affect the global respiratory carbon loss by zooplankton, the proportion of primary production required to compensate for it, and the carbon vertical flux, and (2) The changes in the contribution of zooplankton excretion to the N and P required by phytoplankton, and the N:P proportion of the excreted products. These are basic questions to answer in a future scenario where krill-salps fluctuations will be more frequent and salps are predicted to substitute krill.

MATERIALS AND METHODS
STUDY AREA AND ZOOPLANKTON STRUCTURE AND BIOMASS

The study was made in the framework of the ATOS research project in January-February 2009 on board the R/V "Hespérides"

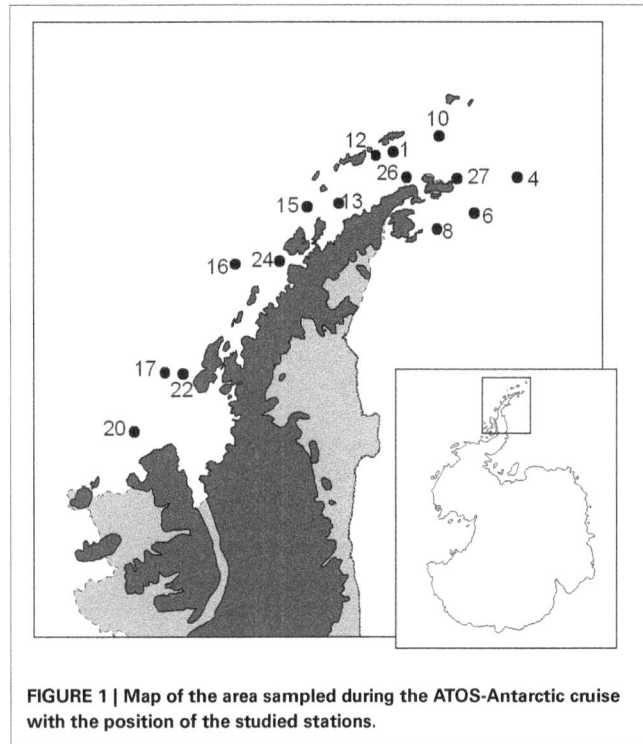

FIGURE 1 | Map of the area sampled during the ATOS-Antarctic cruise with the position of the studied stations.

during the ATOS Antarctic cruise (ATOS-II), as part of the Spanish contribution to the International Polar Year. In a network of stations located in the vicinity of the Antarctic Peninsula (**Figure 1**), the abundance, community composition, and individual biomass of zooplankton was analyzed on samples obtained with a double WP-2 net hauled vertically between 200 m depth (or less in shallower stations) and surface. The volume of water filtered was measured with a back-stop General Oceanics Flow-Meter® placed at the mouth of the net at a distance from the holding ring equivalent to 1/3 of its diameter. The samples corresponding to the two nets were mixed and homogenized in a container, concentrated and fixed in 4% formalin in seawater (final concentration) for abundance, taxonomic and biomass studies.

Crustacean zooplankton abundance and biomass as carbon ("*in situ*" and in the experimental bottles) was estimated according to the biovolume (BV)—zooplankton carbon (C_{zoo}) factor (Alcaraz et al., 2003). The number of organisms and biovolume (BV) determinations were made with the free-user program for image analysis ZooImage® (http://www.sciviews.org/zooimage) on scanned images of preserved organisms made with an EPSOM 4990 Photo scanner at 2400 dpi. Organisms were previously stained in a 0.05% eosin-Y aquatic solution for 24 h. The BV-C_{zoo} factor conversion used was that given by Alcaraz et al. (2003, 2010, 2014) for Arctic zooplankton: $1\,mm^3\,BV = 0.08\,mg\,C_{zoo}$. In the case of *S. thompsoni* blastozoids, the individual C contents (C_S) was calculated by two methods: according to the relationships between atrial-oral length and C_S given by Huntley et al. (1989), and by the relation between the nucleus volume (N_V) and C_S (Alcaraz et al., 2003). The correlation coefficient between the C_S obtained by the two methods was $r = 0.99$. Krill biomass was measured with a Simrad® EK60 multifrequency echosounder, and the data taken from Ruiz-Halpern et al. (2011), where more details can be obtained.

The taxonomic composition of zooplankton was analyzed automatically on the scanned samples using appropriate shape identification algorithms and specific training sets (Fernandes et al., 2009; Saiz et al., 2013) for 10 main taxons or categories of Antarctic zooplankton chosen after the study of selected samples: Two groups of adult copepods, Calanoids and *Oithona*; nauplius; adult and juvenile euphausiids (furcilia); polychaets, chaetognaths, salps, foraminifers and a group of unidentified organisms. The percentage error of automatic classification as compared to manual classification under stereomicroscope in paired samples ranged from 0 (chaetognaths) to less than 6% for nauplii and copepods.

PHYTOPLANKTON BIOMASS, PRIMARY PRODUCTION AND ZOOPLANKTON METABOLISM

Chlorophyll *a* (Chl *a*) concentration was determined in the studied stations by filtering 50 mL samples onto 25-mm diameter GF/F filters from the depths where primary production was measured. Chlorophyll extracted by acetone was measured by fluorescence according to Parsons et al. (1984), and Chl *a* transformed into phytoplankton carbon units using a C:Chl *a* ratio of 100 (mg–mg) following Hewes et al. (1990) for relatively poor Antarctic waters.

In situ primary production was measured by the ^{14}C technique (Steemann-Nielsen, 1952) as described in Morán et al. (2001). Water sampled at 3 depths including the surface (1 m), the subsurface (5 m) and the deep chlorophyll maximum (DCM) was transferred into transparent (light) and dark 150 ml polycarbonate bottles, and inoculated with 100 μCi activity of a ^{14}C working solution. Inoculated bottles were suspended at the corresponding depths from a drifting buoy and incubated *in situ* for 4 h at the same time of the day (from 12.00 to 4.00 p.m.), always including noon. At the end of the incubation period duplicated 5 ml aliquots were transferred into 20 ml scintillation vials for the determination of total labeled organic carbon production (TPP). The remaining volume was filtered through 0.22 μm mesh membrane filters (cellulose membrane filters) of 25 mm diameter to determine particulate primary production (PPP > 0.22 μm). Samples were acidified with 100 μl of 10% HCl and shaken for 12 h to remove inorganic ^{14}C. Then, 10 ml of scintillation cocktail (Packard Ultima Gold XR) were added to TPP vials and the disintegrations per minute were counted after 24 h with a scintillation counter (EG&G/Wallace).

As we had no data on irradiance we integrated the solar curve along the daylight hours corresponding to the latitude and date of the study, considered as proportional to the theoretical irradiance without cloud covering. We calculated also the maximum theoretical irradiance, equivalent to the integral of the maximum (noon) irradiance along the duration of the day. The proportion of the maximum total theoretical irradiance that corresponded to theoretical irradiance, multiplied by the duration of the day gave us the factor f to transform hourly primary production rates into daily rates,

$$f = (t' - t) \left\{ \left(\int_{t}^{t'} TI \right) \Big/ \left(\int_{t}^{t'} MTI \right) \right\} \qquad (1)$$

where f is the factor to multiply hourly primary production rates to obtain daily rates, t and t' the hour of sunrise and sunset during the study, TI the solar curve equivalent to the theoretical irradiance, and MTI the maximum theoretical irradiance, equivalent to the irradiance (height of the solar curve) at noon.

Metabolism (respiration and excretion of ammonia and phosphate) was estimated by incubation experiments on copepods, krill juveniles (unidentified furcilia) and adults (*Euphausia superba* and *E. crystallorophias*), and salps (blastozoids of *Salpa thompsoni*), the most significant groups of Antarctic zooplankton. Experimental copepods, furcilia and salp blastozoids were obtained by vertical WP-2 net tows made at a speed of 10 m min^{-1} from 100 m depth to surface, conducted with the same net as for the study of the zooplankton community structure but fitted with a 6-L plastic bag as cod end to avoid damaging the organisms. Adult krill were caught with short (<3 min) horizontal or oblique trawls using an IKMT net provided with a 20 L rigid PVC cod end and a Scanmar® HC4-D net sounder to control the depth of the trawl. When a krill swarm was located with the Simrad® EK60 echosounder, the ship re-traced the course and the haul was made across the previously observed depth and position of the krill swarm.

WP-2 samples were immediately transferred into thermally isolated 10 L containers filled with "*in situ*" water and transported to the laboratory. Salps and furcilia were separated by gently screening the sample using a 5 mm plastic grid submerged in a 2 L jar containing 0.2 μm-filtered seawater at "*in situ*" temperature, and individually sorted and transferred with a plastic spoon into separated 2 L Pyrex® bottles containing 0.2 μm-filtered seawater. The <5 mm size-fraction copepods were repeatedly cleaned, screened and concentrated using a 200 μm netting submerged in filtered seawater in order to discard phyto- and microzooplankton. Adult krill were gently transferred from the IKMT cod-end into 50 L on-deck containers provided with circulating surface water, individually sorted with a hand net provided with a 200 mL plastic bucket as cod-end and kept on separate 10 L thermally isolated containers at *in situ* temperature.

Incubation experiments for simultaneous estimation of respiration and excretion rates (**Table 1**) were made in Pyrex® bottles from 250 mL to 5 L volume, depending on the biomass of experimental organisms. The bottles were closed by silicone stoppers holding the O_2 probes and a syringe needle to compensate for pressure changes as described in Alcaraz et al. (1998, 2010, 2013, 2014) and sketched in Almeda et al. (2011). Experimental organisms (either an aliquot of the <5 mm copepods, or from 2 to 4 individuals in the case of larger organisms) were transferred in less than 1 h after capture to experimental bottles filled with *in situ* seawater obtained with 12 L Niskin bottles from 20 to 40 m depth, depending of the depth of the maximum chlorophyll, filtered by gravity through 0.2 μm Acro-Pack® filters and O_2-saturated. Control bottles contained only filtered seawater. Once confirmed that there were only intact organisms in the experimental bottles (i.e., all the organisms showing normal swimming behavior),

Table 1 | Taxonomic composition (ind m^{-3}) and biomass (μmol C m^{-3}, bold italics) for the main zooplankton groups in the studied stations (see Methods).

ST.	Cal	Oit	Na	Es	Fur	Pol	Ch	Sal	For	Oth	Total
1	76.6	80.0	6.9	–	0.0	0.0	0.0	0.0	2.6	0.0	166.03
	57.3	*13.5*	*0.8*	*387.8**	*0.0*	*0.0*	*0.0*	*0.0*	*0.3*	*0.0*	*531.22*
4	10.8	4.3	5.6	–	0.0	0.0	0.9	0.0	0.0	0.0	21.51
	4.5	*0.5*	*0.8*	*–*	*0.0*	*0.0*	*52.1*	*0.0*	*0.0*	*0.0*	*63.86*
6	132.5	1.7	36.1	–	0.0	0.0	0.9	0.0	0.9	0.0	172.05
	75.9	*0.2*	*5.3*	*2583.0**	*0.0*	*0.0*	*39.1*	*0.0*	*0.1*	*0.0*	*2785.0*
8	277.0	12.0	69.3	–	34.6	0.0	0.0	0.0	0.0	0.0	392.86
	2621	*0.2*	*24.1*	*–*	*159.3*	*0.0*	*0.0*	*0.0*	*0.0*	*0.0*	*5451.35*
10	60.2	27.5	1.7	–	0.0	0.9	0.0	0.0	0.0	1.7	92.05
	77.0	*3.6*	*0.1*	*–*	*0.0*	*1.2*	*0.0*	*0.0*	*0.0*	*1.2*	*163.96*
12	61.9	73.1	11.2	–	0.0	0.0	0.0	0.0	0.9	0.0	147.11
	57.6	*9.5*	*1.4*	*–*	*0.0*	*0.0*	*0.0*	*0.0*	*0.1*	*0.0*	*136.80*
13	86.0	37.9	1.7	–	0.0	0.0	0.0	0.0	5.2	0.0	130.76
	33.2	*7.7*	*0.3*	*6468.8**	*0.0*	*0.0*	*0.0*	*0.0*	*0.4*	*0.0*	*6551.64*
15	757.0	7.7	20.6	–	0.0	0.0	0.0	0.0	2.6	0.0	788.01
	154.4	*0.7*	*1.6*	*–*	*0.0*	*0.0*	*0.0*	*0.0*	*0.2*	*0.0*	*313.49*
16	–	–	–	–	0.0	0.0	0.0	0.0	0.0	0.0	*
	–	*–*	*–*	*11,133.8**	*0.0*	*0.0*	*0.0*	*0.0*	*0.0*	*0.0*	***
17	–	–	–	–	0.0	0.0	0.0	0.0	0.0	0.0	*
	–	*–*	*–*	*2484.4**	*0.0*	*0.0*	*0.0*	*0.0*	*0.0*	*0.0*	***
20	–	–	–	–	0.0	0.0	0.0	0.0	0.0	0.0	*
	–	*–*	*–*	*5194.7**	*0.0*	*0.0*	*0.0*	*0.0*	*0.0*	*0.0*	***
22	488.6	252.	48.2	–	0.0	3.4	0.9	0.0	3.4	0.9	797.47
	135.9	*30.3*	*6.6*	*–*	*0.0*	*8.1*	*27.4*	*0.0*	*0.3*	*2.3*	*383.57*
24	712.3	80.9	37.0	–	0.0	0.0	0.9	0.0	0.9	0.0	831.89
	576.4	*8.3*	*5.1*	*–*	*0.0*	*0.0*	*36.2*	*0.0*	*0.1*	*0.0*	*1215.61*
26	80.9	9.5	6.0	–	0.0	0.9	0.9	0.9	1.7	0.0	100.65
	149.1	*2.9*	*0.9*	*–*	*0.0*	*1.5*	*16.8*	*82.6*	*0.3*	*0.0*	*407.04*
Avg. Ind	249	53.3	22.2	–	2.47	0.37	0.31	0.06	1.29	0.18	330.95
Stdev. Ind.	275	72	22		9.2	0.9	0.4	0.2	1.6	0.5	318.3
Avg. Biom	**771**	**7.0**	**4.3**	**4708.7**	**14.4**	**0.8**	**12.3**	**7.5**	**0.12**	**0.25**	**5109.1**
Stdev. % Biom	**771**	**8.9**	**6.9**	**3814.5**	**(–)**	**2.1**	**18.5**	**22.1**	**0.15**	**0.7**	**3315.6**
	7.2	**0.13**	**0.08**	**92.5**	**0.28**	**0.01**	**0.24**	**0.15**	**–**	**–**	

ST, station number; Cal, Calanoid copepods; Oit, Oithona sp.; Na, Copepod nauplii; Es, Euphausia adults*; Fur, Furcilia; Pol, Polychaet larvae; Ch, Chaetognaths; Sal, Salpa thompsoni; For, Foraminifers; Oth, Unidentified; –, absence of data. The global average values (Avg), standard deviation (Stdev) and percentage contribution to total biomass (% Biom.) are also given.

*Data from Ruiz-Halpern et al. (2011).

experimental and control bottles were stopped without trapping air bubbles and incubated for 12–24 h in thermostatic baths at the 0–200 m depth "*in situ*" integrated average temperature ±0.1°C and dim light.

Zooplankton respiration was estimated as the decreasing rate of dissolved oxygen concentration during the incubation. The analyses were made with an OXY-10 Pre-Sens® oxygen sensor (optodes, Alcaraz et al., 2010) that allowed semi-continuous (every 5 min.) measurements of O_2 concentration using 6–8 O_2 probes for experimental bottles, and 2–4 for control ones. Respiration rates were estimated as the difference between the slopes of the linear regression equations describing the changes in O_2 concentration during the incubations in experimental and control bottles (Alcaraz et al., 2010, 2013). Oxygen consumption was transformed into respiratory C losses using a respiratory quotient (RQ, the molar ratio of CO_2 produced to O_2 consumed) of 0.97 (Omori and Ikeda, 1984).

Excretion rates were estimated in the same incubation experiments as for respiration. Ammonia and phosphate excretion rates were calculated as the difference in the final concentrations in experimental and control bottles. At the end of the incubation water samples were siphoned from the bottles using silicone tubes ending in broad plastic tips enclosed with $100\,\mu m$-mesh in order to avoid extracting zooplankton organisms with the water sample. Ammonia was analyzed by the fluorimetric method described by Kéruel and Aminot (1997), and phosphate according to Grasshoff et al. (1999). At the end of the incubations, experimental zooplankton was transferred to vials and fixed in 4% formalin (final concentration) for further measurement of experimental biomass as zooplankton carbon.

Metabolic rates were normalized to per unit of zooplankton carbon biomass (C- specific metabolic C_R, N_E, and P_E) by dividing daily gross respiration and excretion rates (μmol C, μmol N and μmol P day^{-1}) by the corresponding experimental biomass in μmol C. Specific metabolic data from other authors when expressed in different units have been re-calculated using the wet mass, dry mass, and organic C transform factors given in Harris et al. (2000). The taxonomic composition and individual biomass of experimental organisms were analyzed as described above. The metabolic C_R:N_E, C_R:P_E, and N_E:P_E quotients were calculated as the ratios between the specific corresponding metabolic rates in each individual experiment and expressed in atoms.

ZOOPLANKTON RESPIRATORY C LOSSES, N AND P EXCRETION AND THEIR RELATION TO PRIMARY PRODUCTION

The daily global average respiratory carbon losses and N and P supplied by zooplankton were calculated by the addition of the average respiratory losses and ammonia and phosphate excreted by the different zooplankton groups. These were calculated as the product of the average *in situ* C biomass of each group by their corresponding C-specific metabolic rates,

$$C_L = C_R C_{ZOO}$$
$$N_S = N_E C_{ZOO}$$
$$P_S = P_E C_{ZOO} \tag{2}$$

where C_L, N_S, and P_S are the daily respiratory C loss and N and P excreted by the group, C_R N_E and P_E the corresponding C-specific metabolic rates, and C_{ZOO} the average *in situ* biomass of the corresponding zooplankton group as carbon.

The total theoretical daily carbon ingested by zooplankton and vertical carbon flux in μmol C m^{-3} day^{-1} were calculated respectively by the addition of the carbon ingested and egested by the different groups. The daily carbon respiratory losses of

each group were considered as equivalent to the carbon assimilated. Therefore, the carbon daily ingested and egested can be estimated from the carbon respiratory losses and the assimilation efficiencies of the different groups (0.7 and 0.52 for krill and salps respectively, Pond et al., 1995; Pakhomov et al., 2006) as follows,

$$C_I = \Sigma(C_{LG}/A_{EG}) \tag{3}$$

where C_I is the global carbon ingestion; C_{LG} are the respiratory C losses for the different groups, and A_{EG} are the corresponding assimilation efficiencies.

The theoretical carbon egested was considered as equivalent to the non-assimilated C and equivalent to the vertical carbon flux as fecal pellets. It was calculated by the addition of the daily fecal pellets production (carbon egested) by salps and krill. The carbon egested by copepods and other small zooplankters was not included in the estimations of carbon export as their fecal pellets are mainly recycled in surface waters and therefore their contribution to the vertical C transfer is negligible, and their carbon egestion in the present conditions and for the predicted salps-krill shift would not change,

$$C_{EX} = \Sigma \{(C_{LG}/A_{EG}) (1 - A_{EG})\} \tag{4}$$

where C_{EX} is the global carbon egested as fecal pellets, and C_{LG} and A_{EG} as described above.

The fraction of total and particulate primary production (TPP and PPP) daily ingested by zooplankton to compensate for their C metabolic losses and vertically exported has been expressed as a percentage,

$$C_I\% \text{ (TPP or PPP)} = 100(C_I/\text{TPP or PPP})$$
$$C_{EX}\% \text{ (TPP or PPP)} = 100(C_{EX}/\text{TPP or PPP}) \tag{5}$$

To estimate the consequences of the zooplankton shift for the carbon and nutrient flux we have assumed a change in the proportion of krill and salps biomass from the present situation (a krill-based zooplankton community) to that of a "salp year" (average salps/krill ratio = 10, Huntley et al., 1989; Loeb et al., 1997, 2010; Alcaraz et al., 1998). In terms of biomass the substitution falls in the known range of a "salps years," between 2900 and $6200\,\mu$mol C m^{-3} (Alcaraz et al., 1998; Tanimura et al., 2008). Average biomass and numbers and the corresponding standard deviations (krill excepted) for the different zooplankton groups were calculated globally for the whole stations sampled. The relationships between individual biomass and C-specific metabolic rates or metabolic quotients have been estimated by linear regression on log-transformed data. All the statistical analysis have been made using JMP® 7.0 software.

RESULTS
ZOOPLANKTON COMMUNITY STRUCTURE

The most abundant and frequent zooplankton group in the study area were copepods. Calanoids, *Oithona* sp., and nauplii contributed to 97.7% of zooplankton as numbers. Foraminifera were scarce but present in most of the stations, followed by chaethognaths and polychaets. Furcilia and salps (*Salpa thompsoni*) were observed only in Stations 8 and 26 respectively (**Table 1**). Copepods occurred in all the stations, and reached concentrations up to 430 individuals m^{-3} (Station 8, **Table 1**). We had no data on krill numbers as we estimated krill biomass by acoustic methods, and the nets used to capture experimental animals (WP-2 and IKMT) are not adequate to sample krill.

In terms of biomass (as μmol C m^{-3}) krill dominated the zooplankton community (91.9%), followed by copepods and furcilia sp. (7.2 and 0.35% of total biomass respectively, **Table 1**). The remaining groups had variable importance, salps contributing to less than 0.2% of total zooplankton C. The average zooplankton biomass (>200 μm-size), including krill, accounted for 5109 μmol C m^{-3}, or 12.26 g C m^{-2} (0–200 m depth).

METABOLISM

The best fit of the time changes of O_2 concentration in control and experimental bottles was the negative linear regression, the average determination coefficient being $r^2 = 0.83$. As no short-term decreases in the rate of O_2 consumption were observed indicating a linear trend in the respiration rates, we assumed a similar linear response for ammonia and phosphate excretion. Average respiratory losses (C_R) of copepods and furcilia sp. were similar, 0.0348 and 0.0330 d^{-1} respectively. The respiration rates of salps (*S. thompsoni*) were higher by a factor of 2.5 than for the crustacean zooplankton groups (0.0841 d^{-1}), while the lowest C_R corresponded to adult krill (*Euphausia superba*, 0.0102 d^{-1}). Carbon-specific ammonia (N_E) and phosphate (P_E) excretion rates were also lower for crustaceans than for salps (**Table 2**). The lowest excretion rates corresponded also to Adult *E. superba* ($N_E = 0.0004$, std. 0.0002 μmol NH$_4$-N μmol C_{ZOO}^{-1} d^{-1}, and $P_E = 0.0003$, std. 0.0003 μmol PO$_4$-P μmol C_{ZOO}^{-1} d^{-1}). In the case of salps $N_E = 0.0073$, std. 0.0006 μmol NH$_4$-N μmol C_{ZOO}^{-1} d^{-1}, and $P_E = 0.0017$, std. 0.0004 μmol PO$_4$-P μmol C_{ZOO}^{-1} d^{-1}.

The atomic C_R:N_E quotients for the different groups ranged from 11.5 (salps, *E. thompsoni*) to 28.4 (furcilia sp.), in both cases higher by a factor from 2 to 5 than the expected Redfield ratio, and also higher than the average values from previously recorded data (**Table 2**), while the C_R:N_E atomic metabolic ratios for krill were similar to the average literature values (**Table 2**). C_R:P_E quotients ranged from 43.2 to 103.4 (furcilia sp. and copepods respectively) and fell within the values given in the scarce previous data (**Table 2**). Regarding the N_E:P_E atomic quotients, again the values were lower than the expected Redfield ratios. The lowest values corresponded to krill, with an average N_E:P_E value for the whole group of 2.4, followed by salps, N_E:$P_E = 4.6$. The highest N:P quotient, 8.1, corresponded to copepods (**Table 2**).

INDIVIDUAL BIOMASS, METABOLIC RATES AND C:N:P METABOLIC STOICHIOMETRY

The individual biomass of the experimental groups (**Table 2** and **Figure 2**) spanned six orders of magnitude, from copepods (0.16–3.46 μmol C ind^{-1}) to adult *E. superba* (6833.1–59,676 μmol C ind^{-1}), with intermediate values for developmental stages of krill (furcilia sp., 2.56–6.49 μmol C ind^{-1}), *E. cristallorophias* (176.53–356.66 μmol C ind^{-1}) an salps, *S. thompsoni* (156.0–193.0 μmol C ind^{-1}). There was a significant, negative relationship between

the specific metabolic rates and individual biomass when the whole range of individual biomass data was considered (**Figure 2**). The relationships between individual biomass and respiration, ammonia and phosphate excretion rates as described by the exponents of the equations (salps excluded) were $C_R = -0.199$, $N_E = -0.238$ and $P_E = -0.177$ (**Table 3A**). When considering individually each group, the exponents were still negative, but were only significant for groups with large data sets and/or a broad span in individual biomass, like copepods and krill (data not shown). As expected by their high average specific metabolic rates, salps occupy an outsider position in the graph (**Figure 2**). Regarding the effects of individual biomass on metabolic stoichiometry, C_R:N_E was not related to individual biomass, while C_R:P_E and N_E:P_E metabolic quotients were inversely and significantly related to individual biomass (**Figure 3** and **Table 3B**).

PHYTOPLANKTON CARBON AND PRIMARY PRODUCTION, AND ZOOPLANKTON CARBON REQUIREMENTS, VERTICAL CARBON EXPORT, AND N AND P EXCRETION

The average chlorophyll concentration was 0.985 μg L^{-1} ± 0.237 *SE*, equivalent to 8211.6 μmol C m^{-3} ± 1978.3 *SE* (Ruiz-Halpern et al., 2011). The depth-integrated (0–50 m) total primary production (TPP) ranged from 24.1 mg C m^{-3} h^{-1} to 363.3 μg C m^{-3} h^{-1}, and particulate primary production (PPP) from 13.3 to 207.5 μmol C m^{-3} h^{-1} at St. 16 and 2 respectively (data not shown). The average TPP and PPP (according to Equation 1) were 1624.7 and 758.9 μmol C m^{-3} day^{-1}. We had no data on the assimilation rate of N and P by phytoplankton, therefore we assumed it to agree with a 106:16:1 C:N:P atomic proportion (Redfield et al., 1963), the N and P theoretically required by phytoplankton for TPP thus being 245.1 μmol N m^{-3} day^{-1}, and 15.5 μmol P m^{-3} day^{-1}.

The carbon theoretically ingested by zooplankton to compensate for their daily average C respiratory losses, once corrected for the assimilation efficiency, averaged 110.9 μmol C m^{-3} day^{-1} (**Table 4**), about 1.3% of the phytoplankton biomass (as carbon), and about 6.8 and 14.6% of the daily total primary production (TPP) and particulate primary production (PPP) respectively (**Table 5**). The N and P excreted as ammonia and phosphate for the zooplankton community, 4.88 and 2.41 μmol N and P m^{-3} day^{-1} respectively (**Table 4**), were equivalent to 2 and 14.7% of the N and P required by phytoplankton for total primary production (TPP, **Table 5**). Regarding the vertical flux, the carbon exported accounted for 33.3 μmol C m^{-3} day^{-1}, 2 and 4.4% of TPP and PPP respectively (**Table 5**).

In the case of the predicted substitution of krill by salps, and assuming primary production rates equivalent to those found during our study, the carbon requirements by zooplankton would average 772.8 μmol C m^{-3} day^{-1} (**Table 4**), equivalent to about 10% of the phytoplankton standing stock, and about 47 and 100% of TPP and PPP respectively (**Table 5**). The ammonia and phosphate excreted will be 35.82 and 7.96 μmol N and P m^{-3} day^{-1} respectively (**Table 4**), or 14.6 and 52% of the N and P required by phytoplankton for TPP (**Table 5**). The new vertical carbon flux would increase by a factor of ten, equivalent to around 23 and 49% of TPP and PPP respectively (**Table 5**). The average N:P

Table 2 | Average values and standard deviation (in italics between brackets) of biomass (μmol C m^{-3}), individual biomass range (μm C ind^{-1}), C- specific metabolic rates (C_R, carbon respiration, d^{-1}; N_E, ammonia excretion, μmol NH$_4$-N μmol C^{-1} d^{-1}; P_E, μmol PO$_4$-P μmol C^{-1} d^{-1}), and C:N, C:P and N:P metabolic ratios by atoms for the studied zooplankton groups.

	Cop	Adult krill	Furc	Tot Krill	Salps
μmol C m^{-3}	**369.75**	**4708.72**	**18.27**	**4726.99**	**7.51**
	(771.21) $n = 11$	*(4536.89)* $n = 6$	*(–)*	*(4536.89)* $n = 7$	*(–)*
μmol C ind.$^{-1}$	**0.16–3.46**	**6833–59,676**	**2.56–6.49**	**2.56–59,676**	**112–193**
C_R	**0.0348**	**0.0100**	**0.0330**	**0.0136**	**0.0841**
	(0.0236) $n = 16$	*(0.0096)* $n = 27$	*(0.0100)* $n = 7$	*(0.0123)* $n = 34$	*(0.0117)* $n = 4$
	0.0242[a]	0.0194[b]*	0.0269[c]	–	0.0230[d]
	(0.0037)[a]	*(0.0119)[b]*	*(–)*	*(–)*	*(0.0073)[d]*
N_E	**0.0036**	**0.0003**	**0.0020**	**0.0007**	**0.0073**
	(0.0058) $n = 23$	*(0.0002)* $n = 19$	*(0.0010)* $n = 5$	*(0.0007)* $n = 24$	*(0.0004)* $n = 4$
	0.0039[a]	0.0012[b]*	0.0014[c]	–	0.0025[c]
	(0.002)[a]	*(0.00079)[b]*	*(0.0005)[c]*	*(–)*	*(0.0017)[d]*
P_E	**0.0011**	**0.0002**	**0.0010**	**0.0004**	**0.0017**
	(0.0001) $n = 23$	*(0.0002)* $n = 26$	*(0.0001)* $n = 7$	*(0.0004)* $n = 33$	*(0.0003)* $n = 4$
	0.0011[a]	0.0006[b]*	–	–	0.0005[d]
	(–)	*(0.0004)[b]*	*(–)*	*(–)*	*(0.0005)[d]*
C:N	**18.2**	**17.6**	**28.4**	**20.3**	**11.5**
	(15.5) $n = 16$	*(3.8)* $n = 14$	*(8.6)* $n = 5$	*(9.7)* $n = 24$	*(2.6)* $n = 4$
	6.8[a]	16.7[b]	19.8[c]	–	9.1[d]
	(–)	*(6.8)[b]*	*(4.0)[c]*	*(–)*	*(–)*
C:P	**105.7**	**50.9**	**43.2**	**54.8**	**50.9**
	(64.1) $n = 16$	*(22.9)* $n = 21$	*(11.3)* $n = 7$	*(33.6)* $n = 33$	*(0.7)* $n = 4$
	61.8[a]	80.8[b]	–	–	50.4[d]
	–	–	–	–	–
N:P	**8.2**	**2.8**	**1.7**	**2.8**	**4.6**
	(3.0) $n = 23$	*(1.1)* $n = 14$	*(0.3)* $n = 5$	*(1.4)* $n = 24$	*(0.9)* $n = 4$
	4.8[a]	5.1[b]*	–	–	5.52 [4]
	–	–	–	–	–

*Cop, Copepods. Adult krill: Euphausia superba and E. crystallorophias. Furc, furcilia sp. Tot. Krill: Adult + Furcilia. Salps, blastozoids of S. Thompsoni. Bold types: Data from this study. Normal types: Average literature values. –, absence of data. [a] Ikeda and Mitchell (1982); Alcaraz et al. (1998). [b] Ikeda and Mitchell (1982), Hirche (1983), Meyer et al. (2009, 2010), Meyer and Oettl (2005), Auerswald et al. (2009), Ikeda and Bruce (1986). [c] Frazer et al. (2002). [d] Ikeda and Mitchell (1982); Alcaraz et al. (1998), Iguchi and Ikeda (2004). *Data on krill metabolism by Ruiz-Halpern et al. (2011) and Lehette et al. (2012) corresponding to hourly rates (deduced from the decreasing trend obtained with short-time incubation dynamic series) have not been included (see comments in Section Zooplankton Metabolism).*

atomic quotient of the excreted products by the whole zooplankton community would increase from 2.0 in the present conditions to 4.5 for the predicted shift (**Table 6**).

DISCUSSION

ZOOPLANKTON COMMUNITY STRUCTURE

The zooplankton community during the ATOS-II cruise was dominated by krill, accounting for more than 90% of total zooplankton biomass. The situation therefore corresponded to a non- "salp year" (in the sense of Huntley et al., 1989 and Alcaraz et al., 1998), in which salps (usually *E. thompsoni* and *Ihlea rakovitzai*) are the dominant group at least in terms of biomass (Alcaraz et al., 1998; Le Fèvre et al., 1998; Perissinotto and Pakhomov, 1998),

displace krill as the main grazer, and can reach more than 90% of total zooplankton biomass (Loeb et al., 1997, 2010; Alcaraz et al., 1998; Atkinson et al., 2004).

The average zooplankton biomass ranged between the 2500 μmol C m^{-3} found by Ward et al. (1995) and the more than 12,500 μmol C m^3 given by Pauly et al. (2000) and by Tanimura et al. (2008) in summer 2002–2003. Part of the differences in the zooplankton community with previous data should be attributed to the strong inter-annual variability coupled with the patchy nature of distribution and abundance that characterize zooplankton in general, and especially Southern Ocean krill and salps (Nishikawa et al., 1995; Loeb et al., 1997; Atkinson et al., 2004; Smetacek and Nicol, 2005). The zooplankton abundance

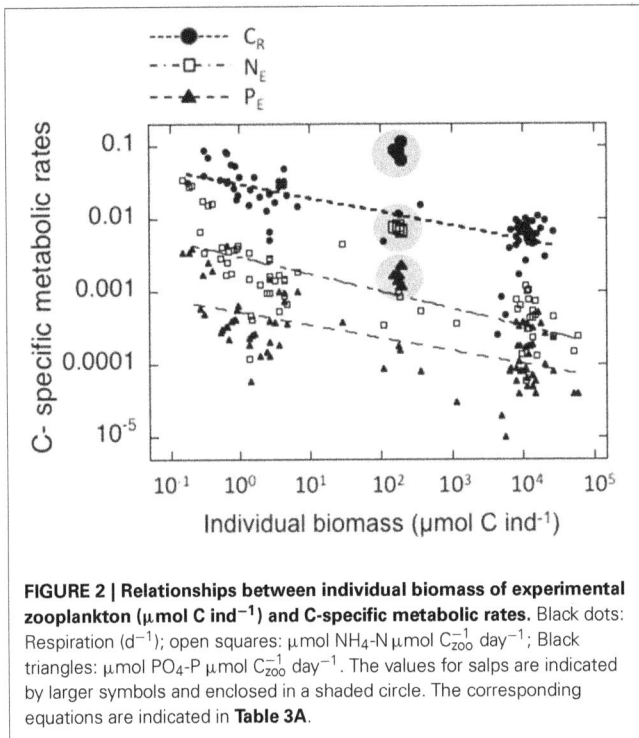

FIGURE 2 | Relationships between individual biomass of experimental zooplankton (μmol C ind^{-1}) and C-specific metabolic rates. Black dots: Respiration (d^{-1}); open squares: μmol NH$_4$-N μmol C$_{zoo}^{-1}$ day^{-1}; Black triangles: μmol PO$_4$-P μmol C$_{zoo}^{-1}$ day^{-1}. The values for salps are indicated by larger symbols and enclosed in a shaded circle. The corresponding equations are indicated in **Table 3A**.

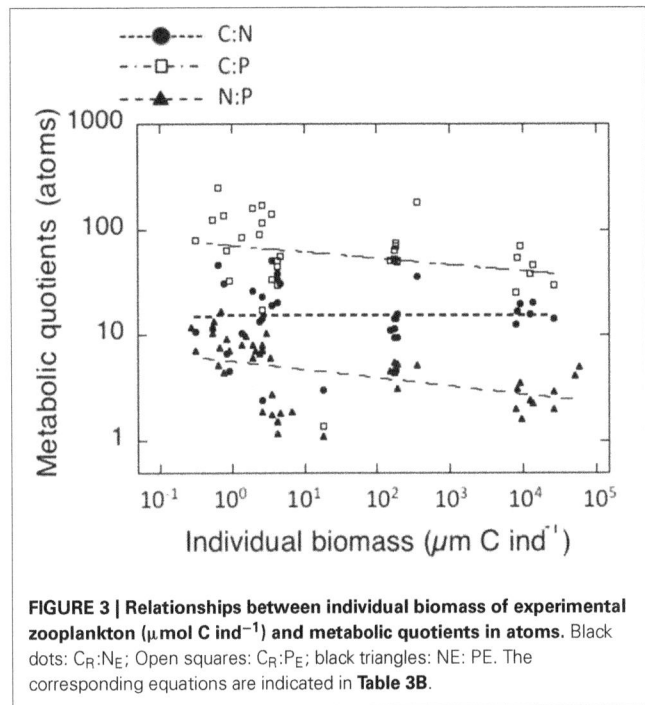

FIGURE 3 | Relationships between individual biomass of experimental zooplankton (μmol C ind^{-1}) and metabolic quotients in atoms. Black dots: C_R:N_E; Open squares: C_R:P_E; black triangles: NE: PE. The corresponding equations are indicated in **Table 3B**.

Table 3 | Equations relating individual zooplankton biomass (C_{zoo}) with (A): C-specific metabolic respiration (C_R), and ammonia and phosphate excretion rates, N_E and P_E respectively, and (B): with C_R:N_E, C_R:P_E and N_E:P_E metabolic quotients (atoms) corresponding to Figures 2, 3.

A				
	$C_R = 0.029*C_{zoo}^{-019}$	$r = -0.784$	$P < 0.01$	$n = 65$
	$N_E = 0.003*C_{zoo}^{-0238}$	$r = -0.614$	$P < 0.01$	$n = 63$
	$P_E = 0.00053*C_{zoo}^{-0177}$	$r = -0.573$	$P < 0.01$	$n = 46$
B				
	C_R:$P_E = 15.1*C_{zoo}^{0.0027}$	$r = 0.155$	(N.S.)	$n = 33$
	C_R:$P_E = 70.9*C_{zoo}^{-0.0599}$	$r = -0.377$	$P < 0.05$	$n = 39$
	N_E:$P_E = 5.68*C_{zoo}^{-0.0077}$	$r = -0.553$	$P < 0.01$	$n = 33$

Salps have been excluded from the calculations. N.S., not significant.

and biomass during summer, strongly dependent from the ice conditions, the position of the circumpolar current, and primary production (Atkinson et al., 2004; Ward et al., 2004; Ross et al., 2008; Montes-Hugo et al., 2009) show strong spatial and temporal changes, the alternative dominance of salps and krill being of different duration but occurring at 3–4 years interval since 1993 (Ross et al., 2008). The progressive tendency to the reduction of krill abundance (Atkinson et al., 2004; Murphy et al., 2007), and differences in the methods of biomass estimation (i.e., nets or echosounders) can also explain the between years differences observed in total zooplankton biomass. Especially in the case of krill, biomass estimations made with different sampling gears are hardly comparable. While echosounders seem to be quite reliable (Ruiz-Halpern et al., 2011), nets clearly induce avoidance reactions in krill that lead to large underestimations (Sameoto et al., 2000).

Table 4 | Zooplankton biomass (μmol C m^{-3}), average metabolic carbon requirements (theoretical carbon ingestion, C_I), ammonia (N_{EX}) and phosphate (P_{EX}) excreted (μmol m^{-3} day^{-1}) for *Present* and *Salp-Krill shift* conditions.

	Cop	Krill	Salps	Other	Total
PRESENT					
Biomass	369.75	4726.99	7.51	17.02	**5121.28**
C_I	12.87	95.99*	1.23*	0.76	**110.85**
N_{EX}	1.34	3.49	0.06	0.07	**4.88**
P_{EX}	0.17	2.23	0.01	0.01	**2.41**
SALP-KRILL SHIFT					
Biomass	369.75	47.35	4687.16	17.02	**5121.28**
C_I	12.87	0.95*	758.3*	0.76	**772.88**
N_{EX}	1.34	0.035	34.37	0.07	**35.82**
P_{EX}	0.17	0.022	7.77	0.01	**7.96**

By groups and total (in bold characters).

*Theoretical carbon ingestion, C_I, calculated from the C-respiratory losses corrected for an average assimilation efficiency of krill and salps (70 and 52% respectively, Pond et al., 1995; Pakhomov et al., 2006).

ZOOPLANKTON METABOLISM

The metabolic rates of zooplankton during our study fell within the range of previous works. The average values for copepods were similar to those given by Ikeda and Mitchell (1982) and Alcaraz et al. (1998), especially regarding ammonia and phosphate excretion (N_E and P_E). In the case of adult krill (*E. superba* and *E. crystallorophias*) our average metabolic rates were similar to those found by Hirche (1983), Auerswald et al. (2009), and Ikeda and Mitchell (1982), but lower than data from Meyer et al. (2009, 2010), Atkinson et al. (2002), Ruiz-Halpern et al.

Table 5 | Allocation and fate of biogenic C, N, and P during our study (*Present*) and in the case of the predicted *Salp-Krill shift*.

	Present			Salp-Krill shift		
	C	N	P	C	N	P
TPP	1624.7	245.1	15.5	1624.7	240	15.5
PPP	758.9	176.2	11.0	758.9	176.2	11.0
Ingest.	110.9	16.7	1.0	772.8	116.6	7.3
(Ing % TPP)	6.8	6.8	6.8	47.6	47.6	47.6
(Ing % PPP)	14.6	14.6	14.6	100	100	100
Supp.	–	4.9	2.4	–	35.8	8.0
(Sup %)	–	2.0	15.4	–	14.9	51.6
Vert.	33.3	5.0	0.3	370.9	56.0	3.5
(Vert. % TPP)	2.0	2.0	2.0	22.8	22.8	22.8
(Vert. % PPP)	4.4	4.4	4.4	48.9	48.9	48.9

The primary production and total zooplankton biomass have been considered to be the same in both situations. TTP and PPP: C, N and P in total and particulate primary production respectively. Ingest.: Total C, N and P ingested by zooplankton. Supp.: N and P supplied by zooplankton excretion. Vert.: C, N and P vertically exported by zooplankton fecal pellets. Data in $\mu mol\ m^{-3}\ day^{-1}$. Ing %: Percentage of TPP and PPP ingested. Supp. %: Percentage of the N and P required by phytoplankton for TPP supplied by zooplankton excretion. Vert. %: Percentage of the TPP vertically exported by salp and krill fecal pellets. Phytoplankton C:N:P ratios as in Redfield et al. (1963). In the case of the salp-krill shift, the relative proportion of krill and salps as in an average "salp year" (Loeb et al., 1997, 2010; Alcaraz et al., 1998). Ingestion and vertical flux derived from respiratory losses corrected for the assimilation efficiency of salps and krill (Pond et al., 1995; Pakhomov et al., 2006).

Table 6 | Average $C_R:N_E$, $C_R:P_E$ and $N_E:P_E$ metabolic quotients (atoms) for the whole zooplankton community.

	Present	Salp-Krill shift
C:N	16.7	11.4
C:P	33.8	51.3
N:P	2.0	4.5

*Present: This study. Salp-Krill shift: For the predicted case of the substitution of krill biomass by salps, as in **Table 5**.*

(2011), and Lehette et al. (2012). The metabolic rates of larval krill (furcilia) were very similar to those given by Frazer et al. (2002) and Meyer and Oettl (2005). Regarding salps, the estimated metabolic rates were higher by a factor of two than those given by Ikeda and Bruce (1986), Iguchi and Ikeda (2004), and Alcaraz et al. (1998), but lower than the data given by Ikeda and Mitchell (1982).

About eight decades ago Marshall et al. (1935) observed a decrease of zooplankton respiration during long incubation experiments in filtered seawater. The decrease is attributable to the combination of multiple factors derived from the experimental conditions like starvation, capture and manipulation stress, food quality and composition, animal crowding and container volume, etc. (Mayzaud, 1973; Ikeda, 1976, 1977; Checkley et al., 1992; Harris et al., 2000). Therefore, the short-time rate

of decrease of metabolic rates along successive measurements, when these are made as close as possible to the starting of the incubation, should allow to estimate metabolic rates at time = 0, considered to be equivalent to the "*in situ*" rates. Recently Ruiz-Halpern et al. (2011) and Lehette et al. (2012) obtained by this method biomass- specific ammonia excretion rates for Antarctic krill significantly higher than previous ones. But although the "*in situ*" rates given by both authors are similar, the rate of decrease given by the exponential model of Ruiz-Halpern et al. (2011) is about twice than that estimated by the potential equation of Lehette et al. (2012). The eventual underestimation of our metabolic rates (12–24 h incubation in filtered seawater) as compared with those of Ruiz-Halpern et al. (2011) would have been 32 and 54%, and from 17 to 40% as compared to Lehette et al. (2012).

In our case the decrease in oxygen concentration in experimental and control flasks displayed a linear trend indicative of a constant respiration rate, a response generally observed in crustaceans for substrate (O_2) concentrations above 70% (Alcaraz, 1974). As the factors responsible for the decrease of metabolic rates affect similarly all the metabolic processes (Mayzaud, 1973; Ikeda, 1976, 1977; Checkley et al., 1992; Harris et al., 2000), we assumed a similar linear trend for excretion rates as for respiration, and therefore the data on ammonia and phosphate excretion were not corrected. Another reason for not correcting the metabolic rates of krill for the possible effects of starvation (or any other laboratory conditions that could modify metabolic rates) was the lack of previous data of their effects on the metabolism of salps, therefore precluding the comparison of the metabolic rates of both zooplankton groups.

Ikeda and Mitchell (1982) and more recently Phillips et al. (2009) reported metabolic rates of *S. thompsoni* as high as those observed here, significantly higher than other groups of similar average individual C content. Large differences in the mass-specific metabolic rates of different zooplankton groups, aside from the effects of individual biomass could be due to the use of inadequate body mass conversion factors when normalizing the units in which the metabolism is expressed (Ikeda and Mitchell, 1982). Schneider (1990) comparing literature data on metabolic rates for crustacean and gelatinous zooplankton found biomass-specific ammonia excretion rates in crustaceans to be about one order of magnitude higher than for gelatinous zooplankton when expressed as per dry mass. However, when using organic C or N as biomass units, the metabolic rates of both groups were equivalent or higher for gelatinous organisms (Schneider, 1990). Differences in the degree of gut fullness (the C gut contents can make from 10 to 60% of their body C, Pakhomov et al., 2006) when salps C-contents is indirectly estimated by inadequate factors relating body carbon with salps dimensions are a complementary source of variability.

The negative relation between individual biomass and C-specific metabolic rates when all the zooplankton groups were included was significant and consistent with the non-similarity theory (Heusner, 1982; Riisgård, 1998). However, salps fell out of range, being clearly outliers. The different relationships between individual biomass (C contents) and C-specific respiration, ammonia and phosphate excretion rates as described by the exponents of the power equations were quite similar to those given by Ikeda and Mitchell (1982) and Ikeda (1985), as were the coefficients of the equations.

C:N:P METABOLIC STOICHIOMETRY

In general, the metabolic quotients were similar to previously reported values for most groups except for copepods (Ikeda and Mitchell, 1982). Average $C_E:N_E$ ratios were higher and $C_E:P_E$ and $N_E:P_E$ ratios lower than Redfield's ones by a factor of at least two, as previously reported in previous studies (see **Table 2**). The deviation of the metabolic ratios from the theoretical Redfield's seems to be general for high latitude zooplankton (Hirche, 1983). Higher than Redfield et al. (1963) $C_E:N_E$ ratios are indicative of the use of carbohydrates and/or lipids as metabolic substrate, of herbivorous feeding (Conover and Corner, 1968; Mayzaud, 1973), or of underestimating ammonia excretion (Ruiz-Halpern et al., 2011; Lehette et al., 2012). In the case of furcilia, aside from the above mentioned reasons the high $C_E:N_E$ values would be consequence of low N_E rates (Meyer and Oettl, 2005) due to metabolic N retention, characteristic of fast growing larval crustaceans (Elser et al., 1996). Contrarily, the $C_E:N_E$ quotient lower than 12 of salps could be due to relatively high N_E rates, indicating the use of N-rich metabolic substrate (Mayzaud and Conover, 1988) or to the differences in the slopes and intercepts of C-scaled specific respiration and excretion rates, similar to those of jellyfish, as discussed by Pitt et al. (2013). The differences in the relationships between individual biomass and specific respiration and ammonia and phosphate excretion rates explain the relation observed between individual body C and $C_E:N_E$, $C_E:P_E$ and $N_E:P_E$ metabolic quotients. It is particularly important in the case of the $N_E:P_E$ ratio, as the variance induced by individual biomass to the quotient can be up to 10% (Ikeda, 1985), that would be added to the effect taxonomic differences. The consequences of changes in the proportion of excreted ammonia and phosphate by zooplankton would be the modification of the N:P stoichiometry of nutrients available for phytoplankton (Elser et al., 1996; Sterner, 1986, 1990).

ZOOPLANKTON METABOLISM IN RELATION TO PHYTOPLANKTON BIOMASS AND PRIMARY PRODUCTION

The phytoplankton C concentration corresponded to what Hewes et al. (1990) qualifies of relatively low Chl a concentration waters. Total primary production (TPP) was in the range of previous estimations for the same area in late summer (from 1163 μmol C m^{-3} day^{-1}, Figueiras et al., 2001, to 2500 μmol C m^{-3} day^{-1}, Morán et al., 2001), but higher than the values observed by Basterretxea and Aristegui (1999) during late spring (558–930 μmol C m^{-3}). These differences, aside from the intrinsic inter annual variability, could be due to changes in the depth range considered in the estimates of primary production. Likely by the same logic the ratio particulate primary production/total primary production (PPP/TPP) was lower than the values given by Morán et al. (2001) corresponding to offshore waters of the same area.

During our cruise zooplankton required a very low percentage of both the phytoplankton standing stock and the particulate carbon produced by phytoplankton (PPP). Similar low impacts on phytoplankton standing stock and primary production in the

Southern Ocean by krill and salps grazing have been reported by Tanimura et al. (2008) with grazing impacts ranging from 0.1 to 1%, exceptionally up to 6% of phytoplankton C. During our study most of the phytoplankton C was required for crustaceans (copepods plus adult krill and furcilia), which needed 98% of the carbon necessary to balance the global respiratory C losses of total zooplankton, while salps required less than 1%. However, there was a radical difference when considering the C requirements of the zooplankton groups during our study as compared to the 1994 "salp year," when crustaceans required only 14% of the carbon allocable to total zooplankton respiratory losses and the remaining 86% corresponded to salps (Alcaraz et al., 1998; Perissinotto and Pakhomov, 1998).

The average supply of ammonia by zooplankton to the N required by phytoplankton for TPP was lower than previous data for a similar area and season of the year (Alcaraz et al., 1998). By groups, krill contributed to more than 70% of the total ammonia excreted, while salps provided only 1.2%. This contrasts with the conditions found during 1994 (a "salp year"), when zooplankton excretion provided up to 7.3% of the N and P required by phytoplankton (Alcaraz et al., 1998), salps alone accounting for 96% of the nutrients excreted. During our study, the total phosphate excreted provided almost 10% of the phytoplankton requirements, again with krill as the main contributors and salps providing less than 0.5% of the P required for TPP. Both the N and P supplied could be roughly 43% higher if theoretical "in situ" metabolic rates had been estimated according to the methods of Ruiz-Halpern et al. (2011) and Lehette et al. (2012), but as discussed above the linear trend in O_2 consumption suggested similarly constant excretion rates, and therefore the theoretical "in situ" rates were not calculated.

ZOOPLANKTON SHIFTS, CARBON CYCLING AND NUTRIENT STOICHIOMETRY

The fraction of PPP required to compensate for the respiratory losses of zooplankton, aside from being an estimator of the relative importance of classical, herbivorous food webs in marine ecosystems (Calbet et al., 1996) is also related to the trophic efficiency of the system (Alcaraz, 1988; Alcaraz et al., 1994) and equivalent to the reciprocal of the quotient Production/Respiration (P/R), considered as a descriptor of the ecosystem's entropy when the respiration of the whole ecosystems is taken into consideration (Odum, 1956; Margalef, 1974).

Assuming the zooplankton shift from krill to salps will lead to a zooplankton community composition equivalent to that of a typical salp year (average salps/krill = 10, Huntley et al., 1989; Loeb et al., 1997, 2010; Alcaraz et al., 1998) the proportion of PPP necessary to compensate for the C respiratory losses of total zooplankton will increase by a factor of 9. At the same time the importance of the so-called regenerative plankton loop will proportionally decrease (Parsons and Lalli, 1988; Miller et al., 1991).

In a future salps-dominated Southern Ocean around half of total primary production (TPP) and roughly 100% of particulate primary production (PPP) will be necessary to compensate for the respiratory zooplankton losses, and near 50% of it will be packed into large, fast sinking fecal pellets (Pakhomov et al.,

2006), thus intensifying the rate of vertical carbon flux to deep waters and increasing the turnover time of biogenic carbon (i.e., from short-lived to long-lived and sequestered, Fortier et al., 1994; Le Fèvre et al., 1998).

Quantitative and qualitative changes in the nutrient environment for phytoplankton due to the climate shift in the Southern Ocean (Montes-Hugo et al., 2009), the lack of metabolic N:P homeostasis (Alcaraz et al., 2013) and the differential regeneration rates of N and P in relation to the food web structure (Elser et al., 1988, 1996) are, besides grazing, prime factors of change in the competitive relations for phytoplankton communities. Although the most direct effects of the predicted shifts in Southern Ocean zooplankton will derive from the different quality as food of krill and salps, the higher specific carbon demand due to the new zooplankton community structure, coupled with higher vertical carbon flux via salps fecal pellets, will strongly affect the characteristics of carbon cycling. At the same time, the increased rate of nutrient re-supply by zooplankton will increase. Coupled with the new $N_E:P_E$ metabolic quotient, the N:P quotient of the nutrient pool will rise by a factor of two, a change that could contribute to modify the structure and function of primary producers (Sterner, 1986, 1990) accelerating the changes already observed in the community structure of phytoplankton by Montes-Hugo et al. (2009).

As a conclusion, a major change in the relative proportion of krill and salps in the Southern Ocean can induce significant variations in the marine food webs. Aside from the direct trophic effects, other major changes will take place via the increase of the metabolic carbon requirements of zooplankton and its vertical export, the higher overall contribution of zooplankton excretion to the nutrient requirements by phytoplankton, and the increasing N:P ratios of the recycled nutrients. Moreover, the zooplankton shift will result in a decrease of the P/R quotient (Odum, 1956; Margalef, 1974), indicator of a potential increase of the trophic efficiency of the system, while paradoxically the regeneration processes in surface waters will decrease.

ACKNOWLEDGMENTS

The authors wish to express their gratitude to the crew of the R/V Hespérides, the technicians of the UTM and two unknown reviewers whose comments contributed significantly to clarify and improve the paper. This work was supported by the Spanish funded projects ATOS (POL 2006-0550/CTM) to Carlos M. Duarte, PERFIL (CTM 2006-12344-C01) to Miquel Alcaraz, and the UE funded project ATP (www.eu-atp.org) contract # 226248 to P. Wassmann.

REFERENCES

Alcaraz, M. (1974). Respiración en crustáceos: influencia de la concentración de oxígeno en el medio. *Inv. Pesq.* 38, 397–411.

Alcaraz, M. (1988). Summer zooplankton metabolism and its relation to primary production in the western Mediterranean. *Oceanol. Acta* 9, 185–191.

Alcaraz, M., Almeda, R., Calbet, A., Saiz, E., Duarte, C. M., Lasternas, S., et al. (2010). The role of arctic zooplankton in biogeochemical cycles: respiration and excretion of ammonia and phosphate during summer. *Polar Biol.* 33, 1719–1731. doi: 10.1007/s00300-010-0789-9

Alcaraz, M., Almeda, R., Saiz, E., Calbet, A., Duarte, C. M., Agustí, S., et al. (2013). Effects of temperature on the metabolic stoichiometry of Arctic zooplankton. *Biogeosciences* 10, 689–697. doi: 10.5194/bg-10-689-2013

Alcaraz, M., Felipe, J., Grote, U., Arashkevich, E., and Nikishina, A. (2014). Life in a warming ocean: thermal thresholds and metabolic balance of Arctic zooplankton. *J. Plankton Res.* 36, 3–10. doi: 10.1093/plankt/fbt111

Alcaraz, M., Saiz, E., Calbet, A., Fernandez, J. A., Trepat, I., and Broglio, E. (2003). Estimating zooplankton biomass through image analysis. *Mar. Biol.* 143, 307–315. doi: 10.1007/s00227-003-1094-8

Alcaraz, M., Saiz, E., and Estrada, M. (1994). Excretion of ammonia by zooplankton and its potential contriobution to nitrogen requirements for primary production in the Catalan Sea (NW Mediterranean). *Mar. Biol.* 119, 69–76. doi: 10.1007/BF00350108

Alcaraz, M., Saiz, E., Fernandez, J. A., Trepat, I., Figueiras, F., Calbet, A., et al. (1998). Antarctic zooplankton metabolism: carbon requirements and ammonium excretion of salps and crustacean zooplankton in the vicinity of the Bransfield strait during January 1994. *J. Mar. Syst.* 17, 347–259. doi: 10.1016/S0924-7963(98)00048-7

Almeda, R., Alcaraz, M., Calbet, A., and Saiz, E. (2011). Metabolic rates and carbon budget of early developmental stages of the marine cyclopoid copepod *Oithona davisae*. *Limnol. Oceanogr.* 56, 403–414. doi: 10.4319/lo.2011.56.1.0403

Atkinson, A., Meyer, B., Stübing, D., Hagen, W., Schmidt, K., and Bathmann, U. (2002). Feeding and energy budgets of Antarctic krill *Euphausia superba* at the onset of winter- II. Juveniles and adults. *Limnol. Oceanogr.* 47, 953–966. doi: 10.4319/lo.2002.47.4.0953

Atkinson, A., Siegel, V., Pakhomov, E., and Rothery, P. (2004). Long-term decline in krill stock and increase in salps within the Southern Ocean. *Nature* 432, 1000–1003. doi: 10.1038/nature02996

Auerswald, L., Pape, C., Stübing, D., Lopata, A., and Meyer, B. (2009). Effect of short-term starvation of adult Antarctic krill, *Euphausia superba*, at the onset of summer. *J. Exp. Mar. Biol. Ecol.* 381, 47–56. doi: 10.1016/j.jembe.2009.09.011

Basterretxea, G., and Aristegui, J. (1999). Phytoplankton biomass and production during late spring (1991) and summer (1993) in the Bransfield Strait. *Polar Biol.* 21, 11–22. doi: 10.1007/s003000050328

Calbet, A., Alcaraz, M., Saiz, E., Estrada, M., and Trepat, I. (1996). Planktonic herbivorous food webs in the Catalan Sea (NW Mediterranean): temporal variability and comparison of indices of phytoplankton-zooplankton coupling based in state variables and rate processes. *J. Planklton Res.* 18, 2329–2347. doi: 10.1093/plankt/18.12.2329

Checkley, D. M., Dagg, M. J., and Uye, S. (1992). Feeding, excretion and egg production by individual and populations of the marine copepods *Acartia* spp. and *Centropages furcatus*. *J. Plankton Res.* 14, 71–96. doi: 10.1093/plankt/14.1.71

Conover, R. J., and Corner, E. D. S. (1968). Respiration and nitrogen excretion by some marine zooplankton in relation to their life cycles. *J. Mar. Biol. Assoc. U.K.* 48, 49–75. doi: 10.1017/S0025315400032410

Constable, A. J., de la Mare, W., Agnew, D. J., Everson, I., and Miller, D. (2000). Managing fisheries to conserve the Antarctic marine ecosystem: practical implementation of the convention on the conservation of Antarctic Marine Resources (CCAMLR). *ICES J. Mar. Sci.* 57, 778–791. doi: 10.1006/jmsc.2000.0725

Constable, A. J., Melbourne-Thomas, J., Corney, S. P., Arrigo, K. R., Barbraud, C., Barnes, D., et al. (2014). Climate change and Southern Ocean ecosystems I: how changes in physical habitats directly affect marine biota. *Global Change Biol.* 20, 3004–3025. doi: 10.1111/gcb.12623

Dagg, M. J., Urban-Rich, J., and Peterson, J. O. (2003). The potential contribution of fecal pellets from large copepods to the flux of biogenic silica and particulate organic carbon in the Antarctic Polar Front region near 170° W. *Deep Sea Res. II* 50, 675–691. doi: 10.1016/S0967-0645(02)00590-8

Duarte, C. M. (ed.) (2008). *Impacts of Global Warming on Polar Ecosystems*. Bilbao: Fundación BBVA.

Duarte, C. M., Agustí, S., Wasmann, P., Arrieta, J. M., Alcaraz, M., Coello, A., et al. (2012). Tipping elements in the Arctic marine ecosystem. *Ambio* 41, 44–55. doi: 10.1007/s13280-011-0224-7

Dubischar, C. D., Pakhomov, E. A., von Harbou, L., Hunt, B. P. V., and Bathmann, U. V. (2012). Salps in the Lazarev Sea, Southern Ocean: II. Biochemical composition and potential prey value. *Mar. Biol.* 159, 15–24. doi: 10.1007/s00227-011-1785-5

Ducklow, H., Clarke, A., Dickhut, R., Doney, S. C., Geisz, H., Huang, K., et al. (2012). "The marine system of the western Antarctic Peninsula," in *Antarctic Ecosystems: An Extreme Environment in a Changing World*, eds A. D. Rogers, N. M. Johnston, E. J. Murphy and A. Clarke (Oxford: Blackwell Publishing Ltd.), 121–159. doi: 10.1002/9781444347241.ch5

Elser, J. J., Dobberfuhl, D. R., MacKay, N. A., and Schampel, J. H. (1996). Organism size, life history, and N:P stoichiometry. *BioScience* 46, 674–684. doi: 10.2307/1312897

Elser, J. J., Elser, M. M., MacKay, N. A., and Carpenter, S. R. (1988). Zooplankton-mediated transitions between N and P limited algal growth. *Limnol. Oceanogr.* 33, 1–14. doi: 10.4319/lo.1988.33.1.0001

Fernandes, J. A., Irigoien, X., Boyra, G., Lozano, J. A., and Albaina, A. (2009). Optimizing the number of classes in automated zooplankton classification. *J. Plankton Res.* 31, 19–29. doi: 10.1093/plankt/fbn098

Figueiras, F. G., Pérez, F. F., Pazos, Y., and Rios, A. F. (2001). Dissolved and particulate primary production and bacterial production in offshore Antarctic waters during austral summer: coupled or uncoupled? *Mar. Ecol. Prog. Ser.* 222, 25–39. doi: 10.3354/meps222025

Fortier, L., Le Fèvre, J., and Legendre, L. (1994). Export of biogenic carbon to fish and to the deep ocean: the role of large planktonic microphages. *J. Plankton Res.* 16, 809–839. doi: 10.1093/plankt/16.7.809

Frazer, T. K., Quetin, L. B., and Ross, R. M. (2002). Energetic demands of larval krill, *Euphausia superba*, in winter. *J. Exp. Mar. Biol. Ecol.* 277, 157–171. doi: 10.1016/S0022-0981(02)00328-3

Grasshoff, K., Kremling, K., and Ehrhardt, M. (1999). *Methods of Seawater Analysis.* Weinheim: Wiley-VCH.

Harris, R. P., Wiebe, P. H., Lenz, J., Skjoldal, H. R., and Huntley, M. (2000). *Zooplankton Methodology Manual.* London: Academic Press.

Heusner, A. A. (1982). Energy metabolism and body size. I. Is the 0.75 mass exponent of Kleiber's equation a statistical artifact? *Respir Physiol.* 48, 1–12. doi: 10.1016/0034-5687(82)90046-9

Hewes, C. D., Sakshaug, E., Reid, F. M. H., and Holm-Hansen, O. (1990). Microbial autotrophic and heterotrophic eucaryotes in Antarctic waters: relationships between biomass and chlorophyll, adenosine triphosphate and particulate organic carbon. *Mar. Ecol. Prog. Ser.* 63, 27–35. doi: 10.3354/meps063027

Hirche, H. J. (1983). Excretion and respiration of the Antarctic krill *Euphausia superba*. *Polar Biol.* 1, 205–209. doi: 10.1007/BF00443189

Huntley, M. E., Sykes, P. F., and Martin, V. (1989). Biometry and trophodynamics of *Salpa thompsoni* Foxton (Tunicata: Thaliacea) near the Antarctic Peninsula in austral summer, 1983-1984. *Polar Biol.* 10, 59–70. doi: 10.1007/BF00238291

Iguchi, N., and Ikeda, T. (2004). Metabolism and elemental composition of aggregate and solitary forms of *Salpa thompsoni* (Tunicata: Thaliacea) in waters off the Antarctic Peninsula during austral summer 1999. *J. Plankton Res.* 26, 1025–1037. doi: 10.1093/plankt/fbh093

Ikeda, T. (1976). The effect of laboratory conditions on the extrapolation of experimental measurements to the ecology of marine zooplankton. I. Effect of feeding conditions on the respiration rate. *Bull. Plankton Soc. Jpn.* 23, 1–10.

Ikeda, T. (1977). The effect of laboratory conditions on the extrapolation of experimental measurements to the ecology of marine zooplankton. IV. Changes in respiration and excretion rates of boreal zooplankton species maintained under fed and starved conditions. *Mar. Biol.* 41, 241–252. doi: 10.1007/BF00394910

Ikeda, T. (1985). Metabolic rates of epipelagic marine zooplankton as a function of body mass and temperature. *Mar. Biol.* 85, 1–12. doi: 10.1007/BF00396409

Ikeda, T., and Bruce, B. (1986). Metabolic activity and elemental composition of krill and other zooplankton from Prydz Bay, Antarctica, during early summer (November-December). *Mar. Biol.* 92, 545–555. doi: 10.1007/BF00392514

Ikeda, T., and Mitchell, A. W. (1982). Oxygen uptake, ammonia excretion and phosphate excretion by krill and other Antarctic zooplankton in relation to their body size and chemical composition. *Mar. Biol.* 71, 283–289. doi: 10.1007/BF00397045

Kéruel, R., and Aminot, A. (1997). Fluorometric determination of ammonia in sea and estuarine waters by direct segmented flow analysis. *Mar. Chem.* 57, 265–275. doi: 10.1016/S0304-4203(97)00040-6

Le Fèvre, J., Legendre, L., and Rivkin, R. B. (1998). Fluxes of biogenic carbon in the Southern Ocean: roles of large microphagous zooplankton. *J. Mar. Syst.* 17, 325–345. doi: 10.1016/S0924-7963(98)00047-5

Lehette, P., Tovar-Sánchez, A., Duarte, C. M., and Hernández-Leon, S. (2012). Krill excretion and its effects on primary production. *Mar. Ecol. Prog. Ser.* 459, 29–38. doi: 10.3354/meps09746

Loeb, V., Hofmann, E. E., Klinck, J. M., and Holm-Hansen, O. (2010). Hydrographic control of the marine ecosystem in the South Shetland-Elephant island and Bransfield strait region. *Deep-Sea Res. II* 57, 519–542. doi: 10.1016/j.dsr2.2009.10.004

Loeb, V., Siegel, V., Holm-Hansen, O., Hewitt, R., Fraser, W., Trivelpiece, W., et al. (1997). Effects of sea-ice extent and krill or salp dominance on the Antarctic food web. *Nature* 387, 897–900. doi: 10.1038/43174

Margalef, R. (1974). *Ecologia.* Barcelona: Omega.

Marshall, S. M., Nicholls, A. G., and Orr, A. P. (1935). On the biology of Calanus finmarchicus VI. Oxygen consumption in relation to environmental conditions. *J. Mar. Biol. Ass. U.K.* 20, 341–346. doi: 10.1017/S0025315400009991

Mayzaud, P. (1973). Respiration and nitrogen excretion of zooplankton. II. Studies on the metabolic characteristics of starved animals. *Mar. Biol.* 21, 19–28. doi: 10.1007/BF00351188

Mayzaud, P., and Conover, R. J. (1988). O:N atomic ratio as a tool to describe zooplankton metabolism. *Mar. Ecol. Prog. Ser.* 45, 289–302. doi: 10.3354/meps045289

Meyer, B., Auerswald, L., Siegel, V., Spahic, S., Pape, C., Fach, B. A., et al. (2010). Seasonal variation in body composition, metabolic activity, feeding, and growth of adult krill *Euphausia superba* in the Lazarev Sea. *Mar. Ecol. Prog. Ser.* 398, 1–18. doi: 10.3354/meps08371

Meyer, B., Fuentes, V., Guerra, C., and Schmidt, K., And others (2009). Physiology, growth and development of larval krill Euphausia superba in autumn and Winter in the Lazarev Sea, Antarctica. *Limnol. Oceanogr.* 54, 1595–1614. doi: 10.4319/lo.2009.54.5.1595

Meyer, B., and Oettl, B. (2005). Effects of short-term starvation on composition and metabolism of larval Antarctic krill *Euphausia superba*. *Mar. Ecol. Prog. Ser.* 292, 263–270. doi: 10.3354/meps292263

Miller, C. B., Frost, B. W., Booth, B., Wheeler, P. A., Landry, M. R., and Welschmeyer, N. (1991). Ecological processes in the subarctic Pacific: iron limitation cannot be the whole story. *Oceanography,* 4, 71–78. doi: 10.5670/oceanog.1991.05

Montes-Hugo, M., Doney, S. C., Ducklow, H., Fraser, W., Martinson, D., Stammerjoh, S. E., et al. (2009). Recent changes in phytoplankton communities associated with rapid regional climate change along the Western Antarctic Peninsula. *Science* 323, 1470–1473. doi: 10.1126/science.1164533

Morán, X. A., Gasol, J. M., Pedrós-Alió, C., and Estrada, M. (2001). Dissolved and particulate primary production and bacterial production in offshore Antarctic waters during austral summer: coupled or uncoupled? *Mar. Ecol. Prog. Ser.* 222, 25–39. doi: 10.3354/meps222025

Murphy, E. J., Watkins, J. L., Trathan, P. N., Reid, K., Meredith, M. P., Thorpe, S. E., et al. (2007). Spatial and temporal operation of the Scotia Sea ecosystem: a review of large-scale links in a centered food web. *Phyl. Trans. R. Soc. B* 362, 1213–1148.

Nishikawa, J., Naganobu, M., and Ichii, T. (1995). Distribution of salps near the South Shetland Islands during austral summer, 1990-1991 with special reference to krill distribution. *Polar Biol.* 15, 31–39. doi: 10.1007/BF00236121

Odum, H. T. (1956). Primary production in flowing waters. *Limnol. Oceanogr.* 1, 102-117. doi: 10.4319/lo.1956.1.2.0102

Omori, M. (1978). Zooplankton fisheries of the world: a review. *Mar. Biol.* 48, 199–205. doi: 10.1007/BF00397145

Omori, M., and Ikeda, T. (1984). *Methods in Zooplankton Ecology.* Cambridge: John Wiley.

Pakhomov, E. A. (2004). Salp/krill interactions in the eastern Atlantic sector of the Southern Ocean. *Deep Sea Res. II* 51, 2645–2660. doi: 10.1016/j.dsr2.2001.03.001

Pakhomov, E. A., Dubischar, C. D., Strass, V., Brichta, M., and Bathmann, U. V. (2006). The tunicate *Salpa thompsoni* ecology in the Southern Ocean. I. Distribution, biomass, demography and feeding ecophysiology. *Mar. Biol.* 149, 609–623. doi: 10.1007/s00227-005-0225-9

Pakhomov, E. A., Froneman, P. W., and Perissinotto, R. (2002). Salp/krill interactions in the Southern Ocean: spatial segregation and implications for the carbon flux. *Deep Sea Res. II* 49, 1881–1907. doi: 10.1016/S0967-0645(02)00017-6

Parsons, T. R., and Lalli, C. M. (1988). Comparative oceanic ecology of the plankton communities of the subarctic Atlantic and Pacific Oceans. *Mar. Biol. A. Rev.* 26, 317–359.

Parsons, T. R., Maita, Y., and Lalli, C.M. (1984). *A Manual of Chemical and Biological Methods for Seawater Analysis.* Cambridge: Pergamonn Press.

Pauly, T., Nicol, S., Higginbottom, I., Hosie, G., and Kitchener, J. (2000). Distribution and abundance of Antarctic krill (*Euphausia superba*) off East Antarctica (80-150° E) during the Austral summer of 1995-1996. *Deep Sea Res. II* 47, 2465–2488. doi: 10.1016/S0967-0645(00)00032-1

Perissinotto, R., and Pakhomov, E. A. (1998). The trophic role of the tunicate *Salpa thompsoni* in the Antarctic marine ecosystem. *J. Mar. Syst.* 17, 361–374. doi: 10.1016/S0924-7963(98)00049-9

Phillips, B., Kremer, P., and Madin, L. P. (2009). Defecation by Salpa thompsoni and its contribution to vertical flux in the Southern Ocean. *Mar. Biol.* 156, 455–467. doi: 10.1007/s00227-008-1099-4

Pitt, K. A., Duarte, C. M., Lucas, C. H., Sutherland, K. R., Condon, R. H., Mianzan, H. et al. (2013). Jellyfish body plans provide allometric advantages beyond low carbon content. *PLoS ONE* 8:e72683. doi: 10.1371/journal.pone.0072683

Pond, D. W., Priddle, J., Sargent, J. R., and Watkins, J. L. (1995). Laboratory studies of assimilation and egestion of algal lipid by Antarctic krill- methods and initial results. *J. Exp. Mar. Biol. Ecol.* 187, 253–268. doi: 10.1016/0022-0981(94)00187-I

Redfield, A. C., Ketchum, B. H., and Richards, F. A. (1963). "The influence of organisms in the composition of seawater," in The Sea, ed M. N. Hill (New York, NY: Interscience), 26–76.

Riisgård, H. U. (1998). No foundation of a "3/4 power scaling law" for respiration in biology. *Ecol. Lett.* 1, 71–73. doi: 10.1046/j.1461-0248.1998.00020.x

Ross, R., Quetin, L. B., Martinson, D. G., Iannuzzi, R. A., Stammerjohn, S. E., and Smith, R. C. (2008). Palmer LTR: Patterns of distribution of five dominant zooplankton species in the epipelagic zone west of the Antarctic Peninsula, 1993-2004. *Deep Sea Res. II* 55, 2086–2105. doi: 10.1016/j.dsr2.2008.04.037

Ruiz-Halpern, S., Duarte, C. M., Tovar-Sanchez, A., Pastor, M., Hortskotte, B., Lasternas, S., et al. (2011). Antarctic krill as a source of dissolved organic carbon to the Antarctic ecosystem. *Limnol. Oceanogr,* 56, 521–528. doi: 10.4319/lo.2011.56.2.0521

Saiz, E., Calbet, A., Isari, S., Antó, M., Velasco, E. M., Almeda, R., et al. (2013). Zooplankton distribution and feeding in the Arctic Ocean during a *Phaeocystis pouchetii* Bloom. *Deep Sea Res I* 72, 17–33. doi: 10.1016/j.dsr.2012.10.003

Sameoto, D., Wiebe, P., Runge, L., Postel, L., Dunn, J., Miller, C., et al. (2000). "Collecting zooplankton," in *Zooplankton Methodology Manual,* ed R. P. Harris, P. H. Wiebe, J. Lenz, H. R. Skjoldal, and M. Huntley (London: Academic Press), 55–81. doi: 10.1016/B978-012327645-2/50004-9

Schneider, G. (1990). A comparison of carbon based ammonia excretion rates between gelatinous and non-gelatinous zooplankton: implications and consequences. *Mar. Biol.* 106, 219–225. doi: 10.1007/BF01314803

Smetacek, V. (2008). "Are declining Antarctic krill stocks a result of a global warming or the decimation of whales?" in *Impacts of Global Warming on Polar Ecosystems,* ed C. Duarte (Bilbao: Fundación BBVA), 45–81.

Smetacek, V., and Nicol, S. (2005). Polar ocean ecosystems in a changing world. *Nature* 437, 362–368. doi: 10.1038/nature04161

Steemann-Nielsen, E. J. (1952). The use of radioactive carbon (^{14}C) for measuring organic production in the sea. *Cons. Perm. Int. Explor. Mer.* 18, 117–140. doi: 10.1093/icesjms/18.2.117

Sterner, R. W. (1986). Herbivores' direct and indirect effect on algal populations. *Science* 231, 605–607. doi: 10.1126/science.231.4738.605

Sterner, R. W. (1990). The ratio of nitrogen to phosphorus resupplied by herbivores: Zooplankton and the algal arena. *Am. Nat.* 136, 209–229. doi: 10.1086/285092

Tanimura, A., Kawaguchi, S., Oka, N., Nishkawa, J., Toczko, S., Takahashi, K. T., et al. (2008). Abundance and grazing impacts of krill, salps and copepods along the 140° E meridian in the Southern Ocean during summer. *Antarct. Sci.* 20, 365–379. doi: 10.1017/S0954102008000928

Tovar-Sanchez, A. C., Duarte, C. M., Hernández-León, S., and Sañudo Wilhemy, S. A. (2007). Krill as a central node for iron cycling in the Southern Ocean. *Geophys. Res. Lett.* 34, L11601, doi: 10.1029/2006GL029096.

Ward, P., Atkinson, A., Murray, A. W. A., Wood, A. G., Williams, R., and Poulet, S. A. (1995). The summer zooplankton community at South Georgia - biomass, vertical migration and grazing. *Polar Biol.* 15, 195–208. doi: 10.1007/BF00239059

Ward, P., Grant, S., Brandon, M., Siegel, V., Sushin, V., Loeb, V., et al. (2004). Mesozooplankton community structure in the Scotia Sea during the CCAMLR 2000 survey: January-February 2000. *Deep Sea Res. II* 51, 1351–1367. doi: 10.1016/j.dsr2.2004.06.016

Wassmann, P., Carroll, J., and Bellerby, R. G. J. (2008). Carbon flux and ecosystem feedback in the northern Barents Sea in an era of climate change. *Deep-Sea Res. PT II,* 55, 2143–2153. doi: 10.1016/j.dsr2.2008.05.025

Wickham, S. A., and Berninger, U. K. (2007). Krill larvae, copepods and the microbial food web: interaction during the Antarc4tic fall. *Aquat. Microbial Ecol.* 46, 1–13 doi: 10.3354/ame046001

Conflict of Interest Statement: The authors declare that the research was conducted in the absence of any commercial or financial relationships that could be construed as a potential conflict of interest.

9

Impact of ocean acidification and warming on the Mediterranean mussel (*Mytilus galloprovincialis*)

Frédéric Gazeau[1,2]*, Samir Alliouane[1,2], Christian Bock[3], Lorenzo Bramanti[4,5], Matthias López Correa[6,7], Miriam Gentile[8], Timo Hirse[3], Hans-Otto Pörtner[3] and Patrizia Ziveri[9,10]

[1] Sorbonne Universités, Université Pierre et Marie Curie Univ Paris 06, Unité Mixte de Recherche 7093, Laboratoire d'Océanographie de Villefranche, Villefranche/Mer, France
[2] Centre National de la Recherche Scientifique, Unité Mixte de Recherche 7093, Laboratoire d'Océanographie de Villefranche, Observatoire Océanologique, Villefranche/Mer, France
[3] Alfred-Wegener-Institute Helmholtz Zentrum für Polar und Meeresforschung, Am Handelshafen 12, D-27570 Bremerhaven, Germany
[4] Sorbonne Universités, Université Pierre et Marie Curie Univ Paris 06, Unité Mixte de Recherche 8222, Laboratorio de Ecofisiología para la Conservación de Bosques, Observatoire Océanologique, Banyuls/mer, France
[5] Centre National de la Recherche Scientifique, Unité Mixte de Recherche 8222, Laboratorio de Ecofisiología para la Conservación de Bosques, Observatoire Océanologique, Banyuls/mer, France
[6] GeoZentrum Nordbayern, Universität Erlangen-Nürnberg, Erlangen, Germany
[7] German University of Technology in Oman, Halban Campus, Muscat, Sultanate of Oman
[8] Consejo Superior de Investigaciones Científicas, Institut de Ciencies del Mar, Barcelona, Spain
[9] Universitat Autònoma de Barcelona, Institut de Ciència i Tecnologia Ambientals, Barcelona, Spain
[10] Institució Catalana de Recerca i Estudis Avançats, Barcelona, Spain

Edited by:
Robinson W. Fulweiler, Boston University, USA

Reviewed by:
Claudia R. Benitez-Nelson, University of South Carolina, USA
Emily Carrington, University of Washington, USA

*Correspondence:
Frédéric Gazeau, Laboratoire d'Océanographie de Villefranche, Centre National de la Recherche Scientifique - Université Pierre et Marie Curie and Unité Mixte de Recherche 7093, 06230 Villefranche-sur-mer, France
e-mail: f.gazeau@obs-vlfr.fr

In order to assess the effects of ocean acidification and warming on the Mediterranean mussel (*Mytilus galloprovincialis*), specimens were reared in aquarium tanks and exposed to elevated conditions of temperature ($+3°C$) and acidity (-0.3 pH units) for a period of 10 months. The whole system comprised a factorial experimental design with 4 treatments (3 aquaria per treatment): control, lowered pH, elevated temperature, and lowered pH/elevated temperature. Mortality was estimated on a weekly basis and every 2 months, various biometrical parameters and physiological processes were measured: somatic and shell growth, metabolic rates and body fluid acid-base parameters. Mussels were highly sensitive to warming, with 100% mortality observed under elevated temperature at the end of our experiment in October. Mortality rates increased drastically in summer, when water temperature exceeded 25°C. In contrast, our results suggest that survival of this species will not be affected by a pH decrease of ~0.3 in the Mediterranean Sea. Somatic and shell growth did not appear very sensitive to ocean acidification and warming during most of the experiment, but were reduced, after summer, in the lowered pH treatment. This was consistent with measured shell net dissolution and observed loss of periostracum, as well as uncompensated extracellular acidosis in the lowered pH treatment indicating a progressive insufficiency in acid-base regulation capacity. However, based on the present dataset, we cannot elucidate if these decreases in growth and regulation capacities after summer are a consequence of lower pH levels during that period or a consequence of a combined effect of acidification and warming. To summarize, while ocean acidification will potentially contribute to lower growth rates, especially in summer when mussels are exposed to sub-optimal conditions, ocean warming will likely pose more serious threats to Mediterranean mussels in this region in the coming decades.

Keywords: ocean acidification, ocean warming, Mediterranean mussels, growth, survival, Mediterranean Sea

INTRODUCTION

During the last 150 years, human activities, through the combustion of fossil fuels (oil, gas, and coal), have led to a dramatic release of carbon dioxide (CO_2) to the Earth's atmosphere. The accumulation of CO_2 impacts the radiative forcing, thereby warming the atmosphere and the ocean. The surface ocean has warmed from 1971 to 2010 by 0.11°C per decade (Rhein et al., 2013), with maximal rates recorded on average in the coastal zone (0.18°C; Lima and Wethey, 2012). Depending on the future emission scenario, surface ocean temperatures are projected to warm in the top 100 m by about 0.6 to 2.0°C by 2100 (Collins et al., 2013). The oceans are not only absorbing a large amount of increased heat, but also about 25% of anthropogenic CO_2 emissions (Le Quéré et al., 2009). This massive CO_2 input greatly impacts seawater chemistry, leaving a surface ocean imprint. These changes are referred to as "ocean acidification" (OA) because increased CO_2 lowers seawater pH (i.e., increases its acidity). The pH in ocean surface waters has already decreased by 0.1 units since the beginning of the industrial era, equivalent to an increased acidity of 26%. According to recent projections, an additional decrease is expected by 2100, ranging from 0.06 to 0.32 units, equivalent to an increased acidity of 15–110%, depending on the considered emission scenario (Ciais et al., 2013).

The effects of OA on marine organisms have been studied over the past 20 years (Gattuso and Hansson, 2011), with particular attention to organisms producing calcareous structures. Indeed, while decreasing pH levels are expected to have profound impacts on the physiology and metabolism of marine organisms through a disruption of intercellular transport mechanisms (Pörtner et al., 2004), the seawater pH decrease will also lead to a decrease in the concentration of carbonate ions (CO_3^{2-}), one of the building blocks of calcium carbonate ($CaCO_3$), and likely alter the ability of calcifying organisms to precipitate $CaCO_3$ (Gazeau et al., 2007). Being both ecologically and economically important species in the coastal zone, the body of literature on mollusks has grown substantially over recent years (see comprehensive review from Gazeau et al., 2013).

To date, very few studies have considered the impacts of OA in synergy with other environmental stressors such as warming, with contrasting results. Moreover, all these studies considered stable pH and/or temperature conditions although large daily and/or seasonal variations in these parameters are common features of coastal sites (Hofmann et al., 2011). For instance, Mackenzie et al. (2014) reported that Atlantic mussels (*Mytilus edulis*) were not impacted by a decrease of pH by 0.4 units, while these organisms were highly sensitive to a temperature increase of 4°C. In contrast, Duarte et al. (2014) found that the Chilean mussel (*Mytilus chilensis*) was able to tolerate an increase in temperature of 4°C for 2 months but that shell growth was significantly affected by a decrease in pH of ~0.2–0.3 units. Additive effects of these two stressors have been highlighted in several studies (Talmage and Gobler, 2011; Hiebenthal et al., 2013) although with large variations between species and even developmental stages (Talmage and Gobler, 2011). Finally, Watson et al. (2012) and Ivanina et al. (2013) reported synergistic effects of these two stressors on the survival of the fluted giant clam (*Tridacna squamosa*), and on the shell hardness of Eastern oysters (*Crassostrea virginica*), and hard clams (*Mercenaria mercenaria*). Interestingly, Ivanina et al. (2013) showed that mortality of oysters was ameliorated under a warmer and more acidified scenario as compared to a warmer scenario only.

The Mediterranean mussel (*Mytilus galloprovincialis*) is one of the most cultivated bivalve species with a global production that has drastically increased in the last 50 years to reach a value of around 1 million tons in 2011, with China and Spain representing the most important producers: ~700,000 and ~200,000 tons, respectively (FAO, 2014). In the Mediterranean Sea, this bivalve is the third-most cultivated species in terms of production, after Seabass and Seabreams. The production is dominated by Italy and Greece with values in 2011 of ~65,000 and ~20,000 tons, respectively. The Spanish production occurs predominantly on the Atlantic coast (i.e., 98%).

In the northwestern Mediterranean Sea, mean maximum summer temperatures have increased by about 1°C between 2002 and 2010 relative to the 1980–2000 average (Marba and Duarte, 2010) and a rapid warming of 2.8 ± 1.1°C is expected by the end of the century (Jorda et al., 2012). In this region, high summer temperature levels already represent a problem for the rearing

of mussels and farmers are obliged to sell their product before summer (Ramón et al., 2005). Moreover, the observed mean rate of OA is similar to that in Atlantic subtropical regions (Orr, 2011), with a 0.05–0.14 decrease in seawater pH since the pre-industrial period (Touratier and Goyet, 2011) and projected to decrease by 0.3 units by the end of the century (Orr, 2011). In addition, Mediterranean coastal marine ecosystems are experiencing the synergistic effects of multiple climatic and non-climatic anthropogenic stressors such as chemical contaminants (The Mermex Group, 2011). In order to assess the potential combined effects of OA and warming on the Mediterranean mussel, we performed laboratory experiments in which mussels were exposed to elevated temperature (+3°C) and acidity (−0.3 pH units) in aquaria flushed with ambient seawater and thus against the background of natural variability for a period of 10 months.

MATERIAL AND METHODS
BIOLOGICAL MATERIAL AND EXPERIMENTAL SET-UP

Mussels with a mean shell length of 45 ± 5 mm were collected in the Delta del Ebro (40°35′59″N; 0°41′21″E) in November 2011 and directly brought to the Institute of Marine Science (Barcelona, Spain). Seawater, pumped from a depth of 10 m at proximity of the institute (300 m from the coast), and filtered onto 50 μm, was continuously supplied to four 100 L header tanks at a minimal rate of 50 L h^{-1}. The whole system comprised a factorial experimental design with 4 treatments (control, lowered pH, elevated temperature, and lowered pH/elevated temperature), and 12 experimental aquaria (30 L) with 3 aquaria per treatment. From each header tank, seawater was delivered by gravity to the three experimental aquaria at a rate of ca. 15 L h^{-1} per aquarium (checked on a weekly basis with a flow meter). A small pump was installed in each aquarium in order to create some water movement. A total of 120 mussels were haphazardly placed in each aquarium. In the perturbation header tanks, pH and temperature were respectively decreased by 0.1 pH units and increased by 1°C/week over 3 weeks, and thereafter maintained at a pH offset of ~ −0.3 and a temperature offset of ~ +3°C for the duration of the experiment (~10 months). The pH offset was controlled by bubbling pure-CO_2 in the corresponding header tanks using a continuous pH-stat system (IKS, Karlsbad, Aquastar). pH electrodes from the pH-stat system were inter-calibrated on a weekly basis using a glass combination electrode (Metrohm, electrode plus) calibrated on the total scale using TRIS buffer solutions with a salinity of 35 (provided by A. Dickson, Scripps Institution of Oceanography, San Diego). The +3°C offset was maintained with thermo-resistances (2 × 500 W per header tank) controlled by a COREMA©temperature-regulation system. Twice a day, the seawater flow was stopped and mussels were fed with a commercial mixture of live *Nannochloropsis oculata*, *Phaeodactylum tricornutum* and *Chlorella* (DT's Live Marine Phytoplankton) at a final concentration of ~4% mussel tissue dry weight, close to ingestion rates measured *in situ* by Galimany et al. (2011) in the area of collection. Mussel average tissue dry weight was estimated at the start of the experiment by weighing 10 randomly selected mussels after 24 h in an oven at 60°C. Food quantity was adjusted in each aquarium based on the remaining number of mussels and considering a constant average tissue dry weight throughout the experiment. Seawater flow was restored once all phytoplankton had been filtered by the mussels (1–2 h). Mortality of the mussels was recorded, dead mussels were removed and aquaria were cleaned on a weekly basis.

Seawater samples for total alkalinity (A_T) measurements were collected once a week in the ambient seawater header tank, filtered on GF/F membranes, immediately poisoned with $HgCl_2$ and analyzed within 2 months at the Laboratoire d'Océanographie de Villefranche (France). A_T was determined potentiometrically using a Metrohm titrator (Titrando 80), and a glass electrode (Metrohm, electrode plus) calibrated using first NBS buffers (pH 4.0 and pH 7.0, to check that the slope was Nernstian) and then using TRIS buffer solutions (salinity 35, provided by A. Dickson, Scripps Institution of Oceanography, San Diego). Triplicate titrations were performed on 50 mL sub-samples at 25°C and A_T was calculated as described by Dickson et al. (2007). Titrations of standard seawater provided by A. Dickson (batch 106) yielded A_T values within 2.4 µmol kg^{-1} of the nominal value ($SD = 1.2$ µmol kg^{-1}, $n = 32$). All parameters of the carbonate chemistry were determined from pH$_T$, A_T, temperature and salinity using the R package seacarb (Lavigne et al., 2014).

BIOMETRICS AND PHYSIOLOGICAL MEASUREMENTS

Every 2 months, various biometrical parameters and physiological processes were measured. Shell length, shell weight and fresh weight were monitored on 20 haphazardly pre-labeled mussels per aquarium. Shell weight was measured using the buoyant weight technique (Spencer Davies, 1989). Net calcification, excretion and respiration rates were measured by incubating 3 haphazardly chosen mussels per aquarium in 300 mL respiration chambers for 3 h. Temperature was maintained at the corresponding level for each treatment by plunging the respiration chambers in tanks with flowing experimental seawater. Mussels were not fed for a period of 6 h before the start of the incubations. Blanks containing only seawater were done at the start and at the end of the 10-month exposure and showed that variations of any of the constituents due to microbial processes were less than 0.5% of the variations observed with mussels (data not shown). Respiration rates were estimated as the rate of oxygen concentrations decrease over the 3 h incubations by means of continuous measurements (1 measurement every 5 s) with fiber-optic oxygen microsensors (PreSens©). Oxygen decrease was always linear during the 3 h incubations and concentrations never dropped below 70% of saturation. Before and after the 3 h incubations, pH$_T$ was measured with a glass electrode calibrated using TRIS buffers (see above) and seawater was sampled for the measurements of ammonium (NH_4) and A_T. pH$_T$ decrease during the incubations averaged 0.14 ± 0.05. For NH_4, 20 mL of seawater were sampled, filtered on 0.2 µm and immediately frozen at -20°C. Measurements were performed within 2 days at the Institute of Marine Science, using an autoanalyzer. For A_T, sampling and measurements were performed as described above. Tissue dry weights of the incubated mussels were measured after 24 h in an oven at 60°C and all rates were expressed as per gram of tissue dry weight (g DW). Net calcification rates were calculated based on the observed difference in A_T before and after incubation, corrected for the effect of excretion on A_T (Wolf-Gladrow et al., 2007), following the equation:

$$G = \frac{E - \Delta A_T}{2}$$

where G are net calcification rates (in µmol $CaCO_3$ g DW^{-1} h^{-1}), E are excretion rates (in µmol N g DW^{-1} h^{-1}) and ΔA_T are the observed A_T changes (in µmol g DW^{-1} h^{-1}).

Within a maximum of 5 days from the metabolism measurements, an analysis of haemolymph and extrapallial fluid pH and pCO_2 was conducted on another batch of mussels ($n = 6$ individuals) from all four groups, except in October when no measurements were performed. A volume of 0.2–1 mL of haemolymph and extrapallial fluid was sampled using a syringe equipped with cannulas of 0.6 × 30 and 0.6 × 80 mm size, respectively. The body fluids were transferred directly to Eppendorf caps and placed into a thermostated water bath (Lauda R100) set to the temperature determined in the aquaria prior to sampling. Body fluid pH was measured using a micro pH electrode (Mettler) combined with a pH meter (WTW pH 3310) and calibrated to the specific temperature. Fluid pCO_2 were determined by calculation (Pörtner et al., 1990, 2010) from measurements of total CO_2 in a carbon dioxide analyzer (Corning 965) operating in a linear range from 0.71 to 11.36 mM using a $NaHCO_3$ standard (1 g L^{-1}) for calibration. The sample volume was 0.1 mL. Total CO_2 was determined in relation to standard solutions set to similar concentrations to compensate for a potential drift of the sensor.

PERIOSTRACUM ANALYSES

At the end of the experimental period, shells from live mussels were sampled for quantification of periostracum cover. As all mussels were dead in the elevated temperature aquaria at the end of the experiment, only shells from the control and the lowered pH treatments were analyzed. Five random specimens from each treatment were used for area quantification and valve dimensions (height, width, and inflation) were additionally measured with a digital caliper as secondary control. Visual inspections of these shells showed areas without periostracum, preferentially in the umbo portions of the shells. In order to quantify systematic differences between the two treatments, the outline of periostracum-free areas and of the shell perimeter were redrawn in CorelDraw from scaled high-resolution pictures, placed on a constant reference area of 1441 × 1441 pixel (equivalent to 10 × 10 cm) and converted into black and white images. Pixel count area quantifications were carried out in ImageJ 1.47v (http://imagej.nih.gov/ij/) and periostracum loss (%) is given relative to the total shell area.

STATISTICS

Growth rates (shell weight, shell length, and fresh weight) were compared based on the analyses of variance of linear regression models performed with the R software. Regarding mortality, metabolic rates, and acid-base parameters (haemolymph and extrapallial fluid pCO_2 and pH), due to the low number of replicates, permutational multivariate analyses of variance (Permanova) were performed using the R package RVAideMemoire (Hervé, 2014) to test for differences between the 4 treatments. These analyses were performed considering two orthogonal fixed factors (pH$_T$ and temperature) and one blocking

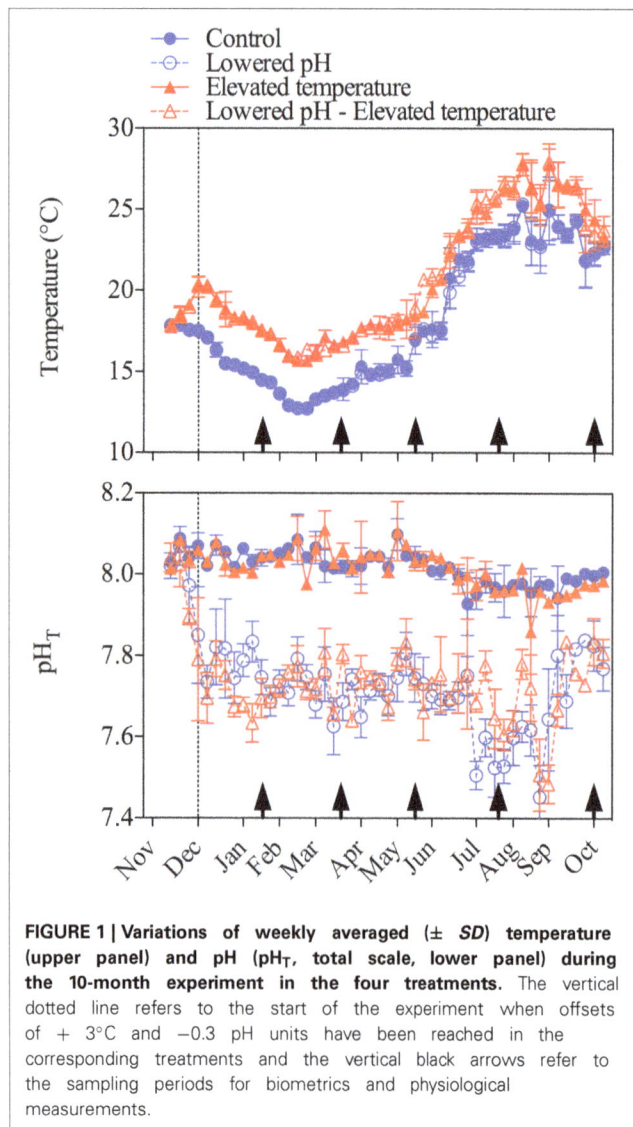

FIGURE 1 | Variations of weekly averaged (± SD) temperature (upper panel) and pH (pH_T, total scale, lower panel) during the 10-month experiment in the four treatments. The vertical dotted line refers to the start of the experiment when offsets of + 3°C and −0.3 pH units have been reached in the corresponding treatments and the vertical black arrows refer to the sampling periods for biometrics and physiological measurements.

factor (time) over 1000 permutations and a significant effect was considered when $p < 0.05$. Differences between control and lowered pH treatments in terms of periostracum cover at the end of the experiment was tested using a non-parametric Mann-Whitney test on R, and considered significant at the risk $\alpha = 5\%$.

RESULTS

Environmental parameters to which the mussels were exposed are presented in **Figure 1** and **Table 1**. Weekly averaged temperature naturally varied between ~12 and ~25°C with minimal levels recorded in February and maximal levels reached in August. Temperature was maintained with mean offsets of respectively 2.7 ± 0.6 and 2.6 ± 0.6°C in the lowered pH/elevated temperature and elevated temperature treatments, as compared to the control. pH_T was relatively constant in the ambient pH treatments (8.01 ± 0.04 and 8.01 ± 0.06 in the control and elevated temperature treatments, respectively). In the lowered pH treatments, offsets of -0.31 ± 0.08 and -0.30 ± 0.07 were observed in the lowered pH and lowered pH/elevated temperature treatments, respectively. However, as can be seen in **Figure 1**, the pH regulation was less efficient in summer (July/August) with observed average offsets of -0.40 ± 0.1 and -0.32 ± 0.1 in the lowered pH and lowered pH/elevated temperature treatments, respectively. Notwithstanding, aragonite saturation states differed among treatments with minimal values in the lowered pH/ambient temperature treatment, but seawater remained oversaturated with respect to this mineral in all treatments during the entire experiment.

Cumulative mortality rates were below 20% for all treatments until the end of June and started to increase abruptly in the elevated temperature treatments when temperature levels reached ~25°C or more (**Figure 2**). Temperature was the main driver for this increased mortality (Permanova, $p < 0.001$, $n = 38$), with no significant effect of pH, either combined, or in isolation (Permanova, $p > 0.05$, $n = 38$, in both cases). In the ambient temperature treatments, regardless of the pH level, mortality rates started to increase after August, when temperature

Table 1 | Parameters of the seawater carbonate chemistry during the 10-month experiment: temperature (Temp, in°C), pH (pH_T, total scale), partial pressure of carbon dioxide (pCO_2, in μatm), total inorganic carbon concentration (C_T, in μmol kg^{-1}), and saturation state of the seawater with respect to aragonite (Ω_a), and calcite (Ω_c).

Treatment	Temp	Δ Temp	pH_T	Δ pH_T	pCO_2	C_T	Ω_a	Ω_c
Control	18.2	–	8.01	–	476	2274	2.9	4.5
	(12.7; 25.2)		(7.93; 8.10)		(375; 601)	(2201; 2326)	(2.5; 3.4)	(3.9; 5.2)
Lowered pH	18.2	−0.1	7.71	−0.31	1090	2422	1.6	2.5
	(12.7; 25.3)	(−1.1; 0.1)	(7.45; 7.84)	(−0.52; −0.14)	(760; 2051)	(2308; 2517)	(1.1; 2.4)	(1.7; 3.7)
Elevated temperature	20.8	2.6	8.01	0.0	479	2252	3.2	4.8
	(15.7; 27.8)	(0.8; 3.6)	(7.86; 8.11)	(−0.10; 0.09)	(363; 706)	(2152; 2334)	(2.5; 3.9)	(3.9; 5.9)
Lowered pH–elevated temperature	20.9	2.7	7.72	−0.30	1060	2401	1.8	2.8
	(15.9; 27.9)	(1.2; 3.6)	(7.48; 7.83)	(−0.49; -0.16)	(772; 1922)	(2292; 2484)	(1.3; 2.8)	(2.0; 4.1)

The difference of the perturbation and the control treatments in terms of pH_T (ΔpH_T) and temperature (Δ Temp) is also reported. For each parameter, the average value is reported as well as the range (min; max). Total alkalinity ranged between 2485 and 2565 μmol kg^{-1} and averaged 2540 μmol kg^{-1}. Salinity ranged between 37.8 and 38.1 and averaged 38.0.

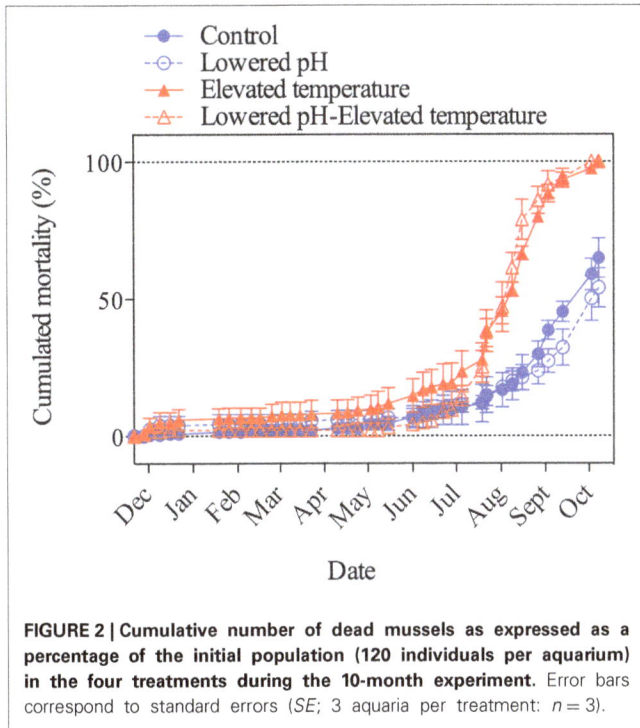

FIGURE 2 | Cumulative number of dead mussels as expressed as a percentage of the initial population (120 individuals per aquarium) in the four treatments during the 10-month experiment. Error bars correspond to standard errors (*SE*; 3 aquaria per treatment: *n* = 3).

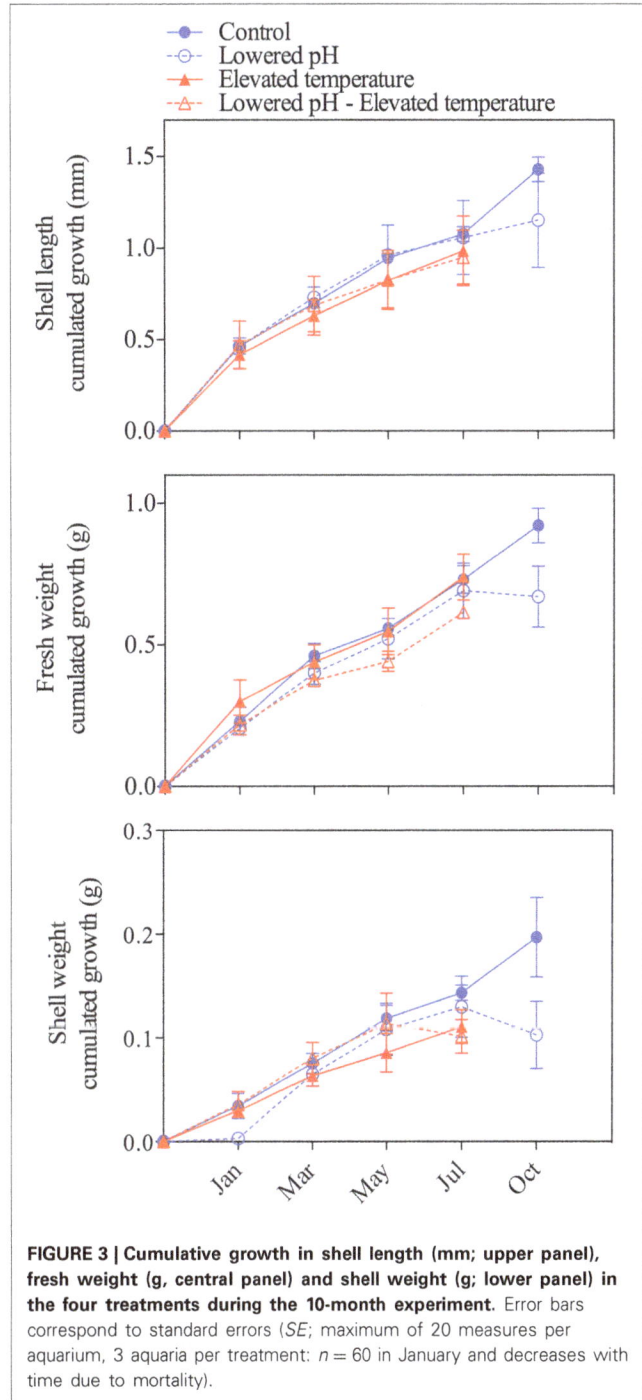

FIGURE 3 | Cumulative growth in shell length (mm; upper panel), fresh weight (g, central panel) and shell weight (g; lower panel) in the four treatments during the 10-month experiment. Error bars correspond to standard errors (*SE*; maximum of 20 measures per aquarium, 3 aquaria per treatment: *n* = 60 in January and decreases with time due to mortality).

levels also reached ∼25°C. A mortality of 100% was recorded by the end of the experiment (October) in the elevated temperature treatments, while in ambient temperature treatments mortality was 60%.

Growth parameters are presented in **Figure 3**. Cumulated growth in terms of shell length, fresh weight, and shell weight were relatively small with maximum growth by the end of the experiment of 1.4 ± 0.1 mm, 0.9 ± 0.1 g, and 0.20 ± 0.1 g, respectively. Note that, as all mussels exposed to elevated temperature levels were dead before the last sampling period in October, no data are available for these treatments. We applied linear models to describe these data, revealing that growth in shell length and in fresh weight was not affected by lowered pH or elevated temperature (**Table 2**). In contrast, growth in shell weight was significantly reduced by both of these perturbations, although without cumulative effects. In October, mussel shell weights even decreased in the lowered pH treatment, a decrease that was accompanied by a decrease in fresh weight and shell length (**Figure 3**).

Respiration, excretion and net calcification rates are presented in **Figure 4**. As no data were collected in October for mussels exposed to elevated temperature levels (no specimen survived), two distinct Permanova tests have been applied: 1) for all treatments and for the period January to July and 2) for ambient temperature treatments (testing only the effect of lowered pH) for the whole experimental period (see **Table 3**). Respiration rates significantly increased at elevated temperature (**Table 3**), and with mean values, for the period January/July, of 21.03 ± 3.3 and $24.2 \pm 4.8 \,\mu\text{mol O}_2\,\text{g DW}^{-1}\,\text{h}^{-1}$ in the ambient and increased temperature treatments, respectively. Excretion rates mean values were $3.3 \pm 1.0 \,\mu\text{mol N g DW}^{-1}\,\text{h}^{-1}$ for all treatments and sampling periods and, as for respiration rates, significantly increased at elevated temperature (2.6 ± 0.7 vs. $3.3 \pm 0.9 \,\mu\text{mol N g}$

$\text{DW}^{-1}\,\text{h}^{-1}$ in the ambient and increased temperature treatments, respectively, for the period January/July). Excretion of ammonium was responsible, on average, for $63.3 \pm 0.2\%$ of the observed A_T variations, and confirmed that a correction of the alkalinity anomaly technique was necessary for this species. Net calcification rates significantly decreased in the lowered pH/elevated temperature treatment. Most of the time, net calcification rates were significantly above 0, except in May and July in the lowered pH/elevated temperature treatment and in October in the ambient pH and lowered pH treatments. In October, negative values

Table 2 | Analysis of variance of the linear models fitted to the growth of mussels exposed to the four treatments ($n = 932$; 5 time points, 2 temperature levels, 2 pH levels).

Source of variation	df	MS	F	p
SHELL LENGTH				
Time	1	635	1761	**<2.2 × 10⁻¹⁶***
Time × Temperature	1	0.24	0.68	0.41
Time × pH_T	1	0.45	1.24	0.27
Time × Temperature × pH_T	1	0.05	0.13	0.72
Residuals	943	0.36		
FRESH WEIGHT				
Time	1	243	2455	**<2.2 × 10⁻¹⁶***
Time × Temperature	1	0.07	0.73	0.39
Time × pH	1	0.14	1.41	0.24
Time × Temperature × pH	1	0.01	0.14	0.71
Residuals	941	0.10		
SHELL WEIGHT				
Time	1	7.80	1084	**<2.2 × 10⁻¹⁶***
Time × Temperature	1	0.03	3.99	**0.04***
Time × pH	1	0.03	4.19	**0.04***
Time × Temperature × pH	1	0.00	0.48	0.49
Residuals	924	0.007		

All regression slopes were significantly different from 0. Asterisks denote significant effects of the considered perturbation at the risk $\alpha = 0.05$.

(net dissolution) were measured in the lowered pH treatment, however as a consequence of the low replication and large associated errors, the difference between ambient and lowered pH treatments were not statistically significant.

Figure 5 depicts the pH changes in haemolymph and extrapallial fluid from January to July. The analysis of variance showed a significant impact of seawater pH only for haemolymph pH (**Table 3**), and no impacts of temperature on both fluids in terms of pH. Except for the mean haemolymph pH of the elevated temperature group, pH in haemolymph and extrapallial fluid started with values below 7.4 in all groups. pH values in all groups increased from January to March, but always remained much lower than seawater pH. After March, pH values in both haemolymph and extrapallial fluid decreased over time, with a larger decrease at lowered pH levels. In July, a large difference (~0.16 units) in extracellular pH of the haemolymph had developed between the two normocapnic groups and the two groups under elevated CO_2. **Figure 6** presents the corresponding changes in pCO_2 in the haemolymph and extrapallial fluid of all groups over time. The changes in pCO_2 correlate with the observed changes in pH in both haemolymph and extrapallial fluid (Pearson correlation: $r = -0.51$, $n = 96$, and $r = -0.37$, $n = 96$, respectively). In contrast to extracellular pH, the analysis of variance showed no significant impact of both temperature and pH for the haemolymph fluid in terms of pCO_2 and large variability between replicates (**Table 3**, **Figure 6**). Temperature appeared as a significant factor with no cumulative effect of pH for extrapallial pCO_2 (**Table 3**).

All valves showed shell areas free of periostracum at or near the umbo, which may spread from there toward progressively

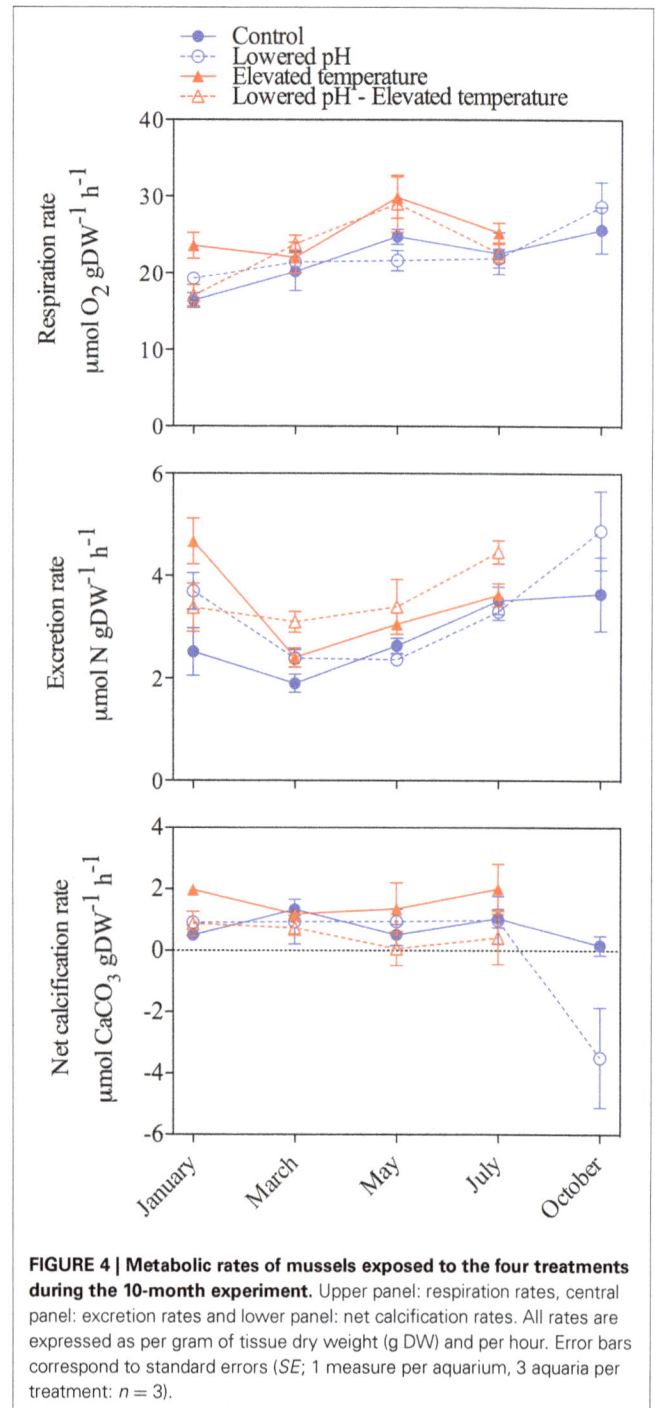

FIGURE 4 | Metabolic rates of mussels exposed to the four treatments during the 10-month experiment. Upper panel: respiration rates, central panel: excretion rates and lower panel: net calcification rates. All rates are expressed as per gram of tissue dry weight (g DW) and per hour. Error bars correspond to standard errors (SE; 1 measure per aquarium, 3 aquaria per treatment: $n = 3$).

younger shell portions (**Figure 7**). Pristine periostracum is rather dark-blue to black, while the periostracum free zones are surrounded by a discolored periostracum with a beige taint. Average periostracum loss for the two treatments for which some mussels were still alive at the end of the experiment was $3.9 \pm 1.5\%$ in the control and $16.9 \pm 6.7\%$ in the lowered pH treatment. Periostracum discoloration was restricted to a narrow fringe around the periostracum free area in the control treatment, but was significantly increased in the lowered pH treatment (Mann-Whitney test, $p = 0.002$, $n = 10$), affecting ~2/3 of the shell area.

Table 3 | Results of the permutational multivariate analyses of variance (Permanova) on metabolic rates and haemolymph/extrapallial fluid pH and pCO_2 ($n = 54$; 4 time points, 2 temperature levels, 2 pH levels).

	Source of variation	df	MS	F	p
RESPIRATION RATES					
All treatments	pH_T	1	11.6	0.84	0.43
(until July)	Temperature	1	116.4	8.45	**0.03***
	$pH_T \times$ Temperature	1	14.03	1.02	0.37
	Residuals	32	10.34		
Control and lowered pH treatments	pH_T	1	3.54	0.35	0.57
(entire experiment)	Residuals	20	12.10		
EXCRETION RATES					
All treatments	pH_T	1	0.58	0.79	0.43
(until July)	Temperature	1	6.24	8.46	**0.01***
	$pH_T \times$ Temperature	1	0.07	0.10	0.75
	Residuals	32	0.27		
Control and lowered pH treatments	pH_T	1	1.77	2.23	0.25
(entire experiment)	Residuals	20	9.93		
NET CALCIFICATION RATES					
All treatments	pH_T	1	3.07	8.92	0.06
(until July)	Temperature	1	0.38	1.09	0.32
	$pH_T \times$ Temperature	1	4.36	12.66	**0.02***
	Residuals	32	0.80		
Control and lowered pH treatments	pH_T	1	3.20	0.73	0.52
(all experiment)	Residuals	20	1.31		
HAEMOLYMPH pH					
	pH_T	1	0.21	23.7	**0.002***
	Temperature	1	0.02	1.7	0.223
	$pH_T \times$ Temperature	1	0.02	2.3	0.161
	Residuals	80	0.01		
EXTRAPALLIAL FLUID pH					
	pH_T	1	0.10	4.59	0.079
	Temperature	1	0.01	0.29	0.586
	$pH_T \times$ Temperature	1	0.01	0.53	0.465
	Residuals	80	0.01		
HAEMOLYMPH pCO_2					
	pH_T	1	0.46	2.08	0.165
	Temperature	1	1.07	4.85	0.059
	$pH_T \times$ Temperature	1	0.00	0.00	0.996
	Residuals	80	0.15		
EXTRAPALLIAL FLUID pCO_2					
	pH_T	1	0.23	3.54	0.085
	Temperature	1	2.09	32.0	**0.001***
	$pH_T \times$ Temperature	1	0.18	2.73	0.149
	Residuals	80	0.11		

Since all mussels exposed to the elevated temperature treatments died before the end of the experiment, separate analyses have been performed considering all treatments until July and considering only the ambient temperature treatments for the whole experimental period. Asterisks and numbers in bold denote significant effects of the considered perturbation at the risk $\alpha = 0.05$.

DISCUSSION

In the present work we tested the effects of OA and warming on growth physiology and survival of the Mediterranean mussel (*Mytilus galloprovincialis*). For the first time, this species has been exposed to pH and temperature levels that are projected for the end of the century by considering constant offsets of -0.3 pH units and $+3°C$ from ambient conditions. Based on these results, there is no doubt that mussels are highly sensitive to a

$3°C$ warming that leads to suboptimal and even lethal temperature levels in summer. Increased mortality rates were recorded in July, when seawater temperature exceeded $\sim25°C$ in the elevated temperature treatment. In this treatment, in August and September, seawater reached $\sim28°C$, and all mussels died within few weeks. It must be stressed that increased mortality was also observed in summer (August and September) in the ambient temperature treatment, corresponding to temperature levels around

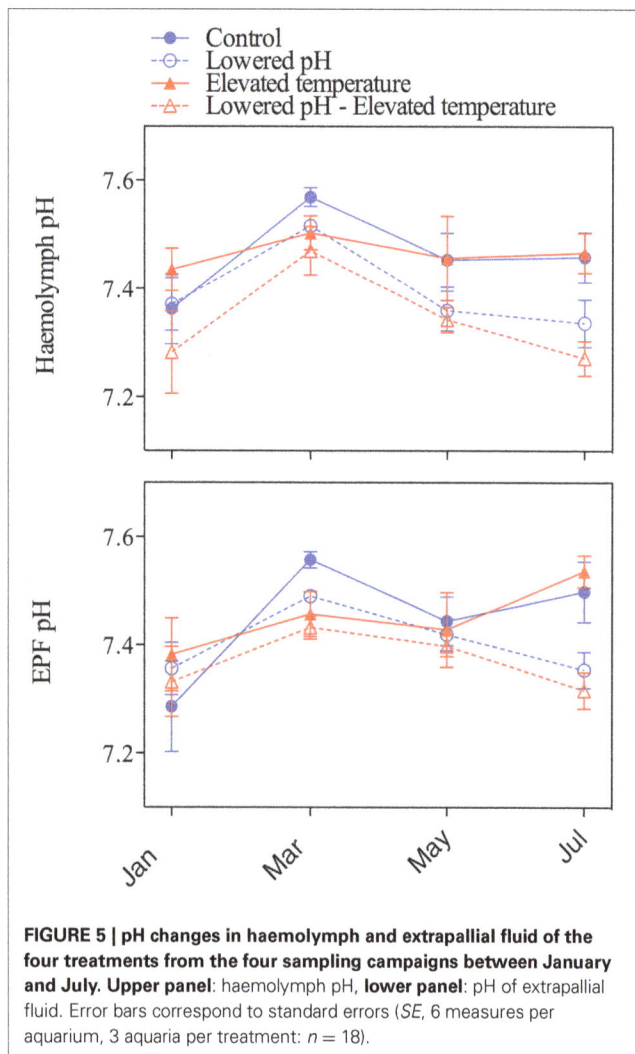

FIGURE 5 | pH changes in haemolymph and extrapallial fluid of the
four treatments from the four sampling campaigns between January
and July. **Upper panel**: haemolymph pH, **lower panel**: pH of extrapallial
fluid. Error bars correspond to standard errors (*SE*, 6 measures per
aquarium, 3 aquaria per treatment: $n = 18$).

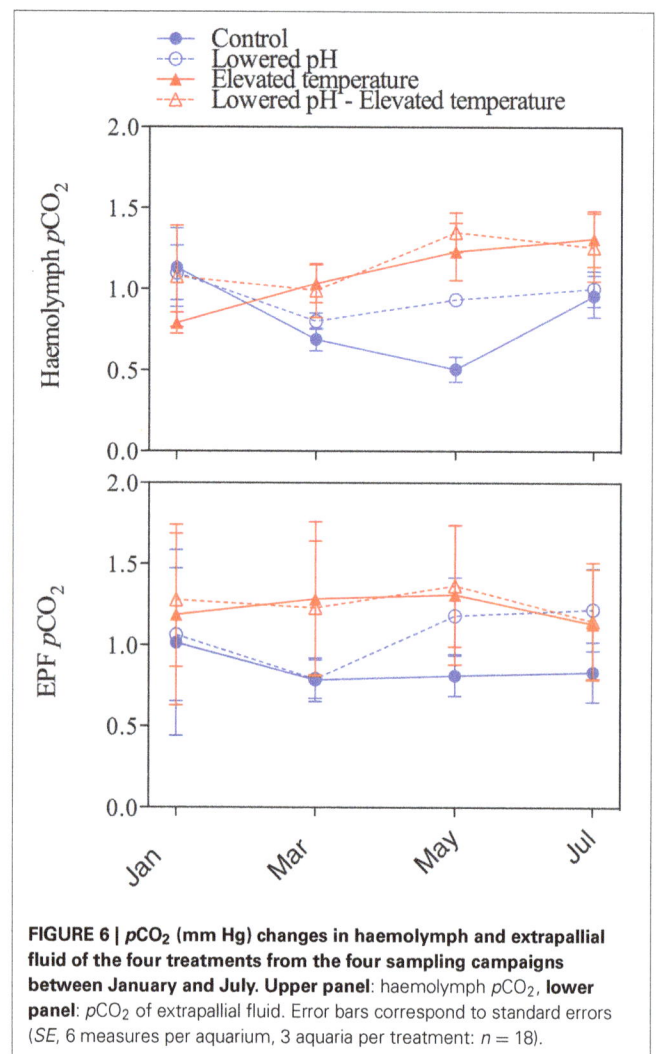

FIGURE 6 | pCO_2 (mm Hg) changes in haemolymph and extrapallial
fluid of the four treatments from the four sampling campaigns
between January and July. **Upper panel**: haemolymph pCO_2, **lower
panel**: pCO_2 of extrapallial fluid. Error bars correspond to standard
errors (*SE*, 6 measures per aquarium, 3 aquaria per treatment: $n = 18$).

25°C. Temperature levels of 24–25°C have been identified, for
this species, as an upper limit for normal physiological activities
(Anestis et al., 2007, 2010) indicating that Mediterranean mussels already live close to their thermal acclimation limits (Anestis
et al., 2007). Total mortality of mussels has been observed in
Fangar Bay, located in the area where mussels used in the present
study were sampled, when seawater temperature reached 28°C
for more than 10 days (Ramón et al., 2007). In the last years,
farmers from this area have already adapted their marketing strategy to increased summer temperature by selling their product
earlier than in the previous years, thus avoiding the period of
heat-induced mortalities (Ramón et al., 2005).

In contrast to warming, our results showed that acidification
alone did not induce higher mortalities. The effects of CO_2 can be
twofold in organisms in environments such as intertidal or shallow subtidal zones. By exerting metabolic depression, CO_2 can
alleviate the level of stress and delay mortality through more efficient exploitation of energy reserves and passive tolerance. This
is not corroborated by our results that do not show any significant delay in mortality at lowered pH levels. Conversely, it has
been hypothesized that OA, as an additional stressor, can narrow

the thermal window of an organism (Pörtner and Farrell, 2008;
Pörtner, 2012), by reducing aerobic performance, i.e., the range of
active tolerance. This effect would become visible once the passive
tolerance range is exploited sooner and faster at high temperature beyond the critical temperature of 25°C. However, again, this
was not observed in our study as the earlier onset in mortality
at lowered pH and elevated temperature was not statistically significant. This absence of acidification effects on mussel survival
is consistent with results reported by Range et al. (2012) who
exposed juveniles of *M. galloprovincialis* to pH offsets of −0.3
and −0.6 for 84 days and also observed no impact on survival.
In contrast, Bressan et al. (2014) observed a slight but significant increase in mortality after 6 months of experiment in the
northern Adriatic Sea. However, it must be stressed that they
exposed mussels to a much larger decrease in pH (0.7) than the
one considered in our study. Our results therefore strongly suggest
that survival of this species will not be affected by a pH decrease
of ~0.3 in the Mediterranean Sea, predicted to occur within the
present century.

According to our results, basic metabolic rates were also not
impacted by OA alone while, as expected, they increased at

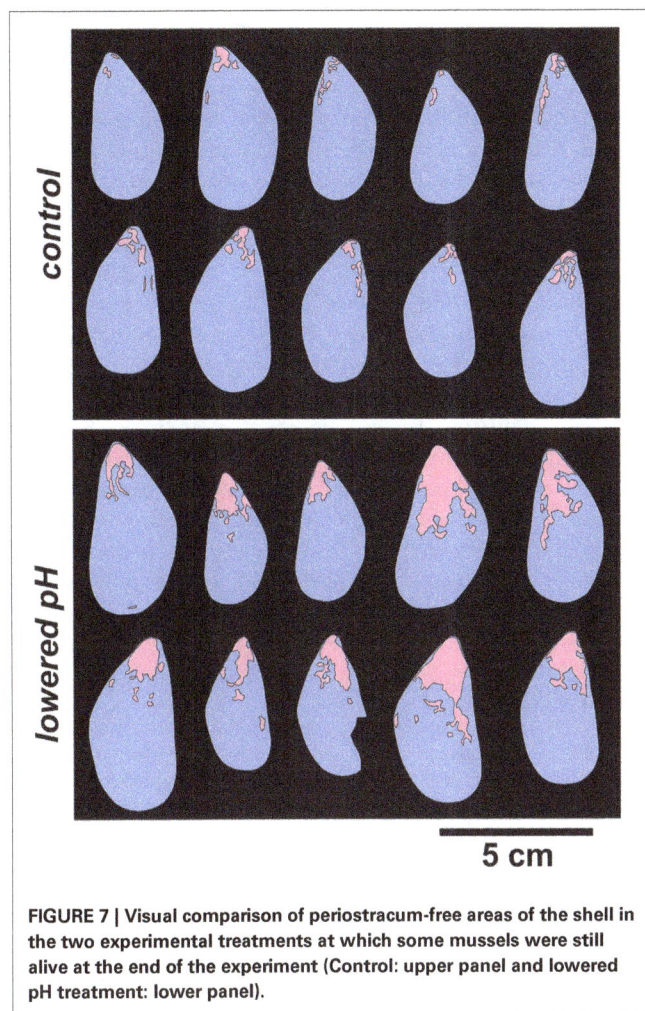

FIGURE 7 | Visual comparison of periostracum-free areas of the shell in the two experimental treatments at which some mussels were still alive at the end of the experiment (Control: upper panel and lowered pH treatment: lower panel).

elevated temperature. This is consistent with results from another experiment conducted on this species in the north Adriatic Sea (Range et al., 2014) that showed no effect of a pH decrease of 0.7 on both respiration and excretion rates. These results are not in line with the general theory of increased amino-acid metabolism and their decreased use in growth-related protein synthesis, which seems to be a typical response in marine invertebrates during hypercapnic acidosis (Pörtner et al., 1998; Michaelidis et al., 2005; Thomsen and Melzner, 2010). In another experiment conducted on Mediterranean mussels from the Atlantic coast in Portugal, respiration rates appeared again insensitive to acidification (water pH lowered by 0.3 and 0.6 compared to control treatment), but excretion rates increased at lowered levels of seawater pH (Fernandez-Reiriz et al., 2012).

The capacity of mollusks to compensate for changes in acid–base status due to elevated CO_2 is believed to be somewhat limited (Melzner et al., 2009). In our experiment, in all groups, patterns of pH and pCO_2 changes were similar in haemolymph and extrapallial fluid, confirming that the carbonate saturation status is similar in both body fluids (Thomsen et al., 2010). All groups displayed similar pH changes between spring and summer, however absolute values were different. The lowest extracellular pH values

were observed in the elevated CO_2 groups, in line with the low capacity of mussel species for extracellular acid base regulation, as previously observed (e.g., Michaelidis et al., 2005; Thomsen et al., 2010). For both body fluids, pH values displayed large seasonal changes in all groups, possibly reflecting shifting set points over time. The up-regulation of pH in spring (March sampling point), observed in all groups, may reflect transition from winter dormancy to enhanced activity, associated with an increase in the capacity for extracellular acid-base regulation, and likely, overall activity. Indeed, mussels reconstitute their energy reserves during the first quarter of the year before undergoing gametogenesis (Lowe et al., 1982). Furthermore, the seawater temperature from March to April was close to the optimum temperature of the species with critical thermal limits close to 25°C (Anestis et al., 2007). Any temperature change outside the optimum range will increase the baseline energy demand of an ectothermic animal and reduce e.g. the growth performance of the organism (oxygen- and capacity-limited thermal tolerance (OCLTT; see Pörtner, 2010). Accordingly, the later sampling points under elevated seasonal temperatures led to a higher baseline energy demand (Guderley and Pörtner, 2010), visible as increased respiration and excretion rates. Moreover, elevated CO_2 concentration caused a significant reduction in pH of the haemolymph in July. An exacerbation of uncompensated extracellular acidosis, in addition to the loss of the periostracum, could explain the negative calcification rates measured in October (see thereafter). Unfortunately, high mortality prevented measurements of extracellular pH during this period. The exacerbation of uncompensated extracellular acidosis, as observed in July, suggests a progressive insufficiency in acid-base regulation capacity at elevated temperatures and pCO_2 levels, conditions that will prevail in the future high CO_2 ocean, affecting the species at the warm end of their distribution range. However, as will be discussed below, the significant decrease of seawater pH in the lowered pH treatment during this summer period could have been also responsible for these lowered compensation capacities and growth.

Somatic and shell growth did not appear highly sensitive to OA and warming during most of the experiment. It must be stressed that growth was very limited and much lower than what can be measured in aquaculture sites in the Mediterranean Sea (i.e., 20–40 mm yr^{-1}; Sarà et al., 1998). Although feeding rates were calculated based on *in situ* ingestion rates, maintaining mussels in the laboratory is challenging and optimal growth rates could not be reached with a feeding protocol using only a few phytoplankton species. It appears essential, in order to improve risk assessment for the Mediterranean aquaculture industry and for designing related adaptation strategies, to conduct future perturbation experiments, for instance using FOCE systems (Gattuso et al., 2014), directly in the field near aquaculture sites. Net calcification rates as estimated by a modified alkalinity anomaly technique showed positive values under all treatments during most of the experiment, except after summer when negative values (net dissolution) were measured in the lowered pH treatment. The visual analysis of the shells indeed showed that mussels exposed to lowered pH levels had much larger areas of their shells free of periostracum (**Figure 7**). This organic layer plays the role of shell protection and can be partially lost as a natural result of abrasion from the bivalve's movements. However, our results suggest, in accordance with findings by Rodolfo-Metalpa et al. (2011), that the periostracum is altered when mussels are exposed to lowered pH conditions. For these mussels, calcitic and aragonitic shell layers were then more exposed to the surrounding seawater and consequently more subject to corrosion, partially explaining the observed net dissolution rates and the decrease of shell weight in low pH treatments after summer. This could reduce the resistance of the shell to mechanical damage, enhancing the risk of predation (Reimer and Tedengren, 1996; Aronhime and Brown, 2009) and of damage by storms and associated wave action (Nehls and Thiel, 1993). Furthermore, although no systematic data have been gathered on this aspect, mussels from all aquaria were able to group and attach to each other with their byssus during most of the experiment, while mussels exposed to low pH conditions after summer were unable to do so and it was possible to easily pick them out of the aquarium individually. This could be due to weaker and less extensible byssal threads for mussels maintained at a lowered pH, as shown by O'donnell et al. (2013) on mussels from the US West coast. Weaker byssal threads negatively affect the ability of mussels to attach to substrate, thereby increasing the probability to be displaced due to hydrodynamic forces generated by storms, which is a significant cause of mortality in the field (Carrington et al., 2009).

Although, as already mentioned, growth was limited in our study, lower net calcification rates, and growth under lowered pH levels in summer have been observed and are consistent with Kroeker et al. (2014) who recently reported significant effects of acidification on *M. galloprovincialis* shell growth after one month exposure at a pH level maintained at an offset of ~-0.3 as compared to control conditions. However, this study, conducted over a relatively short exposure time and at lower temperature levels than in our experiment, showed that warming (from 14 to 20°C) mitigated this negative acidification effect as a consequence of both physiological (performances such as calcification increase if warming occurs at temperatures below the thermal optimum, Pörtner, 2012) and chemical (pH and saturation states increase with temperature) effects. Such mitigation effects of warming were not be observed during our study that was conducted under higher temperature levels. In contrast, and more in line with our results, a decrease in shell calcification for this species has been shown as a consequence of synergistic impacts of elevated pCO_2 and temperature on the same species by Rodolfo-Metalpa et al. (2011). However, based on our experiment, we cannot elucidate if the decrease in net calcification rates and growth as observed after summer is a consequence of lower pH levels during that period (pH offsets of -0.4 vs. -0.3 during the rest of the experiment) or a consequence of a combined effect of acidification and warming as suggested by Rodolfo-Metalpa et al. (2011).

To conclude, this study is the first focused on the combined effects of warming and acidification on the Mediterranean mussel over seasonal exposure time. Our results strongly suggest that ongoing ocean warming and more frequent summer heat waves will be serious threats to this species in the Mediterranean Sea. All in all, OA does not seem to have very serious impacts on this species in this alkaline region although our results suggest lower growth rates, lower acid-base regulation capacities and significant loss of the periostracum cover in summer coinciding with temperatures beyond optimal levels. Based on the present dataset, the question whether decreased growth capacities are due to a combined effect of OA and warming remains however unresolved. More long-term experiments performed at proximity of aquaculture sites and combining these two stressors are still necessary to improve risk assessment for the Mediterranean aquaculture industry and for designing related adaptation strategies.

ACKNOWLEDGMENTS

We like to thank Zora Zittier for her help with the body fluid measurements during the sampling in January, Jean-Pierre Gattuso for help with experimental design and Jean-Olivier Irisson for help with statistics. We wish to thank Claudia R. Benitez-Nelson and Emily Carrington for their constructive reviews. This work was funded by the EC FP7 project "Mediterranean Sea Acidification in a changing climate" (MedSeA; grant agreement 265103), the French program PNEC (Programme national environnement côtier; Institute national des sciences de l'univers) and the EC FP7 project "European Project on Ocean Acidification" (EPOCA; grant agreement 211384).

REFERENCES

Anestis, A., Lazou, A., Portner, H. O., and Michaelidis, B. (2007). Behavioral, metabolic, and molecular stress responses of marine bivalve *Mytilus galloprovincialis* during long-term acclimation at increasing ambient temperature. *Am. J. Physiol. Regul. Integr. Comp. Physiol.* 293, R911–R921. doi: 10.1152/ajpregu.00124.2007

Anestis, A., Pörtner, H. O., Karagiannis, D., Angelidis, P., Staikou, A., and Michaelidis, B. (2010). Response of *Mytilus galloprovincialis* (L.) to increasing seawater temperature and to marteliosis: metabolic and physiological parameters. *Comp. Biochem. Physiol. A* 156, 57–66. doi: 10.1016/j.cbpa.2009.12.018

Aronhime, B. R., and Brown, K. M. (2009). The roles of profit and claw strength in determining mussel size selection by crabs. *J. Exp. Mar. Biol. Ecol.* 379, 28–33. doi: 10.1016/j.jembe.2009.08.012

Bressan, M., Chinellato, A., Munari, M., Matozzo, V., Manci, A., Marèeta, T., et al. (2014). Does seawater acidification affect survival, growth and shell integrity in bivalve juveniles? *Mar. Environ. Res.* 99, 136–148. doi: 10.1016/j.marenvres.2014.04.009

Carrington, E., Moeser, G. M., Dimond, J., Mello, J. J., and Boller, M. L. (2009). Seasonal disturbance to mussel beds: field test of a mechanistic model predicting wave dislodgment. *Limnol. Oceanogr.* 54, 978–986. doi: 10.4319/lo.2009.54.3.0978

Ciais, P., Sabine, C., Bala, G., Bopp, L., Brovkin, V., Canadell, J., et al. (2013). "Carbon and other biogeochemical cycles," in Climate Change 2013: *t*he Physical Science Basis. Contribution of Working Group I to the *Fifth Assessment Report of the Intergovernmental Panel on Climate Change*, eds T. F. Stocker, D. Qin, G.-K. Plattner, M. Tignor, S. K. Allen, and J. Boschung, et al. (Cambridge; New York: Cambridge University Press), 465–570.

Collins, M., Knutti, R., Arblaster, J., Dufresne, J.-L., Fichefe, T., Friedlingstein, P., et al. (2013). "Long-term climate change: projections, commitments and irreversibility," in Climate Change 2013: *t*he Physical Science Basis. Contribution of Working Group I to the *Fifth Assessment Report of the Intergovernmental Panel on Climate Change*, eds T. F. Stocker, D. Qin, G.-K. Plattner, M. Tignor, S. K. Allen, J. Boschung, et al. (Cambridge; New York: Cambridge University Press), 1029–1136.

Dickson, A. G., Sabine, C. L., and Christian, J. R. (2007). *Guide to Best Practices for Ocean CO$_2$ Measurements*. Vol. 3, PICES Special Publication, 191.

Duarte, C., Navarro, J. M., Acuna, K., Torres, R., Manriquez, P. H., Lardies, M. A., et al. (2014). Combined effects of temperature and ocean acidification on the juvenile individuals of the mussel *Mytilus chilensis*. *J. Sea Res.* 85, 308–314. doi: 10.1016/j.seares.2013.06.002

FAO. (2014). *The State of World Fisheries and Aquaculture (SOFIA)*. Rome: Fisheries and Aquaculture Department.

Fernandez-Reiriz, M. J., Range, P., Alvarez-Salgado, X. A., Espinosa, J., and Labarta, U. (2012). Tolerance of juvenile *Mytilus galloprovincialis* to experimental seawater acidification. *Mar. Ecol. Prog. Ser.* 454, 65–74. doi: 10.3354/meps09660

Galimany, E., Ramón, M., and Ibarrola, I. (2011). Feeding behavior of the mussel *Mytilus galloprovincialis* (L.) in a Mediterranean estuary: a field study. *Aquaculture* 314, 236–243. doi: 10.1016/j.aquaculture.2011.01.035

Gattuso, J.-P., and Hansson, L. (2011). "Ocean acidification: background and history," in *Ocean Acidification*, eds J.-P. Gattuso and L. Hansson (Oxford: Oxford University Press), 1–20.

Gattuso, J.-P., Kirkwood, W., Barry, J. P., Cox, E., Gazeau, F., Hansson, L., et al. (2014). Free-ocean CO_2 enrichment (FOCE) systems: present status and future developments. *Biogeosciences* 11, 4057–4075. doi: 10.5194/bg-11-4057-2014

Gazeau, F., Parker, L. M., Comeau, S., Gattuso, J.-P., O'connor, W. A., Martin, S., et al. (2013). Impacts of ocean acidification on marine shelled molluscs. *Mar. Biol.* 160, 2207–2245. doi: 10.1007/s00227-013-2219-3

Gazeau, F., Quiblier, C., Jansen, J. M., Gattuso, J.-P., Middelburg, J. J., and Heip, C. H. R. (2007). Impact of elevated CO_2 on shellfish calcification. *Geophys. Res. Lett.* 34, L07603. doi: 10.1029/2006GL028554

Guderley, H., and Pörtner, H. O. (2010). Metabolic power budgeting and adaptive strategies in zoology: examples from scallops and fish. *Can. J. Zool.* 88, 753–763. doi: 10.1139/Z10-039

Hervé, M. (2014). *RVAideMemoire: Diverse Basic Statistical and Graphical Functions.* Available online at: http://cran.r-project.org/web/packages/RVAideMemoire/index.html

Hiebenthal, C., Philipp, E. E. R., Eisenhauer, A., and Wahl, M. (2013). Effects of seawater pCO_2 and temperature on shell growth, shell stability, condition and cellular stress of Western Baltic Sea *Mytilus edulis* (L.) and *Arctica islandica* (L.). *Mar. Biol.* 160, 2073–2087. doi: 10.1007/s00227-012-2080-9

Hofmann, G. E., Smith, J. E., Johnson, K. S., Send, U., Levin, L. A., Micheli, F., et al. (2011). High-frequency dynamics of ocean pH: a multi-ecosystem comparison. *PLoS ONE* 6:e28983. doi: 10.1371/journal.pone.0028983

Ivanina, A. V., Dickinson, G. H., Matoo, O. B., Bagwe, R., Dickinson, A., Beniash, E., et al. (2013). Interactive effects of elevated temperature and CO_2 levels on energy metabolism and biomineralization of marine bivalves *Crassostrea virginica* and *Mercenaria mercenaria*. *Comp. Biochem. Physiol. Mol. Integr. Physiol.* 166, 101–111. doi: 10.1016/j.cbpa.2013.05.016

Jorda, G., Marba, N., and Duarte, C. M. (2012). Mediterranean seagrass vulnerable to regional climate warming. *Nat. Clim. Change* 2, 821–824. doi: 10.1038/nclimate1533

Kroeker, K. J., Gaylord, B., Hill, T. M., Hosfelt, J. D., Miller, S. H., and Sanford, E. (2014). The role of temperature in determining species' vulnerability to ocean acidification: a case study using *Mytilus galloprovincialis*. *PLoS ONE* 9:e100353. doi: 10.1371/journal.pone.0100353

Lavigne, H., Epitalon, J. M., and Gattuso, J.-P. (2014). *Seacarb: Seawater Carbonate Chemistry with R.* Available online at: http://CRAN.R-project.org/package=seacarb

Le Quéré, C., Raupach, M. R., Canadell, J. G., Marland, G., Bopp, L., Ciais, P., et al. (2009). Trends in the sources and sinks of carbon dioxide. *Nat. Geosci.* 2, 831–836. doi: 10.1038/ngeo689

Lima, F. P., and Wethey, D. S. (2012). Three decades of high-resolution coastal sea surface temperatures reveal more than warming. *Nat. Commun.* 3, 704. doi: 10.1038/ncomms1713

Lowe, D. M., Moore, M. N., and Bayne, B. L. (1982). Aspects of gametogenesis in the marine mussel *Mytilus edulis* L. *J. Mar. Biol. Assoc. U.K.* 62, 133–145. doi: 10.1017/S0025315400020166

Mackenzie, C. L., Ormondroyd, G. A., Curling, S. F., Ball, R. J., Whiteley, N. M., and Malham, S. K. (2014). Ocean warming, more than acidification, reduces shell strength in a commercial shellfish species during food limitation. *PLoS ONE* 9:e86764. doi: 10.1371/journal.pone.0086764

Marba, N., and Duarte, C. M. (2010). Mediterranean warming triggers seagrass (*Posidonia oceanica*) shoot mortality. *Glob. Change Biol.* 16, 2366–2375. doi: 10.1111/j.1365-2486.2009.02130.x

Melzner, F., Gutowska, M. A., Hu, M., and Stumpp, M. (2009). Acid-base regulatory capacity and associated proton extrusion mechanisms in marine invertebrates: an overview. *Comp. Biochem. Physiol. Mol. Integr. Physiol.* 153A, S80–S80. doi: 10.1016/j.cbpa.2009.04.056

Michaelidis, B., Ouzounis, C., Paleras, A., and Pörtner, H. O. (2005). Effects of long-term moderate hypercapnia on acid-base balance and growth rate in marine mussels *Mytilus galloprovincialis*. *Mar. Ecol. Prog. Ser.* 293, 109–118. doi: 10.3354/meps293109

Nehls, G., and Thiel, M. (1993). Large-scale distribution patterns of the mussel *Mytilus edulis* in the Wadden Sea of Schleswig-Holstein: do storms structure the ecosystem? *Netherlands J. Sea Res.* 31, 181–187. doi: 10.1016/0077-7579(93)90008-G

O'donnell, M. J., George, M. N., and Carrington, E. (2013). Mussel byssus attachment weakened by ocean acidification. *Nat. Clim. Change* 3, 587–590. doi: 10.1038/nclimate1846

Orr, J. C. (2011). "Recent and future changes in ocean carbonate chemistry," in *Ocean Acidification*, eds J.-P. Gattuso and L. Hansson (Oxford: Oxford University Press), 41–66.

Pörtner, H. O. (2010). Oxygen- and capacity-limitation of thermal tolerance: a matrix for integrating climate-related stressor effects in marine ecosystems. *J. Exp. Biol.* 213, 881–893. doi: 10.1242/jeb.037523

Pörtner, H.-O. (2012). Integrating climate-related stressor effects on marine organisms: unifying principles linking molecule to ecosystem-level changes. *Mar. Ecol. Prog. Ser.* 470, 273–290. doi: 10.3354/meps10123

Pörtner, H.-O., Bickmeyer, U., Bleich, M., Bock, C., Brownlee, C., Melzner, F., et al. (2010). "Studies of acid-base status and regulation, p137-166," in *Guide to Best Practices for Ocean Acidification Research and Data Reporting*, eds U. Riebesell, V. J. Fabry, L. Hansson, and J.-P. Gattuso (Publications Office of the European Union), 260. Available online at: http://www.epoca-project.eu/index.php/Home/Guide-to-OA-Research/

Pörtner, H.-O., Boutilier, R. G., Tang, Y., and Toews, D. P. (1990). Determination of intracellular pH and pCO_2 after metabolic inhibition by fluoride and nitrilotriacetic acid. *Respir. Physiol.* 81, 255–274. doi: 10.1016/0034-5687(90)90050-9

Pörtner, H. O., and Farrell, A. P. (2008). Physiology and climate change. *Science* 322, 690–692. doi: 10.1126/science.1163156

Pörtner, H. O., Langenbuch, M., and Reipschlager, A. (2004). Biological impact of elevated ocean CO_2 concentrations: lessons from animal physiology and earth history. *J. Oceanogr.* 60, 705–718. doi: 10.1007/s10872-004-5763-0

Pörtner, H. O., Reipschläger, A., and Heisler, N. (1998). Acid-base regulation, metabolism and energetics in *Sipunculus nudus* as a function of ambient carbon dioxide level. *J. Exp. Biol.* 201, 43–55.

Ramón, M., Cano, J., Peña, J. B., and Campos, M. J. (2005). Current status and perspectives of mollusc (*bivalves* and *gastropods*) culture in the Spanish Mediterranean. *Boletin Instituto Español de Oceanografia* 21, 361–373.

Ramón, M., Fernandez, M., and Galimany, E. (2007). Development of mussel (*Mytilus galloprovincialis*) seed from two different origins in a semi-enclosed Mediterranean Bay (NE Spain). *Aquaculture* 264, 148–159. doi: 10.1016/j.aquaculture.2006.11.014

Range, P., Chícharo, M. A., Ben-Hamadou, R., Piló, D., Fernandez-Reiriz, M. J., Labarta, U., et al. (2014). Impacts of CO_2-induced seawater acidification on coastal Mediterranean bivalves and interactions with other climatic stressors. *Reg. Environ. Change* 14, 19–30. doi: 10.1007/s10113-013-0478-7

Range, P., Piloì, D., Ben-Hamadou, R., ChiiCharo, M. A., Matias, D., Joaquim, S et al. (2012). Seawater acidification by CO_2 in a coastal lagoon environment: effects on life history traits of juvenile mussels *Mytilus galloprovincialis*. *J. Exp. Mar. Biol. Ecol.* 424–425, 89–98. doi: 10.1016/j.jembe.2012.05.010

Reimer, O., and Tedengren, M. (1996). Phenotypical improvement of morphological defences in the mussel *Mytilus edulis* induced by exposure to the predator *Asteria rubens*. *Oikos* 75, 383–390. doi: 10.2307/3545878

Rhein, M., Rintoul, S. R., Aoki, S., Campos, E., Chambers, D., Feely, R. A., et al. (2013). "Observations: ocean," in Climate Change 2013: *the Physical Science Basis*. Contribution of Working Group I to the *Fifth Assessment Report of the Intergovernmental Panel on Climate Change*, eds T. F. Stocker, D. Qin, G.-K. Plattner, M. Tignor, S. K. Allen, J. Boschung, et al. (Cambridge; New York: Cambridge University Press), 255–332.

Rodolfo-Metalpa, R., Houlbrèque, F., Tambutté, E., Boisson, F., Baggini, C., Patti, F. P., et al. (2011). Coral and mollusc resistance to ocean acidification adversely affected by warming. *Nat. Clim. Change* 1, 308–312. doi: 10.1038/nclimate1200

Sarà, G., Manganaro, A., Cortese, G., Pusceddu, A., and Mazzola, A. (1998). The relationship between food availability and growth in *Mytilus galloprovincialis* in the open sea (southern Mediterranean). *Aquaculture* 167, 1–15. doi: 10.1016/S0044-8486(98)00281-6

Spencer Davies, P. (1989). Short-term growth measurements of corals using an accurate buoyant weighing technique. *Mar. Biol.* 101, 389–395. doi: 10.1007/BF00428135

Talmage, S. C., and Gobler, C. J. (2011). Effects of elevated temperature and carbon dioxide on the growth and survival of larvae and juveniles of three species of Northwest Atlantic bivalves. *PLoS ONE* 6:e26941. doi: 10.1371/journal.pone.0026941

The Mermex Group. (2011). Marine ecosystems' responses to climatic and anthropogenic forcings in the Mediterranean. *Prog. Oceanogr.* 91, 97–166. doi: 10.1016/j.pocean.2011.02.003

Thomsen, J., Gutowska, M. A., Saphorster, J., Heinemann, A., Trubenbach, K., Fietzke, J., et al. (2010). Calcifying invertebrates succeed in a naturally CO_2-rich coastal habitat but are threatened by high levels of future acidification. *Biogeosciences* 7, 3879–3891. doi: 10.5194/bg-7-3879-2010

Thomsen, J., and Melzner, F. (2010). Moderate seawater acidification does not elicit long-term metabolic depression in the blue mussel *Mytilus edulis. Mar. Biol.* 157, 2667–2676. doi: 10.1007/s00227-010-1527-0

Touratier, F., and Goyet, C. (2011). Impact of the Eastern Mediterranean Transient on the distribution of anthropogenic CO_2 and first estimate of acidification for the Mediterranean Sea. *Deep-Sea Res. I Oceanogr. Res. Papers* 58, 1–15. doi: 10.1016/j.dsr.2010.10.002

Watson, S. A., Southgate, P. C., Miller, G. M., Moorhead, J. A., and Knauer, J. (2012). Ocean acidification and warming reduce juvenile survival of the fluted giant clam, *Tridacna squamosa. Molluscan Res.* 32, 177–180.

Wolf-Gladrow, D. A., Zeebe, R. E., Klaas, C., Körtzinger, A., and Dickson, A. G. (2007). Total alkalinity: the explicit conservative expression and its application to biogeochemical processes. *Mar. Chem.* 106, 287–300. doi: 10.1016/j.marchem.2007.01.006

Conflict of Interest Statement: The authors declare that the research was conducted in the absence of any commercial or financial relationships that could be construed as a potential conflict of interest.

10

Spatial and historic variability of benthic nitrogen cycling in an anthropogenically impacted estuary

Sarah Q. Foster[1] and Robinson W. Fulweiler[1,2]*

[1] Department of Earth and Environment, Boston University, Boston, MA, USA
[2] Department of Biology, Boston University, Boston, MA, USA

Edited by:
Paulina Martinetto, Consejo Nacional de Investigaciones Científicas y Técnicas-Universidad Nactional de Mar Del Plata, Argentina

Reviewed by:
Isaac R. Santos, Southern Cross University, Australia
Iris C. Anderson, College of William and Mary, USA

*Correspondence:
Robinson W. Fulweiler, Department of Earth and Environment, Boston University, 685 Commonwealth Avenue Rm. 130, Boston, MA 02215, USA
e-mail: rwf@bu.edu

Human activities have dramatically altered reactive nitrogen (N) availability in coastal ecosystems globally. Here we used a gradient of N loading found in a shallow temperate estuary (Waquoit Bay, Massachusetts, USA) to examine how key biogeochemical processes respond to environmental change over time. Using a space-for-time substitution we measured sediment oxygen uptake, dissolved inorganic nitrogen, and di-nitrogen (N_2) gas fluxes from sediments collected at four stations. For two stations we compared measurements to those made at the same locations 20 years ago. Spatial variability was not directly correlated to N loading, however the results indicate significant changes in crucial ecosystem processes over time. Sediment oxygen uptake was only 46% of the historic rate and ammonium flux only 34%. The current rate of net denitrification ($36\,\mu mol\ N_2$-N m^{-2} h^{-1}) was also lower than the mean historic rate ($181\,\mu mol\ N_2$-N m^{-2} h^{-1}). Additionally, at one of the stations we measured a negative average N_2 flux rate, indicating that the sediments may be a net source of reactive N. These changes in benthic flux rates are concurrent with a 39% decline in net ecosystem productivity determined from long-term dissolved oxygen data. Although we cannot rule out year-to-year variability we propose that the differences measured between current and historic rates may be explained in part by concurrent changes found in water temperature, precipitation, and freshwater discharge. These regional forcings have the potential to impact N inputs to the estuary, primary producer biomass, and benthic fluxes by altering the supply of organic matter to the sediments. This work highlights the dynamic nature of biogeochemical cycling in coastal ecosystems and underscores the need to better understand long-term changes.

Keywords: benthic nitrogen cycling, sediment oxygen demand, net denitrification, Waquoit Bay, historic variability, spatial variability

INTRODUCTION

Over the past century human activities have dramatically increased both the magnitude and transport rate of nutrients to coastal ecosystems leading to widespread negative impacts including, eutrophication (Bricker et al., 2008; Selman et al., 2008), a rise in toxic phytoplankton blooms (Smayda, 1990; Anderson et al., 2002; Anderson, 2009), increased frequency and duration of low oxygen events (Diaz and Rosenberg, 2008), and losses of biodiversity (Valiela et al., 1992; Heck et al., 1995; Gong and Xie, 2001). However, the impact of excess nutrients and subsequent eutrophication on benthic nutrient cycling is less understood, particularly in terms of how sediment biogeochemistry may be altered under concurrent environmental changes on larger regional and global scales (Duarte et al., 2009). This is important because sediments in shallow coastal ecosystems play a critical role in ecosystem function by supporting organic matter decomposition and subsequent nutrient regeneration, as well as by influencing water column dissolved oxygen (O_2) concentrations (Nixon, 1981; Boynton and Kemp, 1985; Giblin et al., 1997). Furthermore, coastal sediments serve as an important sink for

reactive nitrogen (N) with the capability of removing 20–50% of external N inputs through denitrification [the microbial transformation of nitrate (NO_3^-) to di-nitrogen (N_2) gas] (Seitzinger and Kroeze, 1998; Galloway et al., 2004; Seitzinger et al., 2006). In this way the sediments of estuaries and near-shore ecosystems act as filters, which can mitigate anthropogenic N locally and diminish export to the ocean. Understanding how N pollution alters this critical ecosystem service is key to long-term coastal ecosystem sustainability.

Excess N can alter sediment biogeochemistry through a variety of complex mechanisms. For example, enhanced water column phytoplankton or macroalgae production following nutrient loading leads to declines in seagrass populations (McGlathery, 2001; Hauxwell et al., 2003) while simultaneously increasing deposition of organic matter to the benthos. In turn, this can enhance rates of sediment decomposition and nutrient mineralization (Nixon, 1981). Increased organic matter deposition to the benthos has also been shown to stimulate rates of denitrification (Seitzinger and Giblin, 1996; Cornwell et al., 1999; Fulweiler et al., 2008). In addition, as macroalgae or phytoplankton decompose dissolved O_2 concentrations drop leading to an increase in the overall duration, intensity, and frequency of hypoxia (dissolved O_2 concentrations less than $2\,mg\ O_2\ L^{-1}$; $62.5\,\mu M$; approximately 30% O_2 saturation Vaquer-Sunyer and Duarte, 2008) and/or anoxia. Low oxygen events have important implications for sediment biogeochemistry in general (Middelburg and Levin, 2009) and the N cycle in particular (Lam et al., 2009; Canfield et al., 2010).

In addition to local scale impacts such as N loading and eutrophication, coastal ecosystems are also subject to larger scale regional and global forcings (e.g., warming, sea level rise, increased precipitation, etc., Cloern et al., 2007; Nixon and Fulweiler, 2009). These environmental changes can greatly impact coastal ecosystem biogeochemistry and ecosystem functioning. For example, in Narragansett Bay winter water temperatures have increased by 1.7°C since the 1970s (Nixon et al., 2009). This temperature increase has been linked to a variety of critical ecological changes, including the loss of the winter-spring diatom bloom (Oviatt et al., 2002) and diminished mean annual water column chlorophyll concentrations (Li and Smayda, 1998). In turn, these water column changes led to decreased sediment O_2 uptake, benthic nutrient fluxes, and denitrification rates (Fulweiler and Nixon, 2009; Fulweiler et al., 2010). Thus, the interactions between global, regional, and local scale forcings are complex, non-linear, and often unexpected (Conley et al., 2009; Duarte et al., 2009). However, understanding how coastal systems will respond to these changes is a critical challenge if we hope to manage and even restore these ecosystems.

The objective of this study was to assess how sediment metabolism and nitrogen cycling varies under different N loads within a shallow temperate estuary, Waquoit Bay (Massachusetts, USA). Additionally, we compare our rates to those measured in Waquoit Bay over two decades ago as a way to examine how these ecosystem functions have changed over time.

MATERIALS AND METHODS
SITE DESCRIPTION
Waquoit Bay is a shallow (mean depth 1.8 m, maximum depth 3 m), temperate estuary on the southwestern shore of Cape Cod (**Figure 1**). The surface area of the main basin and the two major tributaries (Childs River and Quashnet River) is approximately $6\,km^2$. River discharge rates range between 0.1 and $0.4\,m^3\ s^{-1}$ and both are approximately 2 km long and 100 m wide with 10 m bridge abutments that restrict flow (Geyer, 1997). The main basin of the estuary opens to Vineyard Sound but water exchange between the two bodies is restricted by barrier beaches along the southern edge of the bay and water flow is through a small (100 m wide) channel. The tidal range is 0.5 m with a weak spring/neap signal (Geyer, 1997).

Over the past century changes in watershed land use have dramatically increased the rate of land-derived reactive N inputs to Waquoit Bay. In a model developed by Bowen and Valiela (2001), the load of N to Waquoit Bay more than doubled from 1938 to 1990. This increase is concurrent with a tenfold increase in the number of houses in the Childs River watershed (Valiela et al., 1992). Residential septic systems are the principal source of N

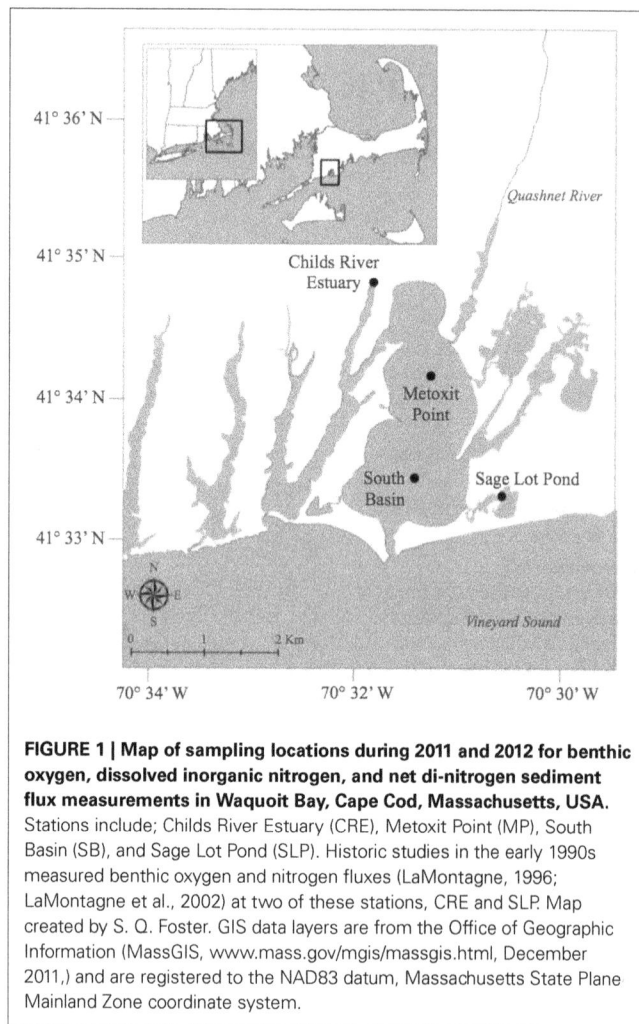

FIGURE 1 | Map of sampling locations during 2011 and 2012 for benthic oxygen, dissolved inorganic nitrogen, and net di-nitrogen sediment flux measurements in Waquoit Bay, Cape Cod, Massachusetts, USA. Stations include; Childs River Estuary (CRE), Metoxit Point (MP), South Basin (SB), and Sage Lot Pond (SLP). Historic studies in the early 1990s measured benthic oxygen and nitrogen fluxes (LaMontagne, 1996; LaMontagne et al., 2002) at two of these stations, CRE and SLP. Map created by S. Q. Foster. GIS data layers are from the Office of Geographic Information (MassGIS, www.mass.gov/mgis/massgis.html, December 2011,) and are registered to the NAD83 datum, Massachusetts State Plane Mainland Zone coordinate system.

because the drainage area has become densely populated and many of the homes are not connected to a sewer system.

Four stations were chosen to capture a range of environmental variability in the Waquoit Bay estuarine system (**Figure 1**). The stations varied in relative external N loading rates, dissolved inorganic nitrogen (DIN) concentrations, chlorophyll a concentrations, sediment characteristics (**Table 1**), and macrophyte biomass (**Table 3**). Childs River Estuary (CRE) is a brackish station located in the tidal section of the river in the northwest region of the Waquoit Bay estuarine system. CRE is one of the two main fresh water sources to Waquoit Bay and receives high N inputs (**Table 1**) (Valiela et al., 1997a, 2000). Two stations in the main basin of Waquoit Bay were also sampled, Metoxit Point (MP) and South Basin (SB). MP is located just south of the Quashnet River, which also carries a high load of N (Valiela et al., 1997a, 2000) from its suburbanized watershed, but this open basin station is more well mixed and has greater tidal exchange than CRE. MP also has high macroalgae biomass and experiences large diel swings in dissolved O_2 concentrations in the summer. SB is located close to the mouth of the estuary where there is enhanced tidal mixing and is presumably less influenced by anthropogenic N inputs. This station receives an average N-load

Table 1 | Characterization of the four sampling stations in Waquoit Bay Massachusetts, USA.

Station (Abbr.)	Latitude (°N, min)	Longitude (°W, min)	Mean Depth (m)	External N load to Sub-Estuary[c] (kg N km^{-2} year^{-1})	Bottom water column characteristics				Sediment characteristics			
					Chl a[d] (µg/L)	Salinity (ppt)	NH$_4^+$ (µmol/L)	NO$_2^-$ + NO$_3^-$ (µmol/L)	Density (g/mL)	Porosity	% C	C:N
Childs River Estuary (CRE)	41° 34.805	70° 31.826	1.3[a]	41.9 * 10^3	29.4 (±5.0)	28.3	8.41 (±1.18)	0.48	1.38 (±0.05)	0.46 (±0.02)	1.12 (±0.32)	10.3 (±0.3)
Metoxit Point (MP)	41° 34.134	70° 31.272	2.1[a]	50.0 * 10^{3f}	4.5 (±1.1)	30.2	3.76 (±0.93)	0.11 (±0.04)	1.06 (±0.02)	0.56 (±0.02)	6.11 (±0.28)	8.6 (±0.2)
South Basin (SB)	41° 33.404	70° 31.442	1.8[b]	n.m.	n.m.	30.6	3.08 (±0.83)	0.14 (±0.09)	1.48 (±0.01)	0.42 (±0.01)	1.02 (±0.07)	8.9 (±0.4)
Sage Lot Pond (SLP)	41° 33.270	70° 30.584	1.1[a]	2.1 * 10^3	7.5 (±2.9)	29.7	1.78 (±0.95)	<MDL	1.07 (±0.02)	0.50 (±0.01)	5.93 (±0.37)	8.9 (±0.5)

Water column grab samples for chlorophyll a (chl a) analysis were taken on a near monthly basis by staff of the Waquoit Bay National Estuarine Research Reserve (WBNERR). The chl a values represent the mean (± standard error) of samples collected July–October in 2012 (n = 4 for the three WBNERR stations. Note that samples were not collected by WBNERR in 2011). Bottom water dissolved inorganic nitrogen (ammonium NH$_4^+$) and nitrite plus nitrate (NO$_2^-$ + NO$_3^-$) samples were collected from cur control cores at the start of each incubation. The water in these cores was pumped and filtered (to 0.2 µm pore size) in the field, approximately 20 h before being sampled for inorganic nitrogen concentrations. Concentrations below analytical minimum detection limits (MDL) are noted. Values that were not measured are indicated by n.m. Values are the mean (± standard error) for incubations conducted July–October 2011 and 2012 (For NH$_4^+$, n = 3 (CRE), n = 5 (MP), n = 5 (SB), n = 3 (SLP). And for NO$_2^-$ + NO$_3^-$, n = 1 (CRE), n = 2 (MP), n = 2 (SB), n = 0 <MDL (SLP)). Sediment porosity and density samples were taken at the end of incubations from each sediment core. Values represent the mean (± standard error) of pooled sediment samples taken at 1 cm increments down to 4 cm depth averaged across all cores taken during the study period (July–October 2011 and 2012) [n = 24 (CRE), n = 54 (MP and SB), n = 40 (SLP)]. Sediment samples for percent carbon (% C) and molar carbon to nitrogen ratios (C:N) were collected from sediment cores at the end of incubations in 2011. These values also represent the mean (± standard error) of pooled samples taken at 1 cm increments down to 4 cm depth [n = 4 (CRE), n = 8 (MP, SB and SLP)].

[a]*NOAA, 2012;*
[b]*Main basin mean depth (D'Avanzo and Kremer, 1994);*
[c]*Valiela et al., 1997a, 2000 (watershed N loads) and D'Avanzo et al., 1996 (sub-estuary surface areas);*
[d]*NOAA, 2012;*
[f]*N load for Metoxit Point is estimated from loading rates to the Quashnet River sub-wa'ershed.*

from multiple Waquoit Bay sub-watersheds, so we assumed an intermediate N load relative to the other stations but no values are available. The final station sampled we sampled was Sage Lot Pond (SLP), a small (surface area 0.17 km^2) lagoon located within a forested watershed of protected state lands, which delivers a low N load relative to the other stations (Valiela et al., 1997a, 2000). SLP is one of the only regions within the Waquoit Bay estuarine system where native eelgrass (*Zostera marina*) stands are still present. The CRE and SLP sampling stations in this study are in the same region as historical stations where benthic O_2 and N fluxes with sealed cores (LaMontagne, 1996; Kirkpatrick and Foreman, 1998; LaMontagne et al., 2002), benthic chambers (Hurlbut et al., 1994), and total ecosystem metabolism (D'Avanzo et al., 1996) were previously measured over the past 20 years. Three of the stations (CRE, MP, and SLP) in this current study correspond to long-term monitoring stations that are part of the National Estuarine Research Reserve (NERR) System-Wide Monitoring Program (SWMP).

NITROGEN LOADING: HISTORIC CALCULATION AND CURRENT ESTIMATE

We calculated historic N loads to Waquoit Bay sub-estuaries for the early 1990s using the total N load to sub-watersheds from the Waquoit Bay Nitrogen Loading Model (NLM) (Valiela et al., 1997a, 2000) divided by the estuary surface areas (D'Avanzo et al., 1996). The Waquoit Bay NLM calculates the sub-watershed N-loads through 1990 (Valiela et al., 1997a, 2000; Bowen and Valiela, 2001) and we assume relative changes in the N load based on well-developed relationships with land use/land cover (Boyer et al., 2002), population (Caraco and Cole, 1999) and housing densities (Valiela et al., 1992). Using the Boyer et al. model (2002) and reclassified land use data (MassGIS), we estimated the current N load to the sub-estuaries. The MassGIS data classifies 21 different types of land-use. We reclassified these land-uses into 3 simple categories: "disturbed," "non-disturbed," or "other." Disturbed land-use categories included all residential, commercial, industrial, transportation, waste disposal areas, and cropland. Non-disturbed included forest, pasture, wetlands, open land, and woody perennial regions (e.g., orchards, nurseries, bogs). Under these classifications, the disturbed area of the Childs River Estuary sub-watershed increased by nearly 90% between the early 1970s and the start of the 2000s. This trend is driven by an increase in residential and commercial regions, a trend that continues into the 2000s (**Figure 2**). In 1999 more than 40% of the CRE sub-watershed was designated as disturbed compared to 20% in 1971. In the Sage Lot Pond sub-watershed, the land area was relatively less disturbed and only showed a small increase (~3%) in disturbed areas between 1971 and 1999 (**Figure 2**). These differences in land use between the sub-watersheds is largely responsible for the observed historic differences in N loads (Valiela et al., 2000) and the estimated current N loads. Although the rate of building and watershed development decreased in the 2000s, there have been no substantial changes in the watershed that would impact the overall N-load budget. Therefore, we assume that based on land use in Waquoit Bay sub-estuaries and well established relationships (Valiela et al., 1992; Caraco and Cole, 1999; Boyer et al., 2002) that the current N-load is relatively higher in

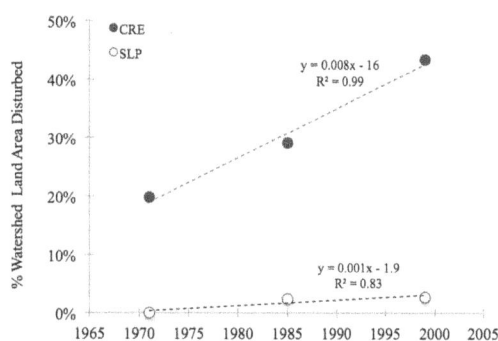

FIGURE 2 | Thirty years of land use/land cover change in two Waquoit Bay sub-watersheds, Childs River Estuary (CRE, closed gray circles) and Sage Lot Pond (SLP, open circles). Disturbed areas were defined as regions under high human influence. This included the following land use categories (data from MassGIS, www.mass.gov): residential, commercial, industrial, transportation, waste disposal and cropland. Equations and the coefficient of determination (R^2) for linear models are displayed next to the curves.

Childs River Estuary compared to Sage Lot Pond. Additionally, Childs River Estuary is likely to have experienced the greatest increase in N-load over the past 40 years.

SEDIMENT CORING AND FIELD MEASUREMENTS

We collected sediment cores on seven occasions for sediment O_2 uptake, benthic nitrogen fluxes, and net sediment N_2 fluxes at the four sampling stations over 2 years (2011 and 2012) in the summer and fall months (July–October). Each station was sampled three to six times during this period (**Table 2**). Triplicate or quadruplicate intact sediment cores were collected using a pull corer, which maintains the vertical structure of the sediment cores. The core tubes (30 cm height, 10 cm diameter) are made of clear polyvinyl chloride (PVC) material. The sediment height in the cores collected ranged from 10 to 18 cm. After collection the cores were capped and kept in the dark in a cooler in order to maintain approximate field temperature until they were brought to an environmental chamber at Boston University, which was set to the average *in situ* bottom water temperature. The time between coring and placement in the environmental chamber was less than 7 h. Dissolved O_2 concentrations, salinity, and water temperature (Hach HQd using LDO101, CDC401 probes) were collected at the surface and bottom at each station.

INCUBATION SET UP AND SAMPLING

After the cores arrived at Boston University they were immediately placed in a water bath in the environmental chamber. The sediment cores were kept in the dark, uncapped, and bubbled gently with ambient air overnight (~12 h) to keep the overlaying water oxygenated without disturbing the surface of the sediment (Fulweiler et al., 2008; Fulweiler and Nixon, 2009). All of our incubations were conducted in the dark. We conducted dark only incubations because we used the N_2/Ar technique for measuring N_2 fluxes which is highly sensitive to bubble formation created by photosynthesis in the light

(Eyre et al., 2002). Additionally, we conducted dark incubations because we wanted to compare these fluxes to those measured historically and all previous incubations were conducted in the dark.

We conducted two separate incubations. The first incubation collected samples for DIN fluxes and the second for the flux of N_2 gas (Fulweiler and Nixon, 2009; Vieillard and Fulweiler, 2012). We separated these incubations in order to minimize the sample volume taken from the cores at each time point thus decreasing potential dilution impacts. The two incubations were separated by 8–12 h (overnight) and cores were again bubbled gently with ambient air. Before each of the core incubations we carefully siphoned off the water overlaying the sediment and replaced it with 0.2 μm filtered, station-specific bottom water. Each core was then sealed with no air headspace with an acrylic gas-tight lid equipped with inflow and outflow ports. This process was done using inflow of the 0.2 μm filtered station water in order to prevent bubbles in the tops of the sealed cores. Magnetic stir bars (8 cm length), fixed to the core tops (Banta et al., 1995; Fulweiler et al., 2010), provided gentle mixing of the overlaying water (~40 revolutions min^{-1}) with minimal re-suspension of sediments. The incubation temperatures ranged from 19.5 to 26°C. Incubation time was determined by balancing the following criteria: achieving a 2 mg L^{-1} drop in dissolved oxygen, five sampling time points at intervals of at least 60 min, and dissolved O_2 concentrations kept above the hypoxic threshold (>2 mg O_2 L^{-1}). Our incubations ranged between 6 and 9 h. The total time between core collection and the end of the second incubation ranged between 48 and 55 h.

At each time point water samples were immediately filtered into 30 ml acid washed and deionized water leached polyethylene bottles using a 60 ml acid washed polypropylene syringe and glass fiber filters (Whatman GF/F, 0.70 μm pore size). The vials were stored in a freezer (at approximately −15°C) until analysis for DIN (ammonium (NH_4^+), nitrite (NO_2^-), and nitrate (NO_3^-)). Dissolved O_2 concentrations were measured at 3 time points (initial, middle, and final) using an optical dissolved oxygen sensor (Hach LDO101). At the start of the second incubation we replaced the overlying water with filtered (0.2 μm) station water and the cores were sealed with the gas tight lids and the incubation proceeded. We collected duplicate water samples in 12 ml Labco Limited Exetainer® vials with gas tight septa for dissolved N_2/Ar gas analysis at five time points and preserved them with 25 μl of saturated zinc chloride solution (Fulweiler and Nixon, 2009, 2011). Sample vials were filled from the bottom with approximately a 12 ml overflow volume, preserved, and then immediately capped in order to minimize atmospheric gas contamination.

At the end of the second incubation we uncapped the cores and bubbled them over night (approximately 12 h). The following day we took sediment sub-samples from each core to analyze for sediment porosity, density (in 2011 and 2012) and sediment percent carbon and molar carbon to nitrogen ratios (in 2011). Sub-core samples were taken using acid washed 60 mL polycarbonate syringes with their ends cut off. We sectioned sub-cores at 1 cm increments up to 4 cm. Sediment samples were then frozen and stored in a −80°C freezer until analysis.

Table 2 | Summary of sediment oxygen (O$_2$), ammonium (NH$_4^+$), nitrite (NO$_2^-$) and di-nitrogen gas per mole nitrogen (N$_2$-N) flux rates measured across the sediment-water interface in 2011 and 2012 (July–October).

Station (Abbr.)	Date	Sal	Temp	O$_2$	NH$_4^+$	NO$_2^-$	N$_2$-N
		ppt	°C	μmol O$_2$ m^{-2} h^{-1}	μmol N m^{-2} h^{-1}	μmol N m^{-2} h^{-1}	μmol N m^{-2} h^{-1}
Sage Lot Pond (SLP)	29-Jul-2011	30.0	25.2	−2394 (±652)	11 (±60)	−2.0 (±0.2)	16 (±8.0)
	24-Aug-2011	29.3	24.5	−3016 (±206)	306 (±86)	−2.4 (±1.6)	64 (±64)
	6-Aug-2012	30.5	30.0	−2120 (±124)	310 (±68)	0.0	n.m.
South Basin (SB)	7-Jul-2011	31.9	26.0	−2551 (±452)	170 (±96)	−0.3 (±0.3)	9.1 (±9.1)
	24-Aug-2011	31.5	24.5	−1164 (±229)	3.0 (±10)	−1.7 (±0.3)	−7.6 (±23)
	11-Oct-2011	29.9	19.5	−1664 (±268)	0.0 (±0.0)	−0.2 (±0.2)	28 (±16)
	11-Jul-2012	31.6	29.0	−1561 (±277)	27 (±21)	−0.2 (±0.4)	44 (±34)
	2-Oct-2012	31.4	20.0	−1117 (±115)	29 (±22)	−1.0 (±0.2)	n.m.
Metoxit Point (MP)	7-Jul-2011	30.5	26.0	−3053 (±378)	225 (±15)	1.6 (±1.6)	−19 (±7.2)
	24-Aug-2011	30.1	24.5	−2233 (±122)	167 (±56)	−0.8 (±0.6)	−28 (±12)
	11-Oct-2011	29.5	19.5	−2908 (±134)	231 (±97)	−1.0 (±0.4)	−4.0 (±26)
	11-Jul-2012	31.3	29.0	−1538 (±144)	38 (±38)	−0.0 (±0.3)	−19 (±39)
	6-Aug-2012	29.5	30.0	−2456 (±281)	106 (±30)	−0.8 (±0.1)	n.m.
	2-Oct-2012	30.0	20.0	−1838 (±182)	365 (±76)	−0.3 (±0.3)	n.m.
Childs River Estuary (CRE)	7-Jul-2011	27.7	26.0	−1589 (±202)	193 (±60)	0.0 (±0.0)	22 (±11)
	29-Jul-2011	28.5	25.2	−1609 (±205)	99 (±34)	−4.6 (±1.3)	18 (±9.8)
	11-Jul-2012	30.1	29.0	−1614 (±318)	−2 (±61)	−0.7 (±0.2)	62 (±3.0)

O$_2$ fluxes are from the first dissolved inorganic nutrient incubation. Positive fluxes represent a net efflux of the analyte from the sediments into the water column, and negative fluxes represent a net influx of the analyte into the sediments from the water column. A net zero flux represents either a balance between analyte production and consumption process in the sediments and/or rates that are below our detection limit. Stations are listed from low to high N loading. Values represent the station means (± standard error) from triplicate or quadruplicate cores. Note that n.m. signifies parameter fluxes for which there is no measurement.

SAMPLING ANALYSIS AND FLUX CALCULATIONS

Dissolved inorganic nitrogen concentrations were determined by high-resolution digital colorimetry on a Seal Auto Analyzer 3 with segmented flow injection using standard techniques (Solorzano, 1969; Johnson and Petty, 1983; Hansen and Koroleff, 1999). Our laboratory mean minimum detection limits (MDL) for the period of this study are 0.164, 0.051, and 0.005 μM for NH$_4^+$, NO$_x$ (NO$_3^-$ + NO$_2^-$), and NO$_2^-$ respectively.

Dissolved gas samples were analyzed for concentrations of N$_2$ and Ar using a quadrupole membrane inlet mass spectrometer (MIMS) using a technique developed by Kana et al. (1994) to measure rates of sediment denitrification (Kana et al., 1998; Eyre et al., 2002; Fulweiler and Nixon, 2009). This method is precise (±0.03% for N$_2$/Ar), rapid (~20–30 samples h^{-1}) and requires a small sample size (<10 ml). N$_2$ concentrations were determined using the measured ratio of N$_2$/Ar multiplied by the calculated theoretical Ar concentration based on solubility constants of Weiss (1970) and given the temperature and salinity of the sample (Colt, 1984; Kana et al., 1998). The N$_2$/Ar technique gives a measure of net N$_2$ flux (i.e., gross denitrification minus gross nitrogen fixation).

Flux rates across the sediment-water interface were determined by first calculating the linear regression slope (R squared value ≥0.65) of the net parameter (e.g., O$_2$, NH$_4^+$, NO$_3^-$, NO$_2^-$, N$_2$) concentration over the incubation time. This rate was then multiplied by the sediment core water column volume and then divided by the sediment core surface area. We report mean flux rates of O$_2$, NH$_4^+$, NO$_2^-$, and N$_2$ from triplicate or quadruplicate cores (± standard error) from a given sampling station and date in consistent units (micromoles per meter squared per hour, μmol m^{-2} h^{-1}). Positive fluxes represent a net efflux of the parameter out of the sediments into the water column, and negative fluxes represent a net influx of the parameter into the sediments from the water column. A net zero flux represents either a balance between analyte production and consumption process in the sediments and/or rates that are below our detection limit.

Concentration values for NO$_x$ were consistently low and in many cases below the MDL of the Seal Auto Analyzer. As a result of this, NO$_x$ concentrations were often less than the measured concentration of NO$_2^-$ yielding a negative value for NO$_3^-$. Therefore, although we were successfully able to calculate fluxes for NO$_2^-$ for most cores we were unable to do so for NO$_3^-$.

Sediment characteristic methods (i.e., sediment porosity, density, percent carbon and nitrogen) followed those outlined in the Protocol Handbook for NICE (Nitrogen Cycling in Estuaries) (Dalsgaard et al., 2000). Analysis of sediment percent carbon and nitrogen was conducted at the Boston University Stable Isotopes Lab, Boston, Massachusetts, USA.

HISTORIC COMPARISON METHODS

In an effort to put our Waquoit Bay sediment flux rates in a larger temporal context we compared our data to all available historical studies for this site. Most of the historical data come from a comprehensive effort in the early 1990s lead by LaMontagne and Valiela (LaMontagne and Valiela, 1995; LaMontagne, 1996;

LaMontagne et al., 2002). In these studies, sediment fluxes of O_2, DIN, and N_2 gas, were measured using similar methods to our current 2011–2012 study. Others also measured DIN fluxes (Kirkpatrick and Foreman, 1998) in Waquoit Bay and we have compared our 2011-2012 rates to those studies as well.

All of the values used in our historic comparisons were derived from near identical field and incubation procedures. For example, sediment cores were collected from similar locations within Waquoit Bay and then transported to laboratories where sealed core incubations were conducted in the dark and at ambient temperatures. Water samples were taken from the cores at discrete time points during the incubations and analyzed for the various parameters. The duration of incubations in the LaMontagne (1996) study, like those we report here, were based on the O_2 uptake rate. The Kirkpatrick and Foreman (1998) study used a set incubation time of 24 h. All studies used the same colorimetric techniques to determine dissolved nitrogen species concentrations.

Methods differ between the studies in the analysis of dissolved gas samples. In the LaMontagne studies (LaMontagne and Valiela, 1995; LaMontagne, 1996; LaMontagne et al., 2002), dissolved gas samples were analyzed for concentrations of O_2 and N_2 using a custom gas stripper connected to a gas chromatograph (GC) (Devol, 1991). This analysis was designed to precisely (<0.5% coefficient of variance) measure dissolved gasses by injecting a fixed volume of liquid, completely stripping the gasses from the water sample into the carrier gas of a GC and separating N_2 and O_2 into discreet peaks. The peaks were quantified with a Perkin Elmer 9310 gas chromatograph equipped with a thermal conductivity detector. In our current study, we measured dissolved O_2 concentrations over the course of the incubation with a Hach LDO101 oxygen sensor and N_2 concentrations were determined by the N_2/Ar technique on a MIMS.

Although the GC gas stripping and the MIMS N_2/Ar methods clearly differ in terms of the instruments used to estimate denitrification, they are similar in that they both are techniques that directly measure the concentration of N_2. Currently, to our knowledge, there are no published simultaneous comparisons between the GC method and the N_2/Ar technique. However, we acknowledge that differences in denitrification rates may be driven, at least in part, by methodical issues.

As part of our historic comparison we also calculated net ecosystem daytime productivity and nighttime respiration using the same change in oxygen concentration method which was first proposed by Odum (1956) and modified according to Thébault (Thébault et al., 2008). We computed rates for our sampling years (2011–2012) at Childs River Estuary, the station where historic measurements were conducted in the early 1990s (D'Avanzo et al., 1996). We used conductivity, temperature, and dissolved oxygen data from the Waquoit Bay National Estuarine Research Reserve (WBNERR) System-Wide Monitoring Program (SWMP) (NOAA, 2012). All input data was measured at 15-min intervals using a YSI model 6600 data logger equipped with a Luminescence Lifetime ProODO optical sensor and is deployed and maintained by WBNERR. The data were carefully examined

and we removed any time points flagged by the WBNERR staff as potentially erroneous. We did not eliminate data based on severe weather events. The historic study (D'Avanzo et al., 1996) used an ENDECO/YSI 1184 rapid-pulsed Clark-type polarographic electrode (membrane technology) to determine dissolved O_2 concentrations. In our 2011–2012 study, the Childs River Estuary data-sonde was a YSI 6600v2 equipped with a 6150 ROX Optical dissolved oxygen probe. Side-by-side simultaneous measurement of dissolved O_2 concentrations using these two sensors has shown agreement within 1% (Thébault et al., 2008).

The basic method for calculating daily productivity and respiration rates uses rates of dissolved O_2 concentration change over a 24-h time period (Odum, 1956). The rate of dissolved O_2 change during the day is net primary production, while at night it is respiration. The relationship is represented by the mass balance equation:

$$dDO/dt = P - R + D \tag{1}$$

Where dDO is the change in concentration of dissolved oxygen (mg O_2 L^{-1}), dt is the change in time (h), P is the rate of gross photosynthesis and R is the absolute value of respiration, and D is the diffusional flux of oxygen between the water-air interface (mg O_2 L^{-1} h^{-1}). The diffusional flux (D) is dependent on wind driven turbulence on the water surface and the concentration gradient between the dissolved oxygen concentration in the water (DO) and the hypothetical saturation concentration at equilibrium with the atmosphere (DO_s). The relationship is described by:

$$D = k(DO_s - DO) \tag{2}$$

In our calculations we set the exchange coefficient (k) equal to 0.5 h^{-1}. While this exchange coefficient is generally representative of estuarine tributaries (Caffrey, 2004), the coefficient can vary with wind conditions and other studies use wind speeds measured concurrently to dissolved oxygen concentrations to determine k (Marino and Howarth, 1993; D'Avanzo et al., 1996; Thébault et al., 2008; Howarth et al., 2014). Therefore, our measure of the diffusional flux is simplistic and may at times be over or underestimated. The DO_s concentration is a function of water salinity and temperature (Benson and Krause, 1984). When the DO concentration in the water exceeds the saturation concentration, the diffusive rate is negative indicating oxygen flux out of the water and when the DO concentration is less than the saturation concentration, the diffusive rate is positive indicating diffusion of oxygen from the atmosphere into the water.

Dissolved O_2 change can also be a result of exchange with other water bodies through advection, but in Waquoit Bay this has been determined to be a negligible process (D'Avanzo et al., 1996; Caffrey, 2003). D'Avanzo et al. (1996) calculated that the tidal excursion for Childs River Estuary was 600 m. This excursion value is less than the length of estuary indicating that tidal advection is likely to be confined within the sub-estuary region during the intervals over which productivity and respiration are measured. In addition, Caffrey (2003) tested the impact of horizontal advection on metabolism calculations for a 6-week period at the end of 1998 in Waquoit Bay using WBNERR sondes at Central Basin and Metoxit Point stations. It was found that production and respiration calculated from these data did not significantly differ between the two stations (Caffrey, 2003). In our calculations we also assumed negligible effects of advection based on the findings of D'Avanzo et al. (1996). Additionally, we did not find a correlation between dissolved O_2 concentrations and salinity ($R^2 = 0.002$), again indicating that dissolved O_2 concentrations where not greatly impacted by tidal advection.

DATA ANALYSIS

All statistical tests were conducted using the JMP software package (version 10.0.0, copyright 2012, SAS Institute Inc.). To determine if flux rates were significantly different from each other we ran two-way analysis of variance (ANOVA) with sampling date and station as the main effects. We then conducted paired student's t test *post-hoc* to further examine where differences existed in our datasets. We determined outliers as points that exceed 1.5 times the interquartile range beyond the first or third quartiles. For all statistical analyses we interpreted a p value ≤ 0.05 to show that the observed trends and correlations were statistically significant.

RESULTS

SEDIMENT OXYGEN UPTAKE

Sediment O_2 uptake from the DIN incubation ranged ($\pm$ standard error) from 1117 ($\pm$115) to 3053 ($\pm$378) μmol O_2 m^{-2} h^{-1} with a mean of 2025 ($\pm$151) μmol O_2 m^{-2} h^{-1} (**Table 2**). O_2 uptake measured from the DIN incubation (conducted first) was significantly lower ($p = 0.04$) than the uptake measured during the incubation for N_2 samples by approximately 14%. In order to be consistent, we use only the O_2 uptake values measured from the first DIN incubation for our comparison analyses. Oxygen uptake values differed significantly across the stations sampled because Metoxit Point and Sage Lot Pond had greater O_2 uptake rates compared to South Basin and Childs River Estuary ($p < 0.01$) (**Figure 3A**). Three dates (7-July-2011, 11-Oct-2011, and 6-Aug-2012) were significantly higher than two other sampling dates (11-July-2012 and 2-Oct 2012) but the data exhibited no correlation with temperature ($R^2 < 0.001$, p $= 0.83$, temperature range 20–30°C).

BENTHIC DISSOLVED INORGANIC NITROGEN FLUXES

We measured positive fluxes of ammonium (NH_4^+) from the sediments at all stations over the course of this study except on one occasion (11-July-2012) where we measured mean NH_4^+ uptake (negative flux) from Childs River Estuary sediments. Additionally,

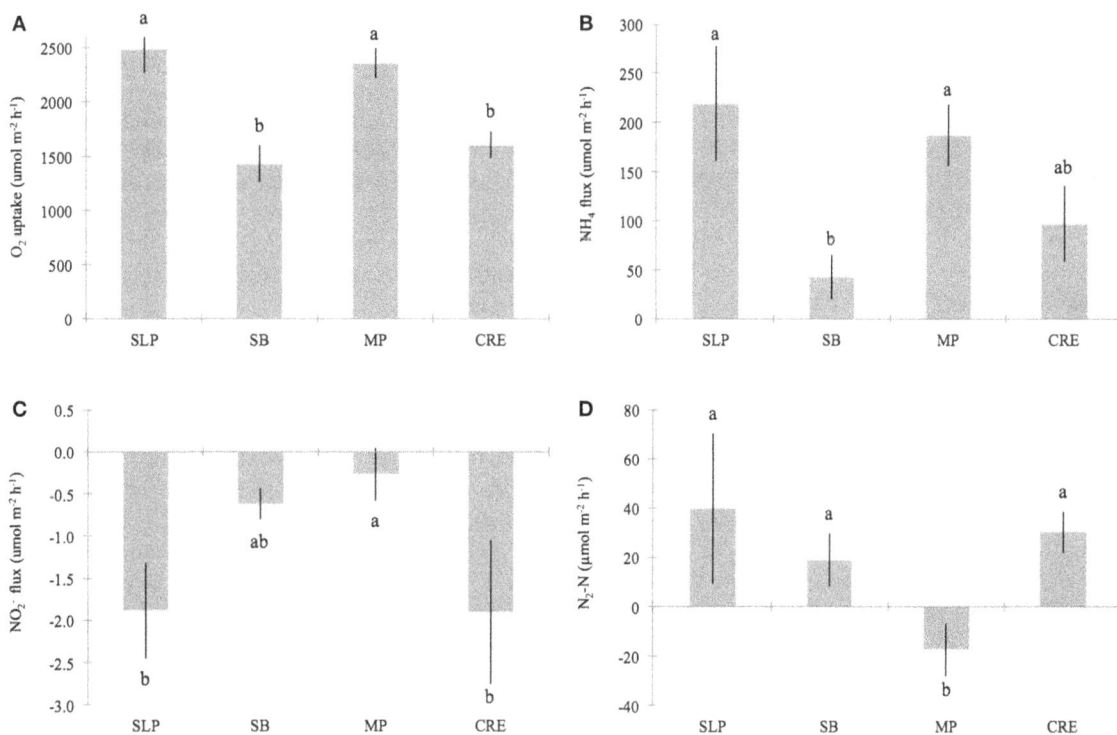

FIGURE 3 | Benthic fluxes for (A) oxygen (O_2) uptake, (B) ammonium (NH_4^+), (C) nitrite (NO_2^-) and (D) di-nitrogen gas per mole nitrogen (N_2-N), across the sediment-water interface in Waquoit Bay. Bars represent the station means ($\pm$ standard error) of all cores measured on seven 2011–2012 sampling dates for O_2 uptake, NH_4^+ and NO_2^- fluxes (plots **A–C**) and five sampling dates 2011–2012 for N_2-N fluxes (plot **D**). Stations organized from relatively low to high external nitrogen load: Sage Lot Pond (SLP, $n = 10$ (plots **A–C**), $n = 6$ (plot **D**)), South Basin (SB, $n = 16$ (plots **A–C**), $n = 13$ (plot **D**)), Metoxit Point (MP, $n = 21$ (plots **A–C**), $n = 12$ (plot **D**)), and Childs River Estuary (CRE, $n = 9$ (plots **A–C**), $n = 8$ (plot **D**)). Note that we did find a significant difference between dates for NO_2^- fluxes ($p < 0.01$), but not for the other flux parameters. Lower case letters above the bars that are not the same indicate fluxes that are significantly different from each other ($\alpha = 0.05$).

on two occasions at the South Basin station and one occasion at the Sage Lot Pond station, individual cores showed NH_4^+ uptake (although the mean of triplicate cores per station was a positive flux). The mean ($\pm$ standard error) NH_4^+ flux rate across all sampling dates and stations was 136 ($\pm$30) μmol N m^{-2} h^{-1} with a range of -2 ($\pm$61) (Childs River Estuary) to 365 ($\pm$76) (Metoxit Point) μmol N m^{-2} h^{-1} (**Table 2**).

Ammonium fluxes were significantly higher at Sage Lot Pond and Metoxit Point than South Basin ($p = 0.002$ for both Sage Lot Pond and Metoxit Point). Childs River Estuary did not differ significantly from the other stations. Additionally, while samples taken on 7-July-2011 had higher rates than samples from 29-July-2011 and 11-July-2012, there was no significant correlation between NH_4^+ flux rates and temperature ($R^2 = 0.004$, p $= 0.650$). Ammonium flux rates were positively correlated with O_2 uptake rates ($R^2 = 0.34$, p $= 0.014$) with high rates of NH_4^+ flux corresponding to high rates of O_2 uptake (**Figures 3A,B**).

In general, we observed low sediment NO_2^- fluxes (**Figure 3C**) with a range of mean rates ($\pm$ standard error) from -4.6 ($\pm$1.3) to 1.6 ($\pm$0.2) μmol N m^{-2} h^{-1} (**Table 2**). Production (positive flux) of NO_2^- was measured only once over the sampling period (Metoxit Point, 7-July-2011, 1.6 ± 1.6 μmol N m^{-2} h^{-1}, $n = 3$). The highest NO_2^- uptake (-4.6 ± 1.3 μmol N m^{-2} h^{-1}, $n = 3$) was from Childs River Estuary sediments collected at the end of July. Nitrite flux rates differed significantly across three sampling dates in 2011 (7-July-2011, 29-July-2011, 24-August-2011) but not in 2012. Fluxes did not correlate significantly with temperature ($R^2 = 0.002$, $p = 0.773$). Metoxit Point had higher flux rates than Childs River Estuary ($p = 0.022$) and Sage Lot Pond ($p = 0.019$) but did not differ significantly from South Basin ($p = 0.363$). Nitrite fluxes did not vary significantly as a function of O_2 uptake or NH_4^+ fluxes for the data set as a whole.

DIRECT MEASURE OF SEDIMENT NET N₂ FLUX

Fluxes of N_2 varied among stations with three stations exhibiting positive (net denitrification) and one station (Metoxit Point) exhibiting negative (net N fixation) fluxes (**Figure 3D**). At Metoxit Point, net N_2 flux was significantly lower than the Sage Lot Pond ($p = 0.012$), South Basin ($p = 0.042$) and Childs River Estuary ($p = 0.021$). Mean net N_2 flux across all stations and sampling dates in 2011-2012 was 14 ($\pm$8.3) μmol N_2-N m^{-2} h^{-1} and did not differ significantly with temperature ($R^2 = 0.002$, $p = 0.766$). Mean net denitrification (positive fluxes only) was 33 ($\pm$7.5) μmol N_2-N m^{-2} h^{-1}, and net N fixation (negative values only) was -16 ($\pm$4.3) μmol N_2-N m^{-2} h^{-1}. We observed the highest rate of net denitrification (190 μmol N_2-N m^{-2} h^{-1}) from a Sage Lot Pond core on 24-Aug-2011. Overall, the N_2 fluxes from Sage Lot Pond had the highest mean N_2 flux rates but also the highest variability between cores compared to the other stations. The highest mean net N fixation rate (-28 ± 12 μmol N_2-N m^{-2} h^{-1}, $n = 3$) was measured in sediments collected from the Metoxit Point during August 2011. In fact, we measured net N fixation in 9 out of 12 Metoxit Point sediment cores incubated over the sampling period and overall mean rates from Metoxit Point. The fluxes of net N_2

from Waquoit Bay sediments were not correlated to sediment O_2 uptake ($p = 0.863$).

DISCUSSION

SPATIAL VARIABILITY

Despite the spatial gradient of N loads in Waquoit Bay the observed benthic fluxes did not follow the trend of N loading (**Figure 3**). Water column and sediment characteristics also did not vary as function of N loading (**Table 1**). Although many studies have found a positive relationship of N load to these benthic flux rates (Seitzinger, 1988; Nowicki, 1994) others have found little to no relationship between nutrient enrichment and net denitrification rates (Oviatt et al., 1995; Fulweiler et al., 2008). Similarly, the historic Waquoit Bay studies (LaMontagne and Valiela, 1995; LaMontagne, 1996; LaMontagne et al., 2002) also did not find a significant relationship between N load and O_2 uptake, NH_4^+, NO_3^-, or net N_2 flux rates. This suggests that in Waquoit Bay factors other than N load may be of equal or greater importance in determining organic matter deposition to sediments and subsequently the rates of benthic metabolism and N cycling dynamics.

SEDIMENT OXYGEN UPTAKE

We measured a significant difference in O_2 uptake rates between the four stations, however the relationship between N load and O_2 uptake was not a positive correlation as has previously been found in other estuarine systems (Nixon, 1981). For example, the Childs River Estuary receives the highest N inputs from its sub-watershed relative to the other stations however O_2 uptake was relatively low. We attribute this pattern to the low concentrations of dissolved O_2 in the Childs River Estuary water column when we collected our sediment cores and station water for incubations, 5.1 ± 0.6 mg O_2 L^{-1} (mean $\pm$ standard error). These concentrations were 2.0–3.6 mg O_2L^{-1} lower than the concentrations measured at the other stations sampled on the same day.

The highest rates of O_2 uptake were measured at Metoxit Point and Sage Lot Pond. The similarity in O_2 uptake rates between these two stations is intriguing as the Metoxit Point station is heavily influenced by anthropogenic N while Sage Lot Pond has the lowest N load (**Table 1**). Large, pervasive, filamentous macroalgae mats of *Cladophora vagabunda* and *Gracilaria tikvahiae* characterize Metoxit Point. In contrast, the dominant primary producer in Sage Lot Pond is the rooted eelgrass, *Zostera marina*, (**Table 3**) which once covered the entire benthos of the Waquoit Bay estuary (Valiela et al., 1992). Although these systems have different N loads and ecological structures, they are both highly productive regions of the bay and have organic rich sediments. The percent carbon content of both of these stations was similar ($6.11 \pm 0.28\%$ and $5.93 \pm 0.37\%$, Metoxit Point and Sage Lot Pond respectively), as was their porosity and density compared to the other two stations (**Table 1**). The South Basin station is moderately impacted by anthropogenic activities because of its location at the mouth of the estuary, which puts it relatively far from land-based N loading and it is the most tidally flushed and well mixed site. At this station we measured lower oxygen demand than both Metoxit Point and Sage Lot Pond but rates similar to

Table 3 | Comparison of mean macrophyte biomass in grams dry weight per meter squared (g d.w. m^{-2}) between surveys conducted in the early 1990s (Hersh, 1996) and the early 2000s (Fox, 2008; Fox et al., 2008).

Sub-Estuary	Macrophyte Biomass (g d.w. m^{-2}) (Hersh, 1996)				Macrophyte Biomass (g d.w. m^{-2}) (Fox, 2008; Fox et al., 2008)			
	Cladophora vagabunda (1990–1992)	*Gracilaria tikvahiae* (1990–1992)	Total Macroalgae (all species) (1990–1992)	*Zostera marina* (1990–1992)	*Cladophora vagabunda* (2002–2003)	*Gracilaria tikvahiae* (2002–2003)	Total Macroalgae (all species) (2002–2003)	*Zostera marina* (1994–1996)
CRE	258	107	383	0	75 ($\pm$14)	49 ($\pm$7)	181 ($\pm$20)	0
SLP	40	39	100	104	15 ($\pm$7)	27 ($\pm$4)	46 ($\pm$9)	44 ($\pm$3)

Both studies conducted surveys at ten stations in each sub-estuary (Childs River Estuary (CRE) and Sage Lot Pond (SLP)) at least once a month. Values reported represent the mean from these surveys ($\pm$ standard error, Fox (2008) and (Fox et al., 2008) for the reported time periods. The Hersh (1996) study did not include an estimate of variability from the mean nor were the data available in any appendices. Since the study was conducted at the same location and using the same sample design and methods as Fox (2008) and Fox et al. (2008) it is likely that the deviation and error about the mean are similar.

Childs River Estuary, the station most heavily influenced by N loading. The relatively low rates at South Basin and Childs River Estuary may be attributed again to their similarity in sediment characteristics. Both these stations have lower percent carbon in their sediments than the other two stations ($1.02 \pm 0.07\%$ and $1.12 \pm 0.32\%$, South Basin and Childs River Estuary respectively) (**Table 1**).

BENTHIC NITROGEN FLUXES

Many estuarine studies of benthic sediment fluxes report high rates of NH_4^+ efflux increasing with N loading, organic matter content, and temperature (Nixon, 1981; Boynton and Kemp, 1985; Nowicki, 1994). In Waquoit Bay, we observed NH_4^+ fluxes that were partially consistent with these previous studies. Metoxit Point, the station with the second highest N load, had significantly greater rates of NH_4^+ production than South Basin, the other main basin station only 1 km away but less impacted by N loading. Most of the NH_4^+ fluxes were positive, indicating a release of NH_4^+ from the sediments. However, from the station least impacted by excess N (Sage Lot Pond) we also measured high rates of NH_4^+ flux relative to South Basin. This is likely a result of the organic rich sediments found in the productive eelgrass beds of Sage Lot Pond (**Table 1**), which can promote microbial decomposition and mineral regeneration (Nixon et al., 1980; Nixon, 1981; Kemp et al., 2005).

Concentrations of NO_3^- and NO_2^- were low and often below the detection limit of our analysis. The historic Waquoit Bay studies also report low NO_3^- and NO_2^- concentrations and fluxes (LaMontagne, 1996; Kirkpatrick and Foreman, 1998; LaMontagne et al., 2002). There is a broad range of literature NO_3^- flux values from other estuaries exhibiting both uptake and release. Kemp et al. (1990) exclusively measured nitrate uptake in the Chesapeake Bay with a flux range from -100 to 0μmol N m^{-2} h^{-1} while other systems were balanced between uptake and release (Gardner et al., 2006; Giblin et al., 2010). Fulweiler et al. (2010) found a maximum nitrate flux of -60μmol N m^{-2} h^{-1} at the Providence River estuary, which corresponded to the highest NH_4^+ release and thus was suggestive of dissimilatory nitrate reduction to ammonium (DNRA).

DIRECT MEASURE OF NET SEDIMENT N$_2$ FLUX

Net sediment N$_2$ fluxes directly measured from water samples were not significantly different from each other except that Metoxit Point was significantly lower than the other stations because of negative N$_2$ fluxes (indicating net N-fixation) measured there. We hypothesize that these negative N$_2$ fluxes are because of the inhibition of coupled nitrification-denitrification at this site. Under frequent and intense low oxygen and sulfidic conditions, nitrification would be diminished thereby decreasing the amount of NO_3^- available for denitrification. This finding is supported by Rysgaard et al. (1994, 1995) who found lowest rates of denitrification in the summer when water column oxygen and nitrate concentrations were at a minimum thus inhibiting coupled nitrification-denitrification in the sediments.

Since Childs River Estuary is also greatly influenced by anthropogenic N loads, we might expect that net N$_2$ flux rates would be more similar to the Metoxit Point station. However, we observed important differences between these two stations. For example, at Childs River Estuary we did not observe extensive macroalgae mats and sulfidic sediments as we did at Metoxit Point. Additionally, Childs River Estuary cores typically had a small, light oxic layer while Metoxit Point sediments were always uniformly dark and likely anaerobic to the surface. These qualitative observations suggest that coupled nitrification-denitrification was less inhibited in Childs River Estuary sediments, which is why rates of net N$_2$ flux were significantly greater than at Metoxit Point.

Although Metoxit Point and Sage Lot Pond had similar sediment O$_2$ uptake, presumably because of their high productivity, they exhibited opposite trends in net N$_2$ fluxes. In fact, at Metoxit Point we only measured net N fixation while the highest rate of net denitrification was measured at Sage Lot Pond. This distinction is an important reminder as to some of the deleterious and perhaps unexpected effects of nutrient pollution and eutrophication. While increased nutrient loading may stimulate organic matter production (Nixon et al., 2001), it also changes the ecosystems biological structure and chemical composition. For example, the loss of eelgrass and associated bioturbating infaunal community is likely to diminish oxic regions within the sediments and therefore decrease areas where coupled nitrification-denitrification can occur. Inhibition of this crucial microbial process is significant as denitrification has the potential, in many coastal systems, to nearly completely filter out reactive N inputs (Seitzinger, 1988). Metoxit Point and Sage Lot Pond may both have high productivity, but the station more strongly altered by anthropogenic N loads (Metoxit Point) is the least capable at providing this important ecosystem service.

HISTORIC COMPARISONS

During the early and mid-1990s sediment O_2 uptake, dissolved inorganic nitrogen, and N_2 sediment flux measurements were made in Waquoit Bay (LaMontagne and Valiela, 1995; LaMontagne, 1996; Kirkpatrick and Foreman, 1998; LaMontagne et al., 2002). These historic studies employed both similar experimental design (i.e., sealed sediment core incubations), as well as similar sample analyses to our current study. LaMontagne (1996) measured O_2 uptake, dissolved inorganic nitrogen, and N_2 fluxes between 1992 and 1994 in two of the stations measured here (Sage Lot Pond and Childs River Estuary).

Sediment O_2 uptake, measured on 33 occasions during this period, ranged from 662 to 7103 μmol O_2 m^{-2} h^{-1}, with the highest rates occurring in the summer (LaMontagne, 1996). The highest O_2 uptake we measured (July-October 2011) was approximately 3016 μmol O_2m^{-2} h^{-1}. Using only measurements between July and October, the historic (1992–1994) mean ($\pm$ standard error) O_2 consumption rates for Sage Lot Pond and Childs River Estuary were 5270 ($\pm$1001) and 4144 ($\pm$346) μmol O_2m^{-2} h^{-1}, respectively (LaMontagne, 1996). These historic rates are 32 and 61% greater than the mean sediment O_2 uptake we measured (2510 ($\pm$265) and 1604 ($\pm$7.6) μmol O_2m^{-2} h^{-1}, $p = 0.009$ and $p = 0.006$, Sage Lot Pond and Childs River Estuary respectively) (**Figure 4**).

Compared to historical NH$_4^+$ flux rates measured in Waquoit Bay our values are again substantially lower for Childs River Estuary, however the difference was not significant for Sage Lot Pond. The average rates we measured ($\pm$ standard error) were 209 ($\pm$99) and 97 ($\pm$56) μmol N m^{-2} h^{-1} for Sage Lot Pond and Childs River Estuary respectively. LaMontagne (1996) published mean ($\pm$ standard error) July–October NH$_4^+$ flux rates of 331 ($\pm$133) and 511 ($\pm$93) μmol N m^{-2} h^{-1} from Sage Lot Pond and Childs River Estuary respectively ($p = 0.530$ (SLP) and $p = 0.024$ (CRE)) (**Figure 4**). Furthermore, values reported for Childs River Estuary by Kirkpatrick and Foreman (1998) are over 6 times higher than those we measured (606 μmol N m^{-2} h^{-1}).

Similar to our findings the historic studies in Waquoit Bay (LaMontagne and Valiela, 1995; LaMontagne, 1996; Kirkpatrick and Foreman, 1998; LaMontagne et al., 2002) also measured extremely low concentrations and fluxes of NO$_x$ and NO$_2^-$ during the summer. This may be a result of assimilation by primary producers. It is also indicative of highly reducing sediments with low nitrification or sediments where nitrification and denitrification are tightly coupled.

Finally, historic net N_2 fluxes from Waquoit Bay sediments were nearly an order of magnitude greater than the net N_2 fluxes we measured when pooling both stations together ($p = 0.012$). The difference was not statistically significant when stations were

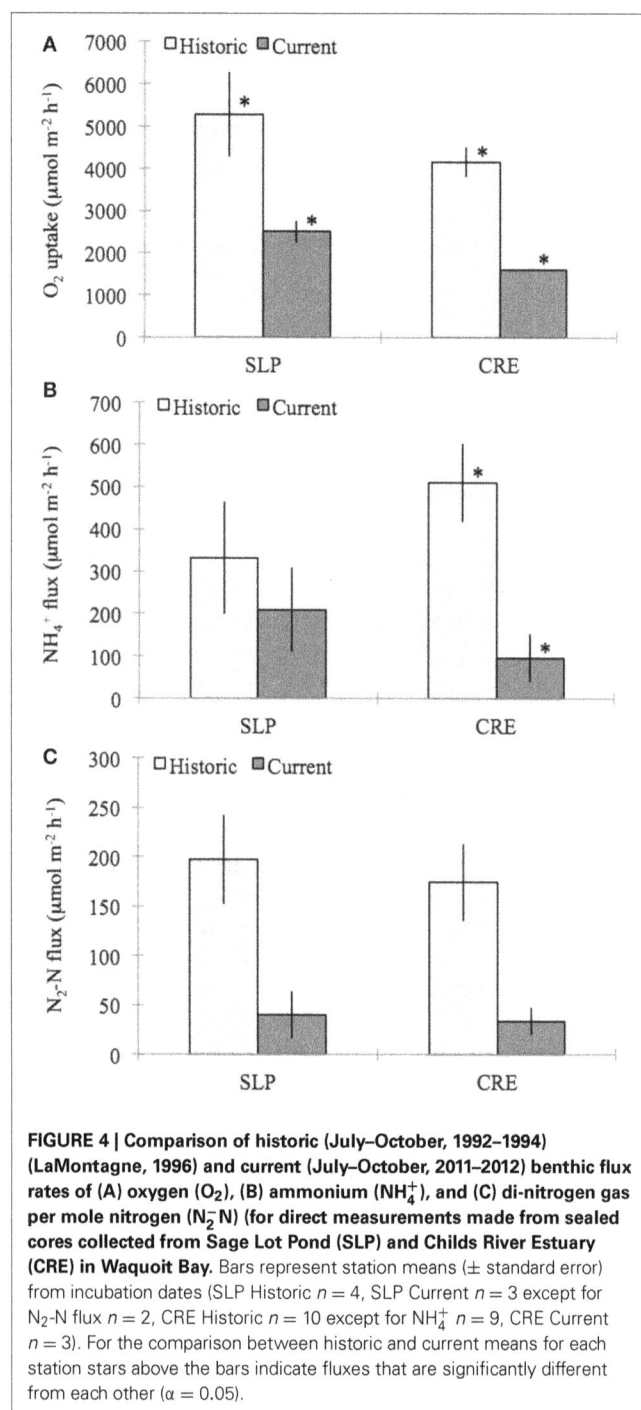

FIGURE 4 | Comparison of historic (July–October, 1992–1994) (LaMontagne, 1996) and current (July–October, 2011–2012) benthic flux rates of (A) oxygen (O_2), (B) ammonium (NH$_4^+$), and (C) di-nitrogen gas per mole nitrogen (N$_2^-$N) (for direct measurements made from sealed cores collected from Sage Lot Pond (SLP) and Childs River Estuary (CRE) in Waquoit Bay. Bars represent station means ($\pm$ standard error) from incubation dates (SLP Historic $n = 4$, SLP Current $n = 3$ except for N$_2$-N flux $n = 2$, CRE Historic $n = 10$ except for NH$_4^+$ $n = 9$, CRE Current $n = 3$). For the comparison between historic and current means for each station stars above the bars indicate fluxes that are significantly different from each other ($\alpha = 0.05$).

separated due to a small sample size in the current study ($n = 3$ and $n = 2$, Childs River Estuary and Sage Lot Pond, respectively). The mean net N_2 flux rates ($\pm SE$) reported between July and October during 1992–1994 by LaMontagne (1996) for Sage Lot Pond and Childs River Estuary are 197 ($\pm$45) and 174 ($\pm$39) μmol N_2-N m^{-2} h^{-1} respectively compared to the current means of 40 ($\pm$24) and 34 ($\pm$14) μmol N_2-N m^{-2} h^{-1} [$p = 0.102$ (SLP) and $p = 0.059$ (CRE)] (**Figure 4**). The historic mean net N_2 flux rate for both stations together is nearly 80%

greater than the mean net N_2 flux rate we observed in our study ($p = 0.012$).

Using July-October net N_2 rates from dark incubations, we scaled up fluxes for the Childs River to the water body surface area of the sub-estuary for an entire year, in order to compare the estimated percentage of anthropogenic N inputs removed. Assuming that these rates are representative of rates for the entire Childs River sub-estuary and throughout the year, we found that the mean historic rate (2818 ($\pm$632) kg N y^{-1}) could have removed 51% ($\pm$11%) of the annual external N inputs (5536 N y^{-1}, Valiela et al., 2000), while the current rate [551 ($\pm$227) kg N y^{-1}] accounts for only 10% ($\pm$4%) of the external N inputs.

The decrease in net N_2 flux rates may be explained by fundamental alterations of the ecological structure and functioning in Waquoit Bay. Alternatively, our observed trend could be a result of high year-to-year variability, methodological differences between our two studies, or a combination of these factors. Although the field and incubation procedures were nearly identical, the methods used to directly measure the N_2 concentration of the water samples differed. Both techniques have similar limitations including gas disequilibria from changing temperatures or from methane ebullition (Cornwell et al., 1999), however the N_2/Ar MIMS method is much more precise than the gas stripping GC method (coefficient of variance 0.03% and 0.5% for N_2/Ar and GC methods, respectively) (Devol, 1991; Kana et al., 1994).

While we propose that the differences observed between our net N_2 flux rates and the historic values are not caused exclusively by the method differences, we are cautious with this interpretation, as denitrification is an extremely difficult process to measure (Cornwell et al., 1999; Groffman et al., 2006). However, the fact that we also see dramatic declines in sediment O_2 uptake, benthic DIN fluxes (which use the same methods) across a range of years in the early 1990s and in our current study suggests that these decreases are caused, at least in part, by larger ecosystem level differences.

In our comparison of net ecosystem daytime production and nighttime respiration, we once again found that current (2011–2012) rates of production and respiration in the Childs River Estuary were lower than historic rates (1991–1992, D'Avanzo et al., 1996) during the July–October sampling period (ANOVA, $p < 0.0001$ for both production and respiration) (**Figure 5**). The mean daily rate ($\pm$ standard error) for daytime net production during this 4-month time period was 4.4 ($\pm$0.3) g O_2 m^{-2} d^{-1} in 2011–2012 compared to 7.2 ($\pm$0.4) g O_2 m^{-2} d^{-1} in 1991–1992. Absolute nighttime respiration rates in 2011–2012 were also lower than the 1991–1992 values [4.6 ($\pm$0.2) and 7.4 ($\pm$0.4) g O_2 m^{-2} d^{-1}, 2011–2012 and 1991–1992 respectively]. During most months of the year, the monthly mean of daytime net production was greater in the historic dataset (above the 1:1 line, **Figure 6A**). The exceptions to this relationship included two winter data points (November, December monthly averages) where historic and current rates were equivalent (on the 1:1 line, **Figure 6A**) and two spring month data points (March, April monthly averages) where the current rates were greater than the historic (below the 1:1 line, **Figure 6A**). Monthly means of daily respiration rates were always greater in the 1991–1992 dataset compared to the 2011–2012 data except for March (**Figure 6B**). Daytime net

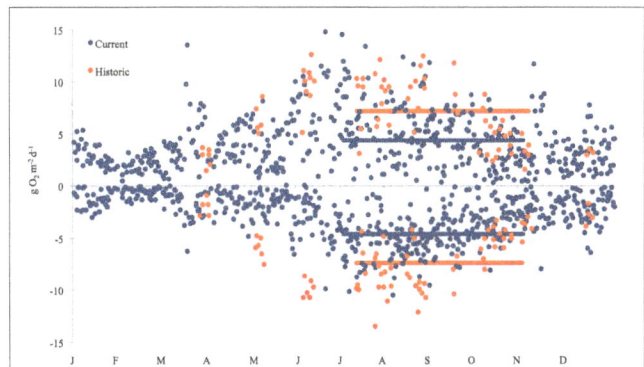

FIGURE 5 | Temporal variation in net daytime primary production (positive values) and nighttime respiration (negative values) rates at the Childs River Estuary station. The current data (blue circles) were calculated using continuous 15-min interval conductivity, temperature, and dissolved oxygen data from deployed WBNERR YSI 6600 data-sondes during 2011 and 2012 (NOAA, 2012). The historic data (red circles) are from D'Avanzo et al. (1996, Figure 1). They include data collected from Childs River Estuary over a year (Jul–Dec 1991, Jan–Jun 1992) in 10 deployments (5–25 days) of automated ENDECO model 1184 meters. The filled bars above zero indicate mean net daytime productivity rates for the sampling time frame (July–October) for the historic 7.18 g O_2 m^{-2} y^{-1} (red bar), and current 4.36 g O_2 m^{-2} y^{-1} (blue bar) time frames, respectively. The filled bars below zero indicate mean nighttime respiration rates for historic −7.39 g O_2 m^{-2} y^{-1} (red bar) and current −4.36 g O_2 m^{-2} y^{-1} (blue bar) time frames respectively. Differences between current and historic daily rates for both production and respiration during the July-October time frames were significant (ANOVA, $p < 0.0001$).

production and respiration monthly mean rates during the 2011–2012 months of sediment core sampling (July–October) were significantly higher in 1991–1992 compared to 2011–2012 ($p = 0.026$ and $p = 0.004$, production and respiration, respectively). While we acknowledge that our calculation of net ecosystem production and respiration is simplistic due to our use of a constant gas exchange coefficient (see Methods section), our aim here was to provide an additional way to examine changes in the system.

In Waquoit Bay, macrophytes are the dominant primary producers (Valiela et al., 1992, 1997b) and drive trends in diel dissolved oxygen concentrations (D'Avanzo and Kremer, 1994). In an effort to explain this decreased production and respiration, we compared macroalgae biomass data from the early 1990s (Hersh, 1996) and 2000s (Fox, 2008; Fox et al., 2008) in Waquoit Bay (**Table 3**). We find that in both the Childs River Estuary and Sage Lot Pond sub-estuaries mean biomass of the dominant macroalgae species (*Cladophora vagabunda* and *Gracilaria tikvahiae*) decreased from the early 1990s compared to the early 2000s. And in Sage Lot Pond the biomass of eelgrass, *Zostera marina*, also decreased. While more extensive analyses including the compilation of all available Waquoit Bay macrophyte survey data and estimates of error would strengthen the validity of this trend, we find this preliminary comparison compelling. Changes in the dominant primary producer and the abundance of this production provide a plausible explanation for decreased benthic sediment oxygen uptake and nitrogen cycling dynamics. Reduced macrophyte biomass would reduce production and respiration for this system and diminish organic matter loading

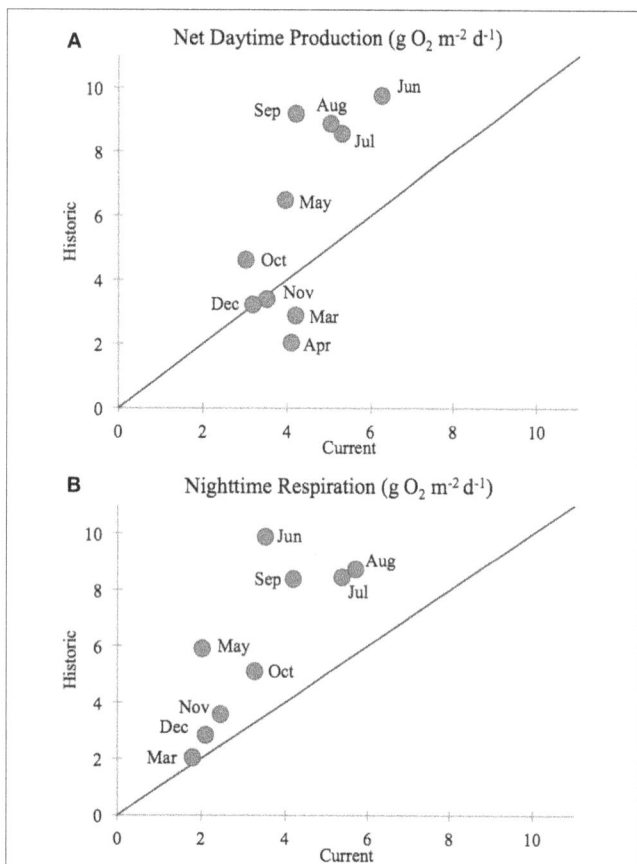

FIGURE 6 | Comparison between Childs River Estuary measurements of (A) daytime net productivity and (B) nighttime respiration rates between the current (2011–2012, this study, WBNERR data) and historic (1991–1992, D'Avanzo et al., 1996) rates. Each data point represents the monthly mean of daily primary production and respiration. The number of days included in those monthly means varies. The number of days included in monthly means was 44 for the current study (2011 and 2012) and 8 for the historic study (1991 and 1992). Monthly averaged daytime production and nighttime respiration values were statistically lower for the current time period compared to the historic (ANOVA $p = 0.026$ and $p = 0.004$, production and respiration means, respectively). Line indicates a 1:1 comparison between the current and historic rates.

to the sediment. We would expect that decreased organic matter availability would result in lower benthic fluxes as correlations between autochthonous and allochthonous carbon supply and benthic sediment oxygen uptake and nutrient regeneration (Nixon, 1981; Hopkinson and Smith, 2005) as well as with N_2 fluxes (Cornwell et al., 1999; Fulweiler et al., 2008, 2013; Eyre et al., 2013) are well established. Further, the findings of decreases in primary production and benthic fluxes mimic those observed in a nearby estuary, Narragansett Bay (further discussion below).

PROPOSED MECHANISMS OF CHANGE

The question remains as to what could be driving the observed differences in system productivity and nitrogen cycling between the early 1990s and 2000s? While at this point we cannot assign a single answer to this question we suggest three plausible explanations, and all may be at least partially contributing to the observed changes.

First, dampened fluxes could be caused by high annual variability. High variability in sediment N_2 fluxes (-257 to $154\,\mu$mol $m^{-2}\,h^{-1}$), directly related to water column chlorophyll a variability, have been observed in a unique 9 year continuous study in Narragansett Bay (Fulweiler and Heiss, 2014). Large seasonal and inter-annual fluctuations of macroalgae and eelgrass biomass have been measured historically in Waquoit Bay (Hersh, 1996; Valiela et al., 1997b; Fox, 2008; Fox et al., 2008) and this variance tends to be more pronounced with higher N loads (Fox, 2008; Fox et al., 2008). It is known that changes in the quantity of organic matter deposited to the benthos will alter decomposition rates and thus rates of nutrient remineralization (Nixon, 1981; Hopkinson and Smith, 2005) and denitrification (Cornwell et al., 1999; Fulweiler et al., 2008, 2013; Eyre et al., 2013). Thus, we could simply be observing the benthic response to year-to-year changes in primary production.

A second explanation could be that dampened benthic fluxes could be the next stage of impacts driven by N pollution and that we are witnessing the birth of "dead zone." Based on land use data from MassGIS (www.mass.gov) and the Boyer (Boyer et al., 2002) model it is possible that N loading to Waquoit Bay could be currently increasing, albeit more slowly than 50 years ago. Nevertheless, ecosystems may reach a threshold or tipping point in N inputs where beyond a certain point the system is fundamentally altered (Scheffer et al., 2001; Webster and Harris, 2004; Duarte et al., 2009). While, at this point we can only suggest a likely scenario based on the information at hand, we propose the following. Take the Childs River Estuary where we measured the most dramatic changes from historic rates as an example: the high macroalgae production and frequent anoxic conditions over time would result in diminished oxygen availability and high sulfide concentrations causing coupled nitrification-denitrification to cease. In time, these conditions would become toxic to bioturbating organisms, aerobic bacteria, and even the macroalgae that once thrived there. Instead, the community would be replaced by slow growing anaerobic organisms, which result in the dampened fluxes we observe. While this explanation may work for the high N load station it does not explain the concurrent decreases observed in the low N load station (Sage Lot Pond).

Thus, other mechanisms driving either high inter-annual variability and/or significant declines of primary production in this system over the past 20 years must include larger scale phenomena. For example, in nearby Narragansett Bay, similar declines in benthic metabolism and nitrogen cycling were reported (Fulweiler et al., 2007; Fulweiler and Nixon, 2009, 2011). The changes in Narragansett Bay were hypothesized to be driven by increased water temperatures, decreased wind speed, and alterations in the timing and magnitude of the winter-spring phytoplankton bloom (Nixon et al., 2009). Although Waquoit Bay is not a phytoplankton-based ecosystem, it is conceivable that changes in meteorological parameters such as temperature, precipitation, and wind speed are driving the changes we observe in Waquoit Bay as well. Perhaps not surprisingly due to their proximity, many meteorological changes observed in Narragansett Bay are also observed in Waquoit Bay. For example, as in Narragansett

Bay surface temperatures in coastal Cape Cod waters have also been increasing over the past 4 decades (Nixon et al., 2004, 2009). Temperature data collected from the Woods Hole Oceanographic Pier shows significant warming from 1970 to 2002 at a rate of 0.04°C per year and an overall mean annual increase of 1.2°C (Nixon et al., 2004). We found that over the past 20 years there has also been a significant increase in total annual rainfall ($p = 0.002$) and discharge from Quashnet River to Waquoit Bay ($p = 0.001$) (**Figure 7**). Finally, decade long wind data from the Waquoit Bay National Estuarine Research Reserve show that while mean annual wind speed from 2002 to 2013 had not significantly changed wind speed between May and October has significantly declined ($p = 0.016$) over this same time period.

Determining how all of these factors combine to ultimately impact benthic fluxes in Waquoit Bay is beyond the scope of this paper, but is the focus of our current and future efforts. Here we can only hypothesize some ways in which these climatic patterns may alter ecosystem function. First, warming temperatures will increase metabolic activity of both autotrophs and heterotrophs (Hopkinson and Smith, 2005; Harris et al., 2006; Anderson et al., 2014). While at first this might lead to an increase in photosynthetic activity ultimately when macroalgae are exposed to temperatures beyond their thermal optimum photosynthesis declines rapidly (Davison, 1991). Additionally, an autotroph's response to

increased temperature may vary when combined with changes in other environmental variables. For example, seagrass response to increased temperature is positive under high light conditions but negative under low light conditions (Bulthuis, 1987). The effect of temperature on eelgrass (*Zostera marina*) in large mesocosm studies was significantly increased under elevated inorganic nutrients and is the proposed mechanism driving major eelgrass declines in the Northeast (Bintz et al., 2003). Of course, primary producers would also be impacted by the observed changes in precipitation, freshwater discharge, and wind speeds of Waquoit Bay. While the data do not exist for the major tributaries of Waquoit Bay it is reasonable to suspect that the increase precipitation and freshwater discharge has also led to an increase in total suspended solids and colored dissolved organic matter which would attenuate light and decrease primary production (Anderson et al., 2014). Finally, decreased wind speeds in May through October (arguably the dominant metabolic period of the year), has likely resulted in decreased vertical mixing and enhanced water column stratification (Nixon et al., 2009). As such, bottom water exchange with the atmosphere would be limited and even though there is evidence of decreased primary production, the lack of water column mixing may increase low oxygen conditions in bottom waters. In turn, this could cause a shift from aerobic to anaerobic respiration, declines in sediment macroafauna abundance and activity, and lower rates of benthic fluxes. Despite not being able to pinpoint the mechanisms driving our observations, the fact that we report similar changes to those found for nearby Narragansett Bay suggests that larger scale climatic patterns are responsible for the decrease in benthic fluxes. Thus, anthropogenic impacts on coastal systems are no longer simply local, instead they are part of the global change that has come to dominate our planet.

CONCLUSIONS

Positioned at the interface between land and sea, estuarine systems are influenced by local stressors such as N loading and also larger, regional and climate scale perturbations. Alterations of the physical and chemical nature of ocean systems can have direct and profound impacts on ecological structure and functioning of estuaries (Cloern et al., 2007; Nixon et al., 2009). In the spatial scope of this study we did not find a clear relationship between the load of N and benthic flux rates across the four sampling stations. However by taking a broader perspective and comparing our results to historic values, we find significantly lower rates than those measured two decades ago in Sage Lot Pond and Childs River Estuary. Rates of net ecosystem production and respiration were 39 and 38% lower than historic rates, respectively ($p < 0.0001$). Sediment O_2 uptake and NH_4^+ flux for both SLP and CRE combined were less than half of historical rates ($p = 0.001$ and $p = 0.023$, O_2 and NH_4^+, respectively). Perhaps one of the most important changes observed is the 80% reduction in net N_2 efflux rates from sediments ($p = 0.012$). If this is truly a long-term trend then this represents a substantial loss in the important ecosystem service of reactive N removal. Alternatively, if this is simply year-to-year variability then these findings suggest that the coastal sediment N cycling in general and N_2 fluxes in particular are more dynamic than previously realized. And finally, a better understanding of how powerful global-scale forcings (e.g.,

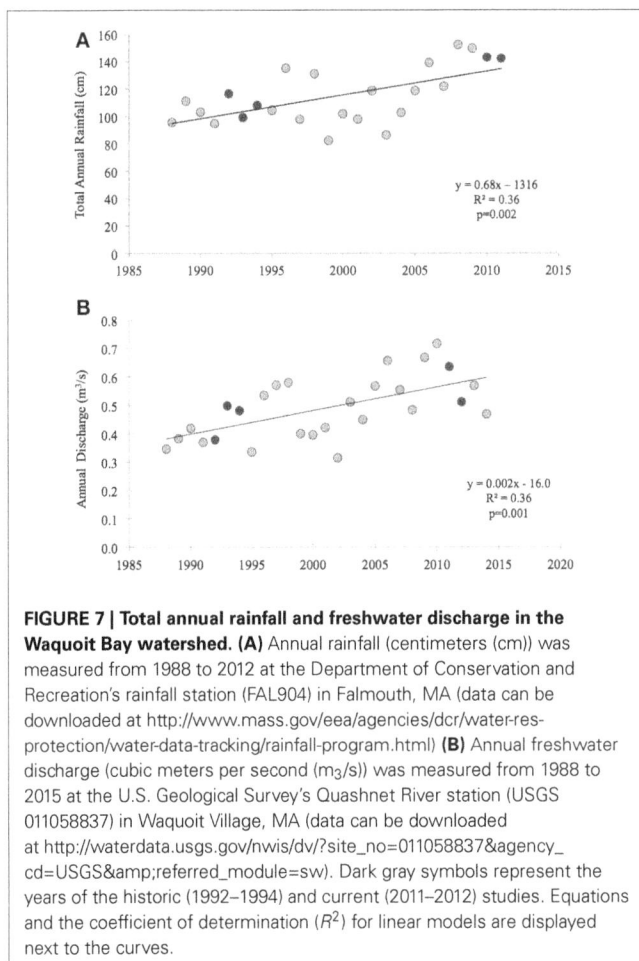

FIGURE 7 | Total annual rainfall and freshwater discharge in the Waquoit Bay watershed. (A) Annual rainfall (centimeters (cm)) was measured from 1988 to 2012 at the Department of Conservation and Recreation's rainfall station (FAL904) in Falmouth, MA (data can be downloaded at http://www.mass.gov/eea/agencies/dcr/water-res-protection/water-data-tracking/rainfall-program.html) **(B)** Annual freshwater discharge (cubic meters per second (m3/s)) was measured from 1988 to 2015 at the U.S. Geological Survey's Quashnet River station (USGS 011058837) in Waquoit Village, MA (data can be downloaded at http://waterdata.usgs.gov/nwis/dv/?site_no=011058837&agency_cd=USGS&referred_module=sw). Dark gray symbols represent the years of the historic (1992–1994) and current (2011–2012) studies. Equations and the coefficient of determination (R^2) for linear models are displayed next to the curves.

changes in temperature, precipitation, sea level, light, and wind) interact with local-scale impacts (e.g., changing nutrient loads) to impact whole system functioning is crucial in our prediction of long-term ecological trajectories and for better management strategies for coastal regions.

ACKNOWLEDGMENTS

This research was supported by a grant to Robinson W. Fulweiler from Woods Hole Sea Grant, by support of Robinson W. Fulweiler from the National Science Foundation Project ICER-1343802, and by the Sloan Foundation in the form of a fellowship to Robinson W. Fulweiler. The Waquoit Bay National Estuarine Research Reserve (WBNERR) System-Wide Monitoring Program (SWMP) water quality data used in this publication is supported by an award from the Estuarine Reserves Division, Office of Ocean and Coastal Resource Management, National Ocean Service, National Oceanic and Atmospheric Administration (NOAA, 2012). We would like to thank the WBNERR personnel, especially C. Weidman and M. K. Fox, who provided us with the boats and facilities to use during our field sampling. We received additional help in the field and lab from E. Heiss (Boston University) and from participants in the Boston University Marine Program: M. K. Rogener, R. Schweiker, A. Banks, D. Forest, S. Andrews, and J. Luthringer. We would also like to thank the following people for their lab analyses: K. Czapla, E. Heiss, and M. K. Rogener. We are grateful to W. Boynton and C. Sperling (Chesapeake Bay Laboratory, University of Maryland Center for Environmental Science) for the use of an automated Excel spreadsheet developed by D. Jasinski for the calculation of ecosystem production and respiration. We would also like to thank J. Sparks for his assistance adjusting the spreadsheet code. Finally, we thank our two reviewers who provided helpful comments and suggestions, which greatly improved our manuscript.

REFERENCES

Anderson, D. M. (2009). Approaches to monitoring, control and management of harmful algal blooms (HABs). Ocean Coast. Manag. 52, 342–347. doi: 10.1016/j.ocecoaman.2009.04.006

Anderson, D. M., Glibert, P. M., and Burkholder, J. M. (2002). Harmful algal blooms and eutrophication: nutrient sources, composition, and consequences. Estuaries 25, 704–726. doi: 10.1007/BF02804901

Anderson, I. C., Brush, M. J., Piehler, M. F., Currin, C. A., Stanhope, J. W., Smyth, A. R., et al. (2014). Impacts of climate-related drivers on the benthic nutrient filter in a shallow photic estuary. Estuaries Coasts 37, 46–62. doi: 10.1007/s12237-013-9665-5

Banta, G. T., Giblin, A. E., Hobbie, J. E., and Tucker, J. (1995). Benthic respiration and nitrogen release in Buzzards Bay, Massachusetts. J. Mar. Res. 53, 107–135. doi: 10.1357/0022240953213287

Benson, B. B., and Krause, D. (1984). The concentration and isotopic fractionation of oxygen dissolved in freshwater and seawater in equilibrium with the atmosphere. Limnol. Oceanogr. 29, 620–632. doi: 10.4319/lo.1984.29.3.0620

Bintz, J. C., Nixon, S. W., Buckley, B. A., and Granger, S. L. (2003). Impacts of temperature and nutrients on coastal lagoon plant communities. Estuaries 26, 765–776. doi: 10.1007/BF02711987

Bowen, J. L., and Valiela, I. (2001). The ecological effects of urbanization of coastal watersheds: historical increases in nitrogen loads and eutrophication of Waquoit Bay estuaries. Can. J. Fish. Aquat. Sci. 58, 1489–1500. doi: 10.1139/f01-094

Boyer, E. W., Goodale, C. L., Jaworski, N. A., and Howarth, R. W. (2002). Anthropogenic nitrogen sources and relationships to riverine nitrogen export in the Northeastern USA. Biogeochemistry 57, 137–169. doi: 10.1023/A:1015709302073

Boynton, W. R., and Kemp, W. M. (1985). Nutrient regeneration and oxygen consumption by sediments along an estuarine salinity gradient. Mar. Ecol. Prog. Ser. 23, 45–55. doi: 10.3354/meps023045

Bricker, S. B., Longstaff, B., Dennison, W., Jones, A., Boicourt, K., Wicks, C., et al. (2008). Effects of nutrient enrichment in the nation's estuaries: a decade of change. Harmful Algae 8, 21–32. doi: 10.1016/j.hal.2008.08.028

Bulthuis, D. A. (1987). Effects of temperature on photosynthesis and growth of seagrasses. Aquat. Bot. 27, 27–40. doi: 10.1016/0304-3770(87)90084-2

Caffrey, J. M. (2003). "Production, respiration and net ecosystem metabolism in US estuaries," in Coastal Monitoring through Partnerships: Proceedings of the Fifth Symposium on the Environmental Monitoring and Assessment Program (EMAP) Pensacola Beach, FL, U.S.A. Vol. 81, eds B. D. Melzian, V. Engle, M. McAlister, and S. S. Sandhu (Netherlands: Springer), 207–219. doi: 10.1007/978-94-017-0299-7_19

Caffrey, J. M. (2004). Factors controlling net ecosystem metabolism in U.S. Estuaries. Estuaries 27, 90–101. doi: 10.1007/BF02803563

Canfield, D. E., Stewart, F. J., Thamdrup, B., De Brabandere, L., Dalsgaard, T., Delong, E. F., et al. (2010). A cryptic sulfur cycle in oxygen-minimum-zone waters off the Chilean coast. Science 330, 1375–1378. doi: 10.1126/science.1196889

Caraco, N. F., and Cole, J. J. (1999). Human impact on nitrate export: an analysis using major world rivers. Ambio 28, 167–170.

Cloern, J. E., Jassby, A. D., Thompson, J. K., and Hieb, K. A. (2007). A cold phase of the East Pacific triggers new phytoplankton blooms in San Francisco Bay. Proc. Natl. Acad. Sci. U.S.A. 104, 18561–18565. doi: 10.1073/pnas.0706151104

Colt, J. E. (1984). Computation of Dissolved Gas Concentrations in Water as Functions of Temperature, Salinity, and Pressure Bethesda, MD: American Fisheries Society Special Publication, 14.

Conley, D. J., Paerl, H. W., Howarth, R. W., Boesch, D. F., Seitzinger, S. P., Havens, K. E., et al. (2009). Controlling eutrophication: nitrogen and phosphorus. Science 323, 1014–1015. doi: 10.1126/science.1167755

Cornwell, J. C., Kemp, W. M., and Kana, T. M. (1999). Denitrification in coastal ecosystems: methods, environmental controls, and ecosystem level controls, a review. Aquat. Ecol. 33, 41–54. doi: 10.1023/A:1009921414151

Dalsgaard, T., Nielsen, L. P., Brotas, V., Viaroli, P., Underwood, J. C., Nedwell, D. B., et al. (2000). "Sediment characteristics," in Protocol Handbook for NICE: Nitrogen Cycling in Estuaries: A Project under the EU Research Programme: MARINE Science and Technology (MAST III), ed T. Dalsgaard (Silkeborg: National Environmental Research Institute), 53–54.

D'Avanzo, C., and Kremer, J. N. (1994). Diel oxygen dynamics and anoxic events in an eutrophic estuary of Waquoit Bay, Massachusetts. Estuaries 17, 131–139. doi: 10.2307/1352562

D'Avanzo, C., Kremer, J., and Wainright, S. C. (1996). Ecosystem production and respiration in response to eutrophication in shallow temperate estuaries. Mar. Ecol. Prog. Ser. 141, 263–274. doi: 10.3354/meps141263

Davison, I. R. (1991). Environmental effects on algal photosynthesis: temperature. J. Phycol. 27, 2–8. doi: 10.1111/j.0022-3646.1991.00002.x

Devol, A. H. (1991). Direct measurement of nitrogen gas fluxes from continental shelf sediments. Nature 349, 319–321. doi: 10.1038/349319a0

Diaz, R. J., and Rosenberg, R. (2008). Spreading dead zones and consequences for marine ecosystems. Science 321, 926–929. doi: 10.1126/science.1156401

Duarte, C. M., Conley, D. J., Carstensen, J., and Sánchez-Camacho, M. (2009). Return to Neverland: shifting baselines affect eutrophication restoration targets. Estuaries Coasts 32, 29–36. doi: 10.1007/s12237-008-9111-2

Eyre, B. D., Rysgaard, S., Dalsgaard, T., and Christensen, P. B. (2002). Comparison of isotope pairing and N2:Ar methods for sediment denitrification - assumptions, modifications, and implications. Estuaries 25, 1077–1087. doi: 10.1007/BF02692205

Eyre, B. D., Santos, I. R., and Maher, D. T. (2013). Seasonal, daily and diel N2 effluxes in permeable carbonate sediments. Biogeosciences 10, 2601–2615. doi: 10.5194/bg-10-2601-2013

Fox, S. E. (2008). Ecological Effects of Nitrogen Loading to Temperate Estuaries: Macrophyte and Consumer Community Structure and Food Web Relationships. Ph.D. thesis, Boston University, Boston, MA.

Fox, S. E., Stieve, E., Valiela, I., Hauxwell, J., and McClelland, J. (2008). Macrophyte abundance in Waquoit Bay: effects of land-derived nitrogen loads on seasonal and multi-year biomass patterns. Estuaries Coasts 31, 532–541. doi: 10.1007/s12237-008-9039-6

Fulweiler, R. W., Brown, S. M., Nixon, S. W., and Jenkins, B. D. (2013). Evidence and a conceptual model for the co-occurrence of nitrogen fixation and denitrification in heterotrophic marine sediments. *Mar. Ecol. Prog. Ser.* 482, 57–68. doi: 10.3354/meps10240

Fulweiler, R. W., and Heiss, E. M. (2014). A decade of directly measured sediment N. *Oceanography* 27, 184. doi: 10.5670/oceanog.2014.22

Fulweiler, R. W., and Nixon, S. W. (2009). Responses of benthic–pelagic coupling to climate change in a temperate estuary. *Hydrobiologia* 629, 147–156. doi: 10.1007/s10750-009-9766-0

Fulweiler, R. W., and Nixon, S. W. (2011). Net sediment N2 fluxes in a southern New England estuary: variations in space and time. *Biogeochemistry* 111, 111–124. doi: 10.1007/s10533-011-9660-5

Fulweiler, R. W., Nixon, S. W., and Buckley, B. A. (2010). Spatial and temporal variability of benthic oxygen demand and nutrient regeneration in an anthropogenically impacted New England Estuary. *Estuaries Coasts* 33, 1377–1390. doi: 10.1007/s12237-009-9260-y

Fulweiler, R. W., Nixon, S. W., Buckley, B. A., and Granger, S. L. (2007). Reversal of the net dinitrogen gas flux in coastal marine sediments. *Nature* 448, 180–182. doi: 10.1038/nature05963

Fulweiler, R. W., Nixon, S. W., Buckley, B. A., and Granger, S. L. (2008). Net sediment N2 fluxes in a coastal marine system—experimental manipulations and a conceptual model. *Ecosystems* 11, 1168–1180. doi: 10.1007/s10021-008-9187-3

Galloway, J. N., Dentener, F., Capone, D. G., Boyer, E. W., Howarth, R. W., Seitzinger, S. P., et al. (2004). Nitrogen cycles: past, present, and future. *Biogeochemistry* 70, 153–226. doi: 10.1007/s10533-004-0370-0

Gardner, W. S., McCarthy, M. J., An, S., and Sobolev, D. (2006). Nitrogen fixation and dissimilatory nitrate reduction to ammonium (DNRA) support nitrogen dynamics in Texas estuaries. *Limnol. Oceanogr.* 51, 558–568. doi: 10.4319/lo.2006.51.1_part_2.0558

Geyer, W. R. (1997). Influence of wind on dynamics and flushing of shallow estuaries. *Estuar. Coast. Shelf Sci.* 44, 713–722. doi: 10.1006/ecss.1996.0140

Giblin, A. E., Hopkinson, C. S., and Tucker, J. (1997). Benthic metabolism and nutrient cycling in Boston Harbor, Massachusetts. *Estuaries* 20, 346–364. doi: 10.2307/1352349

Giblin, A. E., Weston, N. B., Banta, G. T., Tucker, J., and Hopkinson, C. S. (2010). The Effects of salinity on nitrogen losses from an oligohaline estuarine sediment. *Estuaries Coasts* 33, 1054–1068. doi: 10.1007/s12237-010-9280-7

Gong, Z., and Xie, P. (2001). Impact of eutrophication on biodiversity of the macrozoobenthos community in a Chinese shallow lake. *J. Freshw. Ecol.* 16, 171–178. doi: 10.1080/02705060.2001.9663802

Groffman, P. M., Altabet, M. A., Bohlke, J. K., Butterbach-Bahl, K., David, M. B., Firestone, M. K., et al. (2006). Methods for measuring denitrification: diverse approaches to a difficult problem. *Ecol. Appl.* 16, 2091–2122. doi: 10.1890/1051-0761(2006)016[2091:MFMDDA]2.0.CO;2

Hansen, H. P., and Koroleff, F. (1999). "Determination of nutrients," in *Methods of Seawater Analysis*, eds K. Grasshoff, K. Kremling, and M. Ehrhardt (Weinheim: Wiley-VCH Verlag GmbH, D-69469), 159–228.

Harris, L. A., Duarte, C. M., and Nixon, S. W. (2006). Allometric laws and prediction in estuarine and coastal ecology. *Estuaries Coasts* 29, 340–344. doi: 10.1007/BF02782002

Hauxwell, J., Cebrian, J., and Valiela, I. (2003). Eelgrass *Zostera marina* loss in temperate estuaries: relationship to land-derived nitrogen loads and effect of light limitation imposed by algae. *Mar. Ecol. Prog. Ser.* 247, 59–73. doi: 10.3354/meps247059

Heck, K. L. J., Able, K. W., Roman, C. T., and Fahay, M. P. (1995). Composition, abundance, biomass, and production of macrofauna in a New England estuary: comparisons among eelgrass meadows and other nursery habitats. *Estuaries* 18, 379–389. doi: 10.2307/1352320

Hersh, D. (1996). *Abundance and Distribution of Intertidal and Subtidal Macrophytes in Cape Cod: The Role of Nutrient Supply and Other Controls*. Ph.D. thesis, Boston University, Boston, MA.

Hopkinson, C. S., and Smith, E. M. (2005). "Estuarine respiration: an overview of benthic, pelagic, and whole system respiration," in *Respiration in Aquatic Ecosystems*, eds P. A. del Giorgio and P. J. L. Williams (New York, NY: Oxford University Press), 122–146. doi: 10.1093/acprof:oso/9780198527084.003.0008

Howarth, R. W., Hayn, M., Marino, R. M., Ganju, N., Foreman, K., McGlathery, K., et al. (2014). Metabolism of a nitrogen-enriched coastal marine lagoon during the summertime. *Biogeochemistry* 118, 1–20. doi: 10.1007/s10533-013-9901-x

Hurlbut, P., D'Avanzo, C., Sethi, D., and Giulfoyle, K. (1994). Effects of algal biomass on benthic nitrogen flux in nutrient-loaded estuaries of Waquoit Bay, Massachusetts. *Biol. Bull.* 187, 283–284.

Johnson, K. S., and Petty, R. L. (1983). Determination of nitrate and nitrite in seawater by flow injection analysis. *Limnol. Oceanogr.* 28, 1260–1266. doi: 10.4319/lo.1983.28.6.1260

Kana, T. M., Darkangelo, C., Hunt, M. D., Oldham, J. B., Bennett, G. E., and Cornwell, J. C. (1994). Membrane inlet mass spectrometer for rapid high-precision determination of N2, O2, and Ar in environmental water samples. *Anal. Chem.* 66, 4166–4170. doi: 10.1021/ac00095a009

Kana, T. M., Sullivan, M. B., Cornwell, J. C., and Groszkowski, K. M. (1998). Denitrification in estuarine sediments determined by membrane inlet mass spectrometry. *Limnol. Oceanogr.* 43, 334–339. doi: 10.4319/lo.1998.43.2.0334

Kemp, W. M., Boynton, W. R., Adolf, J. E., Boesch, D. F., Boicourt, W. C., Brush, G., et al. (2005). Eutrophication of Chesapeake Bay: historical trends and ecological interactions. *Mar. Ecol. Prog. Ser.* 303, 1–29. doi: 10.3354/meps303001

Kemp, W. M., Sampou, P., Caffrey, J., Mayer, M., Henriksen, K., and Boynton, W. R. (1990). Ammonium recycling versus denitrification in Chesapeake Bay sediments. *Limnol. Oceanogr.* 35, 1545–1563. doi: 10.4319/lo.1990.35.7.1545

Kirkpatrick, J., and Foreman, K. (1998). Dissolved inorganic nitrogen flux and mineralization in Waquoit Bay sediments as measured by core incubations. *Biol. Bull.* 195, 240–241. doi: 10.2307/1542859

LaMontagne, M. G. (1996). *Denitrification and the Stoichiometry of Organic Matter Degradation in Temperate Estuarine Sediments: Seasonal Pattern and Significance as a Nitrogen Sink*. Ph.D. thesis, Boston University, Boston, MA.

LaMontagne, M. G., Astorga, V., Giblin, A. E., and Valiela, I. (2002). Denitrification and the stoichiometry of nutrient regeneration in Waquoit Bay, Massachusetts. *Estuaries* 25, 272–281. doi: 10.1007/BF02691314

LaMontagne, M. G., and Valiela, I. (1995). Denitrification measured by a direct N2 flux method in sediments of Waquoit Bay, MA. *Biogeochemistry* 31, 63–83. doi: 10.1007/BF00000939

Lam, P., Lavik, G., Jensen, M. M., van de Vossenberg, J., Schmid, M., Woebken, D., et al. (2009). Revising the nitrogen cycle in the Peruvian oxygen minimum zone. *Proc. Natl. Acad. Sci. U.S.A.* 106, 4752–4757. doi: 10.1073/pnas.0812444106

Li, Y., and Smayda, T. J. (1998). Temporal variability of chlorophyll in Narragansett Bay, 1973–1990. *ICES J. Mar. Sci.* 55, 661–667. doi: 10.1006/jmsc.1998.0383

Marino, R., and Howarth, R. W. (1993). Atmospheric oxygen exchange in the Hudson River: Dome measurements and comparison with other natural waters. *Estuaries* 16, 433–445. doi: 10.2307/1352591

McGlathery, K. J. (2001). Macroalgal blooms contribute to the decline of seagrass in nutrient—enriched coastal waters. *J. Phycol.* 37, 453–456. doi: 10.1046/j.1529-8817.2001.037004453.x

Middelburg, J. J., and Levin, L. A. (2009). Coastal hypoxia and sediment biogeochemistry. *Biogeosciences* 6, 1273–1293. doi: 10.5194/bg-6-1273-2009

Nixon, S. W. (1981). "Remineralization and nutrient cycling in coastal marine sediments," in *Estuaries and Nutrients*, eds B. Nielsen and L. Cronin (Clifton, NJ: Humana Press), 111–138.

Nixon, S. W., Buckley, B. A., Granger, S. L., and Bintz, J. (2001). Responses of very shallow marine ecosystems to nutrient enrichment. *Hum. Ecol. Risk Assess.* 7, 1457–1481. doi: 10.1080/20018091095131

Nixon, S. W., and Fulweiler, R. W. (2009). "Nutrient pollution, eutrophication, and the degradation of coastal marine ecosystems," in *Global Loss of Coastal Habitats: Rates, Causes and Consequences*, ed C. M. Duarte (Bilbao: Fundacion BBVA), 23–58.

Nixon, S. W., Fulweiler, R. W., Buckley, B. A., Granger, S. L., Nowicki, B. L., and Henry, K. M. (2009). The impact of changing climate on phenology, productivity, and benthic-pelagic coupling in Narragansett Bay. *Estuarine Coast. Shelf Sci.* 82, 1–18. doi: 10.1016/j.ecss.2008.12.016

Nixon, S. W., Granger, S., Buckley, B. A., Lamont, M., and Rowell, B. (2004). A one hundred and seventeen year coastal water temperature record from Woods Hole, Massachusetts. *Estuaries* 27, 397–404. doi: 10.1007/BF02803532

Nixon, S. W., Kelly, J. R., Furnas, B. N., Oviatt, C. A., and Hale, S. S. (1980). "Phosphorus regeneration and the metabolism of coastal marine bottom communities," in *Marine Benthic Dynamics*, eds K. R. Tenore and B. C. Coull (Columbia, SC: University of South Carolina Press), 219–242.

NOAA. (2012). *National Oceanic and Atmospheric Administration, Office of Ocean and Coastal Resource Management, National Estuarine Research Reserve*

System-wide Monitoring Program. Centralized Data Management Office, Baruch Marine Field Lab, University of South Carolina. Available online at: http://cdmo.baruch.sc.edu/data/citation.cfm

Nowicki, B. L. (1994). The Effect of temperature, oxygen, salinity, and nutrient enrichment on estuarine denitrification rates measured with a modified nitrogen gas flux technique. *Estuarine Coast. Shelf Sci.* 38, 137–156. doi: 10.1006/ecss.1994.1009

Odum, H. T. (1956). Primary production in flowing waters. *Limnol. Oceanogr.* 1, 102–117. doi: 10.4319/lo.1956.1.2.0102

Oviatt, C., Doering, P., Nowicki, B. L., Reed, L., Cole, J. J., and Frithsen, J. (1995). An ecosystem level experiment on nutrient limitation in temperate coastal marine enviornments. *Mar. Ecol. Prog. Ser.* 116, 171–179. doi: 10.3354/meps116171

Oviatt, C., Keller, A., and Reed, L. (2002). Annual primary production in Narragansett Bay with no bay-wide winter–spring phytoplankton bloom. *Estuar. Coast. Shelf Sci.* 54, 1013–1026. doi: 10.1006/ecss.2001.0872

Rysgaard, S., Christensen, P. B., and Nielsen, L. P. (1995). Seasonal variation in nitrification and denitrification in estuarine sediment colonized by benthic microalgae and bioturbating infauna. *Mar. Ecol. Prog. Ser.* 126, 111–121. doi: 10.3354/meps126111

Rysgaard, S., Risgaard-Petersen, N., Sloth, N. P., Jensen, K., and Nielsen, L. P. (1994). Oxygen regulation of nitrification and denitrification in sediments. *Limnol. Oceanogr.* 39, 1643–1652. doi: 10.4319/lo.1994.39.7.1643

Scheffer, M., Carpenter, S., Foley, J. A., Folke, C., and Walker, B. (2001). Catastrophic shifts in ecosystems. *Nature* 413, 591–596. doi: 10.1038/35098000

Seitzinger, S., Harrison, J. A., Bohlke, J. K., Bouwman, A. F., Lowrance, R., Peterson, B., et al. (2006). Denitrification across landscapes and waterscapes: a synthesis. *Ecol. Appl.* 16, 2064–2090. doi: 10.1890/1051-0761(2006)016[2064:DALAWA]2.0.CO;2

Seitzinger, S., and Kroeze, C. (1998). Global distribution of nitrous oxide production and N inputs in freshwater and coastal marine ecosystems. *Glob. Biogeochem. Cycles* 12, 93–113. doi: 10.1029/97GB03657

Seitzinger, S. P. (1988). Denitrification in freshwater and coastal marine ecosystems: ecological and geochemical significance. *Limnol. Oceanogr.* 33, 702–724. doi: 10.4319/lo.1988.33.4_part_2.0702

Seitzinger, S. P., and Giblin, A. E. (1996). Estimating denitrification in North Atlantic continental shelf sediments. *Biogeochemistry* 35, 235–260. doi: 10.1007/BF02179829

Selman, M., Greenhalgh, S., Diaz, R., and Sugg, Z. (2008). *Eutrophication and Hypoxia in Coastal Areas: A Global Assessment of the State of Knowledge*. Washington, DC: World Resources Institute Policy Note. Available online at: http://www.wri.org/publication/eutrophication-and-hypoxia-coastal-areas

Smayda, T. J. (1990). "Novel and nuisance phytoplankton blooms in the sea: evidence for a global epidemic," in *Toxic Marine Phytoplankton*, eds E. Gramli, B. Sundstrom, L. Edler, and D. M. Anderson (New York, NY: Elsevier), 29–40.

Solorzano, L. (1969). Determination of ammonia in natural waters by the phenolypochlorite method. *Limnol. Oceanogr.* 14, 799–801. doi: 10.4319/lo.1969.14.5.0799

Thébault, J., Schraga, T. S., Cloern, J. E., and Dunlavey, E. G. (2008). Primary production and carrying capacity of former salt ponds after reconnection to San Francisco Bay. *Wetlands* 28, 841–851. doi: 10.1672/07-190.1

Valiela, I., Collins, G., Kremer, J., Lajtha, K., Geist, M., Seely, B., et al. (1997a). Nitrogen loading from coastal watersheds to receiving estuaries: new method and application. *Ecol. Appl.* 7, 358–380. doi: 10.1890/1051-0761(1997)007[0358:NLFCWT]2.0.CO;2

Valiela, I., Foreman, K., LaMontagne, M. G., Hersh, D., Costa, J., Peckol, P., et al. (1992). Couplings of watersheds and coastal waters: sources and consequences of nutrient enrichment in Waquoit Bay, Massachusetts. *Estuaries* 15, 443–457. doi: 10.2307/1352389

Valiela, I., Geist, M., McClelland, J., and Tomasky, G. (2000). Nitrogen loading from watersheds to estuaries: verification of the Waquoit Bay nitrogen loading model. *Biogeochemistry* 49, 277–293. doi: 10.1023/A:1006345024374

Valiela, I., McClelland, J., Hauxwell, J., Behr, P. J., Hersh, D., and Foreman, K. (1997b). Macroalgal blooms in shallow estuaries: controls and ecophysiological and ecosystem consequences. *Limnol. Oceanogr.* 42, 1105–1118. doi: 10.4319/lo.1997.42.5_part_2.1105

Vaquer-Sunyer, R., and Duarte, C. M. (2008). Thresholds of hypoxia for marine biodiversity. *Proc. Natl. Acad. Sci. U.S.A.* 105, 15452–15457. doi: 10.1073/pnas.0803833105

Vieillard, A. M., and Fulweiler, R. W. (2012). Impacts of long-term fertilization on salt marsh tidal creek benthic nutrient and N2 gas fluxes. *Mar. Ecol. Prog. Ser.* 471, 11–22. doi: 10.3354/meps10013

Webster, I. T., and Harris, G. P. (2004). Anthropogenic impacts on the ecosystems of coastal lagoons: modelling fundamental biogeochemical processes and management implications. *Mar. Freshw. Res.* 55, 67–78. doi: 10.1071/MF03068

Weiss, R. F. (1970). The solubility of nitrogen, oxygen and argon in water and seawater. *Deep Sea Res.* 17, 721–735.

Conflict of Interest Statement: The authors declare that the research was conducted in the absence of any commercial or financial relationships that could be construed as a potential conflict of interest.

11

A seasonal diary of phytoplankton in the North Atlantic

Christian Lindemann * and Michael A. St. John

National Institute of Aquatic Resources, Technical University of Denmark, Charlottenlund, Denmark

Edited by:
Xabier Irigoien, King Abdullah
University for Science and
Technology, Saudi Arabia
Reviewed by:
Sergio M. Vallina, Consejo Superior
de Investigaciones Cientificas, Spain
Xabier Irigoien, King Abdullah
University for Science and
Technology, Saudi Arabia
*Correspondence:
Christian Lindemann, National
Institute of Aquatic Resources,
Technical University of Denmark,
Jægersborg Alle 1, 2920
Charlottenlund, Denmark
e-mail: chrli@aqua.dtu.dk

In recent years new biological and physical controls have been suggested to drive phytoplankton bloom dynamics in the North Atlantic. A better understanding of the mechanisms driving primary production has potentially important implications for the understanding of the biological carbon pump, as it has for prediction of the system in climate change scenarios. However, the scientific discussion regarding this topic has generally failed to integrate the different drivers into a coherent picture, often rendering the proposed mechanisms exclusive to each other. We feel that the suggested mechanisms are not mutually exclusive, but rather complementary. Thus, moving beyond the "single mechanism" point of view, here we present an integrated conceptual model of the physical and biological controls on phytoplankton dynamics in the North Atlantic. Further we believe that the acclimation of physiological rates can play an important role in mediating phytoplankton dynamics. Thus, this view emphasizes the occurrence of multiple controls and relates their variations in impact to climate change.

Keywords: North Atlantic Ocean, phytoplankton spring bloom, abiotic and biotic controls, conceptual model, climate change

INTRODUCTION

Evidence for the impact of climate change on the structure and functioning of the marine environment is becoming increasingly common in for example the North Atlantic, a key region for carbon sequestration (Gruber et al., 2002). In this system, central to our understanding of the biological carbon pump, which is expected to be of increased importance in a high CO_2 ocean (Riebesell et al., 2007), is the role of phytoplankton. In the North Atlantic phytoplankton blooms form an important feature of the annual dynamics of the phytoplankton community and can contribute significantly to carbon export (Allen et al., 2005). Different definitions of the word "bloom" have been used in scientific literature, however it is most commonly referring to as elevate phytoplankton concentrations (Smayda, 1997b). The scientific discussion around the spring bloom has traditionally (Sverdrup, 1953) and more recently (Taylor and Ferrari, 2011b) been dominated by the role of physical drivers. The conditions governing net phytoplankton growth (r) over the seasonal cycle on the other hand have mostly been discussed in the context of grazing control (e.g., Banse, 1992) and food web dynamics, e.g., the PEG (Plankton Ecology Group) model (Sommer et al., 1986, 2012). The "*Dilution-Recoupling-Hypothesis*" (Behrenfeld, 2010) and the extended version thereof, the "*Disturbance-Recovery-Hypothesis*" (Behrenfeld and Boss, 2014) have highlighted the importance of grazing control on spring bloom dynamics. However, publications discussing the onset of the spring bloom have generally highlight one specific process, or as in a recent review (Behrenfeld and Boss, 2014) strongly emphasize one particular control. A more holistic understanding including the importance of the interlinked nature of the biological and physical controls requires the development of an integrated understanding of how these mechanisms act individually and in concert to influence phytoplankton dynamics. Such an understanding should not only consider physical and biological drivers, and their intertwined nature, but also the cells fundamental physiological ability to react to external pressures as for example, acclimation can play an important role in population dynamics and thus help to shape the observed patterns of growth and abundance (Bowler and Scanlan, 2014).

SVERDRUP AND BEYOND

Sverdrup's "*Critical-Depth-Hypothesis*" (Sverdrup, 1953) states that a phytoplankton bloom is initiated when losses including, cell respiration, sinking and grazing are compensated by the integrated light driven growth within the mixed layer, thus leading to a positive net primary production (r) and an increase in phytoplankton concentration. The statement, that positive net production occurs when growth is greater than losses, is inherently true. However, Sverdrup's interpretation, that this situation coincides with the onset of thermal stratification in spring has been called into question. Criticisms can mainly be attributed to different interpretations of the original model as well as a more careful investigation of Sverdrups assumptions.

Sverdrup's first assumption is the existence of a thoroughly mixed surface layer. Typically the mixed layer depth has been defined by density approximated by the homogeneous distribution of water properties such as temperature and salinity, assuming continuous mixing. However, after convective deep winter mixing the onset of stratification can be delayed on the order of days or even longer (Marshall and Schott, 1999; Taylor and Ferrari, 2011b). During this window in time, wind mixing occurs close to the surface, while the water mass is itself still characterized by a deep layer of homogeneous water properties (Chiswell, 2011). Under these conditions the density defined mixed layer depth does not give an accurate representation of the actual mixing depth and surface blooms can occur in the absence of stratification (Townsend et al., 1992; Taylor and Ferrari, 2011b; Ferrari et al., 2014). Taking this physically orientated view, the effect of the interplay between wind mixing and the shutdown of deep convection for the onset of a surface bloom has been proposed in the "*Convection-Shutdown-Hypothesis*" (Ferrari et al., 2014). A more biologically orientated vision describing similar dynamics was put forward by Smayda (1997a), who distinguished between the roles of turbulence and stratification in controlling phytoplankton growth on a cellular and population level due to their eco-physiological impacts.

Sverdrups second assumption also relates to mixing and assumes that within the mixed surface layer the level of turbulence is strong enough to distribute plankton evenly throughout the mixed layer. Assuming homogeneous turbulent diffusion throughout a mixed layer, Huisman et al. (1999) found the existence of a window of turbulence in which phytoplankton are able to achieve positive net growth, thus bloom according to his definition. At levels of turbulence below a certain threshold, the *Critical Turbulence*, phytoplankton growth exceeds losses due to vertical mixing and hence a bloom can develop. On the other hand when turbulence levels become too low, cells will not be maintained in the surface layer by turbulence and will sink out of the mixed layer (Huisman et al., 2002). The "*Critical-Turbulence-Hypothesis*" as defined by Behrenfeld and Boss (2014) applies to active mixing instead of a density-defined mixed layer, thus allowing the combination of these assumptions.

Sverdrup further assumed a constant loss rate, encompassing cell respiration, grazing and sinking. While he himself pointed to the fact that the impact of these variable loss terms can greatly influence the compensation depth, until recently this has largely been ignored. A number of findings have brought this assumption into question. For example, the "*Disturbance-Recovery-Hypothesis*" focuses on the seasonally variable top-down control. During winter, the deepening mixed layer dilutes phyto- and zooplankton concentrations and due to the density dependency of grazing pressure (Landry and Hassett, 1982) lead to a reduction in grazing pressure. This process has been termed the *Decoupling* of phytoplankton biomass from zooplankton grazing pressure (Behrenfeld, 2010) and allows for an increase in *r* and the standing phytoplankton stock, prior to the onset of stratification in early spring (Behrenfeld and Boss, 2014). However, importantly due to mixed layer deepening this is not evidenced as a volumetric increase in biomass. As the water column begins to stratify in early spring the ability of zooplankton to maintain themselves within the stratifying water column, leads to a *Recoupling* of micro-zooplankton with phytoplankton cells, thus increasing grazing pressure (Evans and Parslow, 1985; Behrenfeld and Boss, 2014). This re-stratification also has implications for carbon flux as after the shutdown of deep convection, phytoplankton cells below the depth of spring stratification are *Detrained*, i.e., sink out below the developing summer mixed layer and are lost to the deep ocean and do not contribute to the new spring bloom (Evans and Parslow, 1985; Behrenfeld and Boss, 2014). This phenomenon can potentially

induce an important export to depth prior to the onset of stratification (Körtzinger et al., 2008).

Sverdrup's use of a constant loss rate over a diurnal cycle and over depth has further been criticized on physiological grounds (Smetacek and Passow, 1990). Dark respiration for example is known to be highly variable, not only with regard to species but also depending on growth conditions (Steenmann Nielsen and Hansen, 1959; Falkowski and Owens, 1980; Geider and Osborne, 1989). Within a deep mixed layer cells can experience prolonged periods of darkness, which can result in changes in dark respiration rates. Jochem (1999) found that when placed in darkness within a few days cell can reduce their respiration rate to only a few percent of their respiration rate in light. Further, most taxa can produce resting stages some of which remain photosynthetically active (McMinn and Martin, 2013). French and Hargraves (1980) estimated that a period of 12 h would provide enough energy to sustain some diatom resting spores for 29 days in the dark.

Yet another gap in the Sverdrup Hypothesis is the role of physiology of sinking. Species-specific sinking rates depend on cell size and shape (Smayda, 1970; Miklasz and Denny, 2010) as well as on cell density, which is strongly linked to growth conditions (Anderson and Sweeney, 1977; Bienfang et al., 1982; Brookes and Ganf, 2001) and can exhibit a response time of only a few hours (Waite et al., 1992). Huisman et al. (2002) showed that a certain level of turbulence is required to counteract cell sinking. However, for example Acuña et al. (2010) found that growing diatoms can control their density to achieve positive buoyancy und thus persist even in conditions of low turbulence. Similarly, the existence of predators has the potential to promote the generation of defensive structures such as spines, which not only deter grazers but as well increase surface area thus reducing sinking rates (Nguyen et al., 2011). The potential to adjust defenses against grazers is well recognized (Vos et al., 2004) creating an evolutionary arms race (Smetacek, 2001) as predators also show the ability to respond to changing prey concentration (Mariani et al., 2013). The ability of planktonic organisms to react to increased predator or prey concentrations "increases the relative importance of bottom-up controls" (Vos et al., 2004) thus highlighting the role of physical processes.

Given the above considerations, in the following we present a conceptual model of the seasonal dynamics of phytoplankton in the North Atlantic. With this model we aspire to move toward a more holistic understanding of the different concepts discussed in the literature, highlighting the interplay between bio-physical mechanisms controlling the annual phytoplankton dynamics as well as integrating the cells ability to react to changes in environmental forcing.

PULLING IT TOGETHER: A CONCEPTUAL MODEL OF SEASONAL DYNAMICS

In the winter, the concentration of phytoplankton within the winter mixed layer is low and typically homogeneously distributed (Backhaus et al., 2003; Ward and Waniek, 2007), however the depth integrated standing stock is on the same order of magnitude as during the spring bloom (Backhaus et al., 2003; Behrenfeld, 2010). During this period *Phyto-convection* can maintain a viable phytoplankton stock (Backhaus et al., 2003). *Phyto-convection* has been defined as the ability of convective mixing to sustain a viable phytoplankton population within a deep mixed layer by counteracting cell sinking rates hence frequently returning them into the euphoric zone (Backhaus et al., 1999) (**Figure 1B**). Even though frequently exposed to the euphotic zone and not limited by nutrients during winter, low light levels during winter allow for limited photosynthesis, rendering the respiratory loss term proportionally more important (Sakshaug et al., 1991). Within the winter phytoplankton community a number of processes both physiological and morphological act to determine the composition of the surviving community. The sinking rates of non or slow growing cells have been found to be higher than those of blooming cells (Waite et al., 1992; Acuña et al., 2010). Thus, species, which are low light acclimatized, have the potential to exhibit reduced sinking rates and a greater potential to remain in the deep convective cells. Conversely, cell respiration is reduced relative to fast growing cells (Geider and Osborne, 1989; McMinn and Martin, 2013). Hence, although faster growing cells are favored due to reduced sinking rates, prolonged periods without exposure to light will result in losses due to higher respiration. Micro-zooplankton grazing has been identified as a key control on phytoplankton biomass (e.g., Banse, 1992; Sherr and Sherr, 2002) with changes in importance over the annual cycle. During winter, mixed layer deepening dilutes phytoplankton cells and active grazers resulting in the *Decoupling* of grazing pressure from specific phytoplankton

growth (μ) (Evans and Parslow, 1985) (**Figure 1C**). Therefore, the combined effect of low grazing pressure, low cell respiration and convective mixing are able to compensate for cell sinking and limited light exposure, allowing for a low but positive net phytoplankton growth in the convective mixed layer. As a result, this leads to a slowly increasing standing stock, while the volumetric phytoplankton concentration due to the dilution of the mixed layer stays constant or even may decrease (Backhaus et al., 2003; Behrenfeld and Boss, 2014).

Additionally the ephemeral nature of deep convention can result in periods of reduced turbulence, due to a reduced net surface heat flux in conjunction with low wind stress. During these periods within the still actively mixed part of the upper ocean, levels of turbulence fall below the threshold of *Critical Turbulence* (Huisman et al., 1999) (**Figure 1C**). These periods of quiescence combined with the cells ability to control buoyancy (Waite et al., 1992; Acuña et al., 2010) can maintain cells in the surface layer. However, such events are short lived and therefore do not induce a density stratification of the water column, which enables phytoplankton cells to escape the grazing control of micro-zooplankton (Irigoien et al., 2005). Thus, without *Recoupling* with grazers, this can result in a light induced increase in growth and a subsequent surface bloom in the absence of stratification (Townsend et al., 1992) (**Figure 1C**). Furthermore, mesoscale ocean features such as eddies (Mahadevan et al., 2012) and fronts (Taylor and Ferrari, 2011a) can induce stratification in the absence of net positive

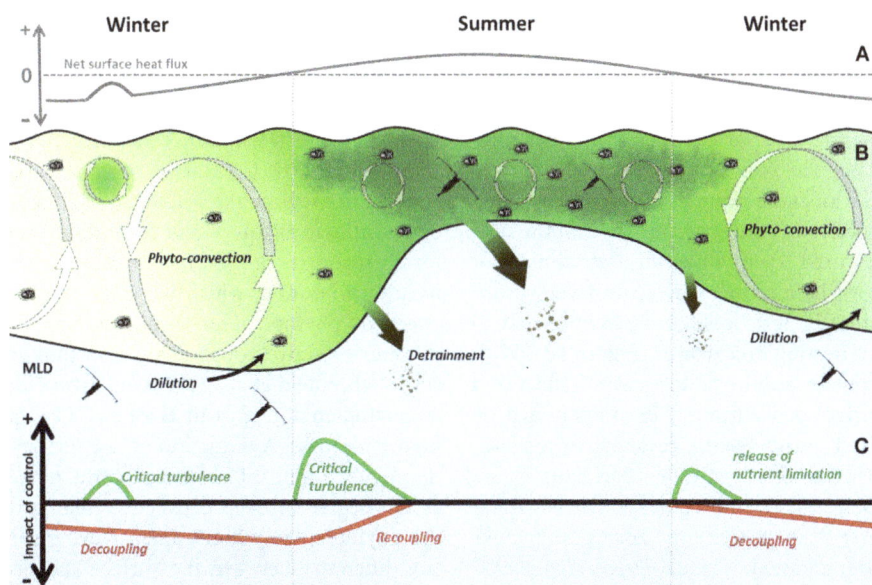

FIGURE 1 | Conceptual model of the physical and biological controls and their impacts on the seasonal cycle of phytoplankton in the open subarctic North Atlantic. (A) Net surface heat flux **(B)** During winter, mixed layer deepening causes the Dilution of plankton leading to a *Decoupling* of grazers from phytoplankton. During this period phytoplankton are sustained by *Phyto-convection* and in combination with ephemeral periods of *Critical Turbulence* result in positive net growth. In early spring the shutdown of deep convection leads to a light driven increase in surface growth. Subsequently re-stratification results in further enhanced growth conditions, i.e., the Critical Depth Model; the *Detrainment* of phytoplankton below the surface mixed layer; and a *Recoupling* with grazers, resulting in a close coupling of biotic and abiotic controls to mixed layer dynamics as observed during summer. In autumn the releases of nutrient limitation due to mixed layer deepening lead to specific growth controlled period before returning to winter conditions. Meso-zooplankton (symbolized by copepods) remain in diapause below the convective mixed layer during winter and migrate into the surface ML in spring. **(C)** Difference of the impact of abiotic (green) and biotic (red) controlling mechanisms on net phytoplankton growth (*r*) relative to the equilibrium dynamics imposed by phytoplankton composition and mixed layer.

heat input, leading to conditions conducive for phytoplankton growth. In the spring, it has been suggested that the initiation of "blooms can begin following winter deep convection, and prior to the vernal development of stratification" (Townsend et al., 1994). Following the "*Convection-Shutdown-Hypothesis*" (Ferrari et al., 2014) during this transitional stage, positive net surface heat flux (**Figure 1A**) leads to the shutdown of deep convection. Stratification does not occur instantaneously resulting in a non-homogeneous mixing throughout the density defined mixed layer (Marshall and Schott, 1999; Taylor and Ferrari, 2011b). Within the upper actively mixed component, turbulence below the threshold of *Critical Turbulence* (Huisman et al., 1999) occurs, thus creating favorable light conditions for a further increase in cell growth. Under conditions of increased cell growth (μ), respiration rates are higher (Falkowski and Owens, 1980; Xue et al., 1996), while the species-specific sinking rates (Waite et al., 1992) decrease and cells may even achieve positive buoyancy (Acuña et al., 2010). This represents a switch in the strength of physiologically driven loss terms, when compared to winter conditions. This change has the potential to be of considerable advantage for cells. Increased light exposure due to the seasonal increase in light and lower mixing depth allows for a higher photosynthetic yield, while the reduction in turbulent mixing increases the impact of cell sinking which is potentially offset by the cells ability to reduce sinking rates. Here, similar to windows of reduced convective mixing in winter, the increase in light exposure, facilitated by adjustment of the physiologically controlled phytoplankton loss rates, allows the increase in specific phytoplankton growth rate. Simultaneously, *Recoupling* with micro-zooplankton gradually increases grazing pressure as stratification sets in. However, during the initial stage after the shutdown of deep convection light-driven phytoplankton growth out-completes the increase in grazing control, thus leading to a further increase in net phytoplankton growth and biomass concentration near the surface (Townsend et al., 1992; Taylor and Ferrari, 2011b).

Furthermore, the reduction of convective mixing and the subsequent shoaling of the mixed layer causes the *Detrainment* of phytoplankton cells below the receding convective layer (Evans and Parslow, 1985; Behrenfeld and Boss, 2014), leaving cells to sink to depth below the retreating mixed layer (**Figure 1B**). With respect to the integrated phytoplankton biomass within the mixed layer, the increase in surface concentration is compensated by the *Detrainment* of cells at greater depth, resulting in a standing stock on the same order of magnitude (Backhaus et al., 2003; Behrenfeld and Boss, 2014). Subsequently, when surface mixing becomes shallower than the euphotic zone phytoplankton dynamics follow the traditional "*Critical-Depth-Hypothesis*" where a combination of mixed layer depth, light availability, and grazing pressure determines net phytoplankton growth and volumetric concentration (Sverdrup, 1953; Banse, 1992). Noteworthy, in the North Atlantic the bulk of the meso-zooplankton community spend the winter in diapause below the permanent mixed layer (Heinrich, 1962), thus removing their grazing pressure independent of mixed layer dynamics. Behrenfeld and Boss (2014) argue that the role of meso-zooplankton has traditionally been overestimated, but despite more recent findings highlighting the importance of micro-zooplankton (e.g., Banse, 1992),

meso-zooplankton still form an important part of the grazing community during spring and early summer. Over the summer, zooplankton grazing, dominated by micro-zooplankton grazing (Banse, 1992) and viral infection (Suttle, 2005), in combination with nutrient limitation, control phytoplankton growth and regulate the collapse and accession of the different phytoplankton groups (Margalef, 1978; Martin, 2012). Species-specific blooms are often reported to be terminated by nutrient limitation, which may induced increased sinking rates. These can lead to the rapid removal of cells from the mixed layer, resulting in carbon export events as has been prominently observed for diatom blooms (Allen et al., 2005). During fall, surface cooling results in a deepening of the pycnocline and an injection of nutrient rich water into the nutrient depleted surface layer as well as a reduction in grazing pressure due to dilution (Sommer et al., 2012; Behrenfeld and Boss, 2014). Both these processes lead to an increase in net phytoplankton growth and can result in an autumn bloom, which potentially terminates with a carbon export event (Findlay et al., 2005). Thereafter the system reverts to winter conditions as described by the "*Disturbance-Recovery-Hypothesis*."

PHYTOPLANKTON DYNAMICS IN A CHANGING WORLD

Unlike subtropical systems, which are relatively stable and dominated by nutrient limitation and the microbial loop (e.g., Sherr and Sherr, 2002) or the more recently identified mixotrophic loop (Mitra et al., 2014) throughout the year, the North Atlantic shows distinct changes over the season. Here, the physical controls shift from light limitation in autumn and winter to nutrient controls during the summer, while the grazing control decreases during the winter and due to *Recoupling* of micro-zooplankton (Behrenfeld, 2010) and the re-emergence meso-zooplankton form diapause (Heinrich, 1962; Sommer et al., 2012) increase sharply in spring. Critical for the North Atlantic phytoplankton community will be the influence of climate change on the processes influencing the winter community and the seed population for the spring bloom. Climate change is expected to increase stratification in winter, which has been proposed to have opposing effects on phytoplankton dynamics. Stratification will reduce the dilution of active zooplankton and thus increase grazing pressure (Behrenfeld et al., 2013) while also reducing light limitation of phytoplankton growth (Doney, 2006) potentially leading to higher biomass. A reduction of net surface heat flux will impact upon the timing of occurrence and retreat, the depth as well as the degree of deep convective mixing thereby impacting on the phytoplankton community. During summer, a greater density difference between the surface and deep layers will reduce the flux of nutrients into the euphotic zone thereby reducing production. As a result the North Atlantic ecosystem is likely to become more similar to lower latitude oligotrophic ecosystems as it moves closer toward a recycling dominated community (Doney, 2006). In the North Atlantic the complex interplay between abiotic and biotic controls on phytoplankton clearly renders our understanding of the impact of climate change anything but straightforward. Given the central role of phytoplankton in marine ecosystems, e.g., in the North Atlantic, moving toward a more integrated approach for understanding and predicting the role phytoplankton in the future is critical. Lewandowska et al. (2014) investigated the combined effects of biological and physical drivers on bloom dynamics focusing on mesocosm experiments and a global modeling approach. They concluded that "the effect of ocean warming on marine plankton depends on the nutrient regime," and thus stratification. However, the temperature dependent growth rate of heterotrophic protists shows a stronger increase than that of phytoplankton. Thus, a scenario of increased temperature could also lead to enhanced trophic coupling, increasing grazing control (Rose and Caron, 2007). In a recent model study Ward et al. (2014) found the total biomass to be depended on nutrient availability, while grazing had a bigger impact on community structure.

Our conceptual model is by no means complete as for example, spatial and temporal small-scale variability of the processes outlined will determine local phytoplankton features. Furthermore, an enhanced understanding on the role of viruses (Suttle, 2005), cellular respiration (Marra, 2009), the adaptive capacities of organisms as well as new insights on the interaction between biophysical processes will continue to modify our perception of phytoplankton dynamics in this ecosystem.

ACKNOWLEDGMENTS

We are thankful to Mark Payne, Adrian Martin, Richard Bellerby, and Artur Palacz for useful comments on an earlier version of the manuscript, fruitful discussions with the participants of the FS Meteor "Deep convection" cruise funded by the Deutsche Forschunggemeinschaft as well the constructive comments of the reviewers how helped to improve this manuscript substantially. Partial financial support was provided by the FP7 program EURO-BASIN (Contract Nr. 264933).

REFERENCES

Acuña, J., López-Alvarez, M., Nogueira, E., and González-Taboada, F. (2010). Diatom flotation at the onset of the spring phytoplankton bloom: an *in situ* experiment. *Mar. Ecol. Prog. Ser.* 400, 115–125. doi: 10.3354/meps08405

Allen, J. T., Brown, L., Sanders, R., Moore, C. M., Mustard, A., Fielding, S., et al. (2005). Diatom carbon export enhanced by silicate upwelling in the northeast Atlantic. *Nature* 437, 728–732. doi: 10.1038/nature03948

Anderson, L., and Sweeney, B. (1977). Diel changes in sedimentation characteristics of Ditylum brightwelli: changes in cellular lipid and effects of respiratory inhibitors and ion-transport modifiers. *Limnol. Oceanogr.* 22, 539–552. doi: 10.4319/lo.1977.22.3.0539

Backhaus, J., Hegseth, E., Wehde, H., Irigoien, X., Hatten, K., and Logemann, K. (2003). Convection and primary production in winter. *Mar. Ecol. Prog. Ser.* 251, 1–14. doi: 10.3354/meps251001

Backhaus, J., Wehde, H., Hegseth, E., and Kämpf, J. (1999). Phyto-convection:the role of oceanic convection in primary production. *Mar. Ecol. Prog. Ser.* 189, 77–92. doi: 10.3354/meps189077

Banse, K. (1992). "Grazing, temporal changes of phytoplankton concentrations, and the microbial loop in the open sea," in *Primary Productivity and Biogeochemical Cycles in the Sea Sea Environmental Science Research*, Vol. 43, eds P. G. Falkowski and A. D. Woodhead (New York, NY: Plenum Press), 409–440.

Behrenfeld, M. J. (2010). Abandoning Sverdrup's critical depth hypothesis on phytoplankton blooms. *Ecology* 91, 977–989. doi: 10.1890/09-1207.1

Behrenfeld, M. J., and Boss, E. S. (2014). Resurrecting the ecological underpinnings of ocean plankton blooms. *Ann. Rev. Mar. Sci.* 6, 167–194. doi: 10.1146/annurev-marine-052913-021325

Behrenfeld, M. J., Doney, S. C., Lima, I., Boss, E. S., and Siegel, D. A. (2013). Annual cycles of ecological disturbance and recovery underlying the subarctic Atlantic spring plankton bloom. *Global Biogeochem. Cycles* 27, 526–540. doi: 10.1002/gbc.20050

Bienfang, P., Harrison, P., and Quarmby, L. (1982). Sinking rate response to depletion of nitrate, phosphate and silicate in four marine diatoms. *Mar. Biol.* 67, 295–302. doi: 10.1007/BF00397670

Bowler, C., and Scanlan, D. J. (2014). Genetics. Being selective in the Prochlorococcus collective. *Science* 344, 366–367. doi: 10.1126/science.1253817

Brookes, J., and Ganf, G. (2001). Variations in the buoyancy response of Microcystis aeruginosa to nitrogen, phosphorus and light. *J. Plankton Res.* 23, 1399–1411. doi: 10.1093/plankt/23.12.1399

Chiswell, S. (2011). Annual cycles and spring blooms in phytoplankton: don't abandon Sverdrup completely. *Mar. Ecol. Prog. Ser.* 443, 39–50. doi: 10.3354/meps09453

Doney, S. C. (2006). Plankton in a warmer world. *Nature* 444, 695–696. doi: 10.1038/444695a

Evans, G., and Parslow, J. (1985). A model of annual plankton cycles. *Biol. Oceanogr.* 3, 327–347. doi: 10.1080/01965581.1985.10749478

Falkowski, P., and Owens, T. (1980). Light—shade adaptation two strategies in marine phytoplankton. *Plant Physiol.* 66, 592–595. doi: 10.1104/pp.66.4.592

Ferrari, R., Merrifield, S. T., and Taylor, J. R. (2014). Shutdown of convection triggers increase of surface chlorophyll. *J. Mar. Syst.* doi: 10.1016/j.jmarsys.2014.02.009. Available online at: http://www.sciencedirect.com/science/article/pii/S0924796314000384

Findlay, H. S., Yool, A., Nodale, M., and Pitchford, J. W. (2005). Modelling of autumn plankton bloom dynamics. *J. Plankton Res.* 28, 209–220. doi: 10.1093/plankt/fbi114

French, F., and Hargraves, P. (1980). Physiological characteristics of plankton diatom resting spores. *Mar. Biol. Lett.* 1, 185–195.

Geider, R. J., and Osborne, B. A. (1989). Respiration and microalgal growth: a review of the quantitative relationship between dark respiration and growth. *New Phytol.* 112, 327–341. doi: 10.1111/j.1469-8137.1989.tb00321.x

Gruber, N., Keeling, C. D., and Bates, N. R. (2002). Interannual variability in the North Atlantic Ocean carbon sink. *Science* 298, 2374–2378. doi: 10.1126/science.1077077

Heinrich, A. K. (1962). The life histories of plankton animals and seasonal cycles of plankton communities in the oceans. *J. Cons. Int. Explor. Mer.* 27, 15–24. doi: 10.1093/icesjms/27.1.15

Huisman, J., Arrayás, M., Ebert, U., and Sommeijer, B. (2002). How do sinking phytoplankton species manage to persist? *Am. Nat.* 159, 245–254. doi: 10.1086/338511

Huisman, J., Oostveen, P., van, and Weissing, F. (1999). Critical depth and critical turbulence: two different mechanisms for the development of phytoplankton blooms. *Limnol. Oceanogr.* 44, 1781–1787. doi: 10.4319/lo.1999.44.7.1781

Irigoien, X., Flynn, K. J., and Harris, R. P. (2005). Phytoplankton blooms: a "loophole" in microzooplankton grazing impact? *J. Plankton Res.* 27, 313–321. doi: 10.1093/plankt/fbi011

Jochem, F. (1999). Dark survival strategies in marine phytoplankton assessed by cytometric measurement of metabolic activity with fluorescein diacetate. *Mar. Biol.* 135, 721–728. doi: 10.1007/s002270050673

Körtzinger, A., Send, U., Lampitt, R. S., Hartman, S., Wallace, D. W. R., Karstensen, J., et al. (2008). The seasonal p CO 2 cycle at 49°N/16.5°W in the northeastern Atlantic Ocean and what it tells us about biological productivity. *J. Geophys. Res.* 113, C04020. doi: 10.1029/2007JC004347

Landry, M. R., and Hassett, R. P. (1982). Estimating the grazing impact of marine micro-zooplankton. *Mar. Biol.* 67, 283–288. doi: 10.1007/BF00397668

Lewandowska, A. M., Boyce, D. G., Hofmann, M., Matthiessen, B., Sommer, U., and Worm, B. (2014). Effects of sea surface warming on marine plankton. *Ecol. Lett.* 17, 614–623. doi: 10.1111/ele.12265

Mahadevan, A., D'Asaro, E., Lee, C., and Perry, M. J. (2012). Eddy-driven stratification initiates North Atlantic spring phytoplankton blooms. *Science* 337, 54–58. doi: 10.1126/science.1218740

Margalef, R. (1978). Life-forms of phytoplankton as survival alternatives in an unstable environment. *Oceanol. acta* 1, 493–509.

Mariani, P., Andersen, K. H., Visser, A. W., Barton, A. D., and Kiørboe, T. (2013). Control of plankton seasonal succession by adaptive grazing. *Limnol. Oceanogr.* 58, 173–184. doi: 10.4319/lo.2013.58.1.0173

Marra, J. (2009). Net and gross productivity: weighing in with 14C. *Aquat. Microb. Ecol.* 56, 123–131. doi: 10.3354/ame01306

Marshall, J., and Schott, F. (1999). Open ocean convection: observations, theory, and models. *Rev. Geophys.* 37, 1–64. doi: 10.1029/98RG02739

Martin, A. (2012). The seasonal smorgasbord of the seas. *Science* 337, 46–47. doi: 10.1126/science.1223881

McMinn, A., and Martin, A. (2013). Dark survival in a warming world. *Proc. R. Soc. B Biol. Sci.* 280:20122909. doi: 10.1098/rspb.2012.2909

Miklasz, K., and Denny, M. (2010). Diatom sinking speeds: improved predictions and insight from a modified Stokes' law. *Limnol. Oceanogr.* 55, 2513–2525. doi: 10.4319/lo.2010.55.6.2513

Mitra, A., Flynn, K. J., Burkholder, J. M., Berge, T., Calbet, A., Raven, J. A., et al. (2014). The role of mixotrophic protists in the biological carbon pump. *Biogeosciences* 11, 995–1005. doi: 10.5194/bg-11-995-2014

Nguyen, H., Karp-Boss, L., Jumars, P. A., and Fauci, L. (2011). Hydrodynamic effects of spines: a different spin. *Limnol. Oceanogr. Fluids Environ.* 1, 110–119. doi: 10.1215/21573698-1303444

Riebesell, U., Schulz, K. G., Bellerby, R. G. J., Botros, M., Fritsche, P., Meyerhöfer, M., et al. (2007). Enhanced biological carbon consumption in a high CO2 ocean. *Nature* 450, 545–548. doi: 10.1038/nature06267

Rose, J., and Caron, D. (2007). Does low temperature constrain the growth rates of heterotrophic protists? Evidence and implications for algal blooms in cold waters. *Limnol. Oceanogr.* 52, 886–895. doi: 10.4319/lo.2007.52.2.0886

Sakshaug, E., Johnsen, G., Andresen, K., and Vernet, M. (1991). Modeling of light-dependent algal photosynthesis and growth: experiments with the Barents sea diatoms Thalassiosira nordenskioldii and Chaetoceros furcellatus. *Deep Sea Res. Part A. Oceanogr. Res. Pap.* 38, 415–430. doi: 10.1016/0198-0149(91)90044-G

Sherr, E. B., and Sherr, B. F. (2002). Significance of predation by protists in aquatic microbial food webs. *Antonie Van Leeuwenhoek* 81, 293–308. doi: 10.1023/A:1020591307260

Smayda, T. (1970). The suspension and sinking of phytoplankton in the sea. *Oceanogr. Mar. Biol. Annu. Rev.* 8, 353–414.

Smayda, T. J. (1997a). Harmful algal blooms: their ecophysiology and general relevance to phytoplankton blooms in the sea. *Limnol. Oceanogr.* 42, 1137–1153. doi: 10.4319/lo.1997.42.5_part_2.1137

Smayda, T. J. (1997b). What is a bloom? A commentary. *Limnol. Oceanogr.* 42, 1132–1136. doi: 10.4319/lo.1997.42.5_part_2.1132

Smetacek, V. (2001). A watery arms race. *Nature* 411, 745. doi: 10.1038/35081210

Smetacek, V., and Passow, U. (1990). Spring bloom initiation and Sverdrup's critical-depth model. *Limnol. Oceanogr.* 35, 228–234. doi: 10.4319/lo.1990.35.1.0228

Sommer, U., Adrian, R., De Senerpont Domis, L., Elser, J. J., Gaedke, U., Ibelings, B., et al. (2012). Beyond the Plankton Ecology Group (PEG) model: mechanisms driving plankton succession. *Annu. Rev. Ecol. Evol. Syst.* 43, 429–448. doi: 10.1146/annurev-ecolsys-110411-160251

Sommer, U., Gliwicz, Z. M., Lampert, W., and Duncan, A. (1986). The PEG-model of seasonal succession of planktonic events in fresh waters. *Arch. Hydrobiol.* 106, 433–471.

Steenmann Nielsen, E., and Hansen, V. (1959). Measurements with the carbon-14 technique of the respiration rates in natural populations of phytoplankton. *Deep Sea Res.* 222–233. doi: 10.1016/0146-6313(58)90015-7

Suttle, C. A. (2005). Viruses in the sea. *Nature* 437, 356–361. doi: 10.1038/nature04160

Sverdrup, H. (1953). On conditions for the vernal blooming of phytoplankton. *J. Cons.* 18, 287–295. doi: 10.4319/lom.2007.5.269

Taylor, J. R., and Ferrari, R. (2011a). Ocean fronts trigger high latitude phytoplankton blooms. *Geophys. Res. Lett.* 38, L23601. doi: 10.1029/2011GL049312

Taylor, J. R., and Ferrari, R. (2011b). Shutdown of turbulent convection as a new criterion for the onset of spring phytoplankton blooms. *Limnol. Oceanogr.* 56, 2293–2307. doi: 10.4319/lo.2011.56.6.2293

Townsend, D., Keller, M., Sieracki, M., and Ackleson, S. (1992). Spring phytoplankton blooms in the absence of vertical water column stratification. *Nature* 360, 59–62. doi: 10.1038/360059a0

Townsend, D. W., Cammen, L. M., Holligan, P. M., Campbell, D. E., and Pettigrew, N. R. (1994). Causes and consequences of variability in the timing of spring phytoplankton blooms. *Deep Sea Res. Part I Oceanogr. Res. Pap.* 41, 747–765. doi: 10.1016/0967-0637(94)90075-2

Vos, M., Verschoor, A., and Kooi, B. (2004). Inducible defenses and trophic structure. *Ecology* 85, 2783–2794. doi: 10.1890/03-0670

Waite, A. M., Thompson, P. A., and Harrison, P. J. (1992). Does energy control the sinking rates of marine diatoms? *Limnol. Oceanogr.* 37, 468–477. doi: 10.4319/lo.1992.37.3.0468

Ward, B. A., Dutkiewicz, S., and Follows, M. J. (2014). Modelling spatial and temporal patterns in size-structured marine plankton communities: top-down and bottom-up controls. *J. Plankton Res.* 36, 31–34. doi: 10.1093/plankt/fbt097

Ward, B., and Waniek, J. (2007). Phytoplankton growth conditions during autumn and winter in the Irminger Sea, North Atlantic. *Mar. Ecol. Prog. Ser.* 334, 47–61. doi: 10.3354/meps334047

Xue, X., Gauthier, D. A., Turpin, D. H., and Weger, H. G. (1996). Interactions between photosynthesis and respiration in the green alga chlamydomonas reinhardtii (characterization of light-enhanced dark respiration). *Plant Physiol.* 112, 1005–1014. doi: 10.1104/pp.112.3.1005

Conflict of Interest Statement: The authors declare that the research was conducted in the absence of any commercial or financial relationships that could be construed as a potential conflict of interest.

12

The R package *EchoviewR* for automated processing of active acoustic data using Echoview

Lisa-Marie K. Harrison[1]*, Martin J. Cox[1,2], Georg Skaret[3] and Robert Harcourt[1]

[1] Marine Predator Research Group, Department of Biological Sciences, Faculty of Science and Engineering, Macquarie University, North Ryde, NSW, Australia
[2] Australian Antarctic Division, Department of the Environment, Australian Government, Kingston, TAS, Australia
[3] Institute of Marine Research, Bergen, Norway

Edited by:
Xabier Irigoien, King Abdullah University of Science and Technology, Saudi Arabia

Reviewed by:
Guillermo Boyra, AZTI, Spain
Anders Røstad, King Abdullah University of Science and Technology, Saudi Arabia

*Correspondence:
Lisa-Marie K. Harrison, Marine Predator Research Group, Department of Biological Sciences, Faculty of Science and Engineering, Macquarie University, Balaclava Road, North Ryde, NSW 2109, Australia
e-mail: lisamarie.k.harrison@gmail.com

Acoustic data is time consuming to process due to the large data size and the requirement to often undertake some data processing steps manually. Manual processing may introduce subjective, irreproducible decisions into the data processing work flow, reducing consistency in processing between surveys. We introduce the R package *EchoviewR* as an interface between R and Echoview, a commercially available acoustic processing software package. *EchoviewR* allows for automation of Echoview using scripting which can drastically reduce the manual work required when processing acoustic surveys. This package plays an important role in reducing subjectivity in acoustic data processing by allowing exactly the same process to be applied automatically to multiple surveys and documenting where subjective decisions have been made. Using data from a survey of Antarctic krill, we provide two examples of using *EchoviewR*: krill biomass estimation and swarm detection.

Keywords: active acoustic, Antarctic krill, data processing, echosounder, Echoview, R package

INTRODUCTION

Active acoustics is a tool widely used for seabed mapping, seabed type classification, underwater tracking and resource monitoring. A suite of active acoustic instruments are available to carry out imaging (e.g., scanning sonars) and more quantitative tasks (e.g., multibeam and scientific echosounders). Echosounders have evolved from being instruments used primarily for mapping and navigation, to precision instruments capable of resolving organisms a few millimeters in length and providing quantitative estimates of, for example, biomass.

This advance has seen widespread use of echosounders to detect organisms in the upper water column of both freshwater and marine environments for commercial fisheries and scientific purposes. In the marine environment, echosounders are routinely used to provide data informing commercial fishery stock assessments (Gerlotto et al., 1999) and to investigate ecological relationships such as predator-prey interactions (Benoit-Bird et al., 2013). Oceanographic applications include seabed habitat mapping (Brown et al., 2004) and environmental monitoring, e.g., oil seep and methane bubble monitoring after the Deepwater Horizon oil spill (Weber et al., 2014). Echosounders are commonly used in conjunction with image/video (McGonigle et al., 2009) and sediment sampling (Van Walree et al., 2005) to verify seabed type, or trawls to verify a biological species' presence, size and target strength (McGonigle et al., 2009).

Echosounder transducers are most commonly embedded in a ship's hull or drop keel, although other platforms such as landers (Johansen et al., 2009), gliders (Guihen et al., 2014), and autonomous underwater vehicles (Brierley et al., 2002) have been used. Regardless of platform, datasets from active acoustics are invariably extremely large and time consuming to process.

In active acoustic surveys, a conventional split-beam echosounder collecting data to a range of 500 m and pinging once per second typically collects around 8 GB of data per day (Note: this depends on settings such as range resolution and pulse repetition rate). This may be compounded by the need to use multiple echosounder frequencies, sometimes more than six, operating simultaneously, further inflating the size of the raw data sets. Moreover, the routine use of broadband systems like the Simrad EK80 on board scientific and commercial vessels is not far away. The amount of data from such systems vastly exceeds those from conventional sounders, and will again push storage and processing capacity. With advances in data storage capacity, data storage is no longer a significant constraint and enhanced computational power has enabled the development of powerful acoustic data processing software.

There are several software packages suitable for the processing of echosounder data e.g., Echoview (Myriax, Hobart; www.echoview.com), LSSS (MAREC, Christian Michelsen Research, Norway, http://www.cmr.no/index.cfm?id=421565) and Sonar5-Pro (University of Oslo, Norway, http://folk.uio.no/hbalk/sonar4_5/). However, processing acoustic data remains time consuming and frequently requires subjective, often undocumented, decisions to be made by the user, such as removal of noise or bad data and allocation of backscatter to targets. Subjective decisions can potentially bias outputs from processed active acoustic data, for example biomass estimates.

Here we present the R package *EchoviewR* as a tool to: (1) reduce the processing time requiring a human operator, (2) document processing steps thereby generating reproducible methodology, and (3) provide a framework within which additional functionality can be built by members of the acoustics community, so reducing the number of subjective decisions. The *EchoviewR* package is an interface between the widely used and freely available R program (http://www.R-project.org/) and Echoview (Myriax, Hobart; www.echoview.com). The methods used are generic and can be transferred to other acoustic processing software with scripting options, but the package as such is incompatible with other acoustic software.

EchoviewR uses Component Object Model (COM) scripting to run Echoview using R. This removes a large portion of the manual processing time and enables entire acoustic surveys to be mostly processed automatically. It also increases consistency in processing because the same methods and thresholds can be applied in exactly the same way to multiple data sets. Hence *EchoviewR* provides a reproducible and transparent automated method for processing acoustic data using Echoview. Some examples of its use include filtering of data, automated biomass estimation and detection of krill swarms.

Using two examples, we illustrate *EchoviewR* functionality. Both examples are based on data collected during surveys of Antarctic krill (*Euphausia superba*; herein krill) using a Simrad EK60 echosounder (Horten, Norway) with downward facing hull-mounted transducers. The first example estimates regional krill biomass, and the second example detects krill swarms.

EchoviewR is intended to speed up processing of already clean acoustic data and is not currently capable of removing false bottom effects, time varied gain or noise spikes although the package can access Echoview virtual variables to do some of these tasks, e.g., "Background noise removal algorithm" virtual variable (De Robertis and Higginbottom, 2007). The package is intended only as a method of automating processing using Echoview and is not a standalone method for processing acoustic data.

METHODS
IMPLEMENTATION AND DEPENDENCIES

EchoviewR was created using R 3.1 (R Development Core Team, 2014; available from http://cran.r-project.org/) with R-Studio 0.98.932 (Rstudio, 2014; available from http://www.rstudio.com/), and Echoview 6.1 (Myriax, 2015; available from http://www.echoview.com/). Both R and Echoview are required to use the package. COM objective handling is achieved using the *RDCOMClient* package. Additional *EchoviewR* functionality uses the *sp, lubridate, geosphere, maptools,* and *rgeos* R libraries (Pebesma and Bivand, 2005; Grolemund and Wickham, 2011; Hijmans, 2014; Bivand and Lewin-Koh, 2014; Bivand and Rundel, 2014). To run Echoview via COM the following modules are required: base, bathymetric, analysis export, and scripting. Worked example one also requires the virtual echogram module and worked example two requires the virtual echogram and schools detection modules.

The *EchoviewR* package is available open source on the GitHub repository (https://github.com/lisamarieharrison/EchoviewR) and can be downloaded and installed as an R package using the "install from.zip file" option in R, or via devtools:install_github().

EXPECTED DATA INPUT FOR THE PACKAGE AND WORKED EXAMPLES

EchoviewR can work with any data type accommodated in Echoview that is accessible via COM. The worked examples provided here have been built using data collected using a Simrad EK60 echosounder (www.simrad.com/ek60). In itself, *EchoviewR* does not create Echoview templates or calibration files, but can use both of these via COM.

FUNCTIONS OF THE PACKAGE

There are 46 functions available in *EchoviewR*, which are described in **Table 1**. A working example for each of these functions is given in the package documentation in the Supplementary Material. Not all Echoview functions are currently available in the package; however any functionality in Echoview that has COM accessibility could be added by the user.

EXAMPLES

Here we present two examples using *EchoviewR*: (1) krill biomass estimation, and (2) krill swarm detection and classification. The purpose of these examples is to demonstrate that these analyses can be run automatically using *EchoviewR* and to show how Echoview output can be seamlessly linked to analyses carried out using R. Both examples assume that the reader is familiar with Echoview and are not intended to be a tutorial on Echoview. It is also assumed that the reader is familiar with R and programming concepts such as for loops.

The data are a subset of the EK60 split-beam data collected during the Krill Acoustics and Oceanography Survey (KAOS) carried out from RV *Aurora Australis*. The KAOS survey was undertaken in January–March 2003 off North Eastern Antarctica. Data from 38, 120, and 200 kHz were written to RAW files. For clarity in the worked examples, we have used the 38 and 120 kHz data because these frequencies are the most useful for detecting and identifying the example species, Antarctic krill.

To demonstrate that biomass estimation and swarm detection can be automatically run on multiple transects where the data are too large to practically read in to Echoview at once, as is the case for most acoustic surveys, segments of six KAOS transects are provided and each 10–20 km transect segment is processed separately (**Figure 1**).

Both these examples have been tested using R Studio and 0.98.932 and Echoview 6.1.32.26088. The data to run these examples are available at the Australian Antarctic Division Data Centre [doi: 10.4225/15/54CF081FB955F]. An example of the data flow for the template used in this example is available as **Figure S1** in the Supplementary Material.

Before running each example some pre-processing is demonstrated to get the data in to a convenient format for analyzing each transect in a separate .EV file. In this pre-processing phase, the six transects are imported separately into Echoview and the following tasks are performed:

1. Create a new .EV file for the transect using the Echoview template file;

Table 1 | Functions available in *EchoviewR*.

Function	Description
EVOpenFile	Opens an existing .EV file
EVSaveFile	Saves an existing .EV file
EVSaveFileAs	Saves an existing .EV file to a new file name
EVCloseFile	Closes an open .EV file
EVNewFile	Creates a new .EV file
EVCreateFileset	Creates a new fileset
EVFindFilesetByName	Finds a fileset by name
EVAddRawData	Adds .RAW files to a fileset
EVCreateNew	Creates a new .EV file from a template
EVminThresholdSet	Sets the minimum dB threshold for an acoustic variable
EVSchoolsDetSet	Sets schools detection parameters
EVAcoVarNameFinder	Finds an acoustic variable by name
EVRegionClassFinder	Finds a region class by name
EVSchoolsDetect	Runs schools detection on an acoustic variable
EVIntegrationByRegionExport	Exports integration by region for an acoustic object
msDateConversion	Converts an Echoview date to readable format
EVAddCalibrationFile	Adds a calibration file to an .EV file
EVFilesInFileset	Finds the names of all .RAW files in the fileset
EVClearRawData	Clears all .RAW files from a fileset
EVFindFilesetTime	Finds the start and end date and time of a fileset
EVNewRegionClass	Creates a new region class
EVImportRegionDef	Imports a regions definition file
EVExportRegionSv	Exports Sv data for a region
EVAdjustRegionBitmap	Adjusts the settings of a region bitmap object
EVFindLineByName	Finds an Echoview line by name
EVChangeVariableGrid	Changes the horizontal and vertical grid for an acoustic variable
EVExportIntegrationByCells	Exports integration by cells for an acoustic variable
EVAddNewAcousticVar	Adds a new acoustic variable
EVShiftRegionDepth	Changes the depth of a region
EVShiftRegionTime	Changes the time of a region
EVGetCalibrationFileName	Finds the calibration file name
EVNewLineRelativeRegion	Creates a new line relative region
EVNewFixedLineDepth	Creates a new fixed depth line
EVDeleteLine	Deletes a line object
EVRenameLine	Renames a line object
EVExportRegionDef	Exports region definitions for a single region
EVFindRegionByName	Finds a region object by name
EVFindRegionClass	Finds a region class by name
EVExportRegionDefByClass	Exports region definitions for an entire region class
EVIntegrationByRegionByCellsExport	Exports integration by region by cells for an acoustic variable

(Continued)

Table 1 | Continued

Function	Description
lawnSurvey	Generate coordinates for a rectangular lawn survey design
zigzagSurvey	Generate coordinates for a zig-zag survey design
centreZigZagOnPosition	Centers a zig-zag survey on a given position
centreLawnOnPosition	Centers a lawn survey on a given position
exportMIF	Write a map information file for import into Echoview
EVImportLine	Imports an Echoview Line object

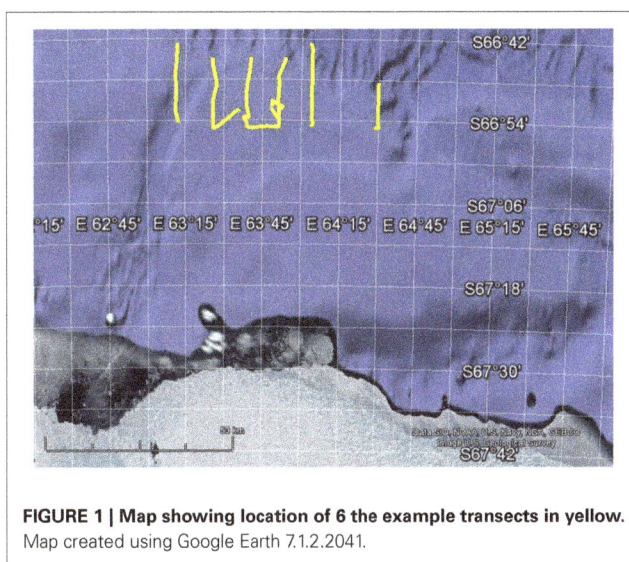

FIGURE 1 | Map showing location of 6 the example transects in yellow. Map created using Google Earth 7.1.2.2041.

2. Import the EK60 .RAW data files for that transect;
3. Add an Echoview.ecs calibration file;
4. Import .evr region definitions files to remove off effort data;
5. Import a seabed exclusion line (lineKAOS .evl);
6. Close and save the file and repeat for remaining transects.

These steps and the code to run them are demonstrated in the "Read data using the R package EchoviewR to control Echoview via COM" pdf vignette that is available with the Supplementary Material. Pre-processing must take place before examples 1 and 2 are run.

EXAMPLE 1—KRILL BIOMASS ESTIMATION

Automated biomass estimation of krill is demonstrated by processing the six transects separately in Echoview and exporting the data into R for density and biomass calculation. For each transect, the following steps are taken in Echoview:

1. Open the transect's .EV file.
2. Set the grid for 38 and 120 kHz noise removed values to 50 ping * 5 m depth.
3. Export integration by cells for 38 and 120 kHz noise removed values.

This produces two .csv files for each transect, one containing 38 kHz and one containing 120 kHz integrated data (i.e., a mean volume backscattering strength value for each cell. Then, the following steps are taken in R.

1. Import the 38 and 120 kHz files for the transect.
2. Remove no data values (set -999 and 999 dB as NA) and depths <0.
3. Calculate the krill difference window of 120–38 kHz for each integration cell using the following formula:

$$\Delta Sv_{ij} = Sv_{120_{ij}} - Sv_{38_{ij}}$$

where $Sv_{120_{ij}}$ = mean 120 kHz backscattering strength for cell at interval j at depth i and $Sv_{38_{ij}}$ = mean 38 kHz backscattering strength for cell at interval j at depth i.

4. Apply the dB difference technique (e.g., Watkins and Brierley, 2002) by setting $Sv_{120_{ij}}$ values outside the survey-specific dB difference range of $1.04 >= \Delta Sv_{ij} <= 14.75$ dB to NA as these windows are unlikely to contain krill.
5. Convert the backscattering strength, $Sv_{120_{ij}}$ for each cell to linear scale, $sv_{120_{ij}}$ (Echoview uses a log scale by default):

$$sv_{ij} = 10^{\frac{Sv_{ij}}{10}}$$

6. Calculate mean volume backscattering strength (MVBS) across all depths for each 50 ping integration interval using the following formula:

$$MVBS_j = 10\log_{10}\frac{1}{n_j}\sum_{i=0}^{n_j} sv_{120_{ij}}$$

where j = integration interval, n = maximum depth within integration interval j and $sv_{120_{ij}}$ = backscattering strength at 120 kHz for interval j at depth i.

7. Calculate estimates of krill density, $\hat{p}_j$, for each integration interval:

$$\hat{p}_j = n_j^* 10^{\left\{\frac{MVBS_j - TS}{10}\right\}}$$

where n_j = maximum depth of integration interval j, $MVBS_j$ = mean volume backscattering strength for interval j as calculated above and TS = target strength for 1 kg of krill at 120 kHz.

8. Calculate the overall transect density, $\hat{p}_k$ for transect k:

$$\hat{p}_k = \frac{1}{s_k}\sum_{j=1}^{s_k} \hat{p}_j$$

where j = integration interval, k = transect and s_k = number of integration intervals within transect k.

9. The full survey density is then estimated using the Jolly and Hampton (1990) method, which uses the weighted density of each transect by length to calculate total survey density. Note

that the formula has been modified to remove stratum as no strata were used in the KAOS example survey design:

$$\hat{p} = w_k \hat{p}_k$$

where k = transect, $w_k = \frac{L_k}{L}$, L_k = length of transect k in km, L = length of all survey transects in km and $\hat{p}_k$ = estimated density for transect k.

10. The full survey biomass estimate, $\hat{b}$, is then calculated by multiplying the weighted survey density by survey area:

$$\hat{b} = \hat{p}A$$

where $\hat{p}$ = estimated survey biomass and A = survey area in km^2.

Both the Echoview and R components above are run within loops to allow each transect to be run separately. This is done to demonstrate how looping over transects or days of a large survey is possible, rather than manually loading and processing each set of files. The *EchoviewR* and R code for the above analysis is shown in the "Biomass estimation using the R package EchoviewR to control Echoview via COM" pdf vignette that is available with the Supplementary Material. **Table 2** shows the estimated density, length and biomass for the sample transects and survey area.

Example 1 has demonstrated the use of *EchoviewR* to automatically process and extract data by transect from Echoview. Krill density and biomass are then calculated in R using the extracted .csv files.

EXAMPLE 2—SWARM DETECTION AND CLASSIFICATION

Automated swarm detection and classification of krill aggregations is demonstrated here using *EchoviewR*. The code for this example is available in the "Schools detection using the R package EchoviewR to control Echoview via COM" pdf vignette file available with the Supplementary Material. Each transect is processed separately to demonstrate how a full survey can be processed automatically using loops. Schools detection is run in Echoview and then detected aggregations are classified and clustered in R. The following steps are undertaken in Echoview using *EchoviewR*:

Table 2 | Estimated transect krill areal density and survey biomass for the six example transects.

Transect number	Mean estimated density, gm^{-2}	Transect length, km	Biomass, tones
1	3.26	13	42
2	20.66	22	454
3	43.74	15	656
4	22.57	22	467
5	6.66	18.5	123
6	4.99	21.5	107
Full survey area	16.79	112	43, 497

1. Open the transect's .EV file.
2. Run schools detection on the variable *120 7x7 convolution*, assigning all detected schools to the region class "aggregations."
3. Export 120 and 38 kHz data for regions of class "aggregations" to a .csv file using the *EVIntegrationByRegionExport* function. This exports a single mean Sv for each aggregation.

In this example, all detected aggregations are exported. However, it is also possible to export only aggregations classified as krill using the *120-38 aggregation dB difference filter* variable included in the template. The filter sets the *Krill aggregations* data to NULL if the *120-38 aggregation dB difference* value for that cell is outside the [1.04, 14.75] dB difference window for the KAOS survey.

The exported aggregations can now be classified and clustered in R. Each transect is run separately using a loop:

1. Import the 120 and 38 kHz export by regions files.
2. Remove null values (-999).
3. Calculate the 120–38 kHz difference window and subset data to only include difference values between [1.04, 14.75].
4. If no aggregations were classified as krill, exit here and move to next transect.
5. If krill aggregations are found, run cluster analysis using the *ClusterSim* library using selected metrics.
6. Print a summary table of the number of aggregations assigned to each identified cluster. **Table 3** shows the number of krill swarms identified and the number of clusters detected for each transect.

This example has demonstrated how school detection, data export and cluster analysis can be run automatically for an entire acoustic survey.

DISCUSSION AND FUTURE DIRECTIONS

EchoviewR is a free interface between R and Echoview that provides automated acoustic data processing. It drastically decreases manual processing time and reduces subjectivity by providing an easy way to implement exactly the same method across surveys. This package enables reproducible methodology, which is a vital part of the scientific method. We have given examples of automated krill biomass estimation and school detection using *EchoviewR* that demonstrate the use of the package on a subset of the KAOS survey. This method can easily be extended to run a full survey by transect, day or any other subset required.

There are a number of limitations to the package. Currently it is only available for use for single and split beam echo sounder data. *EchoviewR* is also unable to handle removal of noise and false bottom effects, which must be completed prior to using the package. Not all functions in Echoview are currently available using *EchoviewR*, however any COM functionality in Echoview can be implemented in R. The COM hierarchy help page is a useful starting point for those wishing to add extra functions.

EchoviewR is accessible as free software from the *EchoviewR* GitHub repository (https://github.com/lisamarieharrison/EchoviewR) and is readily available for community development. An important next step is the implementation of false bottom and noise removal using *EchoviewR*, and it is our hope that the acoustic community will take the tools that we are providing and extend the package to include the functionality that they require. We also underline that the methods described here are generic, and hope the work can inspire the implementation of scripting interface in other acoustic processing software.

ACKNOWLEDGMENTS

We would like to thank Echoview for their support of this project. This research is a contribution to Australian Antarctic Division science programme Project 4104 and project 4102. MC is funded by Australian Research Council grant FS11020005. LH is funded by a Macquarie University Research Excellence Scholarship.

Table 3 | Number of unique krill aggregation clusters identified for each transect.

Transect number	Number of krill swarms	Number of clusters
1	0	0
2	37	9
3	105	6
4	64	3
5	8	3
6	14	6

REFERENCES

Benoit-Bird, K. J., Battaile, B. C., Heppell, S. A., Hoover, B., Irons, D., Jones, N., et al. (2013). Prey patch patterns predict habitat use by top marine predators with diverse foraging strategies. *PLoS ONE* 8:e53348. doi: 10.1371/journal.pone.0053348

Bivand, R., and Lewin-Koh, N. (2014). *maptools: Tools for Reading and Handling Spatial Objects. R Package Version 0.8-29.* Available online at: http://CRAN.R-project.org/package=maptools

Bivand, R., and Rundel, C. (2014). *rgeos: Interface to Geometry Engine - Open Source (GEOS). R package version 0.3-4.* Available online at: http://CRAN.R-project.org/package=rgeos

Brierley, A. S., Fernandes, P. G., Brandon, M. A., Armstrong, F., Millard, N. W., McPhail, S. D., et al. (2002). Antarctic krill under sea ice: elevated abundance in a narrow band just south of ice edge. *Science* 295, 1890–1892. doi: 10.1126/science.1068574

Brown, C. J., Hewer, A. J., Meadows, W. J., Limpenny, D. S., Cooper, K. M., and Rees, H. L. (2004). Mapping seabed biotopes at hastings shingle bank, eastern English Channel. Part 1. Assessment using sidescan sonar. *J. Mar. Biol. Ass. U.K.* 84, 481–488. doi: 10.1017/S002531540400949Xh

De Robertis, A., and Higginbottom, I. (2007). A post-processing technique to estimate the signal-to-noise ratio and remove echosounder background noise. *ICES. J. Mar. Sci.* 64, 1282–1291. doi: 10.1093/icesjms/fsm112

Gerlotto, F., Soria, M., and Fréon, P. (1999). From two dimensions to three: the use of multibeam sonar for a new approach in fisheries acoustics. *Can. J. Fish. Aqua. Sci.* 56, 6–12. doi: 10.1139/cjfas-56-1-6

Grolemund, G., and Wickham, H. (2011). Dates and times made easy with lubridate. *J. Stat. Softw.* 40, 1–25.

Guihen, D., Fielding, S., Murphy, E. J., Heywood, K. J., and Griffiths, G. (2014). An assessment of the use of ocean gliders to undertake acoustic measurements of zooplankton: the distribution and density of Antarctic krill (Euphausia superba) in the Weddell Sea. *Limnol. Oceanogr.* 12, 373–389. doi: 10.4319/lom.2014.12.373

Hijmans, R. J. (2014). *geosphere: Spherical Trigonometry. R Package Version 1.3-8.* Available online at: http://CRAN.R-project.org/package=geosphere

Johansen, G. O., Godø, O. R., Skogen, M. D., and Torkelsen, T. (2009). Using acoustic technology to improve the modelling of the transportation and distribution of juvenile gadoids in the Barents Sea. *ICES. J. Mar. Sci.* 66, 1048–1054. doi: 10.1093/icesjms/fsp081

Jolly, G., and Hampton, I. (1990). A stratified random transect design for acoustic surveys of fish stocks. *Can. J. Fish. Aquat. Sci.* 47, 1282–1291. doi: 10.1139/f90-147

McGonigle, C., Brown, C., Quinn, R., and Grabowski, J. (2009). Evaluation of image-based multibeam sonar backscatter classification for benthic habitat discrimination and mapping at Stanton Banks, UK. *Estuar. Coast. Shelf Sci.* 81, 423–437. doi: 10.1016/j.ecss.2008.11.017

Myriax. (2015). *Echoview.* Hobart, TAS. Available online at: http://www.echoview.com/

Pebesma, E. J., and Bivand, R. S. (2005). *Classes and methods for spatial data in R. R News 5 (2),* Available online at: http://cran.r-project.org/doc/Rnews/

R Development Core Team. (2014). *R: A Language and Environment for Statistical Computing.* Vienna: R Foundation for Statistical Computing.

Rstudio. (2014). *RStudio: Integrated Development Environment for R (version 0.98.932).* Boston, MA. Available online at: http://www.rstudio.com/

Van Walree, P. A., Tęgowski, J., Laban, C., and Simons, D. G. (2005). Acoustic seafloor discrimination with echo shape parameters: a comparison with the ground truth. *Cont. Shelf Res.* 25, 2273–2293. doi: 10.1016/j.csr.2005.09.002

Watkins, J. L., and Brierley, A. S. (2002). Verification of the acoustic techniques used to identify Antarctic krill. *ICES. J. Mar. Sci.* 59, 1326–1336. doi: 10.1006/jmsc.2002.1309

Weber, T. C., Jerram, K., and Mayer, L. (2014). "Acoustic sensing of gas seeps in the deep ocean with split-beam echosounders," in *Proceedings of Meetings on Acoustics* (Edinburgh, EB), 17.

Conflict of Interest Statement: The authors declare that the research was conducted in the absence of any commercial or financial relationships that could be construed as a potential conflict of interest.

13

Deep-sea litter: a comparison of seamounts, banks and a ridge in the Atlantic and Indian Oceans reveals both environmental and anthropogenic factors impact accumulation and composition

Lucy C. Woodall[1]*, Laura F. Robinson[2], Alex D. Rogers[3], Bhavani E. Narayanaswamy[4] and Gordon L. J. Paterson[1]

[1] Department of Life Sciences, The Natural History Museum, London, UK
[2] School of Earth Sciences, University of Bristol, Bristol, UK
[3] Department of Zoology, University of Oxford, Oxford, UK
[4] The Scottish Association for Marine Science, Ecology Department, Scottish Marine Institute, Oban, UK

Edited by:
Alex Ford, University of Portsmouth, UK

Reviewed by:
Adolphe Debrot, Institute for Marine Research and Ecosystem Studies Wageningen University and Research Center, Netherlands
Francois Galgani, Institut Français de Recherche pour l'Exploitation de la Mer, France
Christopher Kim Pham, IMAR-Institute of Marine Research, Portugal

*Correspondence:
Lucy C. Woodall, Department of Life Sciences, The Natural History Museum, Cromwell Road, London SW7 5BD, UK
e-mail: l.woodall@nhm.ac.uk

Marine litter is a global challenge that has recently received policymakers' attention, with new environmental targets in addition to changes to old legislation. There are no global estimates of benthic litter because of the scarcity of data and only patchy survey coverage. However, estimates of baseline abundance and composition of litter are vital in order to implement litter reduction policies and adequate monitoring schemes. Two large-scale surveys of submarine geomorphological features in the Indian and Atlantic Oceans reveal that litter was found at all locations, despite their remoteness. Litter abundance was patchy, but both surveyed oceans had sites of high litter density. There was a significant difference in the type of litter found in the two oceans, with the Indian Ocean sites being dominated by fishing gear, whereas the Atlantic Ocean sites displayed a greater mix of general refuse. This study suggests that seabed litter is ubiquitous on raised benthic features, such as seamounts. It also concludes that the pattern of accumulation and composition of the litter is determined by a complex range of factors both environmental and anthropogenic. We suggest that the tracing of fishing effort and gear type would be an important step to elucidate hotspots of litter abundance on seamounts, ridges and banks.

Keywords: litter, debris, seamount, benthic, fishing gear

INTRODUCTION

Marine litter has been identified as a significant and growing global problem (UNEP, 2009; UNGA, 2012; GOC, 2014). Estimates suggest 6.4 million tons of litter enter the marine ecosystem annually (UNEP, 2009). Litter items, defined as; persistent, manufactured or processed solids that have been disposed of or abandoned, deliberately or unintentionally (UNEP, 2005), are present in all marine environments, including remote locations such as Antarctica (Barnes et al., 2009; Ivar Do Sol et al., 2011) and in the deep sea (Galgani et al., 2000; Ramirez-Llodra et al., 2011; Pham et al., 2014). However, the source and fate of marine litter is little understood (Derraik, 2002). Observations of litter have revealed direct impacts on megafauna through entanglement and ingestion, and on habitats through smothering, transporting alien species, and altering benthic community structure (Gregory, 2009). In addition, plastics can fragment to microplastics which also have potential impacts on the environment and biota, both physical and biochemical (Teuten et al., 2009; Andrady, 2011; Wright et al., 2013; Bakir et al., 2014).

A plethora of studies have reported on abundance and composition of debris in surface water and on beaches (e.g., Ryan et al., 2009). The deep sea, however, is logistically challenging and expensive to survey, therefore few studies have been conducted. Most of these have concentrated on small areas of the seabed, mostly on the continental shelf (Ramirez-Llodra et al., 2013), but there are a few studies that have reported deep-sea litter in more extreme locations e.g., the Ryukyu Trench; [7216 m depth (Miyake et al., 2011)], Molloy Hole; [up to 5500 m; (Galgani and Lecornu, 2004)], and Kuril-Kamchatka Trench (Fischer et al., 2015) and have assessed trends in litter composition and abundance (Bergmann and Klages, 2012; Schlining et al., 2013). To date, records have shown that deep-sea litter is not evenly distributed, with environmental and anthropogenic factors both influencing litter abundance (Schlining et al., 2013; Pham et al., 2014). More specifically, near-shore canyons may accumulate more litter than expected (Pham et al., 2014), and some regions of these canyons can have greater litter abundance than others e.g., more rugose parts of the Monterey Canyon had greater litter densities (Schlining et al., 2013), and thus marine litter assessments may have underestimated the true figure.

The main constituent of beach, seabed and surface water litter is plastic (Derraik, 2002). This is because it has a low degeneration and degradation rate, and production has increased annually since the 1950's (Thompson et al., 2009). Studies suggest litter items arrive in the deep sea from the shore, offshore installations, shipping and fisheries activities (Pham et al., 2014). The proportional contribution of different litter sources is likely to result from the complex interactions of oceanographic processes, geography and local anthropogenic activity (Ramirez-Llodra et al., 2013). Some submarine features are disproportionately affected by some types of litter. For example, in Europe, fishing gear contributes over 70% of the litter found on seamounts, banks and mounds and is also the greatest constituent of litter found in one ocean ridge study (Pham et al., 2014). This large contribution of fishing gear is not unexpected as, seamounts and other geomorphological features are often a focus for fisheries (Clark and Koslow, 2007).

To date, there have been no studies, using a consistent methodology, that survey multiple seamounts or other submarine features within the same ocean basin. The objective of this study was 3-fold; (1) to determine the amount and composition of litter on remote Atlantic Ocean and south-west Indian Ocean submarine features (seamounts, banks and a ridge), and to compare litter within and between the regions; (2) to infer the relative importance of geographical, geomorphological, biological and anthropogenic factors on the patterns of litter abundance and composition (e.g., distance from land, benthic rugosity, and shipping activity) and (3) to discuss results in the context of current legislation.

MATERIALS AND METHODS
STUDY AREAS
Data were collected during two research cruises aboard *R.R.S. James Cook* during 2011 and 2013. During each cruise, 16 remotely operated vehicle (ROV) dives were conducted. Submarine features (seamounts and banks) along the south-west Indian Ocean ridge were observed during JC66 using the ROV *Kiel 6000*. Later, an east-west transect of the equatorial Atlantic Ocean was conducted as part of JC94, observing seamounts and a fracture zone area using the ROV *ISIS*. Ten submarine features were surveyed in total, five on each cruise (**Table 1** and **Figure 1**).

All submarine topographic features had different morphologies (Rogers and Taylor, 2012; Hoy et al., 2014), with different slope pitch and summit area. The ROV dives surveyed

features of 200–3000 m and 100–1500 m depth for the Atlantic Ocean and Indian Ocean, respectively. The benthic profile was detected using the ship-mounted multibeam echo sounder (EM-120, Kongsberg-Simrad), was processed in CARIS, HIPS, and SIPS (http://www.caris.com/products/hips-sips), and 200 m depth contours added in ArcMap. All the features surveyed were over 600 km from land and most were more than 1000 km away (**Table 1**).

DATA RECORDING
The primary purposes of the ROV dive transects were to video benthic habitat and sample specific megafauna. Dives always started at depth and progressed to shallower water. Typically three dives were performed on each submarine feature, but this varied from five to one (**Table 1**). ROV geographic position and depth were recorded using Ocean Floor Observation Protocol (OFOP) software (Huetten and Grienert, 2008). Both ROVs were fitted with parallel scaling lasers to calibrate target size; and an average field of view was worked out for each dive (2 m).

VIDEO ANALYSIS
Video transects from the main color HD camera were played in real-time through Video Annotation and Reference System software (VARS; Schlining and Jacobsen Stout, 2006), which recorded time, geographic location and depth when the observer noted a litter item. All videos were watched twice by the same observer to ensure no objects were missed. The other HD video recordings (downward looking and pan and tilt) were used when possible to confirm the identity of litter items. In addition, the OFOP observer text, made onboard ship during the ROV dives, was queried for annotations that may relate to marine litter such as "trash," "anthropogenic," "fishing," and "plastic" and the video footage was checked. Litter items (**Figure 2**) were placed into five broad categories which included: "fishing gear," "plastic," "metal," "glass" or "other." Unidentified objects, as well as those made from material that were not plastic, metal or glass, were classified as "other" (Table S1). The fishing gear was mainly made from plastic, but the separate designation of "fishing gear" was retained to ensure that the source of these items was recorded in order to determine if there was any correlation between litter type

Table 1 | Sampling data including geomorphological feature surveyed, number of remotely operated vehicle (ROV) dives, area of ROV survey, depth range surveyed, distance from land, shipping activity proxy and litter abundance.

Region	Location	Feature type	ROV Dives	Area covered (ha)	Depth range surveyed (m)	Distance from land (km)	Shipping activity proxy	Items (ha^{-1})
Atlantic	Carter	Seamount	5	2.78	200–2800	630	0.12	12.23
Atlantic	Knipovich	Seamount	3	2.18	600–2800	1360	0.23	2.29
Atlantic	VEMA	Fracture zone	3	1.80	600–3000	1040	0.07	5.56
Atlantic	Vayda	Seamount	3	3.10	400–2300	1170	0.01	1.94
Atlantic	Gramberg	Seamount	2	1.70	900–2200	940	0.21	0.59
Indian	Coral	Seamount	5	2.04	500–1500	1610	0.01	1.47
Indian	Melville	Bank	4	1.35	100–1300	1440	0.00	13.33
Indian	Middle of What	Seamount	3	0.41	1000–1400	1460	0.00	2.44
Indian	Sapmer	Seamount	1	0.46	300–700	1390	0.00	17.39
Indian	Atlantis	Bank	3	1.34	700–1200	1320	0.10	0.75

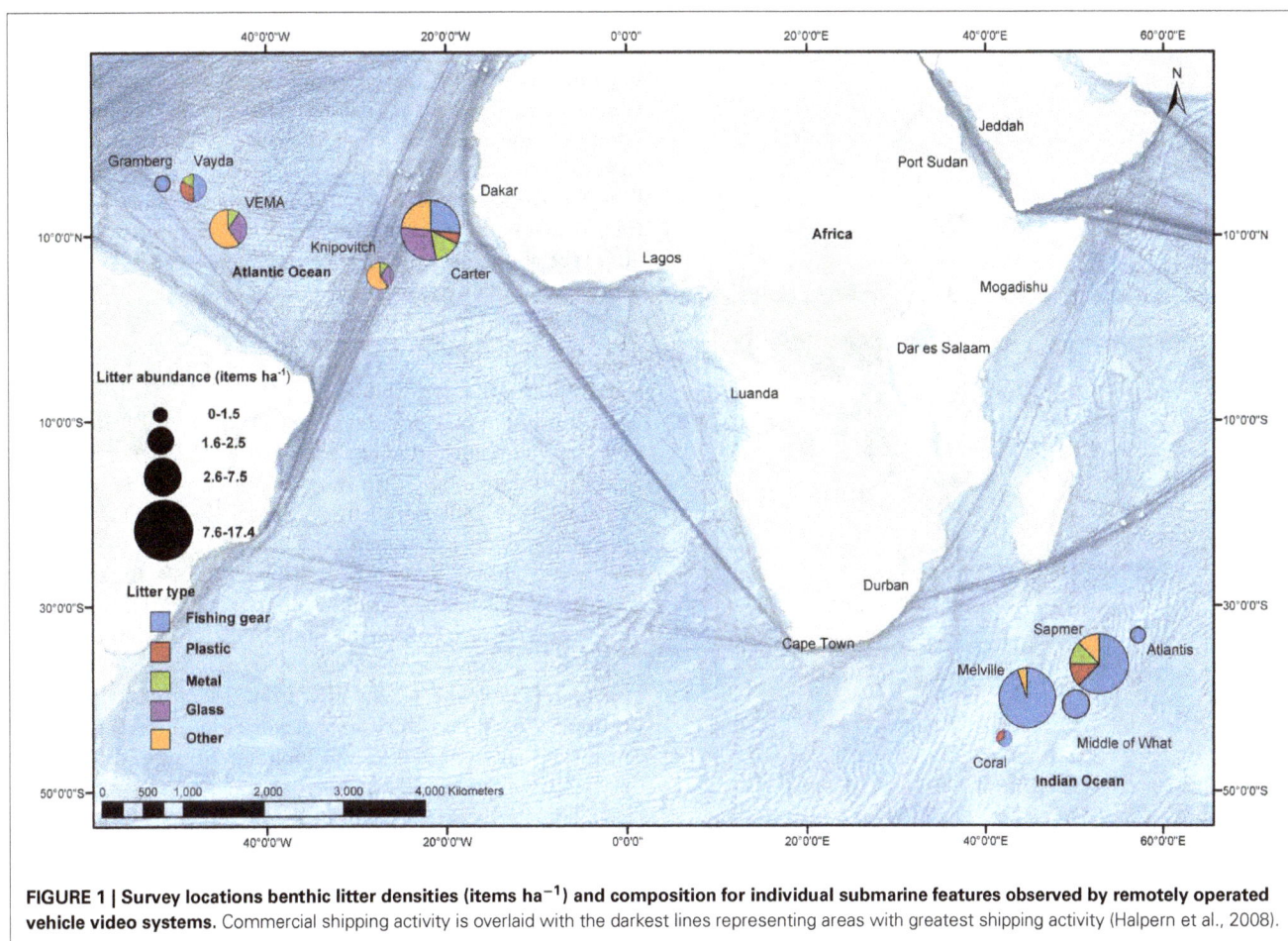

FIGURE 1 | Survey locations benthic litter densities (items ha⁻¹) and composition for individual submarine features observed by remotely operated vehicle video systems. Commercial shipping activity is overlaid with the darkest lines representing areas with greatest shipping activity (Halpern et al., 2008).

and habitat, this also follows other studies (Miyake et al., 2011; Schlining et al., 2013; Pham et al., 2014). Any debris items that were attached to each other were classed as one item for analysis, as they would have resulted from one littering episode.

In the absence of a standardized reporting system, the coverage extent of the litter item on the benthos was recorded using two parameters, shape and size. For the first of these, shape, items were classified as either elongate or oblate; for example rope was classified as elongate and bottles as oblate. Objects were grouped by size into four categories that increase by an order of magnitude <20 cm, 20 cm–2 m, 2 m–20 m, and >20 m for the elongate objects, and <10 cm², 10 cm²–1 m², 1 m²–10 m² and >10 m² for oblate objects. These size categories were chosen to make the measurement simple, using the width between the two lasers (10 cm) and average field of view during the dives (2 m). The impact of the item was then represented by size category with litter type subdivided into either elongate or oblate. When both elongate and oblate items were present e.g., fishing nets (net and the top and bottom ropes), both types of impact were reported separately. Finally any interactions of litter with benthic fauna were recorded, be they through entanglement or use of the litter as substratum.

SPATIAL ANALYSIS

Litter location was plotted using ArcGIS v10.2 (ERSI, 2014) together with bathymetry and ROV bottom tracks. ROV bottom

tracks were smoothed in OFOP and obvious erroneous points corrected by eye. The area surveyed was calculated using ROV bottom track and the average field of view. The depth at which the litter items were found was converted to either "summit" or "flank." Summit was used when the depth was within the shallowest 15% of the submarine feature; otherwise it was categorized as flank. This broad categorization was chosen over absolute depth as the summits of the submarine features varied, with those in the Atlantic Ocean being deeper than those in the Indian Ocean. The areas surveyed were then recalculated to include percentage of the survey areas in summit and flank regions, respectively.

To calculate rugosity, bathymetry raster files for the 10 surveyed features were combined in ArcGIS, using the "mosaic to raster" function. Rugosity was then determined using the terrain ruggedness feature in the Benthic Terrain Modeler (BTM) extension set at five (Wright et al., 2012). This measure of rugosity relies on the resolution of the bathymetry survey and thus the rugosity had 50 and 100 m resolution of both the Indian and Atlantic Oceans, respectively. This rugosity was then assigned to six categories (0–0.9, 1–1.9, 2–2.9, 3–3.9, 4–4.9, 5+). The habitat where the litter was found was used as a proxy for rugosity at finer resolution. The habitat was assigned to one of five categories that increased in rugosity from sand/silt flat areas, to deep rock crags (**Figure S1**).

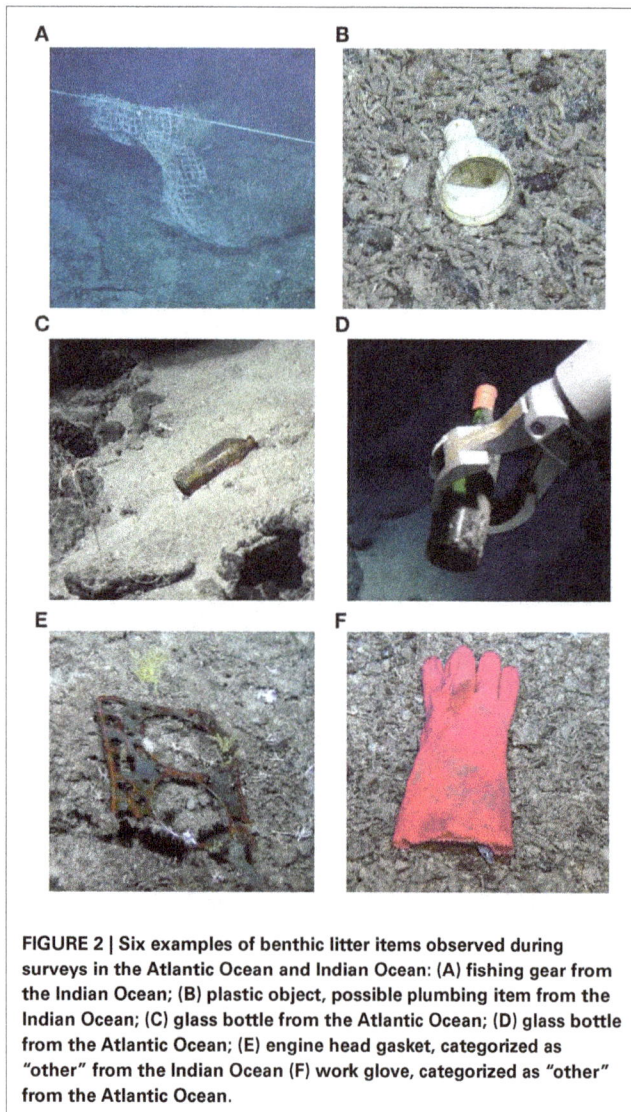

FIGURE 2 | Six examples of benthic litter items observed during surveys in the Atlantic Ocean and Indian Ocean: (A) fishing gear from the Indian Ocean; (B) plastic object, possible plumbing item from the Indian Ocean; (C) glass bottle from the Atlantic Ocean; (D) glass bottle from the Atlantic Ocean; (E) engine head gasket, categorized as "other" from the Indian Ocean (F) work glove, categorized as "other" from the Atlantic Ocean.

Commercial shipping activity was calculated using a proxy derived from the World Meteorological Organization Voluntary Observing Ships Scheme data Oct 2004–Oct 2005 with 1 km resolution (Halpern et al., 2008). At each submarine feature, the mean shipping activity proxy was calculated from the values of the closest $4 \, km^2$. The index runs from 0 to 0.68, zero was recorded when no ships passed through the selected $4 \, km^2$ region during the reporting year, and 0.68 indicates a busy shipping lane such as the Gibraltar or Dover straits.

DATA ANALYSIS
Abundance of litter was calculated for each seamount per hectare surveyed (items ha^{-1}). Non-parametric tests were applied as data were not normally distributed (Ryan-Joiner, $p < 0.05$), but variances were not significantly different (Levene's test $p > 0.05$). Litter patchiness was computed using Lloyd's index implemented in Passage v2, where $n > 1$ means aggregation (Lloyd, 1967). Data were analyzed in Minitab v17 using the Mann-Whitney

tests when comparing differences in litter abundance between two categories (i.e., oceans), Kruskal-Wallis tests when comparing litter abundance between more than two categories (i.e., seamounts) and Spearman's Rank order was used to assess correlation. Multivariate analysis (Analysis of similarity, ANOSIM) was conducted in PRIMER v6.0 (Clark and Gorley, 2006), using Bray Curtis similarity of litter composition following a $log(x − 1)$ transformation of the data to elucidate the relationships between litter type abundance and the other parameters such as shipping activity and distance from land.

RESULTS
LITTER ABUNDANCE AND UBIQUITY
Litter was found on every one of the 10 submarine features surveyed in the Indian and Atlantic Oceans. A total of 56 items were found in the Atlantic Ocean over a survey area of 11.6 ha, and 31 items in the Indian Ocean over 5.6 ha (**Figure 1**). Litter was present in depths ranging from 209 to 2318 m in the Atlantic Ocean, and 112–1278 m in the Indian Ocean. The differences in the depth at which litter was observed reflected the bathymetry of the features surveyed (**Table 1**). In the Indian Ocean litter items, of all size categories, were seen for both shape categories (elongate and oblate), whereas the litter from the Atlantic Ocean was just from the smallest three size categories for both shape types. The greatest amount of litter was found on Sapmer Seamount in the Indian Ocean (17.39 items ha^{-1}), and the least (0.59 items ha^{-1}) on Gramberg Seamount in the Atlantic Ocean. It was not possible to compare litter between different types of topographic features, i.e., bank, ridge, seamount, as the sample size was too small.

There was great variation in the abundance of items between submarine features within the same ocean, with one seamount, Carter, in the Atlantic Ocean and two features (Melville and Sapmer) in the Indian Ocean having a high density of litter (12.23–17.39 items ha^{-1}), while the rest of the sites had much lower densities (0.59–5.56 items ha^{-1}). The mean litter abundance was greatest in the Indian Ocean, but the oceans had similar variance (Atlantic: 4.52 items ha^{-1} SE ± 2.09; Indian: 7.07 items ha^{-1} SE ± 3.45) (**Table 1**). However, the litter had a patchy distribution across all locations ($P = 3.4$) according to Lloyd's index.

There was no significant difference between the number of items ha^{-1} found in each ocean (Mann-Whitney: $W = 25.0$, $\eta_1 = −15.1 \, \eta_2 = 9.8, p = 0.8$). There was also no significant difference between the amount of litter observed on flanks compared with summits of the topographic features ($W = 6$, $\eta_1 = −8.0$ $\eta_2 = −29.0 \, p = 0.7$). But because of sample size it was not possible to compare litter abundance between types of submarine features.

Rugosity, as calculated by BTM and habitat both negatively correlate with litter abundance [BTM: $\rho = −0.9, p < 0.01$; habitat: $\rho = −1.0, p < 0.01$]. Most litter items were found in areas with the flattest rugosity ratings (77% Atlantic Ocean, 69% Indian Ocean). In addition we explored the data to determine if there was correlation of litter abundance with distance from land and shipping activity, however, probably because of the low level of replication these test results were not significant and are not presented.

LITTER TYPE

The relative litter composition of the two oceans was significantly different [1-way ANOSIM $R = 0.242$, $p < 0.05$]; litter in the Indian Ocean was dominated by fishing gear (84%) whereas in the Atlantic Ocean, the litter was a mix of fishing gear, glass and other debris, each comprising about 25% of the total litter, with metal and plastic objects making up the final quarter (**Figure 3**). Objects classed as "other items" in the Atlantic Ocean were as diverse as a pottery urn and machinery gaskets. In the Indian Ocean, this category also included gaskets as well as a work glove. Fishing gear was seen at the most number of different sites, with all locations in the Indian Ocean and three of the five sites in the Atlantic Ocean having at least one occurrence. Plastic items were only seen in two sites in each ocean (Carter and Vayda - Atlantic Ocean; Coral and Sapmer - Indian Ocean) and were of low abundance. The pattern of plastic litter distribution differed from that of glass debris which was seen at three Atlantic sites and was highly abundant at one site (Carter), but less so at the other two (Knipovitch, VEMA) (**Figure 1**).

Multivariate analyses were not successful in elucidating patterns of correlation as a result of the small sample sizes at some sites. The litter was reanalyzed by re-categorizing it as fishing gear or non-fishing litter. Subsequent analyses were conducted to determine correlation between distance from shore, shipping activity, depth category and rugosity; the only significant relationship was a negative one between fishing gear items and shipping activity [Spearman's rank order $\rho - 0.72$, $p < 0.05$].

LITTER COVERAGE EXTENT

There was a significant difference in the coverage extent of the litter items between oceans and between areas according to their shipping activity [1-way ANOSIM Ocean $R = 0.27$, $P < 0.01$; Shipping activity $R = 0.95$ $p < 0.01$], but not between seamounts or distance from shore. SIMPER analysis showed that the difference between the oceans was mainly driven by the dissimilarity between the coverage extent of fishing gear (elongate 33.7% and oblate 20.2%).

When this litter coverage extent was re-categorized as either fishing gear or non-fishing litter, there was no significant correlation between litter coverage extent and rugosity calculated by habitat, however, there is a significant negative correlation between coverage extent of litter and habitat calculated by BTM [Spearman's rank-order $\rho = -0.94$, $p < 0.01$].

ASSOCIATED ORGANISMS

Very few faunal associations were seen, with most being associated with items from the Indian Ocean. Encrusting organisms were observed on 18% of litter items, all of which were fishing gear, these items were often entirely covered and heavily encrusted (**Figure 2A**). The identification of associated taxa was difficult as most litter items were not brought up to the surface, and often

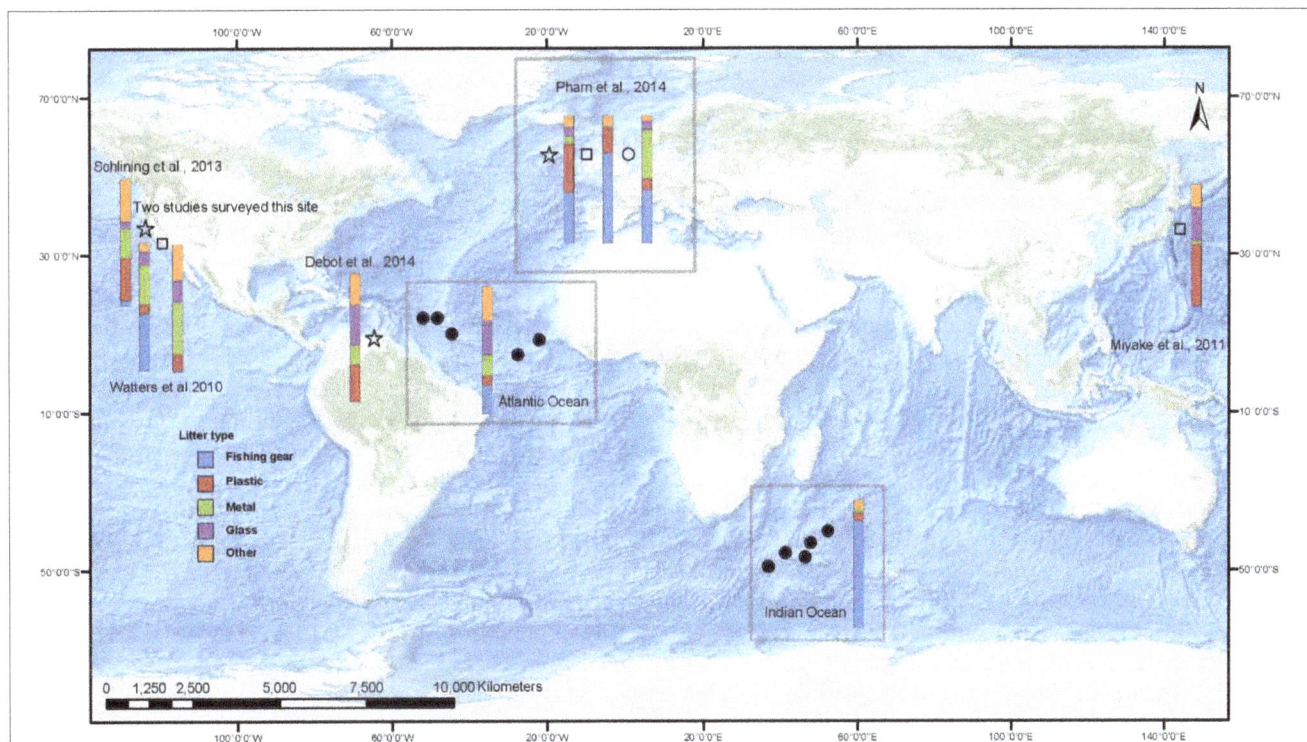

FIGURE 3 | Composition of benthic litter estimated from seabed observations. Circles represent studies on raised features such as seamounts, stars represent canyon studies and other topography is represented by a square. Filled symbols are the summary data from the current study and labeled Indian and Atlantic Ocean. Open symbols are data from previous studies and references are given. Watters et al. (2010) data is given for two areas along the California coast. European data, reviewed by Pham et al. (2014) is summarized for the three physiographic types.

the ROV was unable to get close to fishing gear for operational and safety reasons. However, where identification was possible, coral and hydroids were seen encrusting the gear, whilst fish, crinoids, anemones, sea urchins, and brittle stars were seen using the items as habitat. Entanglement was obvious in four fishing gear items; two of these were also encrusted. These entangled organisms comprised the broadest range of taxa, including coral, sponge, fish and crustacea. Finally, a further three items were used as substrata for organisms to hide under (**Figure 2F**), to lay eggs on, or to use as a holdfast.

DISCUSSION
DISTRIBUTION AND ABUNDANCE

The most notable finding of this study is that litter was found at all deep-sea sites surveyed. The ubiquity of the litter on seamounts, banks and ridges has previously not been as explicit because the features were focused on individually. Litter abundance on these submarine features was patchy, with high and low densities of litter reported within each ocean. However, all litter abundances were within those previously reported in the North Atlantic Ocean for seamounts, banks, mounds and ridges (Pham et al., 2014) and for coastal waters (Galgani et al., 2000; Mifsud et al., 2013), but greater than that previously observed in more remote locations (Antarctic Peninsula and Scotia Arc; Barnes et al., 2009). Different methodologies, including sampling techniques, may account for some of the differences observed, as Barnes et al. (2009) collected data by trawl rather than video systems such as employed in this current study, but could also reflect the more remote areas surveyed by Barnes et al. (2009). This current study highlights the extent of the unseen and thus unreported litter in the oceans.

The surveyed area on each geomorphic feature was limited. This is common in the deep-sea, even when a region is well-studied e.g., Monterey Bay (Schlining et al., 2013). Even though this current study sampled multiple sites, the survey area for each separate feature was within the range that others had reported (0.9–5.6 ha) (Pham et al., 2014); with the exception of two seamounts in the Indian Ocean (Middle of What; Sapmer) which were slightly less, as a result of operational limitations. Deep-sea benthic studies are more time consuming, logistically challenging and expensive than surface water or beach surveys and therefore suffer with a paucity of data (Barnes et al., 2009; Ramirez-Llodra et al., 2011). This current study will have under-reported the litter density, as very small items were not visible with our video system and we chose to class items attached to each other as one item. Sampling for smaller items would have required additional sampling methodologies, e.g., coring or trawling. Indeed much smaller plastic "microplastics (<0.5 cm)" are known to exist in deep-sea sediments, including from some of the same sites as the present study, in greater densities to those found in beach and coastal sediments (Woodall et al., 2014) and therefore this small-size fraction may be an important constituent of marine litter.

To extrapolate our data to provide global estimates of litter on seamounts is challenging, as we only surveyed <1% of each submarine feature, the depth ranges we surveyed are not typical of seamounts generally, and we focused on remote locations.

However, we have extrapolated from existing data (Yesson et al., 2011) on non-overlapping seamounts that are within 500 km of our study locations and that have similar depths (summits > 3000 m). We estimate that over 32 million and 38 million litter items are present on the seamounts of the Atlantic and south-west Indian Oceans, respectively. These figures are derived from the mean abundance (2.15 items ha^{-1}) when the three outlier sites with greatest abundance are removed, and using the seamount data of Yesson et al. (2011). The areas on which this extrapolation is based are about 10% of the seamount area calculated for the FAO regions that contain the areas of interest in our study (Yesson et al., 2011). These extrapolations take account of all data currently available, however, this dataset comes from a very small area compared with the size of the ocean and so our estimates should be used as a guide. Specifically we show that the accumulation of litter on the seabed is patchy, therefore further surveys are required to confirm the average litter abundance on the ocean floor. In addition we have not taken into account the impact of tides, currents and the morphology of the topographic features. Future studies to model the desposition and accumulation of litter would be an important contribution to this field of study, but paucity of data currently prevents this analysis.

Previous studies reveal that certain geomorphological features accumulate more litter than others, with canyons having the highest litter loads (Galgani et al., 2000; Wei et al., 2012; Ramirez-Llodra et al., 2013; Pham et al., 2014, **Figure 3**). These studies suggest hydrodynamic effects act to cause the canyons to function as conduits for the litter. No data on hydrodynamics at depth is available for the areas surveyed in this latest study, and neither the environmental (distance from shore) nor the anthropogenic (shipping index) parameters were good predictors of the density of litter observed or the depth region where litter occurred. However, a decreased rugosity did result in greater litter abundance. This was in contrast to other studies which suggested that distance from land (Barnes et al., 2009; Pham et al., 2014) correlated with a decreasing abundance of litter, and increased rugosity correlated with more litter (Schlining et al., 2013). The correlation with distance seen in other studies could result from the fact that the majority of these studies were carried out close to land, so the decrease in debris could result from the decreasing component of the land as a litter source. In contrast, there was no evidence that land-based litter contributed to the abundances seen in the current study. The difference in the apparent influence of rugosity could be an artifact of low sample size or indeed sampling, because over flatter terrain it is easier to pilot the ROV, giving better quality video footage and so making any litter easier to see. Also in rugged terrain, turbulence may increase and change local hydrodynamics which could, in turn, lead to the patchy accumulation of litter not seen so much in flatter areas.

The survey sites in this study were outside the subtropical gyres which accumulate surface litter (Moore et al., 2001). A previous study reported floating small sized plastic concentrations were low close to some of our survey sites, compared with other ocean locations (Cozar et al., 2014). However, it is not possible to directly compare the abundance as there were very few sightings (6/442) within 500 km of our survey sites and they were reported by weight rather than by number which is our chosen method of reporting. In addition, models predict little floating litter at all sites (Lebreton et al., 2012; Fischer et al., 2015), although the patchiness of surface plastics has been shown (Goldstein et al., 2013). Current evidence suggests that the sites surveyed in this study were not in areas that have high litter accumulation, and thus we could conclude that the abundance seen here are unlikely to be the greatest present on deep-sea seamounts, especially considering the distance from shore. However, without further observations of the seabed, it is hard to predict if benthic litter reflects the patterns of abundance of surface litter. The processes involved in the transportation of litter items from the source to the seabed are poorly known, but must rely on factors including sinking velocity and litter degradation rate. Litter deposition is also likely to be affected by the direction and speed of the circulation of water masses that differ depending on depth and geographic location (Emery and Meincke, 1986). Surface litter accumulations are driven by oceanographic processes and winds (Moore, 2008; Eriksen et al., 2014) and at a regional level highest abundances correspond with high population density (Cozar et al., 2014). Once at depth, benthic litter density is affected by geomorphology (Pham et al., 2014), distance to the coast (Mordecai et al., 2011), hydrology (Galgani et al., 1996), and anthropogenic activities (Bergmann and Klages, 2012; Ramirez-Llodra et al., 2013). It is therefore unlikely that surface litter accumulation is a good predictor of benthic litter deposition. It would be valuable to constrain relationships between the two as benthic litter is more logistically challenging to record than surface litter. Outside the NE Atlantic very few litter estimates from video capture are available (**Figure 3**) and remote sensing technologies that can be used for surface litter (Mace, 2012) are not applicable to sea bed studies. Observations of benthic litter therefore remain an expensive and time-consuming proposition for just one research proposal. However, a coordinated international program that could utilize all global video footage, and analysis by citizen scientists through crowdsourcing, could be ideal mechanisms to address this important question.

TYPE AND IMPACT OF LITTER

A clear difference in the composition of the litter items was found between the two oceans surveyed (**Figure 3**). Fishing gear, the predominant litter type on Indian Ocean submarine features, is also the most abundant litter type on seamounts and other deep-sea features in European waters (Pham et al., 2014), in canyon shelf locations in central Califonia (Watters et al., 2010) and similarly from seabed trawl data in the East China Sea and southern Yellow Sea (Lee et al., 2006). The Atlantic Ocean, on the other hand, displayed a predominance of general litter items specifically associated with food packaging. General litter (i.e., plastic and glass) was also the most common in other deep-sea litter studies focusing on the abyssal plain and canyons, although many of these sites were only surveyed with trawls (Galgani et al., 1996; Koutsodendris et al., 2008; Barnes et al., 2009; Keller et al., 2010; Miyake et al., 2011; Mordecai et al., 2011; Ramirez-Llodra et al., 2011; Bergmann and Klages, 2012; Schlining et al., 2013; Debrot et al., 2014). The difference in litter composition between the two oceans reflects local anthropogenic activity. Seamounts are

targeted by fisheries as they are highly productive areas (Clark and Koslow, 2007). The Indian Ocean features were in waters rarely used by shipping, but have been exploited by fisheries since the 1970's (Clark, 2009), however, the Atlantic Ocean survey sites were mainly situated in waters that experience heavier shipping traffic (Halpern et al., 2008). Indeed the difference in litter impact between the two oceans is driven by the large areas covered by fishing gear in the Indian Ocean as opposed to the elongate aspect of the fishing gear seen in the Atlantic Ocean. This highlights the importance of recording not just presence of litter items, but size and shape as well. Just shipping activity had a significant relationship to litter type and litter coverage extent: fishing gear showed negative correlation with the shipping activity index. It is not currently possible to determine fishing effort for the survey areas, but in the future, data may become available through the use of satellite tracking of vessels through Vessel Monitoring Systems (VMS), Automated Identification Systems (AIS) combined with data from other sensors such as Synthetic Aperture Radar (SAR). The record of fishing effort, including the different gear and technology used, is important to fully understand the patterns of litter composition driven by these industries. The difference in litter abundance within the Indian Ocean may be explained by the Indian Ocean Voluntary Benthic Protected Areas (VBPA) which were established in 2006 (Shotton, 2006) and include the survey sites on Coral Seamount and Atlantis Bank. These were the two sites with the lowest litter abundance in that ocean but this may be coincidental. This illustrates that litter monitoring may provide an opportunity to identify historical fishing activities. The negative correlation of broad habitat rugosity and greatest fishing gear impact found in this study may indicate that vessels are actively avoiding the more rugged areas that are most likely to catch gear. If active avoidance of these areas is practiced, then the technology and gear available on vessels is important to estimate hotspots of litter impact.

In this study few associations between litter and organisms were recorded as (1) most litter was not sampled by the ROV and therefore small encrusting organisms were not seen, and (2) because of the challenges of navigating the ROV close to litter in the most rugged terrain, especially when loose fishing gear was present. However, the variety of taxa, either encrusting or entangled in fishing gear, was diverse and congruent with previous studies (Laist, 1997). Entanglement was present in three forms, simple, benthic scraping and ghost fishing. The simple entanglement of linear fishing gear around coral colonies was evident in both this study and in Pham et al. (2013). Benthic scraping is the abrasion of the seabed by fishing gear, resulting in the accumulation of reef-building coral and other taxa in the gear (Anderson and Clark, 2003; Chiappone et al., 2005). Ghost fishing, is when gear continues to catch organisms despite it no longer being monitored by fishers. This includes self-baiting, which is where caught organisms then become bait for subsequent scavengers which then also become entangled. The self-baiting phenomenon was evident and was at least partly responsible for the change in composition over time of organisms entrapped in ghost fishing gear in previous temporal studies (Kaiser et al., 1996; Arthur et al., 2014). The ultimate impact of such gear is hard to determine, but is fast becoming a research priority (Gilman et al., 2013).

CONCLUSION

There are few legal instruments that regulate waste being dumped at sea. Commercial shipping dumping is regulated by the London Protocol and the recently revised International Convention for the Prevention of Pollution from Ships (MARPOL 73/78), Annex V, and particular regions are covered by specific agreements (NE Atlantic; OSPAR Convention) and reviewed by UNEP (2005). However, legislation is not a panacea for positive change as fishing gear debris rates did not fall when MARPOL was initially implemented in 1989 (Henderson, 2001). This is most likely since some fishing vessels, because of their relatively small size (<400 gross tons), are not covered by MARPOL regulations and policing legislation in the high seas is extremely challenging. The accumulation of fishing gear debris can be attributed to accidental loss as well as deliberate dumping of old and damaged nets. Accidental loss is likely to depend on fishing effort and rugosity of the region, therefore VPBAs, such as those in the southern Indian Ocean, may help. In addition, local schemes such as the "Fishing for litter" initiative (OSPAR, 2007), and port reception facilities for spent fishing gear, may help to reduce gear on the seabed and provide data on location and type. Global and regional programs of marine debris monitoring and litter reduction policies are being implemented (UNEP global initiative on marine litter; UNEP, 2009; Marine Strategy Framework Directive; Galgani et al., 2013). However, with most seamounts, mounds, banks and ridges being by nature out of these jurisdictions, and with litter on these features coming from marine sources, such programs and schemes are unlikely to provide a comprehensive solution to the deep-sea litter challenge. The International Seabed Authority set up to implement UNCLOS beyond the EEZ is focused on regulation of natural resources and currently does not have a policy on marine litter. Thus, novel initiatives will need to be implemented to cut further deposition of debris.

Strandline litter and accumulation of debris in surface water are well-reported in the media, and citizens actively monitor and lobby about these issues. The issue of the un-seen benthic litter has recently attracted more attention (Galgani et al., 2000; Barnes et al., 2009; Mifsud et al., 2013; Schlining et al., 2013; Pham et al., 2014; Woodall et al., 2014). This current study is the first of its kind to assess litter on numerous deep-sea raised topological features, using the same techniques including the same observer, in two oceans with different local anthropogenic activities. The results from this study agree with previous reports that show benthic litter is ubiquitous in the ocean. The abundance and composition of the litter seen in this study appears to be strongly influenced by local marine-based anthropogenic activity. The intensity of these activities may be useful as predictors of benthic litter density and composition in high-seas regions where ground-truthed data is sparse (or vice versa), but this should be tested when such data from the shipping and fishing industries becomes available.

ACKNOWLEDGMENTS

This work was funded by NERC grant NE/F005504/1, ERC grant 278705 and the Philip Leverhulme Trust. We would like to thank participants of JC66 and JC94, especially E. Muller, S. Hoy, and V. Huvenne for bathymetry data, L. Marsh, M. Taylor, N. Serpetti, and M. Packer for their support at sea and the ROV teams and captain and crew of the R.R.S. James Cook for their assistance.

REFERENCES

Anderson, O. F., and Clark, M. R. (2003). Analysis of bycatch in the fishery for orange roughy, *Hoplostethus atlanticus*, on the South Tasman Rise. *Mar. Freshwater Res.* 54, 643–652. doi: 10.1071/MF02163

Andrady, A. L. (2011). Microplastics in the marine environment. *Mar. Pollut. Bull.* 62, 1596–1605. doi: 10.1016/j.marpolbul.2011.05.030

Arthur, C., Sutton-Grier, A. E., Murphy, P., and Bamford, H. (2014). Out of sight but not out of mind: harmful effects of derelict traps in selected U.S. coastal waters. *Mar. Pollut. Bull.* 86, 19–28. doi: 10.1016/j.marpolbul.2014.06.050

Bakir, A., Rowland, S. J., and Thompson, R. C. (2014). Enhanced desorption of persistent organic pollutants from microplastics under simulated physiological conditions. *Environ. Pollut.* 185, 16–23. doi: 10.1016/j.envpol.2013.10.007

Barnes, D. K., Galgani, F., Thompson, R. C., and Barlaz, M. (2009). Accumulation and fragmentation of plastic debris in global environments. *Philos. Trans. R. Soc. Lond. B. Biol. Sci.* 364, 1985–1998. doi: 10.1098/rstb.2008.0205

Bergmann, M., and Klages, M. (2012). Increase of litter at the Arctic deep-sea observatory HAUSGARTEN. *Mar. Pollut. Bull.* 64, 2734–2741. doi: 10.1016/j.marpolbul.2012.09.018

Chiappone, M., Dienes, H., Swanson, D. W., and Miller, S. L. (2005). Impacts of lost fishing gear on coral reef sessile invertebrates in the Florida Keys National Marine Sanctuary. *Biol. Conserv.* 121, 221–230. doi: 10.1016/j.biocon.2004.04.023

Clark, K. R., and Gorley, R. N. (2006). *Plymouth Routines in Mulitvariate Ecological Research – PRIMER v6.* (Plymouth: PRIMER-E).

Clark, M. R. (2009). Deep-sea seamount fisheries: a review of global status and future prospects. *Lat. Am. J. Aquat. Res.* 37, 501–512. doi: 10.3856/vol37-issue3-fulltext-x

Clark, M. R., and Koslow, J. A. (2007). "Impacts of fisheries on seamounts," in *Seamounts: Ecology, Fisheries & Conservation*, eds T. J. Pitcher, T. Morato, P. J. B. Hart, M. R. Clark, N. Haggan, and R. S. Santos (Oxford: Blackwell Publising), 413–441. doi: 10.1002/9780470691953.ch19

Cozar, A., Echevarria, F., Gonzalez-Gordillo, J. I., Irigoien, X., Ubeda, B., Hernandez-Leon, S., et al. (2014). Plastic debris in the open ocean. *Proc. Natl. Acad. Sci. U.S.A.* 111, 10239–10244. doi: 10.1073/pnas.1314705111

Debrot, A. L., Vinke, E., van der Wende, G., Hylkema, A., and Reed, J. K. (2014). Deepwater marine litter densities and composition from submersible video-transects around ABC-islands, Dutch Caribbean. *Mar. Pollut. Bull.* 88, 361–365. doi: 10.1016/j.marpolbul.2014.08.016

Derraik, J. G. B. (2002). The pollution of the marine environment by plastic debris: a review. *Mar. Pollut. Bull.* 44, 842–852. doi: 10.1016/S0025-326X(02)00220-5

Emery, W. J., and Meincke, J. (1986). Global water masses: summary and review. *Oceanol. Acta* 9, 383–391.

Eriksen, M., Lebreton, L. C. M., Carson, H. S., Thiel, M., Moore, C. J., Borerro, J. C., et al. (2014). Plastic pollution in the world's oceans: more than 5 trillion plastic pieces weighing over 250,000 tons afloat in at sea. *PLoS ONE* 9:e111913. doi: 10.1371/journal.pone.0111913

ERSI (2014). *ArcGIS Desktop. V.10.* Redland, CA: Environmental Systems Research Institute.

Fischer, V., Elsner, N. O., Brencke, N., Schwabe, E., and Brandt, A. (2015). Plastic pollution of the Kuril–Kamchatka Trench area (NW Pacific). *Deep-sea Res. II Top. Stud. Oceanogr.* 111, 399–405. doi: 10.1016/j.dsr2.2014.08.012

Galgani, F., and Lecornu, F. (2004). Debris on the seafloor at "Hausgarten": in the expedition ARKTIS XIX/3 of the research vessel POLARSTERN in 2003. *Rep. Polar Mar. Res.* 488, 260–262.

Galgani, F., Hanke, G., Werner, S., and De Vrees, L. (2013). Marine litter within the European Marine strategy framework directive. *ICES J. Mar. Sci.* 70, 1055–1064. doi: 10.1093/icesjms/fst122

Galgani, F., Leaute, J. P., Moguedet, P., Souplet, A., Verin, Y., Carpentier, A., et al. (2000). Litter on the sea floor along European coasts. *Mar. Pollut. Bull.* 40, 516–527. doi: 10.1016/S0025-326X(99)00234-9

Galgani, F., Souplet, A., and Cadiou, Y. (1996). Accumulation of debris on the deep seafloor off the French Mediterranean coast. *Mar. Ecol. Prog. Ser.* 142, 225–234. doi: 10.3354/meps142225

Gilman, E., Suuronen, P., Hall, M., and Kennelly, S. (2013). Causes and methods to estimate cryptic sources of fishing mortality. *J. Fish Biol.* 83, 766–803. doi: 10.1111/jfb.12148

GOC (2014). *From Decline to Recovery: A Rescue Package for the Global Ocean.* Oxford: Global Ocean Commission.

Goldstein, M. C., Titmus, A. J., and Ford, M. (2013). Scales of spatial heterogeneity of plastic marine debris in the northeast pacific ocean. *PLoS ONE* 8:e80020. doi: 10.1371/journal.pone.0080020

Gregory, M. R. (2009). Environmental implications of plastic debris in marine settings: entanglement, ingestion, smothering, hangers-on, hitch-hiking and alien invasions. *Philos. Trans. R. Soc. Lond. B. Biol. Sci.* 364, 2013–2025. doi: 10.1098/rstb.2008.0265

Halpern, B. S., Walbridge, S., Selkoe, K. A., Kappel, C. V., Micheli, F., D'agrosa, C., et al. (2008). A global map of human impact on marine ecosystems. *Science* 319, 948–952. doi: 10.1126/science.1149345

Henderson, J. R. (2001). A pre- and post-MARPOL Annex V summary of Hawaiian Monk Seal entanglements and marine debris accumulation in the north-western Hawaiian Islands, 1982–1998. *Mar. Pollut. Bull.* 42, 584–589. doi: 10.1016/S0025-326X(00)00204-6

Hoy, S. K., Huvenne, V. A., and Robinson, L. F. (2014). *EM-120 Multibeam Swath Bathymetry Collected During James Cook Cruise JC094.* University of Brisiol, School of Earth Sciences. doi: 10.1594/PANGAEA.832836

Huetten, E., and Grienert, J. (2008). Software controlled guidance, recording and post-processing of seafloor observations by ROV and other towed devices: the software package OFOP. *Geophys. Res. Abstr.* 10.

Ivar Do Sol, J. A., Barnes, D. K. A., Costa, M. F., Convey, P., Costa, E. S., and Campos, L. (2011). Plastics in the Antarctic environment: are we looking only at the tip of the iceberg? *Oecologia Austalis* 15, 150–170. doi: 10.4257/oeco.2011.1501.11

Kaiser, M. J., Bullimore, B., Newman, P., Lock, K., and Gilbert, S. (1996). Catches in 'ghost fishing' set nets. *Mar. Ecol. Prog. Ser.* 145, 11–16. doi: 10.3354/meps145011

Keller, A. A., Fruh, E. L., Johnson, M. M., Simon, V., and McGourty, C. (2010). Distribution and abundance of anthropogenic marine debris along the shelf and slope of the US West Coast. *Mar. Pollut. Bull.* 60, 692–700. doi: 10.1016/j.marpolbul.2009.12.006

Koutsodendris, A., Papatheodorou, G., Kougiourouki, O., and Georgiadis, M. (2008). Benthic marine litter in four Gulfs in Greece, Eastern Mediterranean; abundance, composition and source identification. *Estuar. Coast. Shelf Sci.* 77, 501–512. doi: 10.1016/j.ecss.2007.10.011

Laist, D. W. (1997). "Impacts of marine debris: entanglement of marine life in marine debris including a comprehensive list of species with entanglement and ingestion records," in *Marine Debris- Sources, Impacts and Solutions,* eds J. M. Coe and D. B. Rogers (New York, NY: Springer), 99–139. doi: 10.1007/978-1-4613-8486-1_10

Lebreton, L. C., Greer, S. D., and Borrero, J. C. (2012). Numerical modelling of floating debris in the world's oceans. *Mar. Pollut. Bull.* 64, 653–661. doi: 10.1016/j.marpolbul.2011.10.027

Lee, D.-I., Cho, H.-S., and Jeong, S.-B. (2006). Distribution characteristics of marine litter on the sea bed of the East China Sea and the South Sea of Korea. *Estuar. Coast. Shelf Sci.* 70, 187–194. doi: 10.1016/j.ecss.2006.06.003

Lloyd, M. (1967). Mean crowding. *J. Anim. Ecol.* 36, 1–30.

Mace, T. H. (2012). At-sea detection of marine debris: overview of technologies, processes, issues, and options. *Mar. Pollut. Bull.* 65, 23–27. doi: 10.1016/j.marpolbul.2011.08.042

Mifsud, R., Dimech, M., and Schembri, P. J. (2013). Marine litter from circalittoral and deeper bottoms off the Maltese islands (Central Mediterranean). *Mediterr. Mar. Sci.* 14, 298–308. doi: 10.12681/mms.413

Miyake, H., Shibata, H., and Furushima, Y. (2011). "Deep-sea litter study using deep-sea observation tools," in *Interdisciplinary Studies on Environmental Chemistry—Marine Environmental Modeling & Analysis,* eds K. Omori, X. Guo, N. Yoshie, N. Fujii, I. C. Handoh, A. Isobe, and S. Tanabe (Tokyo: Terrapub), 261–269.

Moore, C. J. (2008). Synthetic polymers in the marine environment: a rapidly increasing, long-term threat. *Environ. Res.* 108, 131–139. doi: 10.1016/j.envres.2008.07.025

Moore, C. J., Moore, S. L., Leecaster, M. K., and Weisberg, S. B. (2001). A comparison of plastic and plankton in the North Pacific central gyre. *Mar. Pollut. Bull.* 42, 1297–1300. doi: 10.1016/S0025-326X(01)00114-X

Mordecai, G., Tyler, P. A., Masson, D. G., and Huvenne, V. I. (2011). Litter in submarine canyons off the west coast of Portugal. *Deep-sea Res. II Top. Stud. Oceanogr.* 58, 2489–2496. doi: 10.1016/j.dsr2.2011.08.009

OSPAR (2007). *Background Report on Fishing-For-Litter Activities in the OSPAR Region.* London: OSPAR Comission.

Pham, C. K., Gomes-Pereira, J. N., Isidro, E. J., Santos, R. S., and Morato, T. (2013). Abundance of litter on Condor seamount (Azores, Portugal, Northeast Atlantic). *Deep-sea Res. II Top. Stud. Oceanogr.* 98, 204–208. doi: 10.1016/j.dsr2.2013.01.011

Pham, C. K., Ramirez-Llodra, E., Alt, C. H., Amaro, T., Bergmann, M., Canals, M., et al. (2014). Marine litter distribution and density in European seas, from the shelves to deep basins. *PLoS ONE* 9:e95839. doi: 10.1371/journal.pone.0095839

Ramirez-Llodra, E., De Mol, B., Company, J. B., Coll, M., and Sardà, F. (2013). Effects of natural and anthropogenic processes in the distribution of marine litter in the deep Mediterranean Sea. *Prog. Oceanogr.* 118, 273–287. doi: 10.1016/j.pocean.2013.07.027

Ramirez-Llodra, E., Tyler, P. A., Baker, M. C., Bergstad, O. A., Clark, M. R., Escobar, E., et al. (2011). Man and the last great wilderness: human impact on the deep sea. *PLoS ONE* 6:e22588. doi: 10.1371/journal.pone.0022588

Rogers, A. D., and Taylor, M. L. (eds.). (2012). *Benthic Biodiversity of Seamounts in the Southwest Indian Ocean: Cruise Report – R/V James Cook 066 - November 7th - December 21st 2011.*

Ryan, P. G., Moore, C. J., Van Franeker, J. A., and Moloney, C. L. (2009). Monitoring the abundance of plastic debris in the marine environment. *Philos. Trans. R. Soc. Lond. B. Biol. Sci.* 364, 1999–2012. doi: 10.1098/rstb.2008.0207

Schlining, B. M., and Jacobsen Stout, N. (2006). "MBARI's video annotation and reference system," in *Proceedings of the Marine Technology Society/Institute of Electrical and Electronics Engineers Ocean Conference* (Boston, MA). doi: 10.1109/OCEANS.2006.306879

Schlining, K., Von Thun, S., Kuhnz, L., Schlining, B., Lundsten, L., Jacobsen Stout, N., et al. (2013). Debris in the deep: using a 22-year video annotation database to survey marine litter in Monterey Canyon, Central California, USA. *Deep-sea Res. I Top. Stud. Oceanogr.* 79, 96–105. doi: 10.1016/j.dsr.2013.05.006

Shotton, R. (2006). "Management of demersal fisheries resources of the southern Indian Ocean," in *FAO Fisheries Circular 1020,* (Rome: FAO), 90.

Teuten, E. L., Saquing, J. M., Knappe, D. R., Barlaz, M. A., Jonsson, S., Bjorn, A., et al. (2009). Transport and release of chemicals from plastics to the environment and to wildlife. *Philos. Trans. R. Soc. Lond. B. Biol. Sci.* 364, 2027–2045. doi: 10.1098/rstb.2008.0284

Thompson, R. C., Moore, C. J., Von Saal, F. S., and Swan, S. H. (2009). Plastics, the environment and human health: current consensus and future trends. *Philos. Trans. R. Soc. Lond. B. Biol. Sci.* 364, 2153–2166. doi: 10.1098/rstb.2009.0053

UNEP (2005). *Marine Litter: An Analytical Overview.* Nariobi: UNEP. Available online at: http://www.unep.org/regionalseas/marinelitter/publications/docs/anl_oview.pdf

UNEP (2009). *Marine Litter: A Global Challenge.* Nariobi: UNEP.

UNGA (United Nations General Assembly) (2012). *"The Future We Want"-Resolution Adopted by the General Assembly A/RES/66/288.*

Watters, D. L., Yoklavich, M. M., Love, M. S., and Schroeder, D. M. (2010). Assessing marine debris in deep seafloor habitats off Califonia. *Mar. Pollut. Bull.* 60, 131–138. doi: 10.1016/j.marpolbul.2009.08.019

Wei, C. L., Rowe, G. T., Nunnally, C. C., and Wicksten, M. K. (2012). Anthropogenic "Litter" and macrophyte detritus in the deep Northern Gulf of Mexico. *Mar. Pollut. Bull.* 64, 966–973. doi: 10.1016/j.marpolbul.2012.02.015

Woodall, L., Sanchez-Vidal, A., Canals, M., Paterson, G., Coppock, R., Sleight, V., et al. (2014). The deep sea is a major sink for microplastic debris. *R. Soc. Open Sci.* 1:140317. doi: 10.1098/rsos.140317

Wright, D. P., Pendleton, M., Boulware, J., Walbridge, S., Gerlt, B., Eslinger, D., et al. (2012). *"ArcGIS Benthic Terrain Modeler (BTM)." v3.0: Environmental Systems Research Institute, NOAA Coastal Services Center, Massachusetts Office of Coastal Zone Management.*

Wright, S. L., Thompson, R., and Galloway, T. (2013). The physical impacts of microplastics on marine organisms: a review. *Environ. Pollut.* 178, 483–492. doi: 10.1016/j.envpol.2013.02.031

Yesson, C., Clark, M. R., Taylor, M. L., and Rogers, A. D. (2011). The global distribution of seamounts based on 30 arc seconds bathymetry data. *Deep-sea Res. I Top. Stud. Oceanogr.* 58, 442–453. doi: 10.1016/j.dsr.2011.02.004

Conflict of Interest Statement: The authors declare that the research was conducted in the absence of any commercial or financial relationships that could be construed as a potential conflict of interest.

14

Acute survivorship of the deep-sea coral *Lophelia pertusa* from the Gulf of Mexico under acidification, warming, and deoxygenation

Jay J. Lunden¹†, Conall G. McNicholl¹, Christopher R. Sears¹, Cheryl L. Morrison² and Erik E. Cordes¹*

¹ Department of Biology, Temple University, Philadelphia, PA, USA
² United States Geological Survey, Leetown Science Center, Kearneysville, WV, USA

Edited by:
Paul E. Renaud, Akvaplan-niva AS, Norway

Reviewed by:
Erik Caroselli, Alma Mater Studiorum - University of Bologna, Italy
Jon Havenhand, University of Gothenburg, Sweden

*Correspondence:
Jay J. Lunden, Department of Ecology, Evolution and Marine Biology, University of California, Santa Barbara, Building 520 Rm. 4001 Fl 4L, Santa Barbara, CA 93106-6150, USA
e-mail: jay.lunden@lifesci.ucsb.edu

†**Present address:**
Jay J. Lunden, Department of Ecology, Evolution and Marine Biology, University of California, Santa Barbara, Santa Barbara, USA

Changing global climate due to anthropogenic emissions of CO_2 are driving rapid changes in the physical and chemical environment of the oceans via warming, deoxygenation, and acidification. These changes may threaten the persistence of species and populations across a range of latitudes and depths, including species that support diverse biological communities that in turn provide ecological stability and support commercial interests. Worldwide, but particularly in the North Atlantic and deep Gulf of Mexico, *Lophelia pertusa* forms expansive reefs that support biological communities whose diversity rivals that of tropical coral reefs. In this study, *L. pertusa* colonies were collected from the Viosca Knoll region in the Gulf of Mexico (390 to 450 m depth), genotyped using microsatellite markers, and exposed to a series of treatments testing survivorship responses to acidification, warming, and deoxygenation. All coral nubbins survived the acidification scenarios tested, between pH of 7.67 and 7.90 and aragonite saturation states of 0.92 and 1.47. However, net calcification generally declined with respect to pH, though a disparate response was evident where select individuals net calcified and others exhibited net dissolution near a saturation state of 1. Warming and deoxygenation both had negative effects on survivorship, with up to 100% mortality observed at temperatures above 14°C and oxygen concentrations of approximately $1.5\,ml\cdot l^{-1}$. These results suggest that, over the short-term, climate change and OA may negatively impact *L. pertusa* in the Gulf of Mexico, though the potential for acclimation and the effects of genetic background should be considered in future research.

Keywords: climate change, ocean acidification, *Lophelia pertusa*, survivorship, net calcification, Gulf of Mexico

INTRODUCTION

Human activities are driving noticeable alterations to the Earth's oceans, primarily via warming, deoxygenation, and acidification (Hoegh-Guldberg and Bruno, 2010). As a result of the oceans' uptake of anthropogenic CO_2, drastic changes are expected to occur to the marine environment throughout the current century and beyond (Solomon et al., 2009). Sea surface temperatures increased on average by 0.6°C over the past 100 years, and ongoing warming may result in future increases of as much as 4°C (Solomon et al., 2007; IPCC, 2013). Through processes such as meridional transport and downwelling (Bryan, 1982; Hall and Bryden, 1982), heat and oxygen from photosynthesis are transported to the deep ocean (Barnett et al., 2001, 2005; Levitus, 2005). However, deep-water ventilation may be reduced as ocean temperatures increase, resulting in an expansion of oxygen minimum zones in the deep sea (Keeling et al., 2010; Stramma et al., 2010). Furthermore, absorption of anthropogenic CO_2 by the oceans alters the equilibrium of dissolved inorganic carbon thereby decreasing pH and calcium carbonate saturation. At the current rate of CO_2 emissions, surface seawater pH is projected

to decrease by 0.4 units by the year 2100 (Caldeira and Wickett, 2003). In addition to decreasing surface seawater pH, anthropogenic CO_2 is transported to deeper waters by the same processes listed above, driving the shoaling, or upward movement, of the aragonite saturation horizon (Orr et al., 2005; Guinotte et al., 2006) with potentially negative effects on organisms across a range of depths and latitudes.

The emerging consequences of global climate change (GCC) described above are manifest across the spectrum of biological organization, with a variety of responses observed at the individual, population, community, and ecosystem levels (Walther et al., 2002; Peck, 2011). This may result in a drastic reduction in global biodiversity with deleterious impacts to environmental stability, ecosystem function, and society (Cooley and Doney, 2009; Turley et al., 2010). Despite these changes, many extant species possess physiological and behavioral mechanisms to either cope with or escape from the effects of GCC, including acclimatization, adaptation, and range or habitat shifts through migration. Since these mechanisms operate principally at the individual and population levels, it is prudent to investigate the impacts of GCC at these scales. Furthermore, the role of individual plasticity in acclimatization is important to consider while predicting species and population persistence on a rapidly changing planet.

One of the most abundant and widely distributed deep-sea corals is the scleractinian *Lophelia pertusa* (Linnaeus, 1758), a reef-forming species that occurs on hard substrata at depths from 40 to 3300 m (Strømgren, 1971; Zibrowius, 1980) across the globe (Roberts et al., 2009). The distribution of *L. pertusa* is largely controlled by environmental factors such as temperature, oxygen saturation, food supply, and carbonate chemistry (Roberts et al., 2009; Georgian et al., 2014). *L. pertusa* is typically found in areas with hard substrata where temperatures range from 4 to 12°C, dissolved oxygen ranges from 3 to 5 $ml·l^{-1}$, and the aragonite saturation state is greater than 1 (Freiwald et al., 2004; Guinotte et al., 2006). However, the specific environmental conditions experienced by individual colonies of *L. pertusa* are dependent on both bathymetric and geographic location. The best documented occurrences of *L. pertusa* are in the Northeast Atlantic Ocean surrounding Norway, the United Kingdom, and Ireland at depths primarily from 200 to 1000 m where temperatures range from 6 to 8°C (Zibrowius, 1980; Frederiksen et al., 1992; Freiwald et al., 2004) and dissolved oxygen ranges from 3 to 6 $ml·l^{-1}$ (Freiwald, 2002; Wisshak et al., 2005). Outside of this area, *L. pertusa* occurs in the Mediterranean Sea at depths below 200 m and is associated with higher temperatures from 12.5 to 14°C (Tursi et al., 2004; Freiwald et al., 2009). Populations of *L. pertusa* also occur along the Atlantic coast of the United States and in the Gulf of Mexico. In the Gulf of Mexico, *L. pertusa* reefs are typically found at depths from 300 to 600 m at intermediate temperatures of 8–12°C and relatively low aragonite saturation states of 1.2–1.5 (Cordes et al., 2008; Mienis et al., 2012; Lunden et al., 2013). The range in pH experienced by *L. pertusa* is described only in a few studies, and reports range from 7.94 to 8.08 in the North Atlantic (Form and Riebesell, 2012; Brooke and Ross, 2014) and from 7.85 to 8.03 in the Gulf of Mexico (Lunden et al., 2013). Dissolved oxygen values at *L. pertusa* habitats in the Gulf of Mexico are typically in the 2.7–2.8 ml l^{-1} range (Davies et al., 2010; Schroeder, 2002), although values as low as 1.5 $ml·l^{-1}$ have been observed near *L. pertusa* colonies (Georgian et al., 2014). However, these values are likely to be episodic and long-term (i.e., monthly to annual) data are currently lacking.

Previous studies of global change impacts on *L. pertusa* have predominately been performed on individuals from the Northeast Atlantic and Mediterranean populations. Dodds et al. (2007) found that metabolic rate in *L. pertusa* from the Mingulay Reef Complex was sensitive to ocean deoxygenation despite surviving short periods of anoxia. Naumann et al. (2014) obtained evidence of thermal acclimation to decreased temperature in *L. pertusa* from the Mediterranean, but it is not presently known if this capacity exists for ocean warming. The effects of ocean acidification on *L. pertusa* metabolism were recently investigated by Maier et al. (2013a), and the authors observed no significant decreases in respiration rate in response to elevated pCO_2. Studies on calcification in *L. pertusa* from the Northeast Atlantic and the Mediterranean have mixed results, with observations of short-term decreases in net calcification followed by acclimation to increased pCO_2 in the Northeast Atlantic (Form and

Riebesell, 2012) and conflicting observations of no change in calcification rate due to increased pCO_2 over several months from the Mediterranean (Maier et al., 2013b). The only published laboratory experiment on *L. pertusa* from the Gulf of Mexico identified an upper thermal limit of 15°C after 7 days (Brooke et al., 2013).

This study investigated the effects of ocean warming, deoxygenation, and acidification on survivorship and calcification in *L. pertusa* from the Gulf of Mexico. Due to anthropogenic changes to the physical and chemical properties of seawater, *L. pertusa* is likely to experience increased energetic demands associated with homeostasis and calcification, and therefore its persistence may be compromised in a changing ocean. Here, in a series of short-term experiments in the laboratory, we measured survivorship of *L. pertusa* colonies exposed to varying regimes of temperature, pH, and dissolved oxygen; furthermore, we measured net calcification of *L. pertusa* as a function of pH. The purpose of these experiments was to obtain physiological response data on *L. pertusa* from the Gulf of Mexico.

MATERIALS AND METHODS
COLLECTION SITE DESCRIPTION
The principal collection sites for this study are located in the Viosca Knoll leasing area designated by the U.S. Bureau of Ocean Energy Management (BOEM). Two expansive *Lophelia* reefs occur within two lease blocks of the VK area, VK906 (385–400 m depth, 29.07°N 88.38°W) and VK826 (390–550 m depth, 29.15°N 88.01°W). The physical environment of the two sites is similar: temperatures at VK906 range from 8 to 12.5°C and from 6.5 to 11.6°C at VK826, while salinity at both sites ranges from 34.9 to 35.4 (Mienis et al., 2012; Georgian et al., 2014; Lunden et al., 2014). Observations of dissolved oxygen concentrations from the two sites range from 1.5 to 3.4 $ml·l^{-1}$, with mean dissolved oxygen near 3 $ml·l^{-1}$ (Davies et al., 2010; Georgian et al., 2014). The carbonate chemistry parameters at the two sites are also similar, with pH ranging from 7.85 to 8.03 and aragonite saturation state (Ω_{arag}) ranging from 1.3 to 1.6 (Lunden et al., 2013). Gene flow between the two sites is likely high given that *L. pertusa* populations in this area of the Gulf are considered panmictic (Morrison et al., 2011).

CORAL COLLECTION, PREPARATION, AND MAINTENANCE
Forty-one nubbins of *L. pertusa* used in the experiments were collected in November 2010 on the NOAA Ship *Ronald H. Brown* with ROV *Jason II* as part of the "Lophelia II" project jointly sponsored by the Bureau of Ocean Energy Management and the NOAA Office of Ocean Exploration and Research in the Gulf of Mexico (GoM). Permits for the collection of corals were obtained from the U.S. Department of the Interior prior to any collection activities. Spatially discrete coral branches were collected with the ROV and placed in temperature-insulated bioboxes (volume = ~20 l) at depth. Upon return to the surface, corals were kept alive in 20 l aquaria in the ship's constant-temperature room. Partial water changes were made regularly while at sea. Upon return to port, corals were immediately transported overnight to the laboratory on wet ice.

In the laboratory, corals were maintained in one of two 570 liter recirculating aquaria systems at temperature 8°C and salinity 35 ppt (Lunden et al., 2014). Regular partial water changes (15–20%) were performed with seawater made using Instant Ocean® sea salt. Submersible power heads were placed in each holding tank to ensure water movement and turbulence sufficient to cause swaying of coral polyps. Corals were fed three times weekly using a combination of MarineSnow® Plankton Diet (Two Little Fishies, Miami Gardens, FL) and freshly hatched *Artemia* nauplii.

Prior to experimental manipulations, coral nubbins were fixed to 1" PVC male adapters using HoldFast epoxy forming coral "nubbins" to minimize handling effects. Monofilament line (diameter = 0.30 mm) was looped through the base of each PVC adapter to allow for buoyant weighing of the coral branch. All coral nubbins weighed less than 60 g in seawater. Weights were corrected for the buoyant weight of the base to which they were attached.

For population genetic analyses, small fragments of 240 *L. pertusa* colonies from nine localities in the GoM (Garden Banks, Green Canyon, Mississippi Canyon 751, Gulf Oil and Gulf Penn shipwrecks, VK906, VK862, VK826, West Florida Slope) were collected during four GoM cruises between August 2009 and November 2010. Sampling occurred during August-September of 2009 and October-November 2010, aboard the R/V *Ronald H. Brown* (NOAA) using the remotely operated vehicle (ROV) *Jason II* (Woods Hole Oceanographic Institute); aboard the R/V *Seward Johnson* using the *Johnson-Sea-Link II* (Harbor Branch Oceanographic Institute) submersible in September 2009, as well as aboard the R/V *Cape Hatteras* using the *Kraken II* ROV (University of Connecticut) in September-October 2010. Additionally, 16 samples from the eastern Atlantic Ocean (Sula Ridge and Nordleska Reef) were collected aboard the R/V *Poseidon* in September 2011. Once onboard the vessel, small tissue samples were preserved in 95% ETOH and FTA® Technology Classic card (Whatman®).

CORAL GENOTYPING

All nubbins were swabbed for genotyping prior to experimental assignment and no mortality resulted from swabbing. Genotyping was conducted to ensure that multiple genotypes of corals were present in the experimental cohorts, i.e., that one set of experiments did not include an entire set of clones and that each treatment contained a mixture of genotypes. Coral nubbins were haphazardly assigned to each series of experiments prior to genotyping results. Total DNA was extracted from swabs taken from experimental *L. pertusa* nubbins and FTA cards using the PureGene DNA extraction kit (Gentra Systems Inc., Minneapolis, Minnesota). Six *L. pertusa* microsatellite loci (*LpeA5, LpeC44, LpeC151, LpeC142, LpeD3*, Morrison et al., 2008; *Lpeg62, Molecular Ecology Resources* database entry 51059) were amplified from 41 swab extracts in 20 μL PCR reactions following the conditions in Morrison et al. (2008). Additionally, 240 *L. pertusa* samples from 9 localities in the GoM and 16 samples from the eastern Atlantic Ocean (Sula Ridge and Nordleska Reef) were genotyped at 8 microsatellite loci (*LpeA5, LpeC44, LpeC52, LpeC61, LpeC142, LpeC151, LpeD3*, and *LpeD5*, Morrison et al.,

2008). Fluorescent DNA fragments were electrophoresed on an ABI 3130xl Genetic Analyzer with GeneScan-500 ROX size standard. Alleles were scored using GeneMapper v. 4.1 fragment analysis software (Applied Biosystems). Individuals with identical multilocus genotypes (MLGs) were identified and the probability of identity (PI; i.e., the probability of two individuals sharing the same MLG) was calculated using GenAlEx 6.501 software (Peakall and Smouse, 2006, 2012). To assess inter-regional patterns of connectivity, MLG data from 195 *L. pertusa* individuals collected at seven localities in the northwestern Atlantic Ocean, off the southeastern U.S. (SEUS), two New England Seamounts (Manning and Rehoboth), and five eastern North Atlantic populations (Rockall Bank, Mingulay Reef, Sula Ridge, Trondheimfjord and Nordleska), plus an additional 108 *L. pertusa* samples from 4 localities in the GoM, were included in several analyses (see Table S1 and Morrison et al., 2011 for details).

The probability of identity (PI, the probability of two unrelated individuals sharing the same genotype) was calculated for increasing locus combinations using GenAlEx. A Bayesian model-based clustering approach (Pritchard et al., 2000) implemented in STRUCTURE v. 2.3.2 (Hubisz et al., 2009) was used to describe genetic relationships among individuals. This method infers the number of genetic clusters (*K*) from MLG data by minimizing Hardy-Weinberg and linkage disequilibrium among loci within groups, assigning individuals (probabilistically) to each cluster. Because models utilizing collection location information as priors are useful for small data sets and weak structuring (Hubisz et al., 2009), locality designations were included as priors. Settings for all runs also included an admixture model (i.e., individuals may have mixed ancestry), correlated allele frequencies (Falush et al., 2003), and 20,000 Markov Chain Monte Carlo (MCMC) iterations after a burn-in of 10,000 iterations. Twenty independent chains were run to test each value of *K* from *K* = 1–23. The optimum number of clusters was determined by evaluating the values of *K* as the highest mean likelihood of the probability of the number of clusters given the data observed (LnP(D), Pritchard et al., 2000), and comparing that with ΔK (Evanno et al., 2005) as compiled and graphed using STRUCTURE Harvester v.0.56.1 (Earl, 2009). Each cluster identified in the initial STRUCTURE run was analyzed separately using the same settings to identify potential within-cluster structure since detection of fine-scale structuring can be limited with large data sets (see Jakobsson et al., 2008). The software CLUMPP v.1.2 (Jakobsson and Rosenberg, 2007) was used to merge the results of the 20 runs per *K*. DISTRUCT v.1.1 (Rosenberg, 2004) were used to visualize merged run results.

EXPERIMENTAL DESIGN AND GENERAL SETUP

Three single-factor experiments were conducted in this study: pH ("acidification"), temperature ("warming"), and dissolved oxygen ("deoxygenation"). Each experiment consisted of 3–5 treatments (**Table 1**). All experiments were conducted in a constant-temperature room in the laboratory (see Lunden et al., 2014, for complete description of experimental aquaria). Three 75-l aquaria ("tall" type: 61 × 33 × 43 cm) with individual Hagen® AquaClear® 30 filtration units (Drs. Foster & Smith, Rhinelander,

Table 1 | Tank conditions for the different experiments.

Experiment	Temperature (°C)	pH$_T$	TA (μmol·kg^{-1})	Ω_{arag}	DO (ml·l^{-1})
ACIDIFICATION					
Ambient 1	8.6 ± 0.5	7.90 ± 0.06	2320 ± 85	1.47 ± 0.17	∼6
Low 1	7.8 ± 0.4	7.80 ± 0.07	2352 ± 32	1.18 ± 0.18	∼6
Very Low 1	8.4 ± 0.5	7.67 ± 0.16	2371 ± 9	0.97 ± 0.40	∼6
Ambient 2	8.6 ± 0.5	7.90 ± 0.08	2316 ± 23	1.47 ± 0.23	∼6
Low 2	7.7 ± 0.5	7.78 ± 0.04	2340 ± 66	1.11 ± 0.1	∼6
Very Low 2	8.8 ± 0.2	7.67 ± 0.1	2309 ± 24	0.92 ± 0.23	∼6
WARMING					
8°C	8.4 ± 0.2	7.92 ± 0.04	2343 ± 11	1.53 ± 0.14	∼6
10°C	9.9 ± 0.5	7.92 ± 0.04	2241 ± 13	1.54 ± 0.12	∼6
12°C	11.9 ± 0.3	7.89 ± 0.07	2389 ± 18	1.68 ± 0.22	∼6
14°C	14.0 ± 0.3	7.93 ± 0.03	2370 ± 38	1.92 ± 0.14	∼6
16°C	16.0 ± 0.5	7.95 ± 0.04	2283 ± 18	2.08 ± 0.18	∼6
DEOXYGENATION					
High	8.61 ± 0.49	8.11 ± 0.08	3535 ± 81	3.43 ± 0.51	5.32 ± 0.28
Ambient	8.79 ± 0.33	8.07 ± 0.06	3553 ± 18	3.23 ± 0.38	2.92 ± 0.21
Low	8.5 ± 0.48	8.28 ± 0.12	3570 ± 45	4.88 ± 1.12	1.57 ± 0.28

All data are mean values ± SD over the 7-day experiments. TA, total alkalinity; Ω_{arag}, the aragonite saturation state; and DO, dissolved oxygen.

WI) and Pinpoint® pH controllers (American Marine Inc., Ridgefield, CT) were used for each treatment (3 to 5 depending on experiment) within each experiment. Each treatment lasted for a total of 15 days, with an initial 8-day conditioning period to allow the corals to acclimate to the experimental tank conditions. Treatments were separated by 3-week "recovery" periods in all experimental series.

The acidification experiment consisted of three separate treatments and was conducted in sequence at ambient pH (7.90), low pH (7.75), and very low pH (7.60) from April 2011 to January 2012. Experimental incubations were maintained at temperature 8°C and salinity 35 ppt. pH was controlled by injection of CO_2 using a Pinpoint® pH controller (American Marine Inc., Ridgefield, CT). pH electrodes were calibrated weekly using Tris-HCl and AMP-HCl buffers (Nemzer and Dickson, 2005; Dickson et al., 2007). All of the corals used in the acidification experiment were collected from VK906. Two trials were performed for the acidification experiment with two separate groups of corals (see Results).

The warming experiment consisted of five separate treatments and was conducted in sequence at 8, 10, 12, 14, and 16°C from February to July 2012. Experimental incubations were maintained at pH 7.90 and salinity 35 ppt. To reach the desired temperature for each treatment, the thermostat of the constant-temperature room was adjusted at a rate of 2°C per day. This rate of change is well within the range experienced by *L. pertusa* in its natural environment, where temperature can change as rapidly as 2.3°C per hour (Brooke et al., 2013). Temperature in each aquarium was recorded daily with a digital thermometer. All of the corals used in the warming experiment were collected from VK906. During the recovery period between the 14 and 16°C treatments, all corals experienced 100% mortality. Because of this, a new set of experimental corals (also from VK826) had to be used for the 16°C experiment. Survivorship was recorded as cumulative

survivorship, or the proportion of the original nine coral nubbins surviving.

The deoxygenation experiment consisted of three separate treatments and was conducted in sequence under high DO (5 ml·l^{-1}), ambient DO (3 ml·l^{-1}), and low DO (1ml·l^{-1}) from July to October 2012. Experimental incubations were maintained at temperature 8°C and salinity 35 ppt. In order to manipulate dissolved oxygen concentrations, oxygen-free nitrogen (OFN) gas was bubbled into each tank through an Aqua Medic 1000 CO_2 reactor (Drs. Foster & Smith, Rhinelander, WI). Flow of nitrogen was controlled with CGA 580 regulators (Airgas, Inc., Berwyn, PA). Dissolved oxygen concentration in each aquarium was recorded daily with an Orion 5 Star DO/pH meter that was calibrated each day. All of the corals used in the deoxygenation experiment were collected from VK826.

SEAWATER PREPARATION AND ANALYSES

Experimental seawater was prepared using Instant Ocean® sea salt at a salinity of 35 ppt. For the acidification and temperature experiments, further modifications were necessary in order to manipulate pH. Since Instant Ocean® produces seawater with a total alkalinity of approximately 3600 μmol·kg^{-1} (1.5X that of natural oceanic values), 12.1 N HCl was added to reduce the total alkalinity to 2300 μmol·kg^{-1} (mean total alkalinity at GoM *L. pertusa* reefs, Lunden et al., 2013). The seawater was then bubbled with oxygen for ∼24 h to drive off excess CO_2 and to restore pH to the ambient value of 7.90 and then further manipulated for the other treatments (fully described in Lunden et al., 2014). This reduction in total alkalinity of the Instant Ocean seawater® was not performed in the deoxygenation experiments. Because the addition of nitrogen gas removed both O_2 and CO_2 from the experimental aquaria, dissolved CO_2 was reduced in the aquaria and therefore an elevation in pH occurred. Since pH could not be controlled under this scenario, total alkalinity was not manipulated in the deoxygenation experiments.

Total alkalinity was measured twice weekly in each aquarium by potentiometric open-cell titration using a Mettler-Toledo DL15 automatic titrator (Fisher Scientific, Waltham, MA) according to SOP3b (Dickson et al., 2007) with certified reference materials courtesy of A. Dickson (Scripps Institute of Oceanography, La Jolla, CA). pH (total hydrogen scale) was recorded daily using the Pinpoint® pH controller (American Marine Inc., Ridgefield, CT) calibrated against Tris/HCl and AMP/HCl buffers (Nemzer and Dickson, 2005). The aragonite saturation state was calculated using CO2SYS (Pierrot et al., 2006) with total alkalinity, pH, temperature, and salinity as input variables. Nutrient concentrations (ammonia [NH_3], nitrate [NO_3^-], and nitrite [NO_2^-]) were measured weekly using API® aquarium test kits (Drs. Foster & Smith, Rhinelander, WI).

SURVIVORSHIP MEASUREMENTS

Survivorship was assessed by daily observations of polyp tissue presence and behavior. Final survivorship counts were taken 3 to 4 days following the end of each treatment after transfer to the maintenance tank. Survivorship is reported as percent cumulative mortality.

CALCIFICATION MEASUREMENTS

Net calcification was measured using the buoyant weight technique (Davies, 1989). Coral nubbins were buoyantly weighed at the start and end of each experimental period (days eight and fifteen) using a Denver Instruments SI-64 analytical balance ($d = 0.1$ mg, Fisher Scientific, Waltham, MA). A weighing chamber was constructed using ½" plexiglass to prevent disturbances from air movement during weighing. Each coral nubbin was transported individually from its respective aquarium to the weighing chamber in a four-liter Pyrex® beaker and suspended from the balance. The buoyant weight was recorded after the coral nubbin stabilized, typically 2 min. Each coral nubbin was weighed three times to determine measurement precision ($\sim$2–3 mg). Seawater density was determined in each aquarium by buoyantly weighing a 2.5 cm^2 aluminum block with known density (2.7 g·cm^{-3}).

Coral weight in air (i.e., dry weight) was calculated by the following equation:

$$W_a = \frac{W_w}{1 - \frac{D_w}{SD}}$$

Where

W_a = coral weight in air (dry weight)
W_w = coral weight in water (buoyant weight)
D_w = density of seawater
SD = coral skeletal density (= 2.82 g·cm^{-3}, Lunden et al., 2013).

Coral growth rate is reported as percent growth per day (%·d^{-1}), which was calculated by the equation:

$$G_t = 100 \times \frac{M_{t2} - M_{t1}}{M_{t1}(T_2 - T_1)}$$

Where

G_t = growth rate as %·d^{-1}
M_{t2} = mass (mg, dry weight) at time 2 (end of experimental period, day 15)
M_{t1} = mass (mg, dry weight) at time 1 (start of experimental period, day 8)
T_2 = time 2 (end of experimental period, day 15)
T_1 = time 1 (start of experimental period, day 8).

STATISTICAL ANALYSES AND TANK EFFECTS

All statistical analyses were performed using JMP10® statistical software. In this study, both parametric and non-parametric statistics were employed. Non-parametric statistics were employed where assumptions of normality were not met, despite endeavors to transform the data (log, square-root). Normality was tested using the Shapiro–Wilk W Test. A coral "individual" is defined as a coral nubbin weighing less than 60 g irrespective of genotype. We elect this definition due to lack of genotypic replication in our original experimental design.

The effect of individual experimental tanks ($n = 3$) on coral growth was tested by one-way Kruskal–Wallis test of growth rate against tank for each treatment. All but one treatment in this study yielded $p > 0.5$ for tank effects. The only treatment to fail this threshold was the very low pH treatment for the group 1 corals, where $p > 0.05$. We note that "accepted practices" typically set $p > 0.25$ for tank effects, but that exceptions to this rule exist (e.g., Dufault et al., 2012). Therefore, in our subsequent analyses, we ignored the effect of experimental tanks and treated all coral nubbins within the 3 tanks as individual replicates.

RESULTS
CORAL GENOTYPING AND POPULATION GENETICS

For the 41 experimental L. pertusa samples genotyped, the number of alleles per locus ranged from 3 (Lpeg62) to 11 (LpeC142), with a mean of 7.8. The PI for the six microsatellite loci was 0.000269, or in other words, this combination of microsatellite loci provides adequate power to distinguish among close relatives and clones. There were 30 unique MLGs and 17 samples had identical MLGs (four MLGs represented twice, and two MLGs represented 4–5 times), suggesting that these samples were clones.

In total, 36 of the 41 genotyped coral nubbins were haphazardly assigned to groups and treatments in the laboratory experiments (Table S2). The 36 coral nubbins represented 27 of the 30 unique MLGs, or "genets" identified in the microsatellite analysis. Each of the three experimental series included multiple genetically distinct individuals. For the acidification experiment, two separate groups of 9 coral nubbins were used. The first group of nubbins ("group 1") consisted of 5 genets and 4 clones, and the second group of nubbins ("group 2") consisted of 9 genets and no clones. For the warming experiment, one group of 7 genets and 2 clones was used for the 8, 10, 12, and 14°C treatments. A separate group of 9 genets and no clones was used for the 16°C treatment. For the deoxygenation experiment, one group was used and consisted of 9 genets and no clones.

Our regional population genetic analysis of L. pertusa included 601 unique MLGs from 23 localities in the Gulf of Mexico and North Atlantic Ocean (Table S1). These 8 microsatellite loci were highly polymorphic, with a mean of 12.87 alleles per locus across sampling localities (Table S1). Observed heterozygosities were high in the majority of localities (mean 0.654, Table S1, HO), however, heterozygote deficits were detected in most localities (positive FIS, Table S1).

Population structuring was evident at several hierarchical levels based upon successive STRUCTURE analyses (**Figure 1**). Two population clusters corresponding to the Gulf of Mexico and the North Atlantic Ocean were detected in the full data set (**Figure 1A**, First round). Additional STRUCTURE runs performed on each initial cluster detected sub-structuring only in the North Atlantic between the Western Atlantic populations off the southeastern U.S. coast and the New England Seamounts plus the Eastern Atlantic (**Figure 1B**, Second round), but not within the Gulf of Mexico populations examined. Additional sub-structuring was detected within the North Atlantic cluster among the seamounts, the eastern Atlantic, and Sula Ridge populations (**Figure 1C**, Third round).

ACIDIFICATION EXPERIMENTS

The experimental values for pH and additional carbonate chemistry parameters are reported in **Table 1**. Mean pH varied significantly across all three treatments for both group 1 corals (Kruskal–Wallis test, $H = 19.7$, $p < 0.001$) and group 2 corals (Kruskal–Wallis test, $H = 30.1$, $p < 0.001$). Temperature was significantly different in both the group 1 experiment (Kruskal–Wallis test, $H = 32.03$, $p < 0.0001$) and in the group 2 experiment (Kruskal–Wallis test, $H = 17.9$, $p < 0.0001$), and was lower by roughly 0.8–0.9°C from the control experiments. Survivorship was 100% for all treatments and no mortality was observed during the 3-week recovery phases between the acidification treatments.

In the acidification experiments, calcification rate data were collected from eight coral nubbins from each group, for a total of 16 coral nubbins for the entire experiment. One coral nubbin from each group became detached from its PVC fitting during

the experiment and was excluded from the analysis. There were no significant differences between group 1 and group 2 corals in number of live polyps (Kruskal–Wallis test, $H = -0.464$, $p = 0.643$) or mass (Kruskal–Wallis test, $H = -0.893$, $p = 0.372$).

In the ambient pH treatment, all corals exhibited positive net calcification over the experimental period. Net calcification rates among all corals in the ambient pH treatment ranged from 0.002 to 0.091% day^{-1}, with a mean rate of $0.025 \pm 0.006\%$ day^{-1} ($n = 16$). Group 1 corals exhibited a mean net calcification rate of $0.011 \pm 0.004\%$ day^{-1} ($n = 8$), and group 2 corals exhibited a mean net calcification rate of $0.039 \pm 0.01\%$ day^{-1} ($n = 8$).

Calcification rate was highly variable among all coral nubbins in the low pH treatment; eight coral nubbins exhibited net dissolution, and eight coral nubbins exhibited positive net calcification. Calcification rate in the low pH treatment for all coral nubbins ranged from -0.04 to 0.02% day^{-1}, with a mean of $-0.003 \pm 0.003\%$ day^{-1} ($n = 16$). Despite the slightly decreased temperature in the low pH treatments, this did not have a significant effect on the net calcification of the individuals tested (Two-Way ANOVA, $t = 1.61$, $p = 0.158$). There was no significant interaction between pH and temperature ($p = 0.055$). The two coral groups responded differently in the low pH treatment. Group 1 corals calcified at a rate of $0.01 \pm 0.003\%$ day^{-1} ($n = 8$). However, group 2 corals exhibited net dissolution in the low pH treatment at a rate of $-0.01 \pm 0.004\%$ day^{-1} ($n = 8$).

In the very low pH treatment, 15 of 16 coral nubbins exhibited net dissolution. Calcification rate for all coral nubbins in the very low pH treatment ranged from -0.024 to 0.012% day^{-1}, with a mean rate of $-0.007 \pm 0.002\%$ day^{-1} ($n = 16$). Group 1 corals dissolved at a rate of $-0.008 \pm 0.003\%$ day^{-1} ($n = 8$) and group 2 corals dissolved at a rate of $-0.005 \pm 0.003\%$ day^{-1} ($n = 8$).

The relationship between net calcification and pH was assessed with both a local regression (LOESS, **Figure 2**) and linear

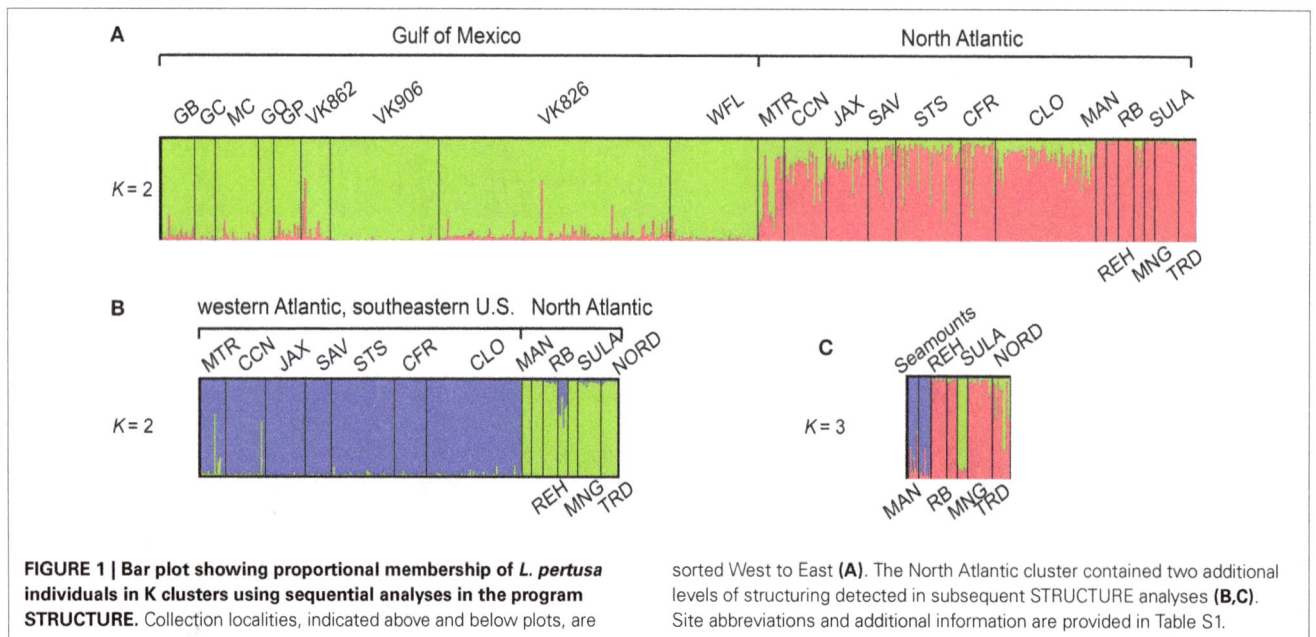

FIGURE 1 | Bar plot showing proportional membership of *L. pertusa* individuals in K clusters using sequential analyses in the program STRUCTURE. Collection localities, indicated above and below plots, are sorted West to East **(A)**. The North Atlantic cluster contained two additional levels of structuring detected in subsequent STRUCTURE analyses **(B,C)**. Site abbreviations and additional information are provided in Table S1.

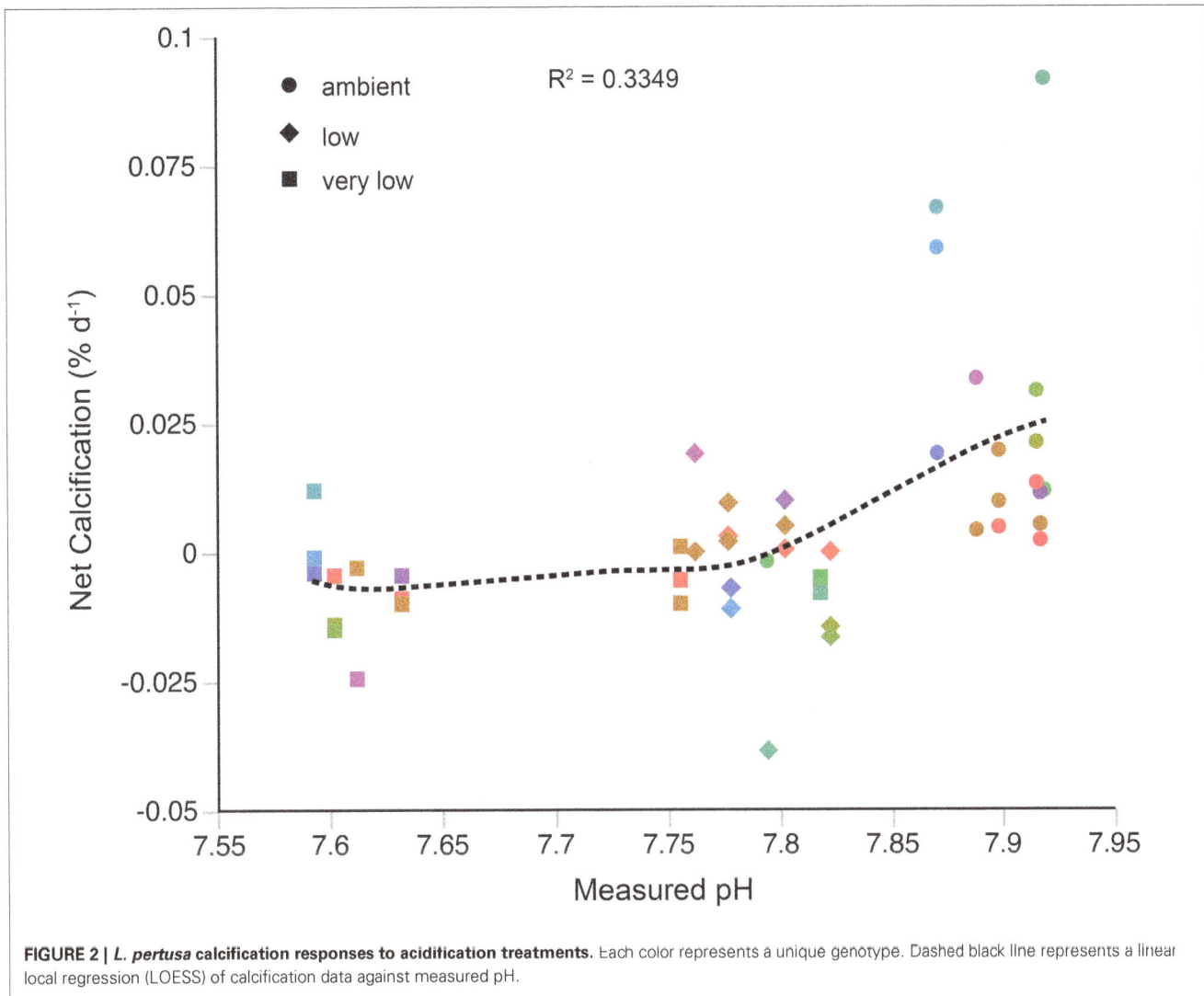

FIGURE 2 | L. pertusa calcification responses to acidification treatments. Each color represents a unique genotype. Dashed black line represents a linear local regression (LOESS) of calcification data against measured pH.

regression. The local regression ($R^2 = 0.3349$) shows two apparent responses to changes in pH, a sharp decline in calcification from the ambient pH treatment to the low pH treatment, and a slight decline in calcification from the low pH treatment to the very low pH treatment. Results from the linear regression ($R^2 = 0.238$, data not shown) suggest an average threshold pH of 7.73 ± 0.06 and aragonite saturation state of 1.05 ± 0.2 for net calcification to occur in the individuals examined.

WARMING EXPERIMENTS

In the warming experiments, temperature was significantly different across all treatments (Kruskal–Wallis test, $H = 92.1$, $p < 0.001$). Mean pH across all temperature treatments was 7.92 ± 0.05. In the 16°C experiment, pH was significantly higher than all other temperature treatments (mean of 7.95 compared to means of 7.89–7.93 for other treatments, Kruskal–Wallis test, $H = 10.7$, $p = 0.0303$). There was no significant interaction between pH and temperature in this set of experiments (Two-Way ANOVA, $t = -0.30$, $p = 0.817$). Ω_{arag} varied significantly among treatments (Kruskal–Wallis test, $H = 63.5$, $p < 0.001$) and was

not statistically different between the 8 and 10°C experiments (Mann–Whitney test, $U = 0.028$, $p = 0.978$). Mean temperatures and other relevant variables for each treatment are reported in **Table 1**.

Cumulative survivorship of L. pertusa differed significantly among temperature regimes after 7 days at each treatment (**Figure 3A,B**, Kruskal–Wallis test, $H = 33.97$, $p < 0.001$). Survivorship was 100% in the control treatment (8°C) and decreased in each successive treatment. At 10°C, survivorship was $86.7 \pm 6.2\%$ (mean $\pm$ SE, $n = 9$). At 12°C, survivorship was $69.9 \pm 6.1\%$ (mean $\pm$ SE, $n = 9$). At 14°C, survivorship was $53.6 \pm 8.5\%$ (mean $\pm$ SE, $n = 9$), though all individuals suffered 100% mortality during the recovery phase following this treatment (within 3 weeks of the conclusion of the 14°C treatment). Survivorship of the second set of nine experimental corals was 0% in the 16°C treatment.

For this set of experiments, several corals were used from the acidification experiments described above (Table S2). To test for carry-over effects on individuals used from the acidification experiments, a "previous treatment" factor was included in the

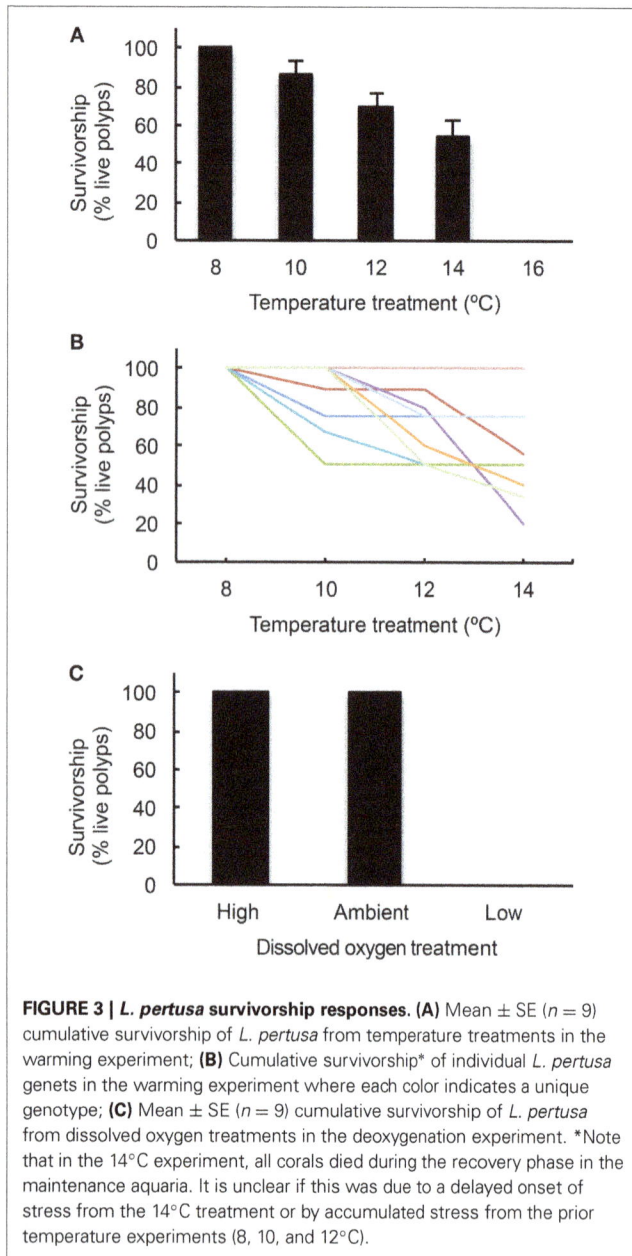

FIGURE 3 | L. pertusa survivorship responses. (A) Mean ± SE (n = 9) cumulative survivorship of *L. pertusa* from temperature treatments in the warming experiment; **(B)** Cumulative survivorship* of individual *L. pertusa* genets in the warming experiment where each color indicates a unique genotype; **(C)** Mean ± SE (n = 9) cumulative survivorship of *L. pertusa* from dissolved oxygen treatments in the deoxygenation experiment. *Note that in the 14°C experiment, all corals died during the recovery phase in the maintenance aquaria. It is unclear if this was due to a delayed onset of stress from the 14°C treatment or by accumulated stress from the prior temperature experiments (8, 10, and 12°C).

analysis and there was a significant effect on coral survivorship (Two-Way ANOVA, $t = 3.49$, $p = 0.002$). Even though they were not used in any previous experiments, all of the corals exposed to 16°C suffered complete mortality.

DEOXYGENATION EXPERIMENTS

In the deoxygenation experiments, DO was significantly different across all treatments (Kruskal–Wallis test, $H = 474.7$, $p < 0.001$) and all treatment values are reported in **Table 1**. Temperature was not significantly different across treatments in this experiment (Kruskal–Wallis test, $H = 2.41$, $p = 0.3$). Survivorship of *L. pertusa* was 100% at both the high and ambient dissolved oxygen treatments. However, survivorship decreased to 0% in the low treatment (DO = 1.57 ± 0.28 ml·l^{-1}, **Figure 3C**). No mortality

was observed during the recovery phases between the deoxygenation treatments. The pH in the low deoxygenation treatment was significantly higher than the high and ambient treatments (Kruskal–Wallis test, $H = 22.96$, $p < 0.001$). This is likely due to the addition of nitrogen gas, which removed dissolved CO_2 (and O_2) and consequently elevated pH.

DISCUSSION

In the present study, the survivorship responses of the deep-sea coral *L. pertusa* were tested against experimental perturbations simulating those projected to occur in the near future from ongoing GCC and ocean acidification in the Gulf of Mexico. Survivorship and calcification responses of individual *L. pertusa* colonies to ocean acidification were tested and the corals exhibited a variable response at pH ∼7.75 whereby some individuals net calcified at rates similar to the controls and others exhibited net dissolution. However, all but one of the individuals exhibited net dissolution at the lowest pH tested (pH 7.60). Exposure to temperatures above 14°C for 7 days led to eventual mortality, while exposure to oxygen concentrations of 1.5 ml·l^{-1} proved fatal to the corals after 7 days. It is important to note that while these results agree well with prior studies on *L. pertusa* that were conducted in a similar manner (e.g., Dodds et al., 2007; Brooke et al., 2013), we cannot rule out the effects of cumulative exposures on mortality in these experiments (as described above in the Methods). As the impacts of GCC and ocean acidification continue to proliferate, it will be necessary to investigate potential mechanisms that species possess in order to cope with or escape from the associated stresses of GCC.

The present study is among the first to explore the impacts of global ocean change on a species of deep-sea coral from the Gulf of Mexico. While deep-sea corals are widely distributed throughout the deep Gulf of Mexico (e.g., Cordes et al., 2008; Quattrini et al., 2014), little is known about their responses to projected future ocean changes. Related studies have investigated the impacts of changes in temperature, oxygen, and pH on *L. pertusa*, but these have been entirely restricted to areas within the Northeast Atlantic Ocean (Dodds et al., 2007; Maier et al., 2009; Form and Riebesell, 2012) and the Mediterranean Sea (Maier et al., 2012, 2013a,b). Existing studies and the population genetics work presented here support the existence of regional populations of *L. pertusa* across the North Atlantic Ocean with restricted gene flow (Morrison et al., 2011). Genetic diversity and connectivity among coral populations is dependent on several factors, including life history, geographic location, and physical environment (Huston, 1985; Hughes, 1989; Selkoe et al., 2010; Maina et al., 2011). High genetic diversity may be found in populations that experience frequent or intense disturbances while more clonal populations are generally found in relatively stable environments due to the success of locally adapted genotypes and lack of space for subsequent recruitment (Hunter, 1993); however, this relationship may vary with respect to species (Coffroth and Lasker, 1998). Long-term monitoring of *L. pertusa* habitats in the Northeast Atlantic reveals relatively stable environmental conditions, with recorded temperature variations typically near 1°C, but episodic variations near 3°C (Mienis et al., 2007). A similar dataset from the Gulf of Mexico shows a wider temperature range of up to 5°C (Mienis et al., 2012). As would be predicted, the structure of the Northeast Atlantic *L. pertusa* population appears to be highly clonal (Waller and Tyler, 2005) and genetically isolated from the Gulf of Mexico population (Morrison et al., 2011). Therefore, the results presented here offer new insight into the responses of *L. pertusa* from a region that is genetically discrete and yet to be rigorously studied in the context of global ocean change.

Results from the acidification experiment show considerable variation in calcification rate among individuals. The most pronounced variability was observed in the ambient pH treatment (**Figure 2**), where calcification rates spanned a range of ~0.08% day^{-1}. In general, growth rates of *L. pertusa* are highly variable (reviewed in Roberts et al., 2009), and are typically much slower than zooxanthellate corals due to the lack of photosynthetic input to calcification (Al-Horani et al., 2003, but see Orejas et al., 2011) and temperature differences of their natural environments. Several studies of *L. pertusa* growth rate show allometric patterns, the differential growth of select parts of a colony relative to the whole (e.g., Mortensen, 2001; Gass and Roberts, 2006; Brooke and Young, 2009), with younger polyps growing significantly faster than older polyps (Maier et al., 2009). Furthermore, periods of active and arrested growth within individual colonies have also been reported (Mortensen, 2001), suggesting episodic growth. While sampling, we exerted great effort to obtain terminal ends of *L. pertusa* branches, and used entirely live colonies in the experiments. However, the combination of allometric and episodic growth patterns in *L. pertusa* may still explain some of the highly variable growth rates obtained at ambient pH in the present study, and our results agree well with previous studies under similar carbonate chemistry conditions (Form and Riebesell, 2012; Maier et al., 2012, 2013b). While a linear response to acidification may be more intuitive (but also see Ries et al., 2010), the high variability in net calcification rates at ambient conditions is better captured by the LOESS curve (**Figure 2**). Furthermore, the LOESS curve suggests varied responses to changes in pH, where calcification initially declines sharply below ambient conditions, and then exhibits a slow decline from the low pH treatment to the very low treatment. This response may indicate a biological "tipping point" for acidification that has been observed in other calcification-dependent invertebrates (Dorey et al., 2013). At this point in time, there are no high-resolution temporal pH data for *L. pertusa* habitats, so it is unknown how much variability in pH this species experiences naturally. If *L. pertusa* does not naturally experience wide variability in pH in the Gulf of Mexico, it may explain the sharp decline in calcification below ambient conditions observed here.

Previous studies of *L. pertusa*'s sensitivity to ocean acidification have revealed a variety of responses. Using specimens from the North Atlantic in a long-term experiment, Form and Riebesell (2012) exposed *L. pertusa* to varying levels of pCO$_2$ ranging from 600 to 980 μatm, corresponding to a pH range of 7.94 to 7.76. They observed a short-term shock response manifest as a reduction in *L. pertusa* calcification rate over the short-term (8 days), but found that after 6 months *L. pertusa* accreted new skeletal material in the high CO$_2$ treatment at comparable rates to the control, implying acclimation to high CO$_2$. In similar studies,

L. pertusa from the Mediterranean Sea was used in experiments with pCO$_2$ treatments ranging from 380 to 930 μatm, spanning a pH range of 8.14 to 7.73, and found no differences in respiration rate (Maier et al., 2013a) or calcification rate (Maier et al., 2013b).

The findings of the present study may reconcile these disparate results regarding calcification rates. Neither Form and Riebesell (2012) nor Maier et al. (2013b) performed genotypic analyses on their *L. pertusa* samples, meaning clones may have been used in the experiments, particularly if samples were collected from a limited number of locations. Prior work on *L. pertusa* in the Northeast Atlantic supports the existence of a highly clonal population within this area (LeGoff-Vitry et al., 2004; Waller and Tyler, 2005). Furthermore, if the genetic variability within a single panmictic population is sufficient to generate a variable response to OA, as may be occurring in the Gulf of Mexico, then one would expect to observe more significant differences among genetically isolated populations such as the North Atlantic and Mediterranean (**Figure 1**). The existence of "OA-hardy" genotypes has been observed in other taxonomic groups (Iglesias-Rodriguez et al., 2008; Langer et al., 2009; Pistevos et al., 2011; Parker et al., 2012), and such genotypes may also occur within the Gulf of Mexico population of *L. pertusa* (**Figure 2**), though further studies are required. Calcification responses of corals and other taxa to ocean acidification are generally complex (Langdon and Atkinson, 2005; Ries et al., 2009), and some of this complexity may be attributed to the inherent genetic diversity within experimental populations.

Temperature is one of the most important abiotic controllers of species' distributions including cold-water corals (Roberts et al., 2009; Davies and Guinotte, 2011). Individual sensitivity to thermal stress is tightly linked to ephemeral physiological mechanisms such as the heat shock response, which permits tolerance to short-term heat (and other) stress through the actions of molecular chaperones (Feder and Hofmann, 1999). The limit for onset of the heat shock response is coupled to an individual's thermal history (O'Donnell et al., 2009). Although it inhabits relatively stable thermal regimes compared to tropical coral species, *L. pertusa* experiences episodic short-term temperature excursions of 2 to 5°C in the Gulf of Mexico (Mienis et al., 2007, 2012; Davies et al., 2010). Previous studies of individuals from the Gulf of Mexico population suggest that *L. pertusa* is able to tolerate short-term (24 h) temperature stress at 15°C without any noticeable effects on survivorship, but prolonged exposure (7 days) at 15°C induces significant mortality (Brooke et al., 2013). The present study agrees to an extent with these results, but the data here indicate that some of the individuals of *L. pertusa* examined are sensitive to prolonged (7 days) exposure to temperatures sustained at 10°C and greater. The presence of a significant effect of previous treatment in the temperature experiment suggests that carry-over effects from the acidification experiments may be influencing the corals' responses to temperature stress. A potential driver of this significant effect of previous treatment may be that all corals used in the 16°C treatment suffered complete mortality. However, all of the corals used in the 16°C treatment had not been used previously, suggesting that the significant effect of previous treatment was biased. In order to further resolve this, future experiments on *L. pertusa* could avoid using corals in multiple experiments. However, from a logistical perspective, this may be challenging, as deep-sea corals require considerable effort and resources to collect and maintain. Our data also suggest that some individuals have wider tolerances to temperature (**Figure 3B**), which may be linked to differential expression of the heat shock response pathways. Future research should also explore the molecular basis of this tolerance and its role in ocean warming, and if this tolerance is linked to specific genotypes of *L. pertusa*.

Like temperature, dissolved oxygen concentration also plays a significant role in controlling *L. pertusa* distribution (Dodds et al., 2007; Georgian et al., 2014). Previous work has explored the metabolic tolerance of *L. pertusa* to various oxygen concentrations, and found that *L. pertusa* is unable to maintain aerobic respiration at oxygen concentrations less than $3.26 \, \text{ml·l}^{-1}$ at 9°C (Dodds et al., 2007). However, this work was performed on samples from the Northeast Atlantic, where the mean local oxygen concentration was $6.10 \, \text{ml·l}^{-1}$. Oxygen concentrations ranging from 1.5 to $3.2 \, \text{ml·l}^{-1}$ have been reported from the Gulf of Mexico surrounding *L. pertusa* mounds (Schroeder, 2002; Davies et al., 2010; Georgian et al., 2014), which suggests that *L. pertusa* from the Gulf of Mexico may possess a lower oxygen threshold for aerobic respiration compared to North Atlantic individuals. The experimental results presented here show that long-term exposure (7 days) at sustained hypoxic conditions near $1.57 \, \text{ml·l}^{-1}$ results in complete mortality. This is despite the unavoidable concomitant increase in pH, which would make it energetically favorable for skeletal precipitation. Although in their environment, these exposures to low O_2 may not last long enough to inflict significant mortality. At the present time, the *L. pertusa* populations of the Gulf of Mexico are surviving on the edge of their dissolved oxygen niche. The frequency and duration of hypoxic conditions may increase in the future due to warming and also expansion of the seasonally oxygen-depleted surface layers of the Gulf of Mexico (Rabalais et al., 2002).

Our work here contributes to the growing body of experimental evidence of deep-sea species sensitivity to ocean acidification, warming, and deoxygenation. The observed responses to climate change-related stressors in *L. pertusa* from the Gulf of Mexico raise new questions related to its persistence in the Anthropocene. Future work should explore variation in response at the individual level, and if certain genotypes possess innate resilience to these stressors. Furthermore, the potential variability of environmental conditions within deep-sea coral habitats should be explored, as this could facilitate the potential for physiological plasticity in species' responses to warming, deoxygenation, and acidification. Such work will enable better management of these habitats as the impacts of global change continue to manifest across the ocean environment.

ACKNOWLEDGMENTS

We thank N. Remon and S. Campellone for help with animal husbandry and S. Georgian and A. Demopoulos for assistance at sea. We are also very grateful for the technical skills of the ROV *Jason* group and the captain and crew of the *Ronald H. Brown*. A special thanks goes to J. Brooks and TDI-Brooks International, Inc., for coordinating logistics for this project, including cruises. Additionally, we thank S. Ross and S. Brooke who coordinated cruises aboard the R/V *Seward Johnson* and R/V *Cape Hatteras* that contributed samples for the population genetics work, and M. Springmann who generated the microsatellite data. J. M Roberts and S. Hennige donated samples from their R/V *Poseidon* cruise in the eastern Atlantic Ocean. We thank two anonymous reviewers for helpful comments on the manuscript. This study was supported by several agencies and awards: the Bureau of Ocean Energy Management (BOEM) and the National Oceanic and Atmospheric Administration (NOAA) Office of Ocean Exploration and Research [contract M08PC20038 to T.D.I. Brooks International, subcontracted to Erik E. Cordes] and the National Science Foundation [OCE-1220478 to Erik E. Cordes]. Financial support for the genetics work was provided by the USGS Outer Continental Shelf Ecosystem Program, sponsored by BOEM. Jay J. Lunden received partial support from the National Science Foundation Bridge to the Doctorate fellowship program. Any use of trade, product, or firm names is for descriptive purposes only and does not imply endorsement by the U.S. government.

REFERENCES

Al-Horani, F. A., Al-Moghrabi, S. M., and de Beer, D. (2003). The mechanism of calcification and its relation to photosynthesis and respiration in the scleractinian coral *Galaxea fascicularis. Mar. Biol.* 142, 419–426. doi: 10.1007/s00227-002-0981-8

Barnett, T. P., Pierce, D. W., Achuta Rao, K. M., Gleckler, P. J., Santer, B. D., Gregory, J. M., et al. (2005). Penetration of human-induced warming into the world's oceans. *Science* 309, 284–287. doi: 10.1126/science.1112418

Barnett, T. P., Pierce, D. W., and Schnur, R. (2001). Detection of anthropogenic climate change in the world's oceans. *Science* 292, 270–274. doi: 10.1126/science.1058304

Brooke, S., and Ross, S. W. (2014). First observations of the cold-water coral *Lophelia pertusa* in mid-Atlantic canyons of the USA. *Deep-Sea Res. Pt. II.* 104, 245–251. doi: 10.1016/j.dsr2.2013.06.011

Brooke, S., Ross, S. W., Bane, J. M., Seim, H. E., and Young, C. M. (2013). Temperature tolerance of the deep-sea coral *Lophelia pertusa* from the southeastern United States. *Deep-Sea Res. Pt. II.* 92, 240–248. doi: 10.1016/j.dsr2.2012.12.001

Brooke, S., and Young, C. M. (2009). *In situ* measurement of survival and growth of *Lophelia pertusa* in the northern Gulf of Mexico. *Mar. Ecol. Prog. Ser.* 397, 153–161. doi: 10.3354/meps08344

Bryan, K. (1982). Poleward heat transport by the ocean: observations and models. *Annu. Rev. Earth Pl. Sci.* 10, 15–38. doi: 10.1146/annurev.ea.10.050182.000311

Caldeira, K., and Wickett, M. E. (2003). Anthropogenic carbon and ocean pH. *Nature* 425, 365. doi: 10.1038/425365a

Coffroth, M. A., and Lasker, H. R. (1998). Population structure of a clonal gorgonian coral: the interplay between clonal reproduction and disturbance. *Evolution* 52, 379–393. doi: 10.2307/2411075

Cooley, S. R., and Doney, S. C. (2009). Anticipating ocean acidification's economic consequences for commercial fisheries. *Environ. Res. Lett.* 4:024007. doi: 10.1088/1748-9326/4/2/024007

Cordes, E. E., McGinley, M., Podowski, E. L., Becker, E. L., Lessard-Pilon, S., Viada, S. T., et al. (2008). Coral communities of the deep Gulf of Mexico. *Deep-Sea Res. Pt. I.* 55, 777–787. doi: 10.1016/j.dsr.2008.03.005

Davies, A. J., Duineveld, G. C. A., Van Weering, T. C. E., Mienis, F., Quattrini, A. M., Seim, H. E., et al. (2010). Short-term environmental variability in cold-water coral habitat at Viosca Knoll, Gulf of Mexico. *Deep-Sea Res. Pt. I.* 57, 199–212. doi: 10.1016/j.dsr.2009.10.012

Davies, A. J., and Guinotte, J. M. (2011). Global habitat suitability for framework-forming cold-water corals. *PLoS ONE* 6:e18483. doi: 10.1371/journal.pone.0018483

Davies, P. S. (1989). Short-term growth measurements of corals using accurate buoyant weighing technique. *Mar. Biol.* 101, 389–395. doi: 10.1007/BF00428135

Dickson, A. G., Sabine, C. L., and Christian, J. R. (2007). *Guide to Best Practices for Ocean CO$_2$ Measurements*. PICES Special Publication.

Dodds, L. A., Roberts, J. M., Taylor, A. C., and Marubini, F. (2007). Metabolic tolerance of the cold-water coral *Lophelia pertusa* (Scleractinia) to temperature and dissolved oxygen change. *J. Exp. Mar. Biol. Ecol.* 349, 205–214. doi: 10.1016/j.jembe.2007.05.013

Dorey, N., Lancon, P., Thorndyke, M., and Dupont, S. (2013). Assessing physiological tipping point of sea urchin larvae exposed to a broad range of pH. *Glob. Change Biol.* 19, 3355–3367. doi: 10.1111/gcb.12276

Dufault, A. M., Cumbo, V. R., Fan, T.-Y., and Edmunds, P. J. (2012). Effects of diurnally oscillating pCO2 on the calcification and survival of coral recruits. *Proc. Biol. Sci.* 279, 2951–2958. doi: 10.1098/rspb.2011.2545

Earl, D. A. (2009). *Structure Harvester v. 0.56.3*. Available online at: http://taylor0.biology.ucla.edu/struct_harvest/

Evanno, G., Regnaut, S., and Goudet, J. (2005). Detecting the number of clusters of individuals using the software STRUCTURE: a simulation study. *Mol. Ecol.* 14, 2611–2620. doi: 10.1111/j.1365-294X.2005.02553.x

Falush, D., Stephens, M., and Pritchard, J. (2003). Inference of population structure using multilocus genotype data: linked loci and correlated allele frequencies. *Genetics* 164, 1567–1587.

Feder, M. E., and Hofmann, G. E. (1999). Heat-shock proteins, molecular chaperones, and the stress response: evolutionary and ecological physiology. *Annu. Rev. Physiol.* 61, 243–282. doi: 10.1146/annurev.physiol.61.1.243

Form, A. U., and Riebesell, U. (2012). Acclimation to ocean acidification during long-term CO$_2$ exposure in the cold-water coral *Lophelia pertusa*. *Glob. Change Biol.* 18, 843–853. doi: 10.1111/j.1365-2486.2011.02583.x

Frederiksen, R., Jensen, A., and Westerberg, H. (1992). The distribution of the scleractinian coral *Lophelia pertusa* around the Faroe Islands and the relation to internal tidal mixing. *Sarsia* 77, 157–171. doi: 10.1080/00364827.1992.10413502

Freiwald, A. (2002). "Reef-forming cold-water corals," in *Series Ocean Margin Systems*, eds G. Wefer, D. Billett, D. Hebbeln, B. B. Jorgensen, M. Schluter, and T. Van Weering (Berlin: Springer-Verlag), 365–385.

Freiwald, A., Beuck, L., Rüggeberg, A., Taviani, M., and Hebbeln, D. (2009). The white coral community in the central Mediterranean Sea revealed by ROV surveys. *Oceanography* 22, 36–52. doi: 10.5670/oceanog.2009.06

Freiwald, A., Fosså, J. H., Grehan, A., Koslow, T., and Roberts, J. M. (2004). *Cold-water Coral Reefs*. Cambridge, UK: UNEP-WCMC.

Gass, S. E., and Roberts, J. M. (2006). The occurrence of cold-water coral *Lophelia pertusa* (Scleractinia) on oil and gas platforms in the North Sea: colony growth, recruitment and environmental controls on distribution. *Mar. Pollut. Bull.* 52, 549–559. doi: 10.1016/j.marpolbul.2005.10.002

Georgian, S. E., Shedd, W., and Cordes, E. E. (2014). High resolution ecological niche modeling of the cold-water coral *Lophelia pertusa* in the Gulf of Mexico. *Mar. Ecol. Prog. Ser.* 506, 145–161. doi: 10.3354/meps10816

Guinotte, J. M., Orr, J., Cairns, S., Freiwald, A., Morgan, L., and George, R. (2006). Will human-induced changed in seawater chemistry alter the distribution of deep-sea scleractinian corals? *Front. Ecol. Environ.* 4, 141–146. doi: 10.1890/1540-9295(2006)004[0141:WHCISC]2.0.CO;2

Hall, M. M., and Bryden, H. L. (1982). Direct estimates and mechanisms of ocean heat transport. *Deep-Sea Res.* 29, 339–359. doi: 10.1016/0198-0149(82)90099-1

Hoegh-Guldberg, O., and Bruno, J. F. (2010). The impact of climate change on the world's marine ecosystems. *Science* 328, 1523–1528. doi: 10.1126/science.1189930

Hubisz, M. J., Falush, D., Stephens, M., and Pritchard, J. K. (2009). Inferring weak population structure with the assistance of sample group information. *Mol. Ecol. Resour.* 9, 1322–1332. doi: 10.1111/j.1755-0998.2009.02591.x

Hughes, R. N. (1989). *A Functional Biology of Clonal Animals*. New York, NY: Chapman and Hall.

Hunter, C. (1993). Genotypic variation and clonal structure in coral populations with different disturbance histories. *Evolution* 47, 1213–1228. doi: 10.2307/2409987

Huston, M. A. (1985). Patterns of species diversity on coral reefs. *Annu. Rev. Ecol. Evol. S.* 16, 149–177. doi: 10.1146/annurev.es.16.110185.001053

Iglesias-Rodriguez, M. D., Halloran, P. R., Rickaby, R. E. M., Hall, I. R., Colmenero-Hidalgo, E., Gittins, J. R., et al. (2008). Phytoplankton calcification in a high-CO$_2$ world. *Science* 320, 336–340. doi: 10.1126/science.1154122

IPCC. (2013). "Summary for Policymakers," in: *Climate Change 2013: The Physical Science Basis. Contribution of Working Group I to the Fifth Assessment Report of the Intergovernmental Panel on Climate Change*, eds T. F. Stocker, D. Qin, G.-K. Plattner, M. Tignor, S.K. Allen, J. Boschung, et al. Midgley (Cambridge, UK; New York, NY: Cambridge University Press), 1062–1068.

Jakobsson, M., and Rosenberg, N. A. (2007). CLUMPP: a cluster matching and permutation program for dealing with label switching and multimodality in analysis of population structure. *Bioinformatics* 23, 1801–1806. doi: 10.1093/bioinformatics/btm233

Jakobsson, M., Scholz, S. W., Scheet, P., Gibbs, J. R., Van Liere, J. M., Fung, H., et al. (2008). Genotype, haplotype and copy-number variation in worldwide human populations. *Nature* 451, 998–1003. doi: 10.1038/nature06742

Keeling, R., Körtzinger, A., and Gruber, N. (2010). Ocean deoxygenation in a warming world. *Annu. Rev. Mar. Sci.* 2, 199–229. doi: 10.1146/annurev.marine.010908.163855

Langdon, C., and Atkinson, M. J. (2005). Effect of elevated pCO$_2$ on photosynthesis and calcification of corals and interactions with seasonal change in temperature, irradiance and nutrient enrichment. *J. Geophys. Res.* 110, C09SC07. doi: 10.1029/2004JC002576

Langer, G., Nehrke, G., Probert, I., Ly, J., and Ziveri, P. (2009). Strain-specific responses of *Emiliania huxleyi* to changing seawater carbonate chemistry. *Biogeosciences* 6, 2637–2646. doi: 10.5194/bg-6-2637-2009

LeGoff-Vitry, M. C., Pybus, O. G., and Rogers, A. D. (2004). Genetic structure of the deep-sea coral *Lophelia pertusa* in the northeast Atlantic revealed by microsatellites and internal transcribed spacer sequences. *Mol. Ecol.* 13, 537–549. doi: 10.1046/j.1365-294X.2004.2079.x

Levitus, S. (2005). Warming of the world ocean, 1955-2003. *Geophys. Res. Lett.* 32, L02604. doi: 10.1029/2004GL021592

Lunden, J. J., Georgian, S. E., and Cordes, E. E. (2013). Aragonite saturation states at cold-water coral reefs structured by *Lophelia pertusa* in the northern Gulf of Mexico. *Limnol. Oceanogr.* 58, 354–362. doi: 10.4319/lo.2013.58.1.0354

Lunden, J. J., Turner, J. M., McNicholl, C. G., Glynn, C. K., and Cordes, E. E. (2014). Design, development, and implementation of recirculating aquaria for maintenance and experimentation of deep-sea corals and associated fauna. *Limnol. Oceanogr. Meth.* 12, 363–372. doi: 10.4319/lom.2014.12.363

Maier, C., Bils, F., Weinbauer, M. G., Watremez, P., Peck, M. A., and Gattuso, J.-P. (2013a). Respiration of Mediterranean cold-water corals is not affected by ocean acidification as projected for the end of the century. *Biogeosciences Discuss.* 10, 7617–7640. doi: 10.5194/bgd-10-7617-2013

Maier, C., Hegeman, J., Weinbauer, M. G., and Gattuso, J.-P. (2009). Calcification of the cold-water coral *Lophelia pertusa* under ambient and reduced pH. *Biogeosciences* 6, 1671–1680. doi: 10.5194/bg-6-1671-2009

Maier, C., Schubert, A., Berzunza Sánchez, M. M., Weinbauer, M. G., Watremez, P., and Gattuso, J.-P. (2013b). End of the century pCO$_2$ levels do not impact calcification in Mediterranean cold-water corals. *PLoS ONE* 8:e62655. doi: 10.1371/journal.pone.0062655

Maier, C., Watremez, P., Taviani, M., Weinbauer, M. G., and Gattuso, J.-P. (2012). Calcification rates and the effect of ocean acidification on Mediterranean cold-water corals. *Proc. Biol. Sci.* 279, 1716–1723. doi: 10.1098/rspb.2011.1763

Maina, J., McClanahan, T. R., Venus, V., Ateweberhan, M., and Madin, J. (2011). Global gradients of coral exposure to environmental stresses and implications for local management. *PLoS ONE* 6:e23064. doi: 10.1371/journal.pone.0023064

Mienis, F., De Stigter, H. C., White, M., Duineveld, G., De Haas, H., and Van Weering, T. C. E. (2007). Hydrodynamic controls on cold-water coral growth and carbonate-mound development at the SW and SW Rockall Trough Margin, NE Atlantic Ocean. *Deep-Sea Res. Pt. I.* 54, 1655–1674. doi: 10.1016/j.dsr.2007.05.013

Mienis, F., Duineveld, G. C. A., Davies, A. J., Ross, S. W., Seim, H., Bane, J., et al. (2012). The influence of near-bed hydrodynamic conditions on cold-water corals in the Viosca Knoll area, Gulf of Mexico. *Deep-Sea Res. Pt. I.* 60, 32–45. doi: 10.1016/j.dsr.2011.10.007

Morrison, C. L., Eackles, M. S., Johnson, R. L., and King, T. L. (2008). Characterization of 13 microsatellite loci for the deep-sea coral, Lophelia pertusa (Linnaeus 1758), from the western North Atlantic Ocean and Gulf of Mexico. *Mol. Ecol. Resour.* 8, 1037–1039. doi: 10.1111/j.1755-0998.2008.02147.x

Morrison, C. L., Ross, S. W., Nizinski, M. S., Brooke, S., Järnegren, J., Waller, R. G., et al. (2011). Genetic discontinuity among regional populations of *Lophelia pertusa* in the North Atlantic Ocean. *Conserv. Genet.* 12, 713–729. doi: 10.1007/s10592-010-0178-5

Mortensen, P. B. (2001). Aquarium observations on the deep-water coral *Lophelia pertusa* (L., 1758) (Scleractinia) and selected associated invertebrates. *Ophelia* 54, 83–104. doi: 10.1080/00785236.2001.10409457

Naumann, M. S., Orejas, C., and Ferrier-Pagès, C. (2014). Species-specific physiological response by the wold-water corals *Lophelia pertusa* and *Madrepora oculata* to variations within their natural temperature range. *Deep-Sea Res. Pt. II.* 99, 36–41. doi: 10.1016/j.dsr2.2013.05.025

Nemzer, B. V., and Dickson, A. G. (2005). The stability and reproducibility of Tris buffers in synthetic seawater. *Mar. Chem.* 96, 237–242. doi: 10.1016/j.marchem.2005.01.004

O'Donnell, M. J., Hammond, L. M., and Hofmann, G. E. (2009). Predicted impact of ocean acidification on a marine invertebrate: elevated CO_2 alters response to thermal stress in sea urchin larvae. *Mar. Biol.* 156, 439–446. doi: 10.1007/s00227-008-1097-6

Orejas, C., Ferrier-Pagès, C., Reynaud, S., Tsounis, G., Allemand, D., and Gili, J. M. (2011). Experimental comparison of skeletal growth rates in the cold-water coral *Madrepora oculata* Linnaeus, 1758 and three tropical scleractinian corals. *J. Exp. Mar. Biol. Ecol.* 405, 1–5. doi: 10.1016/j.jembe.2011.05.008

Orr, J. C., Fabry, V. J., Aumont, O., Bopp, L., Doney, S. C., Feely, R. A., et al. (2005). Anthropogenic ocean acidification over the twenty-first century and its impact on calcifying organisms. *Nature* 437, 681–686. doi: 10.1038/nature04095

Parker, L. M., Ross, P. M., O'Connor, W. A., Borysko, L., Raftos, D. A., and Pörtner, H.-O. (2012). Adult exposure influences offspring response to ocean acidification in oysters. *Glob. Change Biol.* 18, 82–92. doi: 10.1111/j.1365-2486.2011.02520.x

Peakall, R., and Smouse, P. E. (2006). GENALEX 6: genetic analysis in Excel. Population genetic software for teaching and Research. *Mol. Ecol. Notes* 6, 288–295. doi: 10.1111/j.1471-8286.2005.01155.x

Peakall, R., and Smouse, P. E. (2012). GenAlEx 6.5: genetic analysis in Excel. Population genetic software for teaching and research – an update. *Bioinformatics* 28, 2537–2539. doi: 10.1093/bioinformatics/bts460

Peck, L. (2011). Organisms and responses to environmental change. *Mar. Genomics* 4, 237–243. doi: 10.1016/j.margen.2011.07.001

Pierrot, D., Lewis, E., and Wallace, D. W. R. (2006). *MS Excel program developed for CO_2 system calculations. ORNL/CDIAC-105.* Carbon Dioxide Information Analysis Center, Oak Ridge National Laboratory, U.S. Department of Energy.

Pistevos, J. C. A., Calosi, P., Widdicombe, S., and Bishop, J. D. D. (2011). Will variation among genetic individuals influence species responses to global climate change? *Oikos* 120, 675–689. doi: 10.1111/j.1600-0706.2010.19470.x

Pritchard, J. K., Stephens, M., and Donnelly, P. (2000). Inference of population structure using multilocus genotype data. *Genetics* 155, 945–959.

Quattrini, A. M., Etnoyer, P. J., Doughty, C., English, L., Falco, R., Remon, N., et al. (2014). A phylogenetic approach to octocoral community structure in the deep Gulf of Mexico. *Deep-Sea Res. Pt. II.* 99, 92–102. doi: 10.1016/j.dsr2.2013.05.027

Rabalais, N. N., Turner, R. E., and Wiseman, W. J. Jr. (2002). Gulf of Mexico hypoxia, a.k.a. "the dead zone." *Annu. Rev. Ecol. Syst.* 33, 235–263. doi: 10.1146/annurev.ecolsys.33.010802.150513

Ries, J. B., Cohen, A. L., and McCorckle, D. C. (2010). A nonlinear calcification response to CO_2-induced ocean acidification by the coral *Oculina arbuscula*. *Coral Reefs* 29, 661–674. doi: 10.1007/s00338-010-0632-3

Ries, J. B., Cohen, A. L., and McCorkle, D. C. (2009). Marine calcifiers exhibit mixed responses to CO_2-induced ocean acidification. *Geology* 37, 1131–1134. doi: 10.1130/G30210A.1

Roberts, J. M., Wheeler, A. J., Freiwald, A., and Cairns, S. (2009). *Cold-Water Corals: The Biology and Geology of Deep-Sea Coral Habitats.* New York, NY: Cambridge University Press. doi: 10.1017/CBO9780511581588

Rosenberg, N. A. (2004). DISTRUCT: a program for the graphical display of population structure. *Mol. Ecol. Notes* 4, 137–138. doi: 10.1046/j.1471-8286.2003.00566.x

Schroeder, W. W. (2002). Observations of *Lophelia pertusa* and the surficial geology at a deep-water site in the northeastern Gulf of Mexico. *Hydrobiologia* 471, 29–33. doi: 10.1023/A:1016580632501

Selkoe, K. A., Watson, J. R., White, C., Horin, T. B., Iacchei, M., Mitarai, S., et al. (2010). Taking the chaos out of genetic patchiness: seascape genetics reveals ecological and oceanographic drivers of genetic patterns in three temperate reef species. *Mol. Ecol.* 19, 3708–3726. doi: 10.1111/j.1365-294X.2010.04658.x

Solomon, S., Qin, D., Manning, M., Chen, Z., Marquis, M., Averyt, K. B., et al. (eds.). (2007). "Climate change 2007: the physical science basis," in *Contribution of Working Group I to the Fourth Assessment Report of the Intergovernmental Panel on Climate Change.* Cambridge, UK: Cambridge University Press.

Solomon, S., Plattner, G.-K., Knutti, R., and Friedlingstein, P. (2009). Irreversible climate change due to carbon dioxide emissions. *Proc. Natl. Acad. Sci. U.S.A.* 106, 1704–1709. doi: 10.1073/pnas.0812721106

Stramma, L., Schmidtko, S., Levin, L. A., and Johnson, G. C. (2010). Ocean oxygen minima expansions and their biological impacts. *Deep-Sea Res. Pt. I.* 57, 587–595. doi: 10.1016/j.dsr.2010.01.005

Strømgren, T. (1971). Vertical and horizontal distribution of *Lophelia pertusa* (Linne) in Trondheimsfjorden on the west coast of Norway. *K Norske Vidensk Skr.* 6, 1–9.

Turley, C., Eby, M., Ridgwell, A. J., Schmidt, D. N., Findlay, H. S., Brownlee, C., et al. (2010). The societal challenge of ocean acidification. *Mar. Poll. Bull.* 60, 787–792. doi: 10.1016/j.marpolbul.2010.05.006

Tursi, A., Mastrototaro, F., Matarrese, A., Maiorano, P., and D'Onghia, G. (2004). Biodiversity of the white coral reefs in the Ionian Sea (Central Mediterranean). *Chem. Ecol.* 20, 107–116. doi: 10.1080/02757540310001629170

Waller, R. G., and Tyler, P. A. (2005). The reproductive biology of two deep-water, reef-building scleractinians from the NE Atlantic Ocean. *Coral Reefs* 24, 514–522. doi: 10.1007/s00338-005-0501-7

Walther, G.-R., Post, E., Convey, P., Menzel, A., Parmesan, C., Beebee, T. J. C., et al. (2002). Ecological responses to recent climate change. *Nature* 416, 389–395. doi: 10.1038/416389a

Weir, B. S., and Cockerham, C. C. (1984). Estimating F-statistics for the analysis of population structure. *Evolution* 38, 1358–1360. doi: 10.2307/2408641

Wisshak, M., Freiwald, A., Lundalv, T., and Gektidis, M. (2005). "The physical niche of the bathyal *Lophelia pertusa* in a non-bathyal setting: Environmental controls and palaeoecological implications," in *Cold-Water Corals and Ecosystems*, eds A. Freiwald and J. M. Roberts (New York, NY: Springer) 979–1001.

Zibrowius, H. (1980). Les Scléractiniaires de la Mediterranèe et de l'Atlantique nord-oriental. *Mèm. I. Océanogr, Monaco.* 11, 1–284.

Conflict of Interest Statement: The authors declare that the research was conducted in the absence of any commercial or financial relationships that could be construed as a potential conflict of interest.

15

Wave exposure as a predictor of benthic habitat distribution on high energy temperate reefs

Alex Rattray[1,2], Daniel Ierodiaconou[2]* and Tim Womersley[3,4]

[1] Dipartimento di Biologia, Università di Pisa, Pisa, Italy
[2] Centre for Integrative Ecology, School of Life and Environmental Sciences, Deakin University, Warrnambool, VIC, Australia
[3] DHI Water and Environment Pty Ltd., Perth, WA, Australia
[4] Water Technology Pty Ltd., Melbourne, VIC, Australia

Edited by:
*Christos Dimitrios Arvanitidis,
Hellenic Centre for Marine
Research, Greece*

Reviewed by:
*Ibon Galparsoro, AZTI-Tecnalia,
Spain
Vasilis D. Valavanis, Hellenic Centre
for Marine Research, Greece
Yiannis Issaris, Hellenic Centre for
Marine Research, Greece*

*Correspondence:
*Daniel Ierodiaconou, Centre for
Integrative Ecology, School of Life
and Environmental Sciences, Deakin
University, Princes Hwy.,
Warrnambool, VIC 3280, Australia
e-mail: iero@deakin.edu.au*

The new found ability to measure physical attributes of the marine environment at high resolution across broad spatial scales has driven the rapid evolution of benthic habitat mapping as a field in its own right. Improvement of the resolution and ecological validity of seafloor habitat distribution models has, for the most part, paralleled developments in new generations of acoustic survey tools such as multibeam echosounders. While sonar methods have been well demonstrated to provide useful proxies of the relatively static geophysical patterns that reflect distribution of benthic species and assemblages, the spatially and temporally variable influence of hydrodynamic energy on habitat distribution have been less well studied. Here we investigate the role of wave exposure on patterns of distribution of near-shore benthic habitats. A high resolution spectral wave model was developed for a 624 km^2 site along Cape Otway, a major coastal feature of western Victoria, Australia. Comparison of habitat classifications implemented using the Random Forests algorithm established that significantly more accurate estimations of habitat distribution were obtained by including a fine-scale numerical wave model, extended to the seabed using linear wave theory, than by using depth and seafloor morphology information alone. Variable importance measures and map interpretation indicated that the spatial variation in wave-induced bottom orbital velocity was most influential in discriminating habitat classes containing the canopy forming kelp *Ecklonia radiata*, a foundation kelp species that affects biodiversity and ecological functioning on shallow reefs across temperate Australasia. We demonstrate that hydrodynamic models reflecting key environmental drivers on wave-exposed coastlines are important in accurately defining distributions of benthic habitats. This study highlights the suitability of exposure measures for predictive habitat modeling on wave-exposed coastlines and provides a basis for continuing work relating patterns of biological distribution to remotely-sensed patterns of the physical environment.

Keywords: habitat mapping, multibeam sonar, remote sensing, hydrodynamic modeling, video survey, random forests

INTRODUCTION

A major difficulty faced by managers in developing policy and implementing measures to safeguard ecologically important areas of the oceans is the relative paucity of scientific information available to direct and inform such initiatives. In comparison to terrestrial ecosystems, spatial management of marine ecosystems has been constrained by the lack of high quality, spatially explicit data describing the basic patterns of their biophysical constituents. This is for the most part a function of the inherent difficulties and costs associated with data collection in the marine environment. As a result quantitative spatial information in marine ecosystems is typically sparse, localized and patchily distributed through space and time (Kostylev and Hannah, 2007; Foster et al., 2009).

The emergence of remotely sensed acoustic technologies coupled with the ability to collect seabed information with georeferenced towed camera systems, opens the possibility of surveying large areas of seafloor and producing high resolution maps of topography, subsurface structures, and benthic habitats (Rattray et al., 2009). Acoustic habitat mapping utilizes sonar-derived physical variables as proxies to describe the range of abiotic conditions (e.g., substrate type) and processes (e.g., light availability) that define the realized niche and subsequent distribution of benthic species and assemblages. Commonly, features used to predict the distribution of benthic assemblages are derived directly from topographic information and acoustic backscatter response. Thus, the role of wave exposure on habitat distribution is only indirectly considered through postulated associations with water depth and seafloor orientation (aspect). Wave energy, however, varies spatially and temporally, and is locally modified by factors such as coastline geometry and bottom topography. It is therefore unlikely in shallow coastal zones that depth and orientation of an area of the seafloor are fully indicative of structuring effects of exposure on biological assemblages, especially in areas which are known to experience gradients in wave activity.

The southern Australian coastline is one of the highest energy coastlines in the world (Hemer et al., 2008; Hughes and Heap, 2010). As a result, wave energy is arguably one of the primary variables influencing the morphology, community structure and spatial organization of benthic taxa in the region (Wernberg and Goldberg, 2008; Wernberg and Vanderklift, 2010). The effects of wave energy on the composition, functional morphology and distribution of species and assemblages have been documented in most areas of the shallow marine environment across a wide range of taxonomic groups. The hydrodynamic energy regime has been demonstrated as an important factor controlling the spatial distribution of macroalgae (Pedersen et al., 2012; Thomson et al., 2012), sessile invertebrates (Bell and Barnes, 2000; Chollett and Mumby, 2012), seagrasses (Fonseca and Bell, 1998; Turner et al., 1999), mollusks (Boulding et al., 1999; Pfaff et al., 2011) and fishes (Letourneur, 1996; Friedlander et al., 2003), and has been identified as a key indicator of species abundance and diversity (Denny, 2006).

Wave energy determines benthic habitat availability through a number of direct and indirect processes which exert effects on benthic organisms (Denny, 2006). Sessile benthic taxa are reliant on water circulation for delivery of nutrients and oxygen, timing and dispersal of larvae and propagules, and removal of waste. Hydrodynamic exposure is also an important agent of stress and disturbance through sediment flux processes, specifically abrasion, burial and limitation of light availability (Airoldi, 2003), or mechanical tearing or removal of sessile species from their places of attachment (Thomsen et al., 2004). On shallow rocky reefs dominated by canopy forming kelps, wave energy may also determine canopy size, morphology and spatial patchiness, influencing understory community composition through altering light availability, water motion and direct physical abrasion (Toohey et al., 2004).

There are relatively few studies that use a direct proxy of hydrodynamic exposure as a variable for predictive mapping. Quantitative estimation by cartographic fetch models or more complex mathematical simulations of sea state have been used to derive exposure/organism relationships and also to predict their distributional patterns (Bekkby et al., 2008). At the local scale, cartographic fetch models based on the distance from a given location over which wind waves are able to generate (i.e., distance to barrier) have also been used to quantify a metric of exposure often assigned to a fixed number of ordinal categories (Lindegarth and Gamfeldt, 2005). Fetch-based exposure models have been demonstrated to respond well in enclosed or semi-enclosed areas where coastal perturbations, inlets or islands are the principal mediators of local wave energy (Ekebom et al., 2003; Greenlaw et al., 2011), but are potentially less applicable to open coasts where submarine topography such as offshore banks or reefs are often the significant factors mediating fully-developed wave conditions from remote synoptic events (Chollett and Mumby, 2012). Numerical wave modeling approaches are commonly used in coastal engineering applications and are capable of incorporating the combined effects of complex seabed topography and coastlines as well as spatial variation in wave energy caused by shallow water processes such as refraction, diffraction, wave on wave interactions and energy dissipation due to white-capping and wave breaking. Their use in local-scale ecological studies however has not been widely reported (England et al., 2008). This is potentially due to the computational complexity and expert knowledge required for their implementation (Hill et al., 2010).

While sonar methods have been well demonstrated to provide useful proxies of the relatively static geophysical patterns that reflect distribution of benthic species and assemblages, the spatially and temporally variable influence of hydrodynamic energy on benthic habitat distribution has been less well studied. Given the strong associations between marine taxa and their hydrodynamic environment, it is expected that measures of wave energy will also provide useful information for benthic habitat characterization and mapping. The principal hypothesis under investigation in this study is that a surrogate measure of wave energy can be used to improve the predictive accuracy of acoustic mapping techniques for sublittoral benthic habitat characterization. We investigate the effectiveness of a proxy for wave energy by comparing classified maps and measures of variable importance derived using predictors of depth and seafloor morphology, to those derived inclusive of a model of wave-induced orbital velocity.

MATERIALS AND METHODS
STUDY AREA

The study was conducted on the Otway coast of Victoria, southeastern Australia. The site extends ~95 km from east to west around Cape Otway, the prominent coastal feature of western Victoria (**Figure 1**). Acoustic data for the site were acquired in four survey blocks of approximately equal area using a Reson Seabat 101 multibeam echosounder (MBES) operating at a frequency of 240 kHz aboard the Australian Maritime College vessel R.V. Bluefin. Block 1 was surveyed in November 2005 and blocks 2–4 in November 2007. Together, the four survey blocks encompass 624 km^2 of seafloor ranging in depth from 8 to 79 m. Large sandy embayments characterize the site with topographically complex rocky reef systems extending offshore from major headlands. Areas of shallow reef (10–30 m) were populated by diverse assemblages of macroalgae which are characterized by the canopy forming kelps *Phyllospora comosa* and *Ecklonia radiata*, while deeper reefs were populated by diverse communities of sponges and other sessile invertebrates.

The wave climate at the site, like much of the continental margin of southern Australia, is largely dominated by swell waves propagating from west to east moving low pressure systems in the Southern Ocean (Hemer et al., 2008). The majority of Australia's southern shelf is subject to persistent high energy swells of above 3.5 m 30–50% of the time (Porter-Smith et al., 2004) and annual return significant wave heights of up to 8.7 m (Harris and Hughes, 2012). The orientation of Cape Otway to prevailing swells originating from the south-west quadrant causes a gradient of wave energy across the site from highly exposed on the western side to moderately exposed in the east.

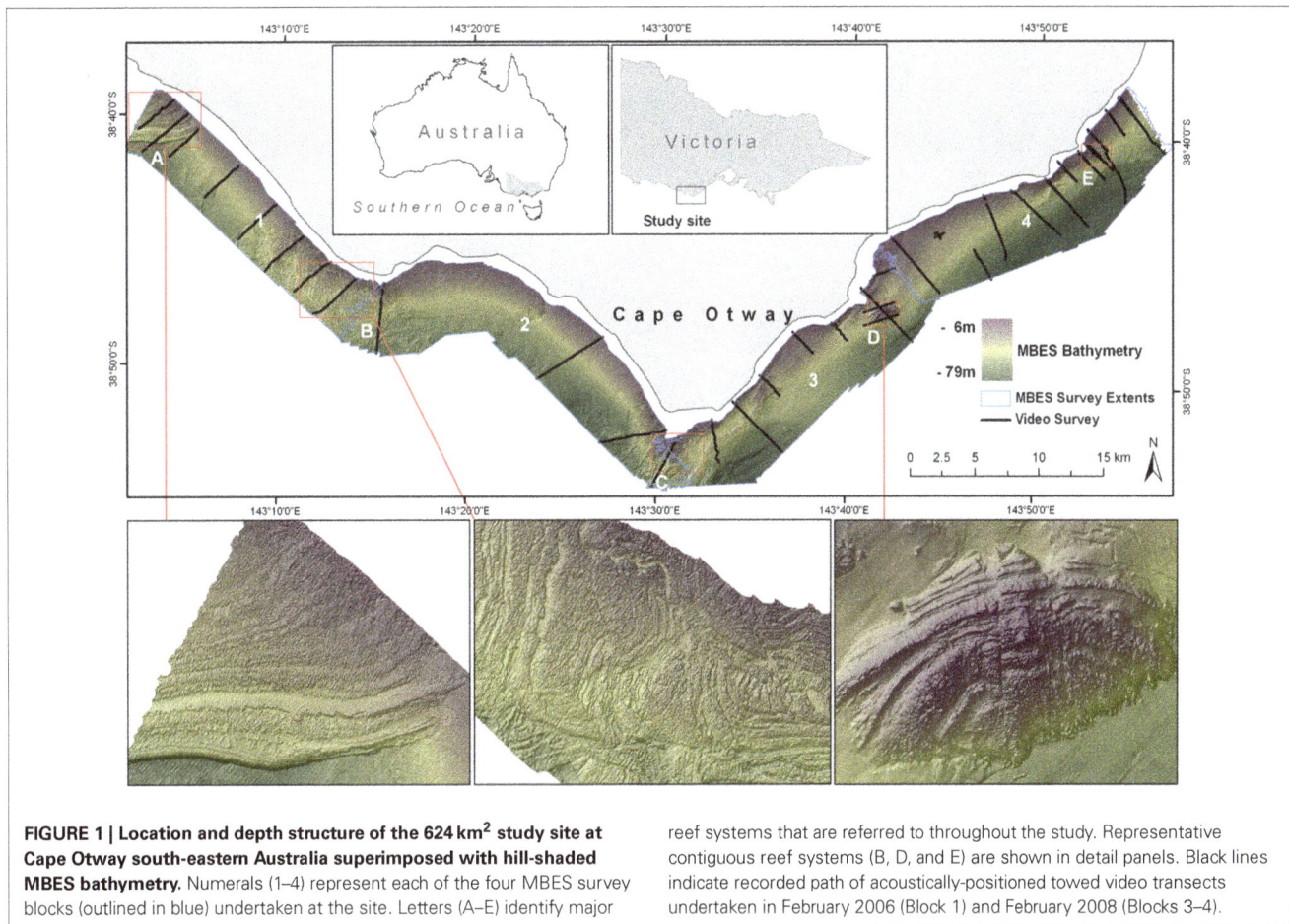

FIGURE 1 | Location and depth structure of the 624 km² study site at Cape Otway south-eastern Australia superimposed with hill-shaded MBES bathymetry. Numerals (1–4) represent each of the four MBES survey blocks (outlined in blue) undertaken at the site. Letters (A–E) identify major reef systems that are referred to throughout the study. Representative contiguous reef systems (B, D, and E) are shown in detail panels. Black lines indicate recorded path of acoustically-positioned towed video transects undertaken in February 2006 (Block 1) and February 2008 (Blocks 3–4).

MBES DATA ACQUISITION AND PROCESSING

Prior to each survey, calibration offsets for pitch, roll, yaw and latency were applied after conducting a detailed patch test. Daily sound velocity profiles were collected at the deepest area of the site during survey period to correct for local variations in sound velocity through the water column during processing. Positioning was accomplished using a real-time differential GPS integrated with a positioning and orientation system for marine vessels (POS MV) for dynamic heave, pitch, roll and yaw corrections ($\pm$ 0.1° accuracy). Navigation, data logging, real-time quality control and display were carried out using Starfix suite 7.1 (Fugro proprietary software). The sounding data were edited on board ship and corrections for tides, sound velocity, vessel draft, settlement, squat and relative position of the transducer head were applied. The raw xyz data were then used to produce a bathymetric grid at 2.5 m horizontal resolution and a range resolution of $\pm$12.5 mm. Backscatter values were corrected for gain and time varied gain using the University of New Brunswick (UNB1) algorithm (Starfix suite 7.1). Backscatter processing also corrected for transmission loss, the actual area of ensonification on the bathymetric surface, source level, and transmit and receive beam patterns. Additionally backscatter was corrected for seafloor bathymetric slope derived from the MBES bathymetry dataset. This resulted in normalized corrected grid (2.5 m resolution) representing relative backscatter intensity (dB).

Processed bathymetry and backscatter grids from each of the 4 survey blocks were combined at their highest resolution of 2.5 m. Edges between each of the survey blocks were normalized whereby overlapping values at a distance of 50 pixels (250 m on ground) from the edge of each block were averaged using a linear ramping technique. In order to minimize misregistration error between MBES products and *in situ* video observations, bathymetry and backscatter images for the entire site were resampled to a resolution of 5 m cell size before further processing.

A suite of environmental data was derived from the MBES datasets using a variety of neighborhood based topographic and spectral methods (see Rattray et al., 2013 for further information regarding derivative products). Prior to analysis, multicollinearity of derivative variables was assessed using a step-wise procedure where at each iteration the variable with the greatest variance inflation factor (VIF) was removed until remaining covariates displayed VIF values less than 10. MBES bathymetry, backscatter and retained derivatives were geographically overlaid to form an image stack of 12 predictor variables (**Table 1**). Further to the previously described set of MBES-derived predictor variables, a model representing energy exposure at the seabed was developed.

Table 1 | Environmental predictor variables used to inform the Random Forests models.

Code	Variable	Analysis Scale (m)	Resolution (m)
Bath	Bathymetry (m)	–	5
Back	Backscatter intensity (dB)	–	5
Slope	Slope (degrees)	3×3	5
Comp	Complexity	3×3	5
Maxc	Maximum Curvature	3×3	5
East	Aspect (Eastness)	3×3	5
North	Aspect (Northness)	3×3	5
Rug	Rugosity	3×3	5
HSlr	HSI_R (Red band)	High pass 11×11, Low pass 5×5	5
HSlg	HSI_G (Green band)	High pass 11×11, Low pass 5×5	5
HSlb	HSI_B (Blue band)	High pass 11×11, Low pass 5×5	5
BPI	Benthic Position Index	Inner radius 10, Outer radius 50	5
Maxu	Maximum Orbital Velocity - u_{max} (m.s^{-1})	–	60

All predictors except maximum orbital velocity (u_{max}) were derived from MBES bathymetry and backscatter intensity.

WAVE ENERGY MODEL

A fine-scale (60 m cell size) estimation of wave-induced orbital velocities at the seabed, used here as a surrogate for wave exposure, was created using a 3 step process:

(1) Results of a global wave hindcast model were downscaled to a regional scale (Victorian coastline) spectral wave model.
(2) A Detailed spectral wave model of the Otway coastline (study area) was created by incorporating local bathymetric variation (MBES and LIDAR derived). The wave boundaries for the detailed spectral wave model were downscaled from the regional wave model based on 1 year of representative annual wave conditions derived from a longer term wave climate assessment undertaken with the regional wave model.
(3) Wave-induced orbital velocities were transferred to the seabed by applying linear wave theory to surface spectral wave conditions.

Regional-scale model parameterization

Numerical wave modeling was accomplished using a MIKE 21 spectral wave (SW) model developed by Water Technology Pty Ltd using the DHI MIKE software suite (DHI, 2012[1]) applied to a bathymetry mesh generated from the LIDAR / Multibeam mosaic and boundary depths derived from the Geoscience Australia 2009 bathymetry grid (0.025°) (Whiteway, 2009). MIKE 21 SW is a 3rd generation spectral wind-wave model capable of simulating wave growth by action of wind, non-linear wave-wave interaction, dissipation by white-capping, dissipation by wave breaking, dissipation due to bottom friction, refraction due to depth variations, and wave-current interaction. The model domain incorporated the western and eastern coastlines of Victoria, Tasmania and adjacent areas of continental shelf including Bass Strait (**Figure 2**). Long-term directional distribution and size of significant wave heights from the Wavewatch 3 model between the years 2000 and 2010 were obtained for the region. We compared the directional distribution and magnitude of significant wave heights for

each year against the 10-year average, and selected the year that displayed the most similarities to the long-term record. Annual variability in significant wave heights and mean wave direction was generally low across the 10-year period. The year 2000 was selected as it contained the fewest extreme values (outliers) with respect to the 10-year average (**Figure 3**). Global hindcast results (10 m u and v wind velocity) from the National Oceanic and Atmospheric Administration (NOAA) Wavewatch 3 model were extracted and linearly interpolated (0.25° spatial, 3-hourly temporal) to provide boundary inputs for the regional scale spectral wave model. Spatially and temporally varying open wave results from the NOAA model provided wave boundary conditions along the western, southern and eastern model boundaries.

The regional spectral wave model was calibrated and validated against measured wave buoy data from Cape Sorell on the west coast of Tasmania (42° 7.2′S 145°E) and Point Lonsdale, south west of Melbourne (38° 18.2′S 144° 34.2′E) for the year 2000. Comparative agreement of hindcast wave conditions (significant wave height and peak period) to measured data was considered appropriate to use this model to assess wave climate along the Victorian coastline.

Local-scale model parameterization

A site-specific spectral wave model was generated for the waters around Cape Otway with western, southern and eastern boundary conditions provided by the regional scale model (**Figure 4**). A spectral wave hindcast was generated using a combination of the 0.0025° bathymetry grid, local MBES bathymetry (5 m) and bathymetric LIDAR (5 m) (Zavalas et al., 2014) to provide depth attenuation inputs in order to accurately propagate waves to uppermost extent of the sublittoral zone.

Modeled wave conditions corresponding to significant wave height and spectral peak period for the year were used to calculate a spatially explicit estimate of maximum instantaneous bottom orbital velocity (u_{max}), used here as a surrogate for exposure to wave-induced energy. Linear wave theory was then used to predict the horizontal component of the wave orbital velocity (u_o) at a particular area on the seabed for small-amplitude, monochromatic waves as follows:

---

[1] Available online at: http://www.dhigroup.com/ [Accessed 20/08 2012].

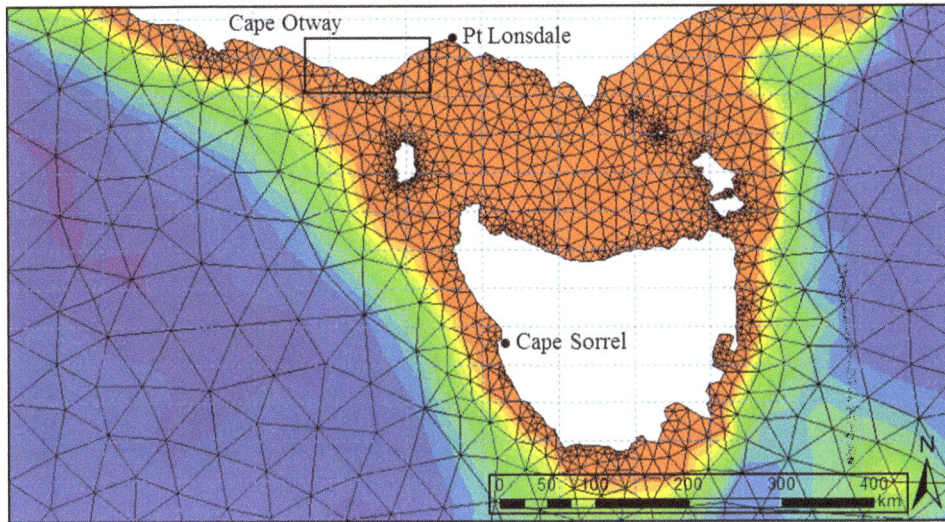

FIGURE 2 | Domain of the regional spectral wave model incorporating the eastern and western coastlines of Victoria, Tasmania and Bass Strait. The triangular irregular network used to inform the numerical wave model was created from the 0.0025° Australian national bathymetry grid. Boundary inputs were obtained from the NOAA Wavewatch III global hindcast model. Wave buoy locations used to calibrate the model are located at Point Lonsdale (Victoria) and Cape Sorell (Tasmania). Extents of study area (**Figure 1**) are shown in box.

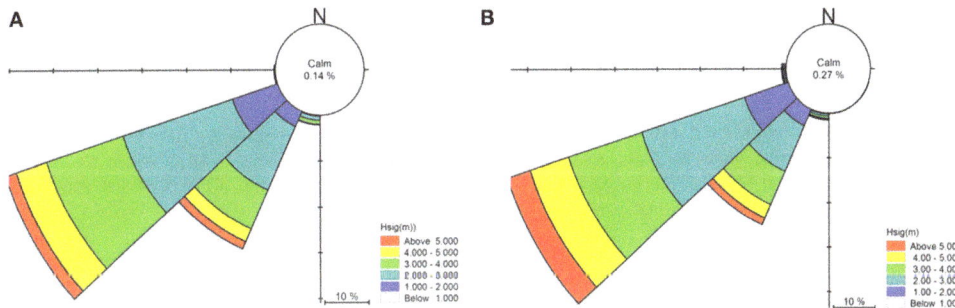

FIGURE 3 | Summary of significant wave heights (Hsig), direction and percentage occurrence for Cape Otway showing prevailing swell conditions for (A) the year 2000 and (B) the 10-year average (2000–2010).

$$u_o = \left[\frac{H\pi}{T \sinh(kd)} \right] \cdot \cos(kx - wt)$$

Where H = wave height (m), T = wave period (s), d = water depth (m), k = wave number, w = radian frequency. As u_o varies sinusoidally through a wave period, the maximum velocity u_{max} occurs when $\cos(kx - wt) = 1$. Instantaneous maximum seabed orbital velocity was calculated for the entire study site at a resolution of 60×60 m and subsequently resampled to a 5×5 m grid to match the resolution of the other physical predictor layers. While resampling did not alter the resolution of the dataset it rendered it compatible with the remaining grids for further processing.

Biological observation data

Observational data were collected using acoustically located towed video in February 2006 (MBES survey block 1) and February 2008 (MBES survey blocks 2–4). The towed video platform was maintained at ~1 m from the seabed by a shipboard operator viewing a real-time video feed via an umbilical control and data cable. An Ultra Short Base Line (USBL) transponder attached to the video unit allowed 3-dimensional positioning of the video unit relative to the vessel's dGPS antenna which was located directly above the pole mount housing the USBL transceiver. Angular rates of roll, pitch and azimuth (±0.1°) at the dGPS antenna were measured and corrected using a KVH motion sensor. A total of 35 video transects covering ~129 linear kilometers of seabed were used to capture the range of depths, topographic and textural diversity at the site determined by visual examination of the MBES bathymetry and backscatter intensity products. Video frames were individually reviewed and assigned to 4 benthic habitat classes (**Table 2**).

Habitat classification

The Random Forests (RF) classification algorithm (Breiman, 2001) was used to quantify relationships between environmental

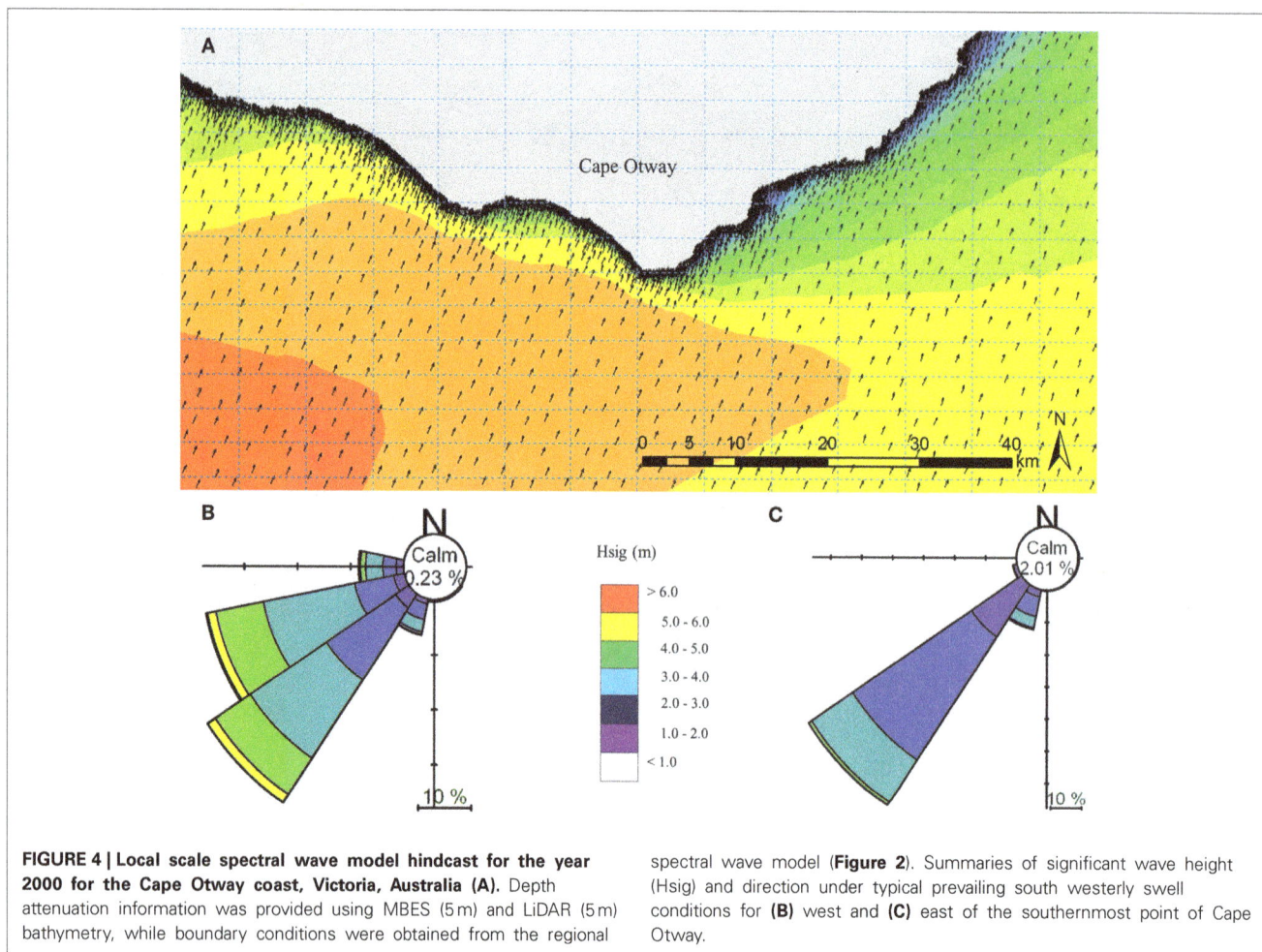

FIGURE 4 | Local scale spectral wave model hindcast for the year 2000 for the Cape Otway coast, Victoria, Australia (A). Depth attenuation information was provided using MBES (5 m) and LiDAR (5 m) bathymetry, while boundary conditions were obtained from the regional spectral wave model (**Figure 2**). Summaries of significant wave height (Hsig) and direction under typical prevailing south westerly swell conditions for **(B)** west and **(C)** east of the southernmost point of Cape Otway.

Table 2 | Summary of the four category classification scheme used in the study.

Habitat Code	Description of habitat attributes
ALGDOM	Dominant canopy forming macroalgal species—small patches of the kelp *Phyllospora comosa* and occasional *Sargassum* and *Cystophora* spp. in the bathymetric highs giving way to dense canopies of the common kelp *Ecklonia radiata* on deeper reefs with sparse to medium understorey of mixed red algae
ALG/INV	Mixed class of generally massive and encrusting sponge forms in a mosaic of patches under a thinning canopy of *E. radiata* with mixed red algal understorey
INVDOM	Dense sponge dominated invertebrate communities displaying high morphological diversity on high profile solid reef to small globular and pedunculate sponges on sand swept pavement reef and in dune troughs
SED	Unconsolidated sandy sediments—Inshore fine sandy sediments with low morphological complexity to coarse shelly sand in irregular dune formations offshore—wavelengths to 30 m

The four habitat classes used in the study are ALDOM, Algae dominant; ALG/INV, Mixed algae and sessile invertebrates; INVDOM, Sessile invertebrates dominant; SED, Unconsolidated sandy sediments.

data layers and video observations. The RF algorithm uses bootstrap samples of the training data and randomly selected subsets of available predictor variables to grow multiple classification trees. At each bootstrap iteration of the RF process the resultant tree is used to predict those data not included in the training process ("out of bag" or OOB observations) and calculate a misclassification rate. Probabilities of membership for the various classes are estimated by the proportions of OOB predictions in each class (Cutler et al., 2007). Each tree provides a unit vote for the most popular class at each input instance and the final classification label is determined by a majority vote of all trees in the ensemble.

Trees are left unpruned (i.e., fully fitted to the training data) in order to diminish potential bias introduced by any stopping rules. The algorithm yields an ensemble that can achieve both low bias and low variance (from averaging over a large ensemble of low-bias, high-variance but low correlation trees).

In this study the RF procedure was applied using a MATLAB implementation (Jaiantilal, 2009) of the code proposed by Breiman and Cutler (available online at http://www.stat.berkeley.edu/users/breiman/). The number of decision trees (ntree) was specified at 500 and variables selected from the pool of predictor variables for splitting at each node (m) was the square root of the number of available predictors, a value which has been commonly used in other implementations of the routine (Breiman, 2001; Cutler et al., 2007). Classification rules and importance measures were obtained from two separate implementations of the RF procedure. The first model included 12 predictor variables derived from and including the primary bathymetry and backscatter products (hereafter referred to as the MBES model). The procedure was run again with the addition of a grid layer representing annual maximum orbital velocity at the seabed (hereafter referred to as the wave energy model). The performance of the RF models was evaluated by comparing each one against a subset (30%) of video observation data that were withheld from the modeling process. Global accuracy of each model was established using confusion matrices (Overall accuracy and κ-statistic), similarly class specific accuracy was derived using metrics of user's and producer's accuracies.

Importance indices from each implementation of RF were obtained by randomly permuting the values for each input variable in the classification in the OOB samples for each tree. Decrease in accuracy caused by effectively removing a particular feature from a tree denotes its relevance to the classification accuracy of that tree. Changes in accuracy as a result of permutation were averaged across all trees in the forest and used to calculate a relative measure of variable importance (permutation importance measure) based on mean decrease in accuracy for each feature used in the classification across all classes.

RESULTS

A model representing maximum bottom orbital velocity (u_{max}) was created using inputs from a global wave model attenuated by a bathymetric surface composed of coarse-scale (~270 m) regional bathymetry and then fine-scale (5 m) local bathymetry. Values of u_{max} ranged from 0.5 to 1.36 m/s (**Figure 5**). The spatial pattern of bottom orbital velocities reflects the bathymetry and orientation to surface wave conditions which arrive predominantly from the south-west quadrant. As a result, highly energetic hydrodynamic conditions at the seabed are evident in the western half of the site reducing to moderate conditions in the eastern portion of the site which is largely sheltered from prevailing wave conditions by Cape Otway.

Cross-validated classification accuracy metrics corresponded well with those obtained from internal validation using the OOB data. The overall accuracy of the model that considered both MBES derived information and wave exposure was found to be higher (93%) than the model considering only the MBES derived predictors (88%). Accuracy as defined by κ was higher for the exposure classification (0.87) than the MBES classification (0.77). A pairwise test for significance of the κ statistic

for each error matrix (Congalton and Green, 2009) revealed a significant difference between the two error matrices (z = 13.3) with the exposure model performing significantly better than the MBES model. User's and producer's accuracies for each habitat class were universally higher for the exposure model. Increase in accuracy was most evident for the ALG/INV class which was commonly misclassified as either ALGDOM or INVDOM in the MBES classification. Producer's accuracy increased from 47 to 76% and user's accuracy increased from 68 to 82% in this class with the addition of the exposure layer to the classification.

Further accuracy assessment was done to assess the relative influence of each set of predictor variables between the sheltered eastern side of the site and the more exposed west. Similar patterns of improved accuracy with the addition of the wave energy variable were observed in all cases. Accuracy gains in the east of the site, however, were comparatively small and non-significant (z = 1.82) (**Table 3**). The greatest increases in overall accuracy and corresponding κ-values were observed in the west of the site. Overall accuracy increased from 84 to 93%, and κ was significantly higher (z = 13.84), increasing from 0.75 to 0.89. Improvements in model accuracy corresponded largely with better discrimination of the ALGDOM and ALG/INV classes in the wave energy model. Producer's accuracy increased from 78% to 93% and 42% to 69% for each class respectively.

Variables identified as most important over all classes for the MBES classification in order of decreasing importance were bathymetry, rugosity, the backscatter derivative HSIr and backscatter intensity (**Figure 6**). These predictors were also most important in varying degrees to the discrimination of individual habitat classes except for the ALG/INV class which was not well resolved by backscatter intensity. Maximum curvature, the variables representing aspect (northness and eastness) and Benthic Position Index (BPI) were the least important predictors across all habitat classes.

The introduction of the wave energy variable to the classification did not appreciably change the relative patterns of contribution of the MBES variables to classification accuracy. The wave energy proxy u_{max} was identified as an important feature (second only to bathymetry) across all habitat classes except for the ALG/INV class where it was of primary importance to the discrimination of that class from all others. The relationship between depth, wave energy and habitat categories east and west of Cape Otway is evident in **Figure 7**. Habitat classes are particularly well partitioned along the wave energy axis into observations made west and east of Cape Otway. Observations occurring west of Cape Otway display higher separability between classes, again along the wave energy axis, than those east of Cape Otway which overlap along the depth axis.

Decision rules derived from the two RF classifications were executed over the full extents of their respective sets of predictor variables to create full coverage habitat maps of the site. Class coverages in the MBES classification were lower for the ALGDOM class (8.7%) and the SED class (3.5%) (**Table 4**) and higher for the INVDOM class (8.8%). Most notably, the area covered by the ALG/INV class was 33.8% greater in the wave energy classification than the MBES classification.

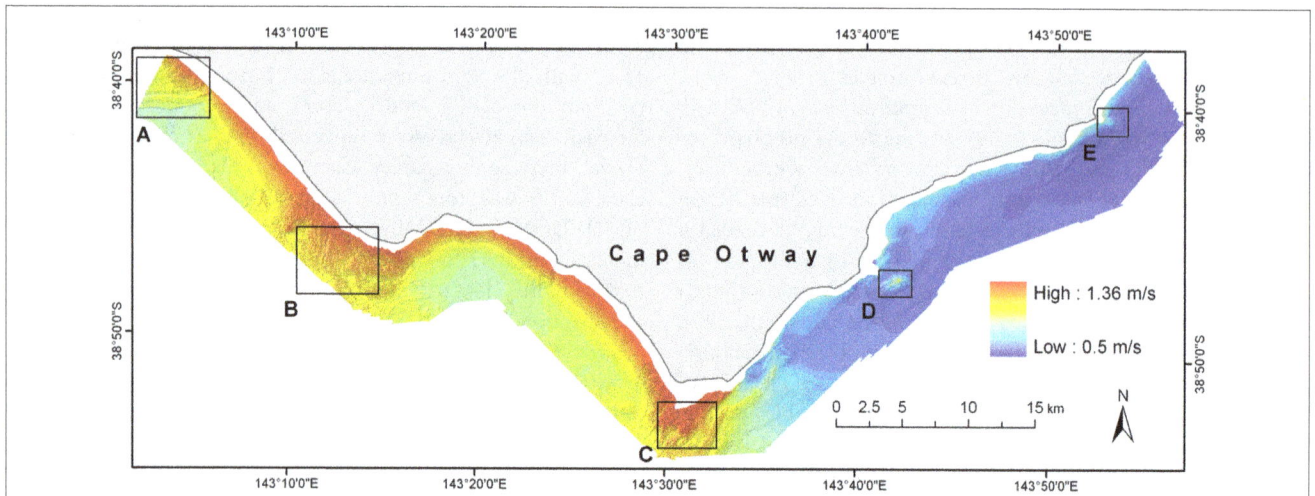

FIGURE 5 | Distribution of modeled maximum orbital velocity (u_max) values across the Cape Otway study site for the year 2000. Maximum bottom orbital velocities were calculated by extending outputs of the local scale numerical wave model to the seabed using linear wave theory. Boxes (A–E) delineate major reef systems at the site and are analogous to those shown in **Figure 1**.

Table 3 | Confusion matrices for the classified images derived from the MBES and wave energy models models east and west of Cape Otway.

	Reference data					%Producer's	%User's
	ALGDOM	**ALG/INV**	**INVDOM**	**SED**	**Total**	**Accuracy**	**Accuracy**
MBES-EAST (OVERALL ACCURACY = 90%; κ = 0.76)							
ALGDOM	*265*	55	27	37	384	92	69
ALG/INV	8	*103*	44	8	163	57	63
INVDOM	7	13	*885*	169	1074	72	82
SED	7	8	268	*4664*	4947	95	94
Total	287	179	1224	4878	6568		
MBES-WEST (OVERALL ACCURACY = 84%; κ = 0.75)							
ALGDOM	*329*	58	7	2	396	78	83
ALG/INV	48	*148*	7	1	204	42	72
INVDOM	36	143	*1425*	104	1708	83	83
SED	5	3	270	*1641*	1919	93	85
Total	418	352	1709	1748	4227		
WAVE ENERGY-EAST (OVERALL ACCURACY = 93%; κ = 0.79)							
ALGDOM	*256*	20	7	18	301	89	85
ALG/INV	11	*124*	24	9	168	69	73
INVDOM	9	22	*973*	125	1129	79	86
SED	11	13	220	*4726*	4970	96	95
Total	287	179	1224	4878	6568		
WAVE ENERGY-WEST (OVERALL ACCURACY = 93%; κ = 0.89)							
ALGDOM	*389*	30	9	1	429	93	90
ALG/INV	21	*280*	18	5	324	79	86
INVDOM	7	42	*1630*	109	1788	95	91
SED	1	0	52	*1633*	1686	93	96
Total	418	352	1709	1748	4227		

Each column corresponds to the ground reference pixels used for accuracy assessment for a single class. The values in the columns indicate the number of those ground observation pixels classified into each class, while the values on the main diagonal (italicized) indicate agreement between validation data and classified maps.

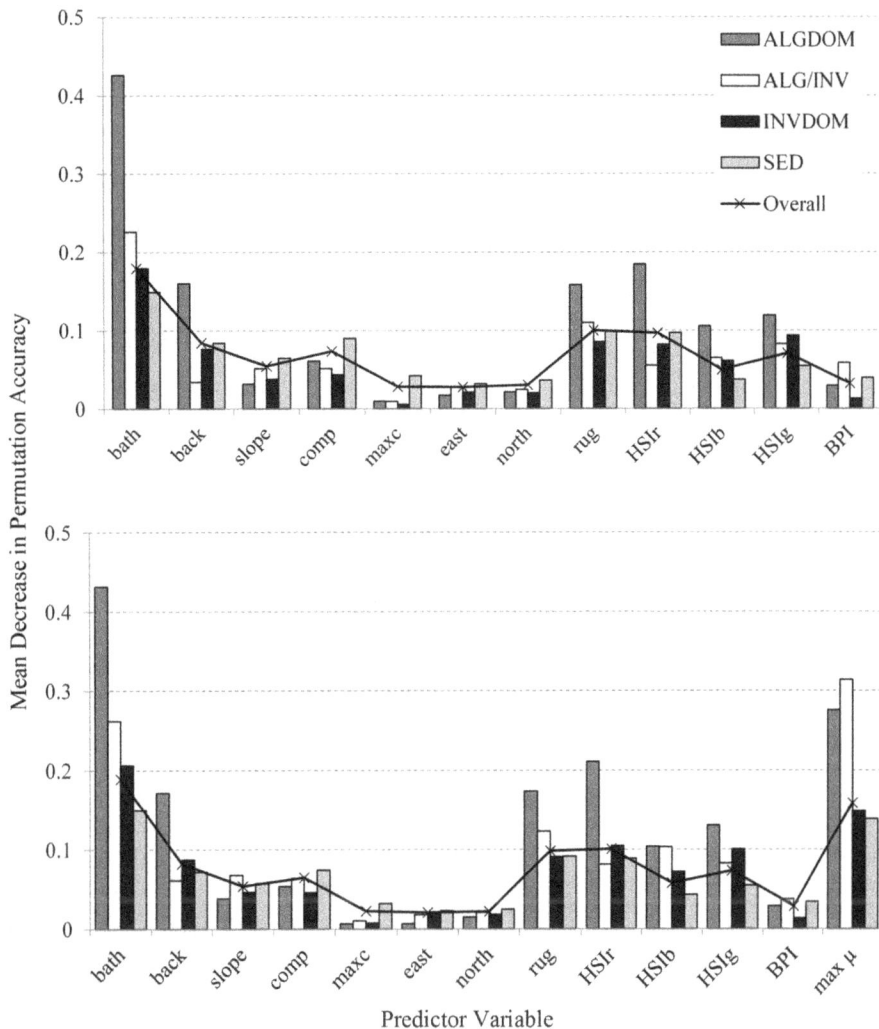

FIGURE 6 | Variable importance assessed by mean decrease in permutation accuracy obtained from the Random Forests classification (y-axis). Mean decrease in permutation accuracy is calculated using the difference between the misclassification rate of the out of bag (OOB) data, and the misclassification rate if values of a given variable are randomly permuted for the OOB observations and passed down the tree to create new predictions. Results from the model incorporating only MBES predictor variables are shown in the top graph, while those incorporating both MBES predictors and modeled maximum bottom orbital velocity are shown at bottom. Variable codes on the x-axis are detailed in **Table 2**.

Major contiguous reef systems at the site, (identified in **Figures 1, 4**), from the exposed western end of the site (Reef A) through to its more sheltered eastern extent (Reef E) showed a clear trend in the zonation of benthic habitat types achieved by each of the models in the study. In the exposed west of the site the MBES classification predicted a zone of change (ALG/INV) between the ALGDOM and INVDOM classes in accordance with other areas of reef at similar depths, although there are no records of that habitat in the ground observations (**Figure 8**). The wave energy classification however, showed an obvious delineation between macroalgal dominated reef and invertebrate dominated reef with the ALGDOM class extending to ~34 m, notably deeper than at any other area of the site.

Reef systems depicted in insets B and C showed an opposing trend. While reef coverage of the ALGDOM class appear very similar, predictions of the ALG/INV class by the wave energy model showed it to cover a considerably more extensive area and extend to greater depths (ca.40 m) than the MBES model (ca.32 m). It also appears that these areas of the site may be the principal source of differences in estimated area of the ALG/INV class between the two classifications.

By contrast, there appear to be only relatively small differences between the two habitat classifications on the sheltered eastern side of the site (Reefs D and E). Both classifications visually show a similar pattern of habitat distributions.

DISCUSSION

This study has demonstrated that a reef habitat classification model incorporating modeled wave-induced orbital velocity performs significantly better than one incorporating indirect proxies derived from depth and seafloor morphology in describing patterns of benthic habitat distribution at a wave-exposed site in

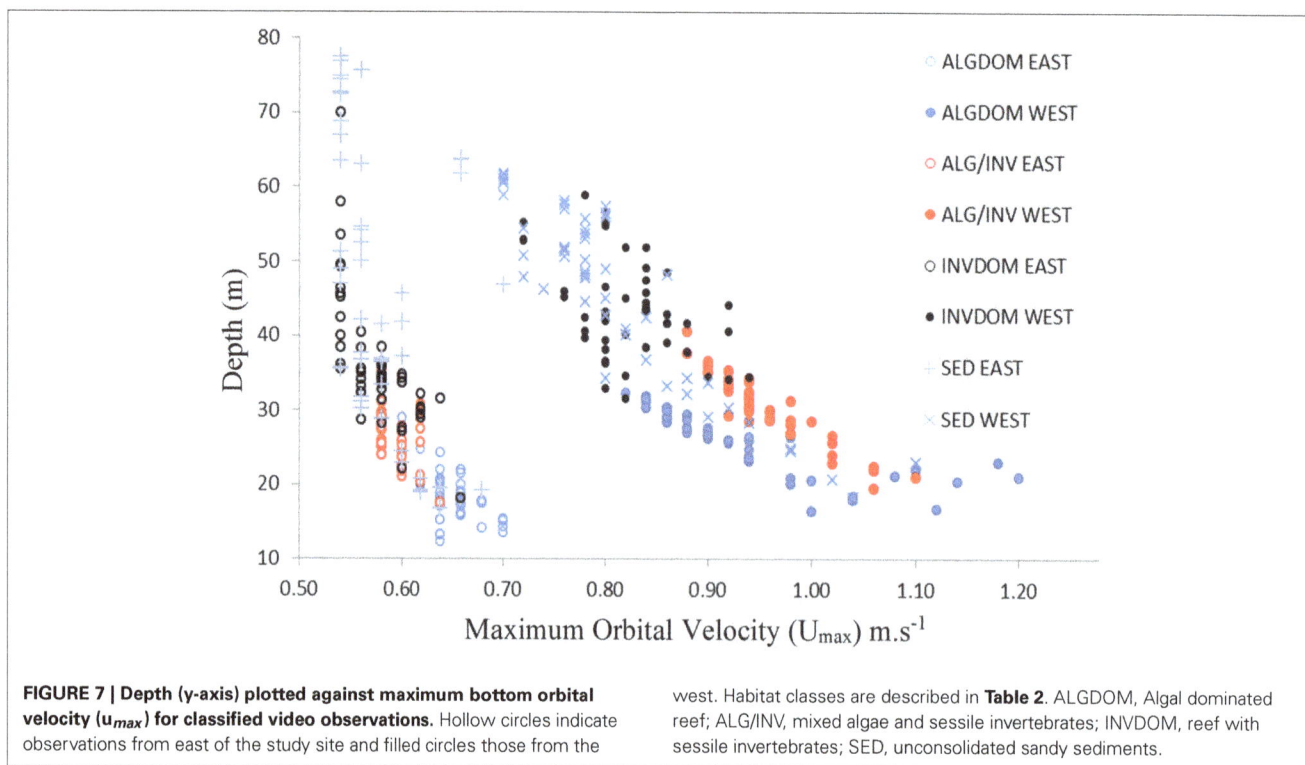

FIGURE 7 | Depth (y-axis) plotted against maximum bottom orbital velocity (u_{max}) for classified video observations. Hollow circles indicate observations from east of the study site and filled circles those from the west. Habitat classes are described in **Table 2**. ALGDOM, Algal dominated reef; ALG/INV, mixed algae and sessile invertebrates; INVDOM, reef with sessile invertebrates; SED, unconsolidated sandy sediments.

Table 4 | Class area estimations and areal differences derived from the model containing only MBES variables (Acoustic) and the model containing both MBES and wave energy variables.

	Habitat area		Areal difference	
	MBES (km²)	Wave energy (km²)	Difference (km²)	Difference (%)
ALGDOM	22.1	20.3	1.8	8.7
ALG/INV	12.6	19.1	6.4	33.8
INVDOM	115.4	126.5	11.0	8.8
SED	473.6	457.8	15.8	3.5
Total	623.7	623.7		

temperate Australia. Improvement in the model was largely due to increased classification capacity of shallow reef habitat types along a gradient of wave energy, allowing their distribution to be more accurately predicted. This pattern corroborated with cross-validation measures which showed an improvement in the classification accuracy for all habitat categories defined in the study. The area covered by each habitat class differed between the two models. Differences were most evident in a transitional habitat between algal dominated reefs and reefs characterized by sponge dominated sessile invertebrates.

In the present investigation, variables of primary importance to classification accuracy of the best performing model were bathymetry, bottom orbital velocity, rugosity, backscatter intensity and HSI backscatter derivatives. A similar pattern is reported by Bekkby et al. (2009) in their distribution modeling study of the kelp Laminaria hyperborea where depth, terrain

curvature, wave and light exposure were found to be the most important geophysical factors explaining the distribution of the species. Similarly depth, slope, wave and light exposure were found to best explain the potential distribution of the fucoid kelp Saccharina latissima in Norwegian waters (Bekkby and Moy, 2011).

Differentiation of benthic habitats in this study was also largely determined by proxies of light availability (bathymetry), availability of suitable substrate for attachment (rugosity) and hydrodynamic energy (seabed orbital velocity). While the contribution of backscatter intensity and its HSI derivatives is more difficult to interpret. It is surmised that these products are important to the classifier in distinguishing textural differences between inhomogeneous substrate types that are indicative of suitable areas for attachment of sessile species.

The high relative contribution of the depth and wave energy variables in explaining habitat distribution, in particular for the two classes defined by the presence of the canopy forming kelp E. radiata (ALG/INV and ALGDOM), is well supported by the ecological relevance of these features in explaining distribution of the species. Bathymetry acts as an indirect mediator for light availability, and limits the depth at which the basic requirements for photosynthesis can be met. The influence of wave energy is also attenuated by depth although it is evident that bathymetry alone did not capture the full influence of wave energy as a variable structuring the distribution of habitats across the site. Differences between classifications can be attributed largely to the contrasting regimes of wave energy on each side of Cape Otway. When analyzed independently, classification accuracy for the eastern section of the site did not improve significantly with the addition of the bottom orbital velocity variable which was

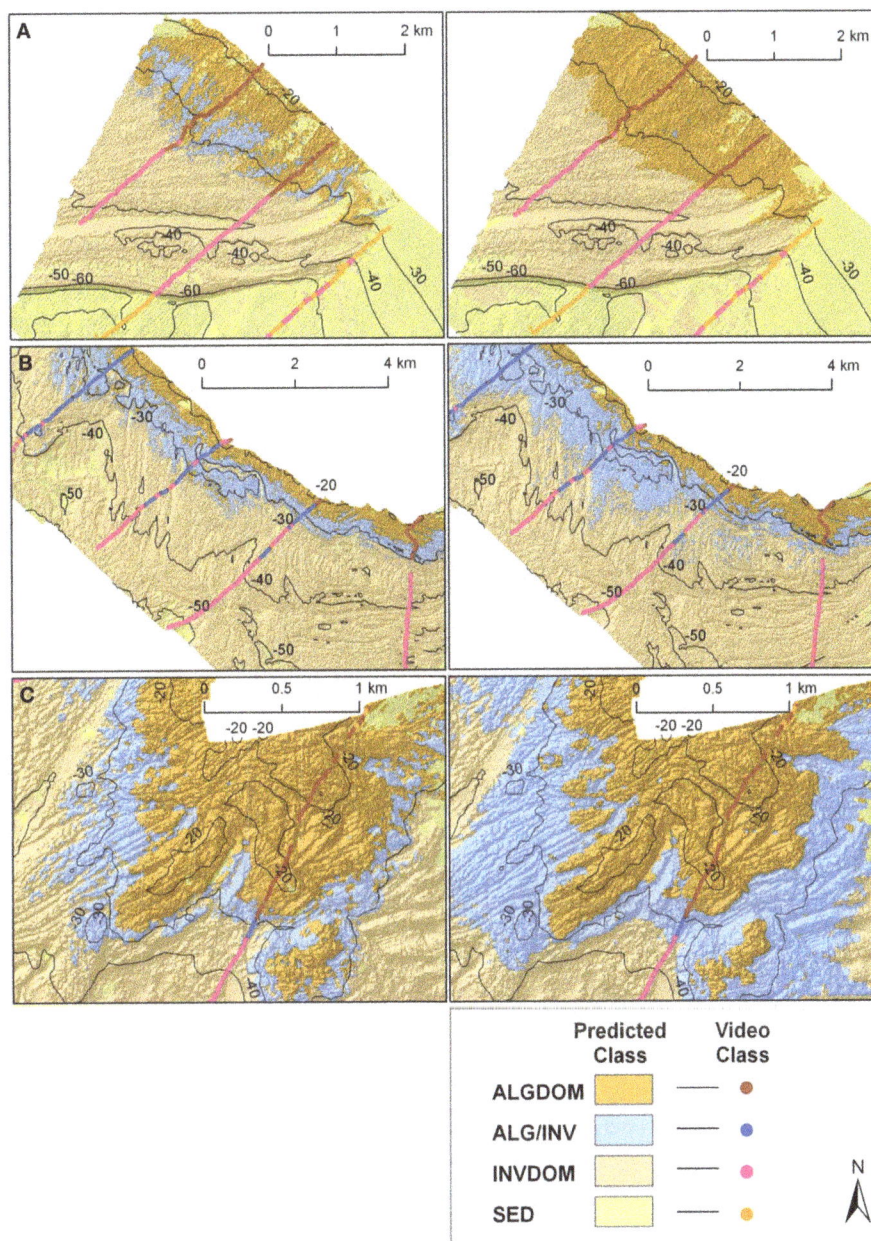

FIGURE 8 | Classified habitat maps of representative reef systems *west* of Cape Otway overlaid with classified observation data from towed video transects. Locations of reefs A, B and C are shown in **Figures 1**, **4**. Classifications on the *left hand side* of the figure were derived using MBES only variables, while those on the *right* were derived using both MBES predictor variables and maximum bottom orbital velocities.

limited in its range from 0.54 to 0.74 m s^{-1}. This indicates that the variable describing bottom orbital velocity improves the predictive capacity of the model only where wave energy is a more restrictive environmental factor. In both models the kelp dominated (ALGDOM) class transitions to invertebrate dominated reef (INVDOM) through a narrow depth band (5–7 m) of the mixed algae and invertebrates class (ALG/INV) (**Figure 9**, reefs D and E). The kelp *E. radiata* is restricted in vertical distribution to depths less than 30 m beyond which invertebrate dominated reef becomes the primary reef habitat type.

West of Cape Otway a different pattern emerges, with observations of *E. radiata* extending to depths of 49 m and the transition zone between algal and invertebrate dominated reef types spanning a greater depth range (~25 m). This east-west variation in depth distributions is captured to some extent by the model incorporating only MBES variables which predicts the ALG/INV class occurring marginally deeper (33 m) than in the east but is better described by the wave energy model which predicts the ALG/INV class to occur both deeper (42 m) and across a greater depth range (~20 m).

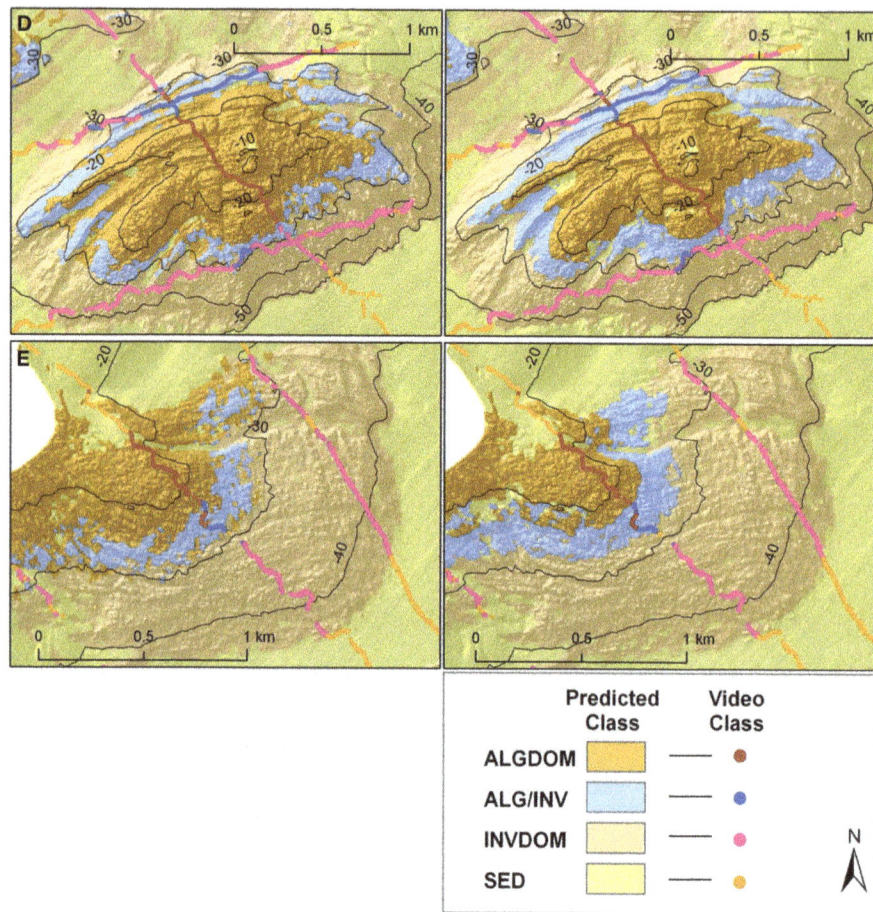

FIGURE 9 | Classified habitat maps of representative reef systems *east* **of Cape Otway overlaid with classified observation data from towed video transects.** Locations of reefs D and E are shown in **Figures 1**, **4**.

Classifications on the *left hand side* of the figure were derived using MBES variables alone, while those on the *right* were derived using MBES variables and maximum bottom orbital velocities.

Differences in patterns of distribution of the ALGDOM class between reef A, and reefs B and C on the western side of the site are potentially caused by a temporal mismatch in collection of observation data which in MBES survey block 1 (**Figure 1**) was collected 2 years prior to the remainder of the study site. Although observation data were collected in February of each year there is evidence to suggest that canopy density of *E. radiata* is temporally variable and largely dependent on the timing of optimum environmental conditions conducive to growth, for example, temperature, nutrient and light availability (Wernberg and Goldberg, 2008). It is therefore conceivable that the observation data associated with survey block 1 represents a different stage on the annual senescence to peak biomass cycle of the species and in that respect is not consistent with observational data from the rest of the site.

Alternatively, these differences may reflect the interaction of incoming wave energy with local reef geometry which is noticeably different between reefs. Reef A displays a relatively steep and regular offshore gradient with little topographic diversity descending to depths of 60 m close to the coast. Reefs depicted in insets B and C by comparison have a shallower offshore gradient

and are characterized by rugged terrain composed of medium to high-profile crests (<1–2 m), troughs and ridges extending farther offshore. While modeled maximum orbital velocity is of similar magnitude for all of these areas, the complexity of the reef surface at areas B and C may provide a wider range of hydrodynamic conditions caused by localized topographic diversity allowing the establishment of invertebrate communities in a mosaic of lower flow areas within the reef. This theory is corroborated by the work of Toohey and Kendrick (2008) who linked greater species richness on reefs with complex topography to a reduction in the structuring effects of *E. radiata* canopy on understorey communities.

Our results suggest that the importance of depth as a predictor of reef habitat distribution is strongly mediated by variation in hydrodynamic energy. This may indicate that light is not the limiting factor in the vertical distribution of *E. radiata* in the east of the site. The limited depths attained by the species compared to the west of the site are potentially the result of competitive interactions with sessile invertebrates for limited hard substrata suitable for attachment. Under this assumption it can also be suggested that stronger wave energy conditions in the west of the site afford some measure of competitive advantage to *E. radiata* allowing the species to successfully occupy space to a greater depth. This contention is supported by the known ecology of the species which exhibits a plastic morphology in response to hydrodynamic stress. Individuals at exposed sites have been reported to display drag reducing morphological characters such as smaller size, narrower laterals and blades as well as thicker holdfasts and stipes (Wernberg and Thomsen, 2005; Wernberg and Vanderklift, 2010). Higher energy conditions may additionally mediate the influence of *E. radiata* on understorey communities through increased effects of direct physical abrasion by fronds (Toohey et al., 2004; Fowler-Walker et al., 2005). There is also evidence to suggest that some kelp species achieve a higher rate of primary productivity, increasing both individual density and canopy biomass in high vs. low wave energy environments (Hurd, 2000). Pedersen et al. (2012) relate this pattern to higher epiphytic load and self-shading in low energy sites and speculate that higher energy conditions may increase light availability to the canopy through continuous and frequent movement.

The results presented have increased our knowledge of the structuring effects of wave energy on subtidal habitats and demonstrated its relevance to benthic habitat mapping. There are however a number of limitations concerning derivation of the wave energy model that should be considered when interpreting these results. Foremost, the temporal resolution of the spectral wave model used to calculate orbital velocity at the seabed is restricted to a single year which may not have fully captured the upper range of extreme wave conditions experienced at the site. Significant wave heights modeled in this study did not exceed 6.2 m for the year 2000 although Hemer et al. (2008) estimate a centennial return significant wave height of 15.51 m for Cape Sorell and cite a 13.2 m event measured by the wave buoy in 1985. Therefore, habitat structuring by wave energy at the site could well be the result of larger wave events occurring outside the temporal resolution of the study.

Secondly, the spatial grain of the wave energy model (60 m) was observed in a small number of cases to cause block artifacts in the habitat classification, predominantly in the areas classified as ALG/INV. This is presumably a function of the value of the wave energy dataset in defining these areas and is of consequence as it potentially masks fine-scale variation in habitat boundaries important in analysis of patch metrics (e.g., Ierodiaconou et al., 2011).

Exposure to hydrodynamic energy is one of the fundamental variables of the coastal environment (Nishihara and Terada, 2010) and has been well demonstrated to play in integral role in the life histories and evolutionary biology of the organisms found there (Hurd, 2000). There is a wealth of evidence to suggest that the degree of adaptation to varying levels of hydrodynamic energy strongly influences the available niche of many species. Despite the evidence linking the distributional ecology of marine taxa to the physical aspects of their hydrodynamic environment, there are relatively few studies (e.g., Galparsoro et al., 2013) that explore the application of these variables for local-scale (10's–100's km^2) predictive distribution modeling. In this study benthic habitats at a wave-exposed site were characterized according to environmental variables obtained from MBES variables only, and compared with a characterization based on the addition of a fine-scale

wave energy model. Measures of classification accuracy obtained with the addition of the wave energy variable to the model were significantly higher overall and contributed to greater resolvability between habitat classes than MBES derived variables alone. Furthermore, an insight was gained into the interaction between the structuring effects of depth (a proxy for light availability) and exposure to wave energy over the full depth range of a foundation kelp species that affects biodiversity and ecological functioning on shallow reefs across temperate Australasia. This study highlights the suitability of exposure measures for predictive benthic habitat modeling on wave-exposed coastlines and provides a basis for continuing research relating patterns of biological distribution to measurable aspects of the physical environment.

ACKNOWLEDGMENTS

DI conceived the project and led the fieldwork. AR led the analyses and writing with contributions from DI and TW. This work was supported by the National Heritage Trust and Caring for Country as part of the Victorian Marine Habitat Mapping Project with project partners Glenelg Hopkins Catchment Management Authority, Department of Environment and Primary Industries, Parks Victoria, University of Western Australia and Fugro Survey. We thank the crew from the Australian Maritime College research vessel Bluefin, which was used for the multibeam data collection. We also thank the crew from Deakin University research vessel Courageous II for assisting DI and AR collecting the towed video data used in this project. Thanks to Richard Zavalas for assistance with the validation matrices. GIS laboratory facilities at Deakin University, Warrnambool, Victoria were used for spatial analyses. We also thank the three reviewers for their constructive suggestions to improve the manuscript.

REFERENCES

Airoldi, L. (2003). The effects of sedimentation on rocky coast assemblages. *Oceanogr. Mar. Biol.* 41, 161–236. doi: 10.1016/S0022-0981(96)02770-0

Bekkby, T., Isachsen, P. E., Isaeus, M., and Bakkestuen, V. (2008). GIS modeling of wave exposure at the seabed: a depth-attenuated wave exposure model. *Mar. Geod.* 31, 117–127. doi: 10.1080/01490410802053674

Bekkby, T., and Moy, F. E. (2011). Developing spatial models of sugar kelp (Saccharina latissima) potential distribution under natural conditions and areas of its disappearance in Skagerrak. *Estuarine Coast. Shelf Sci.* 95, 477–483. doi: 10.1016/j.ecss.2011.10.029

Bekkby, T., Rinde, E., Erikstad, L., and Bakkestuen, V. (2009). Spatial predictive distribution modelling of the kelp species Laminaria hyperborea. *Ices J. Mar. Sci.* 66, 2106–2115. doi: 10.1093/icesjms/fsp195

Bell, J. J., and Barnes, D. K. A. (2000). The distribution and prevalence of sponges in relation to environmental gradients within a temperate sea lough: vertical cliff surfaces. *Divers. Distrib.* 6, 283–303. doi: 10.1046/j.1472-4642.2000.00091.x

Boulding, E. G., Holst, M., and Pilon, V. (1999). Changes in selection on gastropod shell size and thickness with wave-exposure on Northeastern Pacific shores. *J. Exp. Mar. Biol. Ecol.* 232, 217–239. doi: 10.1016/S0022-0981(98)00117-8

Breiman, L. (2001). Random forests. *Mach. Learn.* 45, 5–32. doi: 10.1023/A:1010933404324

Chollett, I., and Mumby, P. J. (2012). Predicting the distribution of Montastraea reefs using wave exposure. *Coral Reefs* 31, 493–503. doi: 10.1007/s00338-011-0867-7

Congalton, R. G., and Green, K. (2009). *Assessing the Accuracy of Remotely Sensed Data: Principles and Practices.* Boca Raton, FL: CRC Press.

Cutler, D. R., Edwards, T. C., Beard, K. H., Cutler, A., and Hess, K. T. (2007). Random forests for classification in ecology. *Ecology* 88, 2783–2792. doi: 10.1890/07-0539.1

Denny, M. (2006). Ocean waves, nearshore ecology, and natural selection. *Aquat. Ecol.* 40, 439–461. doi: 10.1007/s10452-004-5409-8

Ekebom, J., Laihonen, P., and Suominen, T. (2003). A GIS-based step-wise procedure for assessing physical exposure in fragmented archipelagos. *Estuarine Coast. Shelf Sci.* 57, 887–898. doi: 10.1016/S0272-7714(02)00419-5

England, P. R., Phillips, J., Waring, J. R., Symonds, G., and Babcock, R. (2008). Modelling wave-induced disturbance in highly biodiverse marine macroalgal communities: support for the intermediate disturbance hypothesis. *Mar. Freshw. Res.* 59, 515–520. doi: 10.1071/MF07224

Fonseca, M. S., and Bell, S. S. (1998). Influence of physical setting on seagrass landscapes near Beaufort, North Carolina, USA. *Mar. Ecol. Prog. Ser.* 171, 109–121. doi: 10.3354/meps171109

Foster, S. D., Bravington, M. V., Williams, A., Althaus, F., Laslett, G. M., and Kloser, R. J. (2009). Analysis and prediction of faunal distributions from video and multi-beam sonar data using Markov models. *Environmetrics* 20, 541–560. doi: 10.1002/env.952

Fowler-Walker, M. J., Gillanders, B. M., Connell, S. D., and Irving, A. D. (2005). Patterns of association between canopy-morphology and understorey assemblages across temperate Australia. *Estuarine Coast. Shelf Sci.* 63, 133–141. doi: 10.1016/j.ecss.2004.10.016

Friedlander, A. M., Brown, E. K., Jokiel, P. L., Smith, W. R., and Rodgers, K. S. (2003). Effects of habitat, wave exposure, and marine protected area status on coral reef fish assemblages in the Hawaiian archipelago. *Coral Reefs* 22, 291–305. doi: 10.1007/s00338-003-0317-2

Galparsoro, I., Borja, Á., Kostylev, V. E., Rodríguez, J. G., Pascual, M., and Muxika, I. (2013). A process-driven sedimentary habitat modelling approach, explaining seafloor integrity and biodiversity assessment within the European Marine Strategy Framework Directive. *Estuar. Coast. Shelf Sci.* 131, 194–205. doi: 10.1016/j.ecss.2013.07.007

Greenlaw, M. E., Roff, J. C., Redden, A. M., and Allard, K. A. (2011). Coastal zone planning: a geophysical classification of inlets to define ecological representation. *Aquat. Conserv. Mar. Freshw. Ecosyst.* 21, 448–461. doi: 10.1002/aqc.1200

Harris, P. T., and Hughes, M. G. (2012). Predicted benthic disturbance regimes on the Australian continental shelf: a modelling approach. *Mar. Ecol. Prog. Ser.* 449, 13–25. doi: 10.3354/meps09463

Hemer, M. A., Simmonds, I., and Keay, K. (2008). A classification of wave generation characteristics during large wave events on the Southern Australian margin. *Cont. Shelf Res.* 28, 634–652. doi: 10.1016/j.csr.2007.12.004

Hill, N. A., Pepper, A. R., Puotinen, M. L., Hughes, M. G., Edgar, G. J., Barrett, N. S., et al. (2010). Quantifying wave exposure in shallow temperate reef systems: applicability of fetch models for predicting algal biodiversity. *Mar. Ecol. Prog. Ser.* 417, 83–95. doi: 10.3354/meps08815

Hughes, M. G., and Heap, A. D. (2010). National-scale wave energy resource assessment for Australia. *Renewable Energy* 35, 1783–1791. doi: 10.1016/j.renene.2009.11.001

Hurd, C. L. (2000). Water motion, marine macroalgal physiology, and production. *J. Phycol.* 36, 453–472. doi: 10.1046/j.1529-8817.2000.99139.x

Ierodiaconou, D., Monk, J., Rattray, A., Laurenson, L., and Versace, V. L. (2011). Comparison of automated classification techniques for predicting benthic biological communities using hydroacoustics and video observations. *Cont. Shelf Res.* 31, S28–S38. doi: 10.1016/j.csr.2010.01.012

Jaiantilal, A. (2009). *Classification and Regression by Random Forest—MATLAB [Online]*. Available online at: http://code.google.com/p/randomforest-matlab/ [Accessed November 10th 2013].

Kostylev, V. E., and Hannah, C. G. (2007). "Process-driven characterization and mapping of seabed habitats," in *Mapping the Seafloor for Habitat Characterization*, eds B. J. Todd and H. G. Greene (St John's, NL: Geological Society of Canada), 171–184.

Letourneur, Y. (1996). Dynamics of fish communities on Reunion fringing reefs, Indian Ocean.1. Patterns of spatial distribution. *J. Exp. Mar. Biol. Ecol.* 195, 1–30. doi: 10.1016/0022-0981(95)00089-5

Lindegarth, M., and Gamfeldt, L. (2005). Comparing categorical and continuous ecological analyses: effects of "wave exposure" on rocky shores. *Ecology* 86, 1346–1357. doi: 10.1890/04-1168

Nishihara, G. N., and Terada, R. (2010). Species richness of marine macrophytes is correlated to a wave exposure gradient. *Phycological Res.* 58, 280–292. doi: 10.1111/j.1440-1835.2010.00587.x

Pedersen, M. F., Nejrup, L. B., Fredriksen, S., Christie, H., and Norderhaug, K. M. (2012). Effects of wave exposure on population structure, demography, biomass and productivity of the kelp Laminaria hyperborea. *Mar. Ecol. Prog. Ser.* 451, 45–60. doi: 10.3354/meps09594

Pfaff, M. C., Branch, G. M., Wieters, E. A., Branch, R. A., and Broitman, B. R. (2011). Upwelling intensity and wave exposure determine recruitment of intertidal mussels and barnacles in the southern Benguela upwelling region. *Mar. Ecol. Prog. Ser.* 425, 141–152. doi: 10.3354/meps09003

Porter-Smith, R., Harris, P. T., Andersen, O. B., Coleman, R., Greenslade, D., and Jenkins, C. J. (2004). Classification of the Australian continental shelf based on predicted sediment threshold exceedance from tidal currents and swell waves. *Mar. Geol.* 211, 1–20. doi: 10.1016/j.margeo.2004.05.031

Rattray, A., Ierodiaconou, D., Laurenson, L., Burq, S., and Reston, M. (2009). Hydro-acoustic remote sensing of benthic biological communities on the shallow South East Australian continental shelf. *Estuar. Coast. Shelf Sci.* 84, 237–245. doi: 10.1016/j.ecss.2009.06.023

Rattray, A., Ierodiaconou, D., Monk, J., Versace, V. L., and Laurenson, L. J. B. (2013). Detecting patterns of change in benthic habitats by acoustic remote sensing. *Mar. Ecol. Prog. Ser.* 477, 1–13. doi: 10.3354/meps10264

Thomsen, M. S., Wernberg, T., and Kendrick, G. A. (2004). The effect of thallus size, life stage, aggregation, wave exposure and substratum conditions on the forces required to break or dislodge the small kelp Ecklonia radiata. *Botanica Marina* 47, 454–460. doi: 10.1515/BOT.2004.068

Thomson, D. P., Babcock, R. C., Vanderklift, M. A., Symonds, G., and Gunson, J. R. (2012). Evidence for persistent patch structure on temperate reefs and multiple hypotheses for their creation and maintenance. *Estuar. Coast. Shelf Sci.* 96, 105–113. doi: 10.1016/j.ecss.2011.10.014

Toohey, B. D., and Kendrick, G. A. (2008). Canopy-understorey relationships are mediated by reef topography in Ecklonia radiata kelp beds. *Eur. J. Phycol.* 43, 133–142. doi: 10.1080/09670260701770554

Toohey, B., Kendrick, G. A., Wernberg, T., Phillips, J. C., Malkin, S., and Prince, J. (2004). The effects of light and thallus scour from Ecklonia radiata canopy on an associated foliose algal assemblage: the importance of photoacclimation. *Mar. Biol.* 144, 1019–1027. doi: 10.1007/s00227-003-1267-5

Turner, S. J., Hewitt, J. E., Wilkinson, M. R., Morrisey, D. J., Thrush, S. F., Cummings, V. J., et al. (1999). Seagrass patches and landscapes: the influence of wind-wave dynamics and hierarchical arrangements of spatial structure on macrofaunal seagrass communities. *Estuaries* 22, 1016–1032. doi: 10.2307/1353080

Wernberg, T., and Goldberg, N. (2008). Short-term temporal dynamics of algal species in a subtidal kelp bed in relation to changes in environmental conditions and canopy biomass. *Estuar. Coast. Shelf Sci.* 76, 265–272. doi: 10.1016/j.ecss.2007.07.008

Wernberg, T., and Thomsen, M. S. (2005). The effect of wave exposure on the morphology of Ecklonia radiata. *Aquat. Bot.* 83, 61–70. doi: 10.1016/j.aquabot.2005.05.007

Wernberg, T., and Vanderklift, M. A. (2010). Contribution of temporal and spatial components to morphological variation in the kelp Ecklonia (Laminariales). *J. Phycol.* 46, 153–161. doi: 10.1111/j.1529-8817.2009.00772.x

Whiteway, T. (2009). *Australian Bathymetry and Topography Grid, June 2009*. Scale 1:5000000. Canberra, ACT: Geoscience Australia.

Zavalas, R., Ierodiaconou, D., Ryan, D., Rattray, A., and Monk, J. (2014). Habitat classification of temperate marine macroalgal communities using bathymetric LiDAR. *Remote Sens.* 6, 2154–2175. doi: 10.3390/rs6032154

Conflict of Interest Statement: The authors declare that the research was conducted in the absence of any commercial or financial relationships that could be construed as a potential conflict of interest.

16

Seasonal variability in irradiance affects herbicide toxicity to the marine flagellate *Dunaliella tertiolecta*

Sascha B. Sjollema[1]*, Charlotte D. Vavourakis[1], Harm G. van der Geest[1], A. Dick Vethaak[2,3] and Wim Admiraal[1]

[1] Department of Aquatic Ecology and Ecotoxicology, Institute for Biodiversity and Ecosystem Dynamics, University of Amsterdam, Amsterdam, Netherlands
[2] Deltares, Marine and Coastal Systems, Delft, Netherlands
[3] Department Chemistry and Biology, Institute for Environmental Studies (IVM), VU University Amsterdam, Amsterdam, Netherlands

Edited by:
Ram Kumar, Central University of Bihar, India

Reviewed by:
Xiaoshou Liu, Ocean University of China, China
Dongyan Liu, Chinese Academy of Sciences, China

*Correspondence:
Sascha B. Sjollema, Department of Aquatic Ecology and Ecotoxicology, Institute for Biodiversity and Ecosystem Dynamics, University of Amsterdam, Science Park 904, 1098 XH Amsterdam, Netherlands
e-mail: s.b.sjollema@uva.nl

Photosynthetically Active Radiation (PAR) and Ultraviolet Radiation (UVR) of the solar spectrum affect microalgae directly and modify the toxicity of phytotoxic compounds present in water. As a consequence seasonal variable PAR and UVR levels are likely to modulate the toxic pressure of contaminants in the field. Therefore, the present study aimed to determine the toxicity of two model contaminants, the herbicides diuron and Irgarol®1051, under simulated irradiance conditions mimicking different seasons. Irradiance conditions of spring and autumn were simulated with a set of Light Emitting Diodes (LEDs). Toxicity of both herbicides was measured individually and in a mixture by determining the inhibition of photosystem II efficiency (ΦPSII) of the marine flagellate *Dunaliella teriolecta* using Pulse Amplitude Modulation (PAM) fluorometry. Toxicity of the single herbicides was higher under simulated spring irradiance than under autumn irradiance and this effect was also observed for a mixture of the herbicides. This irradiance dependent toxicity indicates that herbicide toxicity in the field is seasonally variable. Consequently toxicity tests under standard light conditions may overestimate or underestimate the toxic effect of phytotoxic compounds

Keywords: seasonal variability, irradiance, herbicide toxicity, microalgae, pulse amplitude modulation (PAM) fluorometry

INTRODUCTION

Microalgae are primary producers and play a key role in aquatic ecosystems due to their position at the base of food webs. Hence, toxicants affecting microalgae potentially affect the carrying capacity of marine and freshwater ecosystems (Hylland and Vethaak, 2011). Herbicides are, due to their specific mode of action, amongst the most harmful contaminants for microalgae. Because herbicides act often directly on the photosynthetic machinery, solar radiation is likely a key factor determining the actual toxicity of herbicides in the field. Numerous studies investigated the effect of light stress on individual microalgal species as well as on natural phytoplankton communities (Vassiliev et al., 1994; Buma et al., 2001; Helbling et al., 2001). Photosynthetically Active Radiation (PAR, 400–700 nm) is captured by the light harvesting pigments in the chloroplasts of the microalgae and is used to convert CO_2 and water into carbohydrates during photosynthesis. PAR will be a limiting factor for algal photosynthesis when radiation is low, while excess radiation causes oxidative stress reducing the photosynthetic capacity (Kirk, 2011). Simultaneously, microalgae are exposed to Ultra Violet Radiation (UVR, <400 nm) which has the potential to damage important biochemical molecules (Larson and Berenbaum, 1988). Excess radiance levels of PAR and UVR may result in the formation of Reactive Oxygen Species (ROS) which may lead to a decreased photosynthetic efficiency or even viability loss (Janknegt et al., 2009a). At the same time, light may also affect compound toxicity in several ways. Solar radiation, especially UVR, may cause

degradation of the contaminant, creating transformation products with a potential different persistence and modified toxicity to microalgae. Additionally, UVR can change the structure of a compound through photomodification or photosensitization which may enhance toxicity, a phenomenon known as photo-enhanced toxicity, described for PAHs (Gala and Giesy, 1992; Wiegman et al., 2001; Southerland and Lewitus, 2004), PCBs (Ruben et al., 1990), TBT (Sargian et al., 2005) as well as pesticides (Guasch and Sabater, 1998; Lin et al., 1999; Pelletier et al., 2006). Considering the described direct and indirect effects of solar radiation on microalgae and contaminants and the fact that intensity of PAR and UVR vary strongly over the year, it is likely that solar irradiance might play an important role in the ultimate toxicity in the field. Consequently, it can be expected that standard toxicity tests may over- or underestimate toxicity of contaminants when compared to the actual toxicity under variable field conditions.

Nevertheless, there has been little attention paid toward seasonal variation in toxicity. Guasch et al. (1997) described a difference in herbicide toxicity between seasons for natural periphyton communities, but as several environmental factors were tested together, it is unclear if the observed difference was caused by a seasonal difference in light conditions. Yet, experimental confirmation on the effect of seasonal variability in solar irradiance on herbicide toxicity is still lacking. Therefore, we aimed to determine herbicide toxicity to microalgae under simulated irradiance conditions mimicking different seasons. Of the total solar radiation 100% of the UVC and 90% of the UVB radiation is blocked by stratospheric ozone, while UVA (320–400 nm) is hardly retained (Coldiron, 1992). As a consequence, microalgae in the water are mainly exposed to UVA (315–400 nm) and PAR (400–700 nm). Additionally, it has been demonstrated that photosynthesis of a natural phytoplankton community is more inhibited by UVA than by UVB (Helbling et al., 2001). Consequently, the present study focuses on the effect of UVA and PAR on herbicide toxicity. Subsurface irradiance at our latitude was simulated in a laboratory setup with Light Emitting Diodes (LEDs) of different wavelengths. We selected two model herbicides, diuron and Irgarol®1051 (Cybutryne), which are both described as priority substances under the European Water Framework Directive (2013/39/EU). Toxicity tests (with both single compounds and with a mixture of both) were performed using PAM fluorometry under the mimicked light conditions to determine the effect of different seasonal light conditions on herbicide toxicity.

MATERIALS AND METHODS
SIMULATING SOLAR RADIATION
Light regimes were simulated in a controlled laboratory setup based on the average light intensity in spring and autumn in The Netherlands. To this purpose, the average light intensity in spring (21 March–21 June 2002–2008) and autumn (21 September–21 December 2002–2008) was obtained from the Royal Netherlands Meteorological Institute resulting in an average intensity of 202 and 53 W/m^2 for spring and autumn, respectively. The average daily light periods were 14:57 h:min and 9:34 h:min, respectively (Royal Netherlands Meteorological Institute). In addition, the solar spectrum was measured outside (Amsterdam, The Netherlands) at the end of May during clear sky with a USB4000 Fiber Optic Spectrometer (Ocean Optics, Dunedin, Florida, USA). This spectrum was normalized to the average calculated PAR intensities of 202 and 53 W/m^2 to obtain a spring and autumn light profile. Since phytoplankton is not exposed to the irradiance levels at the surface, the experimental irradiance levels were attenuated to 10% to simulate subsurface irradiance. The attenuation level was chosen to match the conditions at our field sites in the coastal zone of the North Sea with a mixed water column of ca. 30 m and a light attenuation co-efficient Kd of roughly 0.6 m-1(Delft3D-GEM for the North Sea). The 10% irradiance level of the normalized spectrum (**Figure 1**, solid lines), was simulated in the laboratory with five LEDs emitting light peaking at wavelengths of 450, 475, 522, 632, and 685 nm, corresponding to the absorbance maximum of the main algal pigments. An additional LED with a peak at 370 nm was used to simulate UVA. A total of 16 clusters of six LEDs provided homogeneous light during the experiments. The spectra for both spring and autumn light conditions, hereafter referred to as SPRING and AUTUMN, are depicted in **Figure 1** (dashed lines).

TEST SPECIES AND CHEMICALS
All tests were performed with the marine flagellate *Dunaliella tertiolecta* (Butcher, CCAP 19/27) which was cultured in Erlenmeyer flasks on artificial seawater medium. Artificial seawater (33%) was obtained by dissolving sea salts (Aqua Bio Solutions, Wormerveer, The Netherlands) in de-ionized water (MilliQ) and this was enriched with f/2 medium (Guillard, 1975) (Sigma Aldrich Chemie B.V., Zwijndrecht, The Netherlands). Toxicity tests were performed using diuron (CAS: 330-54-4, analytic standard, Sigma Aldrich, Zwijndrecht, The Netherlands) and Irgarol®1051 (CAS: 28159-98-0, >97%, Ciba Specialty Chemicals Inc., Basle, Switzerland). Stock solutions of both compounds were made in methanol (ULC/MS grade, Biosolve, Valkenswaard, The Netherlands). All presented concentrations are nominal concentrations. Culturing and toxicity experiments were performed under SPRING and AUTUMN light conditions at a constant temperature of 13 ± 0.5°C.

TOXICITY TEST
To investigate the algal growth rate under the different irradiance regimes, cell densities of acclimatized *D. tertiolecta* were counted daily in a Bürker counting chamber. These data were used to

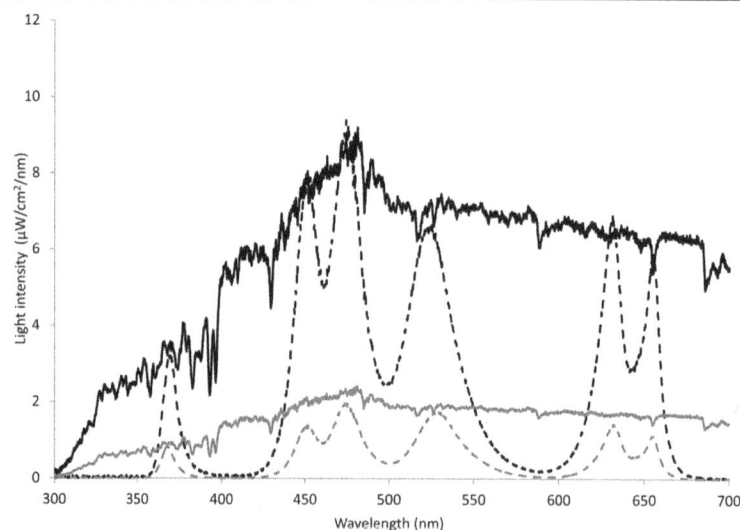

FIGURE 1 | Measured solar radiation (10%) and simulated light regime under spring and autumn light conditions. Light intensity (μW/cm^2/nm) of the LEDs for SPRING (black dotted line) and AUTUMN (gray dotted line) were normalized to 10% of the average solar radiation (black and gray line, respectively).

determine the exponential growth phase under both light regimes and all toxicity experiments were performed with light adapted, exponentially growing populations. The growth rates of *D. tertiolecta,* determined by linear regression of ln-transformed cell densities were 0.8 and 0.3 (day^{-1}) for SPRING and AUTUMN, respectively.

The toxic effects of the Irgarol®1051 and diuron (five concentrations per herbicide, six replicates per concentration and five replicates for the control) on the photosynthetic efficiency of the algae was determined for SPRING and AUTUMN. To this purpose, Pulse Amplitude Modulation (PAM) fluorometry bioassays were performed in which the effective photosystem II efficiency (ΦPSII) was determined after 4.5 h using a WATER-PAM (Heinz Walz GmbH, Effeltrich, Germany). Minimum and maximum fluorescence (F and F'_m, respectively) were determined and ΦPSII was calculated as $[F'_m-F]/F'_m$. The PSII inhibition was expressed as percentage of the corresponding solvent control (% of control). The effect on the photosynthetic efficiency of the algae was expressed as the 50% effect concentration (EC_{50}). Next to the experiments with diuron and Irgarol individually, toxic effect of both herbicides was determined in a mixture for SPRING as well as AUTUMN. The Toxic Unit (TU) concept was applied for the composition of an equitoxic mixture of both herbicides, based on the individual EC_{50} values. The toxicity of this mixture (tested for eight concentrations) was determined using the same bioassay as described above and the results were interpreted according to the Concentration Addition model (Könemann, 1981).

The actinic light used to determine ΦPSII of these PAM measurements consisted of LEDs of 632 nm (chlorophyll *a* fluorescence) which were identical to the 632 nm LEDs used to simulate solar radiation. The same light intensity of these LEDs was used during exposure and measurement to maintain the required simulated field relevant irradiance.

DATA ANALYSIS

The log-logistic dose-response model described by Haanstra et al. (1985) was used to determine the 50% reduction (EC_{50}) in ΦPSII and was calculated as $y = c/(1 + e^{b(log(x)-log(a))})$, where y is the ΦPSII (% control), x is the concentration of the toxicant (μg/L), a is the EC50 value (μg/L), b is the slope of the curve, and c is the ΦPSII of the control. Likelihood ratio tests were applied to compare effect concentrations of SPRING and AUTUMN ($\chi_1^2 > 3.84$, $df = 1$, $p < 0.05$). The analyses were all performed with SPSS (IBM SPSS Statistics 20).

RESULTS

Clear dose-response relationships of the effect of Irgarol and diuron on ΦPSII of *D. tertiolecta* were obtained for SPRING and AUTUMN (**Figure 2**). The observed toxicity was significantly ($p < 0.05$) higher for SPRING (EC_{50}: 0.8 and 3.6 μg/L) compared to AUTUMN (EC_{50}: 1.3 and 4.8 μg/L) for Irgarol and diuron, respectively. The same pattern, with herbicides having a higher toxic effect in SPRING compared to AUTUMN, was observed for the binary mixture of the herbicides (**Figure 3**). The EC_{50} of the equitoxic mixture for SPRING was significantly lower (0.8 TU, 95% C.I. 0.7–0.9) than 1 TU, indicating a more than additive effect for the mixture of both herbicides. In contrast, the EC_{50} of the mixture for AUTUMN was significantly higher (1.2 TU,

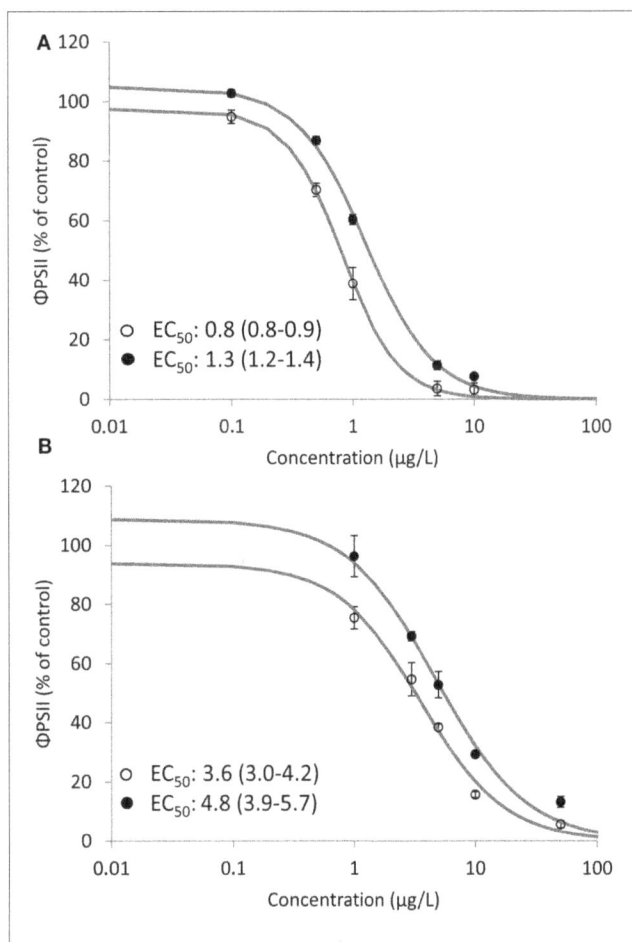

FIGURE 2 | Dose response relationships for Irgarol (A) and diuron (B) on the effective PSII efficiency (ΦPSII) (% of control) of *D. tertiolecta* after 4.5 h of exposure under two seasonal irradiance regimes. EC_{50} values and corresponding 95% confidence intervals (μg/L) are presented for all dose response relationships. Open (SPRING) and closed (AUTUMN) circles represent average ΦPSII and the solid line is the log-logistic model. Error bars represent standard deviation. Concentration is expressed as μg/L. $N = 6$ for treatments; $N = 5$ for control.

95% C.I. 1.1–1.3) than 1 TU, indicating a less than additive effect.

DISCUSSION

We clearly demonstrated a significant difference in the toxicity of Irgarol and diuron to *D. tertiolecta* cultures under simulated spring and autumn irradiance conditions, thereby providing the first experimental confirmation that seasonal variation in irradiance indeed affects herbicide toxicity to microalgae. Results of previous studies on the toxicity of both Irgarol and diuron to microalgae are recently reviewed by Suresh Kumar et al. (2014). It was found that Irgarol, used in antifouling paints on ship hulls, is highly toxic to individual microalgal species as well as to microalgal communities (e.g., Bérard et al., 2003; Devilla et al., 2005; Gatidou and Thomaidis, 2007; Buma et al., 2009; Sjollema et al., 2014). Negative effects of diuron, used in agriculture and antifouling paints, on individual microalgal species and communities have also been described (e.g., Gatidou and Thomaidis, 2007; Magnusson et al., 2008; Knauert et al., 2009; Pesce et al., 2010;

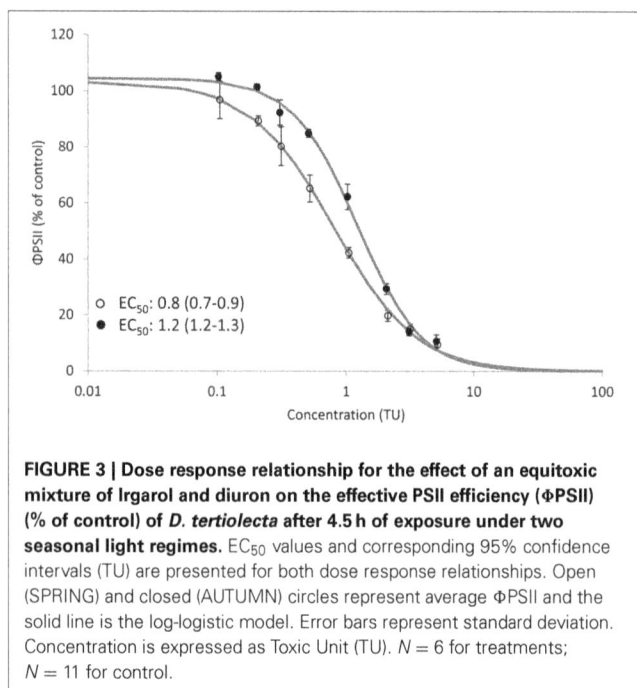

FIGURE 3 | Dose response relationship for the effect of an equitoxic mixture of Irgarol and diuron on the effective PSII efficiency (ΦPSII) (% of control) of *D. tertiolecta* after 4.5 h of exposure under two seasonal light regimes. EC$_{50}$ values and corresponding 95% confidence intervals (TU) are presented for both dose response relationships. Open (SPRING) and closed (AUTUMN) circles represent average ΦPSII and the solid line is the log-logistic model. Error bars represent standard deviation. Concentration is expressed as Toxic Unit (TU). $N = 6$ for treatments; $N = 11$ for control.

Sjollema et al., 2014). From these studies it can be concluded that Irgarol is in general more toxic to microalgae compared to diuron (Gatidou and Thomaidis, 2007; Sjollema et al., 2014), which was confirmed by the present study. However, the majority of the toxicity tests of Irgarol and diuron are performed under standardized laboratory conditions, using fluorescent tubes as a light source. As the present study demonstrated that irradiance conditions affect the toxicity of these herbicides, the question remains how these reported toxicity levels relate to the actual risk of these compounds under field conditions. The observed short term seasonal-specific effect on ΦPSII of these herbicides are indicative for the ecological relevant chronic effect on growth (Magnusson et al., 2008; Buma et al., 2009) and must therefore be taken into account when determining toxic pressure of contaminants in aquatic and marine ecosystems.

Although the differences in toxicity between seasons are relatively small in the present laboratory study, the observations on single herbicides are corroborated by their joint effects under different irradiance regimes. Yet, our observations urge to develop more insight in field relevant irradiance conditions. Algae in the field are exposed to a complex and dynamic light regime with constantly fluctuating light conditions as a result, while this study was performed under constant light conditions, with both seasons based on one solar spectrum measured in spring. Additionally, light intensities will decrease with increasing depth due to absorption and scattering of the light (Kirk, 2011). The main light-absorbing component at wavelength below 500 nm is chromophoric dissolved organic matter (CDOM or yellow substance/humic mater/Gelbstoff), influencing the ratio between UVA, UVB, total UVR, and PAR (Markager and Vincent, 2000). This reduction in light intensity and changes in contribution of UVA, UVB, total UVR, and PAR in the water will affect the herbicide toxicity to microalgae. As the highest CDOM concentrations are found at locations close to direct sources of terrestrial organic

matter (Kowalczuk et al., 2003), it is likely that the phototoxic effects will be location specific. On the other hand, algae can also be present as microphytobenthos on exposed sediments of shallow tidal flats like the Wadden Sea (The Netherlands) where they can be exposed to very high levels of UVR during most of the tidal cycle (Peletier et al., 1996). As a result the photoenhanced toxicity for these benthic species might be higher compared to phytoplankton living in the water column.

Next to the seasonal dynamics in light conditions, concentrations of contaminants can also fluctuate over time. Since diuron is used for agricultural applications and both tested herbicides are used in antifouling products, seasonal fluctuations due to differences in application of these compounds in relation to the growing and boating season are indeed observed (Lamoree et al., 2002). It is likely that peak concentrations coincide with high irradiance levels in spring and summer and consequently microalgae are exposed to higher concentrations of herbicides which also have a higher toxic effect. Additionally, in temperate regions like The Netherlands, the main algal bloom is typically observed in spring (Kaiser et al., 2011), co-occurring with higher contaminant concentrations and high irradiance levels. As a consequence, the hazard and risk for herbicide contamination will be higher during this productive season. Especially in this season, effects on the development of algal populations might potentially affecting higher trophic levels.

The toxic pressure of herbicides in the field will mainly depend on the timing of (1) concentrations of the compounds, (2) presence of algae, and (3) the irradiance spectrum and total light intensity. Additional multi-stress factors which might interfere with the toxicity of the herbicides are nutrient limitation (Hall et al., 1989; Guasch et al., 2004) and temperature (Chalifour and Juneau, 2011). A combination of timing as well as the presence of additional multi-stress factors will determine the herbicide toxicity in the field, while the ultimate effect on the ecosystem will also depend on the ability of the microalgae to recover from damage by solar radiation (Janknegt et al., 2009b) and/or contaminants (Buma et al., 2009; Magnusson et al., 2012). Therefore, standard toxicity test may over- or underestimate the toxic effect when performed under controlled laboratory conditions, thereby misjudging the potential hazard of these compounds in the field.

ACKNOWLEDGMENTS

This research was financed by DELTARES, The Netherlands. We would like to thank the Technology Centre of the University of Amsterdam (particularly Gerrit Hardeman and Theo van Lieshout) for technical assistance.

REFERENCES

Bérard, A., Dorigo, U., Mercier, I., Becker-van Slooten, K., Grandjean, D., and Leboulanger, C. (2003). Comparison of the ecotoxicological impact of the triazines Irgarol 1051 and atrazine on microalgal cultures and natural microalgal communities in Lake Geneva. *Chemosphere* 53, 935–944. doi: 10.1016/S0045-6535(03)00674-X

Buma, A. G., De Boer, M. K., and Boelen, P. (2001). Depth distributions of DNA damage in Antarctic marine phyto-and bacterioplankton exposed to summertime UV radiation. *J. Phycol.* 37, 200–208. doi: 10.1016/j.seares.2008.11.007

Buma, A. G. J., Sjollema, S. B., van de Poll, W. H., Klamer, H.J. C., and Bakker, J. F. (2009). Impact of the antifouling agent Irgarol 1051 on marine phytoplankton species. *J. Sea Res.* 61, 133–139. doi: 10.1046/j.1529-8817.2001.037002200.x

Chalifour, A., and Juneau, P. (2011). Temperature-dependent sensitivity of growth and photosynthesis of Scenedesmus obliquus, Navicula pelliculosa and two strains of Microcystis aeruginosa to the herbicide atrazine. Aquat. Toxicol. 103, 9–17. doi: 10.1016/j.aquatox.2011.01.016

Coldiron, B. M. (1992). Thinning of the ozone layer: facts and consequences. J. Am. Acad. Dermatol. 27, 653–662. doi: 10.1016/0190-9622(92)70233-6

Devilla, R. A., Brown, M. T., Donkin, M., Tarran, G. A., Aiken, J., and Readman, J. W. (2005). Impact of antifouling booster biocides on single microalgal species and on a natural marine phytoplankton community. Mar. Ecol. Prog. Ser. 286, 1–12. doi: 10.3354/meps286001

Gala, W. R., and Giesy, J. P. (1992). Photo-induced toxicity of anthracene to the green alga, Selenastrum capricornutum. Arch. Environ. Contam. Toxicol. 23, 316–323. doi: 10.1007/BF00216240

Gatidou, G., and Thomaidis, N. S. (2007). Evaluation of single and joint toxic effects of two antifouling biocides, their main metabolites and copper using phytoplankton bioassays. Aquat. Toxicol. 85, 184–191. doi: 10.1016/j.aquatox.2007.09.002

Guasch, H., Muñoz, I., Rosés, N., and Sabater, S. (1997). Changes in atrazine toxicity throughout succession of stream periphyton communities. J. Appl. Phycol. 9, 137–146.

Guasch, H., Navarro, E., Serra, A., and Sabater, S. (2004). Phosphate limitation influences the sensitivity to copper in periphytic algae. Freshw. Biol. 49, 463–473. doi: 10.1111/j.1365-2427.2004.01196.x

Guasch, H., and Sabater, S. (1998). Light history influences the sensitivity to atrazine in periphytic algae. J. Phycol. 34, 233–241. doi: 10.1046/j.1529-8817.1998.340233.x

Guillard, R. R. (1975). "Culture of phytoplankton for feeding marine invertebrates," in Culture of Marine Invertebrate Animals, eds W. L. Smith and M. H. Chanley (New York, NY: Plenum Press), 29–60. doi: 10.1007/978-1-4615-8714-9_3

Haanstra, L., Doelman, P., and Voshaar, J. H. O. (1985). The use of sigmoidal dose response curves in soil ecotoxicological research. Plant Soil 84, 293–297. doi: 10.1007/BF02143194

Hall, J., Healey, F. P., and Robinson, G. G. C. (1989). The interaction of chronic copper toxicity with nutrient limitation in two chlorophytes in batch culture. Aquat. Toxicol. 14, 1–13. doi: 10.1016/0166-445X(89)90051-9

Helbling, E. W., Buma, A. G., de Boer, M. K., and Villafañe, V. E. (2001). In situ impact of solar ultraviolet radiation on photosynthesis and DNA in temperate marine phytoplankton. Mar. Ecol. Prog. Ser. 211, 43–49. doi: 10.3354/meps211043

Hylland, K., and Vethaak, A. D. (2011). "Impact of contaminants on pelagic ecosystems Chapter 10," in Ecological Impacts of Toxic Chemicals, eds F. Sánchez-Bayo, P. J. van den Brink, and R. M. Mann (Bentham Science Publishers Ltd), 212–224.

Janknegt, P. J., De Graaff, C. M., Van de Poll, W. H., Visser, R. J., Helbling, E. W., and Buma, A. G. J. (2009a). Antioxidative responses of two marine microalgae during acclimation to static and fluctuating natural UV radiation. Photochem. Photobiol. 85, 1336–1345. doi: 10.1111/j.1751-1097.2009.00603.x

Janknegt, P. J., De Graaff, C. M., Van De Poll, W. H., Visser, R. J., Rijstenbil, J. W., and Buma, A. G. J. (2009b). Short-term antioxidative responses of 15 microalgae exposed to excessive irradiance including ultraviolet radiation. Eur. J. Phycol. 44, 525–539. doi: 10.1080/09670260902943273

Kaiser, M. J., Attrill, M. J., Jennings, S., Thomas, D. N., Barnes, D. K. A., Brierley, A. S., et al. (2011). Marine Ecology: Processes, Systems, and Impacts. Oxford: Oxford University Press.

Kirk, J. T. O. (2011). Light and Photosynthesis in Aquatic Ecosystems. New York, NY: Cambridge university press.

Knauert, S., Dawo, U., Hollender, J., Hommen, U., and Knauer, K., (2009). Effects of photosystem II inhibitors and their mixture on freshwater phytoplankton succession in outdoor mesocosms. Environ. Toxicol. Chem. 28, 836–845. doi: 10.1897/08-135R.1

Könemann, H. (1981). Fish toxicity tests with mixtures of more than two chemicals: a proposal for a quantitative approach and experimental results. Toxicology 19, 229–238. doi: 10.1016/0300-483X(81)90132-3

Kowalczuk, P., Cooper, W. J., Whitehead, R. F., Durako, M. J., and Sheldon, W. (2003). Characterization of CDOM in an organic-rich river and surrounding coastal ocean in the South Atlantic Bight. Aquat. Sci. 65, 384–401. doi: 10.1007/s00027-003-0678-1

Lamoree, M. H., Swart, C. P., van der Horst, A., and van Hattum, B. (2002). Determination of diuron and the antifouling paint biocide Irgarol 1051 in Dutch marinas and coastal waters. J. Chromatogr. A 970, 183–190. doi: 10.1016/S0021-9673(02)00878-6

Larson, R. A., and Berenbaum, M. R. (1988). Environmental phototoxicity. Environ. Sci. Technol. 22, 354–360. doi: 10.1021/es00169a001

Lin, Y., Karuppiah, M., Shaw, A., and Gupta, G. (1999). Effect of simulated sunlight on atrazine and metolachlor toxicity of surface waters. Ecotoxicol. Environ. Saf. 43, 35–37. doi: 10.1006/eesa.1998.1751

Magnusson, M., Heimann, K., and Negri, A. P. (2008). Comparative effects of herbicides on photosynthesis and growth of tropical estuarine microalgae. Mar. Pollut. Bull. 56, 1545–1552. doi: 10.1016/j.marpolbul.2008.05.023

Magnusson, M., Heimann, K., Ridd, M., and Negri, A. P. (2012). Chronic herbicide exposures affect the sensitivity and community structure of tropical benthic microalgae. Mar. Pollut. Bull. 65, 363–372. doi: 10.1016/j.marpolbul.2011.09.029

Markager, S., and Vincent, W. F. (2000). Spectral light attenuation and the absorption of UV and blue light in natural waters. Limnol. Oceanogr. 45, 642–650. doi: 10.4319/lo.2000.45.3.0642

Peletier, H., Gieskes, W. W. C., and Buma, A. G. J. (1996). Ultraviolet-B radiation resistance of benthic diatoms isolated from tidal flats in the Dutch Wadden Sea. Mar. Ecol. Prog. Ser. 135, 163–168. doi: 10.3354/meps135163

Pelletier, É., Sargian, P., Payet, J., and Demers, S. (2006). Ecotoxicological effects of combined UVB and organic contaminants in coastal waters: a review. Photochem. Photobiol. 82, 981–993. doi: 10.1562/2005-09-18-RA-687.1

Pesce, S., Lissalde, S., Lavieille, D., Margoum, C., Mazzella, N., Roubeix, V., et al. (2010). Evaluation of single and joint toxic effects of diuron and its main metabolites on natural phototrophic biofilms using a pollution-induced community tolerance (PICT) approach. Aquat. Toxicol. 99, 492–499. doi: 10.1016/j.aquatox.2010.06.006

Royal Netherlands Meteorological Institute. Measuring station De Kooy. Avilable online at: http://data.knmi.nl (Accessed May, 2013).

Ruben, H. J., Cosper, E. M., and Wurster, C. F. (1990). Influence of light intensity and photoadaptation on the toxicity of PCB to a marine diatom. Environ. Toxicol. Chem. 9, 777–784. doi: 10.1002/etc.5620090612

Sargian, P., Pelletier, E., Mostajir, B., Ferreyra, G. A., and Demers, S. (2005). TBT toxicity on a natural planktonic assemblage exposed to enhanced ultraviolet-B radiation. Aquat. Toxicol. 73, 299–314. doi: 10.1016/j.aquatox.2005.03.019

Sjollema, S. B., MartínezGarcía, G., van der Geest, H. G., Kraak, M. H. S., Booij, P., Vethaak, A. D., et al. (2014). Hazard and risk of herbicides for marine microalgae. Environ. Pollut. 187, 106–111. doi: 10.1016/j.envpol.2013.12.019

Southerland, H. A., and Lewitus, A. J. (2004). Physiological responses of estuarine phytoplankton to ultraviolet light-induced fluoranthene toxicity. J. Exp. Mar. Biol. Ecol. 298, 303–322. doi: 10.1016/30022-0981(03)00364-2

Suresh Kumar, K., Dahms, H., Lee, J., Kim, H. C., Lee, W. C., and Shin, K. (2014). Algal photosynthetic responses to toxic metals and herbicides assessed by chlorophyll a fluorescence. Ecotoxicol. Environ. Saf. 104, 51–71. doi: 10.1016/j.ecoenv.2014.01.042

Vassiliev, I. R., Prasil, O., Wyman, K. D., Kolber, Z., Hanson, A. K., Prentice, J. E., et al. (1994). Inhibition of PS II photochemistry by PAR and UV radiation in natural phytoplankton communities. Photosyn. Res. 42, 51–64. doi: 10.1007/BF00019058

Wiegman, S., van Vlaardingen, P. L., Bleeker, E. A., de Voogt, P., and Kraak, M. H. (2001). Phototoxicity of azaarene isomers to the marine flagellate Dunaliella tertiolecta. Environ. Toxicol. Chem. 20, 1544–1550. doi: 10.1002/etc.5620200718

Conflict of Interest Statement: The authors declare that the research was conducted in the absence of any commercial or financial relationships that could be construed as a potential conflict of interest.

17

A comprehensive model for chemical bioavailability and toxicity of organic chemicals based on first principles

Jay Forrest[1], Paul Bazylewski[1], Robert Bauer[1], Seongjin Hong[2], Chang Yong Kim[3], John P. Giesy[4,5,6,7,8], Jong Seong Khim[2]* and Gap Soo Chang[1]*

[1] Department of Physics and Engineering Physics, University of Saskatchewan, Saskatoon, SK, Canada
[2] School of Earth and Environmental Sciences and Research Institutes of Oceanography, Seoul National University, Seoul, Republic of Korea
[3] Canadian Light Source, Saskatoon, SK, Canada
[4] Department of Veterinary Biomedical Sciences and Toxicology Centre, University of Saskatchewan, Saskatoon, SK, Canada
[5] Department of Zoology, and Center for Integrative Toxicology, Michigan State University, East Lansing, MI, USA
[6] Department of Biology and Chemistry and State Key Laboratory in Marine Pollution, City University of Hong Kong, Kowloon, Hong Kong, China
[7] School of Biological Sciences, University of Hong Kong, Hong Kong, China
[8] State Key Laboratory of Pollution Control and Resource Reuse, School of the Environment, Nanjing University, Nanjing, China

Edited by:
Kyung-Hoon Shin, Hanyang University, South Korea

Reviewed by:
Xiaoshou Liu, Ocean University of China, China
Jongseong Ryu, Anyang University, South Korea

*Correspondence:
Jong Seong Khim, School of Earth and Environmental Sciences and Research Institutes of Oceanography, Seoul National University, Seoul 151-742, Republic of Korea
e-mail: jskocean@snu.ac.kr;
Gap Soo Chang, Department of Physics and Engineering Physics, University of Saskatchewan, Saskatoon, SK S7N 5E2, Canada
e-mail: gapsoo.chang@usask.ca

Here, we present a novel model to predict the toxicity and bioavailability of polychlorinated biphenyls (PCBs) as model compounds based on a first principles approach targeting basic electronic characteristics. The predictive model is based on an initio density functional theory. The model suggests HOMO-LUMO energy gap as the overarching indicator of PCBs toxicity, which was shown to be the primary factor predicting toxicity, but not the only factor. The model clearly explains why chlorination of both para positions is required for maximum toxic potency. To rank toxic potency, the "dipole moment" in relation to the most chemically active Cl-sites was critical. This finding was consistent with the accepted toxic equivalency factor (TEF) model for these molecules, and was also able to improve on ranking toxic potency of PCBs with similar TEFs. Predictions of HOMO-LUMO gap made with the model were consistent with measured values determined by synchrotron based X-ray spectroscopy for a subset of PCBs. HOMO-LUMO gap can also be used to predict bioaccumulation of PCBs. Overall, the new model provides an *in silico* method to screen a wide range of chemicals to predict their toxicity and bioavailability to act as an AhR agonist.

Keywords: bioaccumulation, dioxin-like PCBs, dipole moment, HOMO-LUMO, toxicity

INTRODUCTION

There is a large number of chemicals, both natural and synthetic that can bind to the aryl hydrocarbon receptor (AhR), sometimes referred to as the "*dioxin receptor*" (Giesy et al., 1994a,b). Most research into the toxicity of AhR-agonists and other chemicals has focused on assays with actual biomaterial, which can be time-consuming and require use of live animals. Alternatively, modern molecular techniques and *in silico* studies of quantitative structure-activity relationship (QSAR) methods are being used more and more to predict biological activities of organic molecules (Safe, 1993; Yang et al., 2009, 2010). Such approaches are, however, limited in their ability to explain the inherent nature of bioactivity (viz., potential toxicity) of chemicals because of their case- or compound-specific masking effects (Safe, 1993; Chana et al., 2002). This method is not always accurate and uses information on the physical structure of molecules to calculate toxicity.

Polychlorinated biphenyls (PCBs) were manufactured and used as electrical insulating liquids from 1929 to the late 1970s, when they were voluntarily withdrawn by some manufacturers and subsequently banned in Europe and North America, because of their environmentally persistent, bioaccumulative, and

potential toxic effects on wildlife and humans (DiGiovanni et al., 1987). Some congeners of PCBs, which comprise a group of 209 possible compounds with two chlorinated 6-member carbon (C) rings connected by a C-C bond (Giesy and Kannan, 1998) are potent, chronic toxins (Giesy et al., 2006). Previous researches have shown that, in animals or *in vitro*, PCBs are potentially carcinogenic (DiGiovanni et al., 1987), neurotoxic (Seegal, 1996), and also affect the endocrine system (Birnbaum, 1994).

Toxicity of PCBs is known to be dependent on the pattern of chlorine substitution, viz. their numbers and positions allocated, on the phenyl rings (Giesy et al., 2006). For example, a subgroup of 12 PCBs (non-*ortho*- or mono-*ortho*-chlorinated) exhibit "dioxin-like" activities associated with the AhR. It has also been shown that the energy gap between the highest occupied molecular orbital (HOMO) and the lowest unoccupied molecular orbital (LUMO) is an indication of stability in similar molecules (Lynam et al., 1998) where a larger gap implies a more stable molecule with respect to reactions with biomolecules. The relative stability of more chlorinated PCBs with greater lipophilicity results in widespread distribution in various environments followed by long-term accumulation in wildlife and humans (Alcock et al., 1998). Despite these extensive previous studies, toxicity of PCBs is known with limited accuracy and this makes PCBs an ideal candidate to test the proposed model for the toxicity and bioaccumulation behaviors of AhR-agonists and antagonists (Newsted and Giesy, 1987).

Here, an approach to predict toxicities of AhR-agonists from their physicochemical nature, namely their electronic characteristics, is presented. PCBs were selected as a model system to prove the concept of explaining toxicity based on chemical and electronic structure. The proposed first principles model allows for a fundamental understanding of toxic potency of individual PCB congeners but can be applied to other AhR-agonists, including both natural and synthetic compounds, encompassing both chlorinated and brominated or chloro-bromo-compounds. There are many natural and synthetic chemicals that can interact with the AhR (Giesy et al., 1994a). This model will also allow for the rapid screening and classification of tens of thousands prospective AhR-active chemicals without the need to use animal testing and thus help industry develop the chemical products with minimal environmental/toxic impact while retaining useful properties.

MATERIALS AND METHODS
DENSITY FUNCTIONAL THEORY CALCULATIONS

Ab initio DFT calculations were performed by use of Gaussian 03 software (Frisch et al., 2004). All calculations employed the 3-21G basis set. The exchange correlation functional and basis sets used were B3LYP and 6-21G. After geometry optimizations, density of states (DOS) and electric DM of each PCB congener were determined by calculations of occupied and virtual (unoccupied) molecular orbital (MO) energies. Mulliken charge population analysis and isosurface visualizations were also performed.

CHEMICALS AND SAMPLE PREPARATIONS

For X-ray spectroscopic measurements, selected PCB congeners including PCB-101, PCB-105, PCB-118, and PCB-153 were obtained from Wellington Laboratories (Guelph, Canada). One set of these congeners was then exposed to a rainbow trout microsome and incubated at 37°C for 3 h. The microsome was then extracted, leaving only metabolites of the reaction. Those metabolites were then dissolved in 200 μL hexane. Films of this hexane solution were fabricated on Si (100) wafers using spin-coating at 800 rpm for 20 s.

NEXAFS AND XES MEASUREMENTS

The unoccupied electronic structure of PCB molecules before and after exposure to microsomes was investigated by employing Near Edge X-Ray Absorption Fine Structure (NEXAFS) spectroscopy. Measurements of C 1s NEXAFS were made by use of the Spherical Grating Monochromator (SGM) beam line (Regier et al., 2007) at the Canadian Light Source of the University of Saskatchewan (Saskatoon, SK, Canada). Spectra were measured in total electron yield (TEY) mode. The C 1s NEXAFS spectra were energy calibrated with the π^* (C=C) transition at 285.5 eV of highly ordered pyrolytic graphite (HOPG). Samples were tested for radiation damage because of the sensitivity of carbon-based materials to X-ray radiation. This was performed by comparing results of five repeated measurements of the desired energy range for 1–2 min on the same sample spot. To reduce radiation damage,

concurrent measurements were taken at different locations of the same sample. Measurements made using NEXAFS spectra were intensity-normalized to the incoming photon flux as recorded by a photodiode. Non-resonant C $K\alpha$ ($2p \rightarrow 1s$ transition) XES were taken at Beam line 8.0.1 of the Advanced Light Source in Berkeley, CA (Jia et al., 1995), with excitation energy of 310 eV for all samples.

RESULTS AND DISCUSSION
FIRST PRINCIPLES MODEL OF TOXICITY

PCBs consist of two benzene rings bonded at opposite carbon sites, with varying number and location of chlorine termination. The molecular structure of PCBs and the chlorine termination, indicating nomenclature for the congeners, are shown in **Figure 1**. The 12 dioxin-like PCBs out of a total of 209 congeners have a few structural similarities. These similarity factors are: (i) co-planarity between two benzene rings, (ii) either zero or one Cl-substituted *ortho* position, (iii) both *para* positions chlorinated, and (iv) two or more *meta* positions chlorinated.

The dioxin-like PCBs are known to activate the AhR, otherwise referred as the "dioxin receptor." Since this involves chemical bonding through sharing or transferring of electrons, the above similarities must be associated with the electronic configuration of PCBs (Farmahin et al., 2013). Especially, HOMO and LUMO levels determine how a chemical molecule shares its valence electron in the occupied molecular orbitals and donates (or accepts) electrons to ligand, and thus the HOMO-LUMO energy gap has been suggested as an indicator of the stability of molecules (Lynam et al., 1998). Density functional theory (DFT) numerical calculations were made to obtain HOMO-LUMO energy gaps for all 209 PCB congeners. Calculated HOMO-LUMO gaps, calculated as a function of torsional angle between two benzene rings (α) are given in **Figure 1**.

Most coplanar PCBs (α = 0°) had lesser HOMO-LUMO gap energies than did non-coplanar congeners. The 12 dioxin-like PCB congeners are among those having lesser energy gaps. PCB-77 (3,3′,4,4′-tetrachlorobiphenyl) has an energy gap

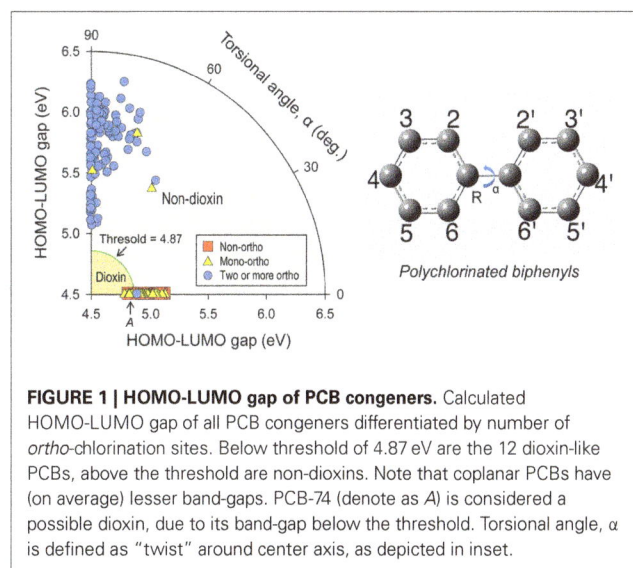

FIGURE 1 | HOMO-LUMO gap of PCB congeners. Calculated HOMO-LUMO gap of all PCB congeners differentiated by number of *ortho*-chlorination sites. Below threshold of 4.87 eV are the 12 dioxin-like PCBs, above the threshold are non-dioxins. Note that coplanar PCBs have (on average) lesser band-gaps. PCB-74 (denote as *A*) is considered a possible dioxin, due to its band-gap below the threshold. Torsional angle, α is defined as "twist" around center axis, as depicted in inset.

of 4.87 eV, which is proposed as a threshold energy gap for toxic potency of dioxin-like PCBs (see **Figure 1**). The results of these predictions are consistent with those of other studies showing that coplanar PCBs are more toxic through AhR-mediated mechanisms than are the di-*ortho*-substituted PCBs (Safe, 1993; Van den Berg et al., 2006). There was a relatively narrow distribution of HOMO-LUMO gaps among coplanar PCBs, with a range of 0.35 eV for all coplanar and only 0.09 eV for dioxin-like congeners. The range was wider for the other non-coplanar PCBs (1.2 eV). A larger range of energy-gaps for non-coplanar PCBs with various α angles and no PCB with a small α angle near the coplanar geometry suggest that dioxin-like PCBs are in a metastable state. Therefore, it is expected that dioxin-like PCBs would transform to non-coplanar configurations with larger HOMO-LUMO gaps and stabilize themselves by changing the configurations of chlorine atoms during reaction with the ligand binding domain (LBD) of the AhR protein.

The HOMO-LUMO gap for PCB-74 (2,4,4′,5-tetrachlorobiphenyl) is less than the suggested threshold for significant binding affinity to the LBD of the AhR. PCB-74 is not included in a subgroup of 12 dioxin-like PCBs to be classified as having two or more *meta*-chlorine atoms but it does satisfy the other similarity factors. The model developed here includes PCB-74 as a dioxin-like PCB, but is classified as a "possible" dioxin-like PCB.

Based on DFT calculations, the HOMO-LUMO energy gap is an overarching indicator of potency of AhR-mediated effects of PCBs that can distinguish the dioxin-like and non-dioxin-like molecules (*Similarity* 1). The trend observed in HOMO-LUMO gap and potency of PCB congeners is not consistent with the values for 2,3,7,8-TCDD toxic equivalency factors (TEFs) of PCBs (**Table 1**). This means that a small HOMO-LUMO energy gap that is less than the threshold is a necessary, but not a sufficient condition for significant potency of molecules and there must be other factors to consider when relating toxicity within the dioxin-like subgroup.

HOMO and LUMO isosurfaces of the most potent dioxin-like PCB-126 (3,3′,4,4′,5-pentachlorobiphenyl) and those for non dioxin-like PCB-107 (2,3,3′,4′,5-pentachlorobiphenyl) did not seem to demonstrate any AhR-mediated potency (**Figure 2**). Although there were no significant differences in spatial distribution of HOMO and LUMO states between two molecules with different chlorination configuration, the chlorinated sites contributed differently to the HOMO and LUMO states. For this reason another site-specific factor representing how sensitive each chlorinated site is to electron transfer (accepting or donating an electron) when a PCB molecule is interacting with the AhR was considered for inclusion in the predictive model.

Examination of the Cl-terminated sites of the 12 dioxin-like PCBs, $\Delta_+(\Delta_-)$ is the absolute difference in partial charge at the Cl-terminated sites between neutral and positively (negatively) ionized molecules. This factor, expressed in units of electron charge, e, is a measure of which sites are most chemically active, and is thus referred to as the "site-specific reactivity" factor. A Mulliken charge population analysis (Frisch et al., 2004) was performed for all dioxin-like PCB congeners in negatively and positively ionized states, as well as an electrically neutral state (**Figure 3**). *Para* positions (4 and 4′ sites) were most likely to either accept or donate more of its partial charge. Especially, when positively ionized, *para* sites have about 50% larger Δ_+ than other chlorinated sites. This observation is consistent with the *para* positions being most favorable sites for interaction with the AhR, followed by *meta* sites then *ortho* sites. Results of the Mulliken analysis were consistent with *similarity* factors 3 (both *para* positions chlorinated) and 4 (two or more *meta* positions chlorinated). Furthermore, Δ_+ was found to be consistently larger than Δ_- across all sites

Table 1 | Structural and electronic configuration parameters of the dioxin-like PCBs and the toxicity ranking predicted by first principles model.

PCB congener #	# of ortho	# of meta	Energy gap eV	$|\vec{p}|$	DM angle θ	DRF	TEF	Rank	Observed toxicity log (1/EC$_{50}$)	Lipophilicity log K$_{ow}$
NON-*ORTHO*										
126	0	3	4.84	0.5553	61.4	4.515	0.1	1	6.89	6.89
169	0	4	4.82	0.0000	N/A	4.021	0.03	2	na	7.42
81	0	2	4.86	0.5542	0	3.589	0.003	3	na	6.36
77	0	2	4.87	1.0000	90	3.107	0.0001	4	6.15	6.36
MONO-*ORTHO*										
189	1	4	4.78	0.2523	135.7	3.046	0.00003	5	na	7.71
157	1	3	4.82	0.6676	121.3	3.012	0.00003	6	5.33	7.18
167	1	3	4.80	0.3275	185.9	2.847	0.00003	7	4.80	7.27
156	1	3	4.83	0.2327	91.4	2.666	0.00003	8	5.15	7.18
105	1	2	4.81	0.3801	98.8	2.182	0.00003	9	5.37	6.65
118	1	2	4.82	0.4007	117	2.084	0.00003	10	4.85	6.74
114	1	2	4.78	0.5726	37.5	1.932	0.00003	11	5.39	6.65
123	1	2	4.83	1.0000	120.8	1.908	0.00003	12	na	6.74
74	1	1	4.84	0.0630	156.7	−1.762	0.00003	13	na	6.20

TEF, The toxic equivalency factor (TEF) values from Van den Berg et al. (2006); The observed toxicity data are AhR receptor binding affinities of PCB congeners from Mekenyan et al. (1996) and Safe et al. (1985); The log K$_{ow}$ values of PCB congeners from Svendsgaard et al. (1997).

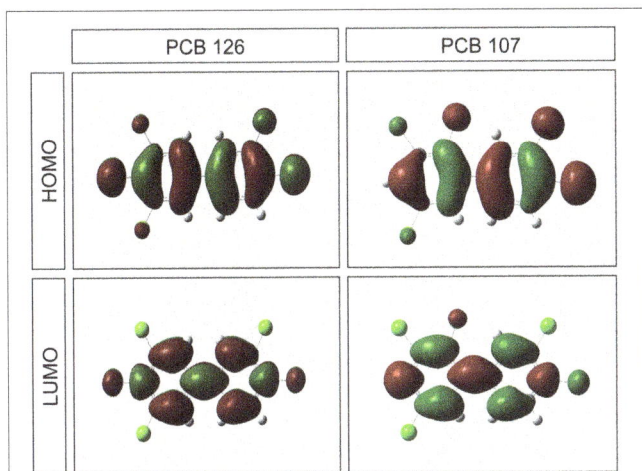

FIGURE 2 | HOMO and LUMO isosurfaces of PCBs. The HOMO and LUMO isosurfaces of PCB 126 and PCB-107.

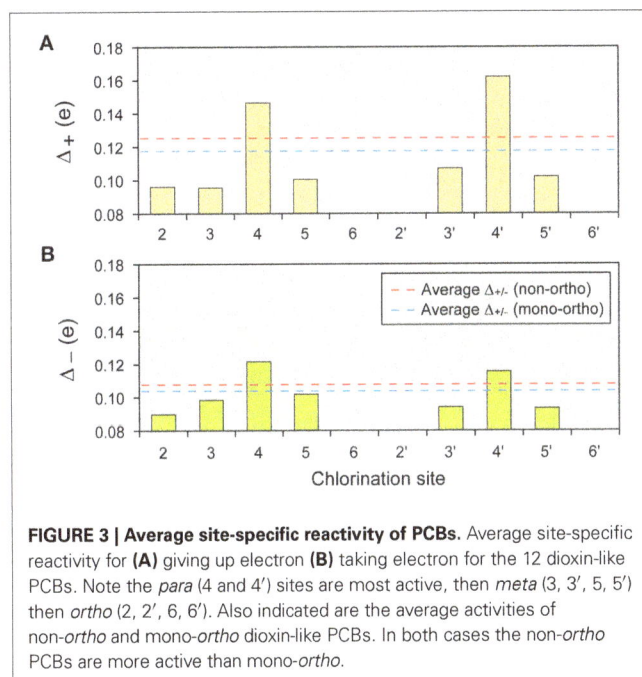

FIGURE 3 | Average site-specific reactivity of PCBs. Average site-specific reactivity for (A) giving up electron (B) taking electron for the 12 dioxin-like PCBs. Note the *para* (4 and 4') sites are most active, then *meta* (3, 3', 5, 5') then *ortho* (2, 2', 6, 6'). Also indicated are the average activities of non-*ortho* and mono-*ortho* dioxin-like PCBs. In both cases the non-*ortho* PCBs are more active than mono-*ortho*.

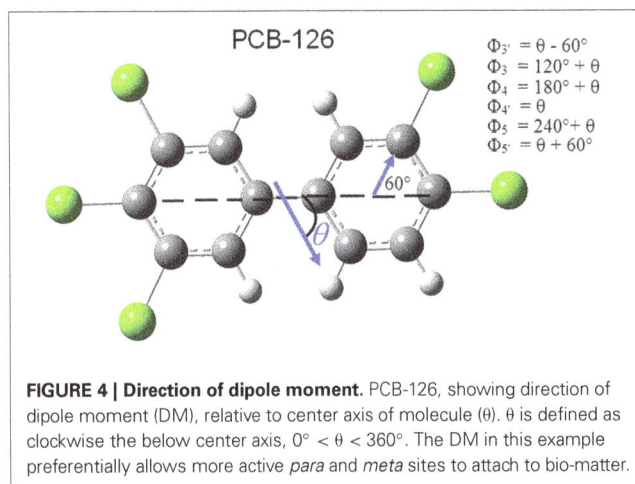

FIGURE 4 | Direction of dipole moment. PCB-126, showing direction of dipole moment (DM), relative to center axis of molecule (θ). θ is defined as clockwise the below center axis, $0° < \theta < 360°$. The DM in this example preferentially allows more active *para* and *meta* sites to attach to bio-matter.

and PCB molecules, which suggests that for a given interaction PCBs preferentially donate electron charge and become positively ionized.

The partial charge differences (Δ_+ and Δ_-) averaged over all chlorinated sites of non-*ortho*-substituted PCBs (red dashed lines) are greater than those of mono-*ortho*-substituted molecules (blue dashed lines). Overall chemical reaction behavior of the dioxin-like PCBs is therefore suppressed when the *ortho* position is chlorinated. The *ortho* chlorinated site has more delocalized HOMO states than do *meta* sites in the same benzene ring (3 and 5 sites). This means that the more the *ortho* positions are chlorinated, the more difficult it is for a PCB molecule to retain its coplanar structure. This is due to the Coulomb repulsion of delocalized π (occupied) orbitals of the two benzene rings.

By considering these effects of the *ortho*-chlorination, the 12 dioxin-like PCBs could be separated into two subgroups, the more potent non-*ortho* PCBs and the less potent mono-*ortho* PCBs (**Table 1**).

The relative toxic potency of each AhR-active PCB in each subgroup was then calculated. The first principles calculations suggested that binding of dioxin-like PCB with a receptor molecule, such as the AhR is not a random event but is strongly influenced by electronic configurations.

The dipole moment (DM) induced by nonhomogeneous charge distribution in a molecule can be a useful parameter for prediction of toxic potency, but the DM of the 12 dioxin-like PCBs varies widely from 0 to 3.3 D with no clear trend (Chana et al., 2002). Therefore, direction as well as magnitude of the DM should be considered relative to affinity of binding and thus potency of toxicity. It was hypothesized that the toxic potency would be greater when the DM direction with respect to the center axis of the molecule (θ as depicted in **Figure 4**) maximizes exposure of chemically reactive chlorinated sites to the AhR LBD. The 4 non-*ortho* PCBs in the more toxic subgroup, PCB-126 (3,3',4,4',5-pentachlorobiphenyl), PCB-169 (3,3'4,4',5,5'-hexachlorobiphenyl), PCB-81 (3,4,4',5-tetrachlorobiphenyl), and PCB-77 (3,3',4,4'-tetrachlorobiphenyl) were considered initially. As given in **Table 1**, PCB-81 has a DM directed along the center axis of a PCB ($\theta = 0°$) and, thus, the molecule can align such that the most reactive *para* sites are preferentially exposed to a neighboring receptor. In the case of PCB-77, the direction of the DM is perpendicular to the center axis ($\theta = 90°$), which results in preferential exposure of the 3 and 3' *meta* sites. This explains why PCB-81 has greater toxic potency than does PCB-77. PCB-126 has a DM directed $\theta \sim 61°$ below the molecule's center (see **Figure 4**). Although the DM alignment in this case has lesser site-specific reactivity contributed by *para* sites, the contribution from more *meta* chlorinated sites in PCB-126 than in PCB-81 and PCB-77 could compensate for the angle of the DM, producing a greater toxicity in PCB-126.

Based on this analysis, a function that describes the directional reactivity factor (DRF), analogous to TEF, was derived (Equation 1).

$$DRF = \sum_{n=1}^{6} \pm \Delta_{+n}(1 + |\vec{p}_n| \, |\cos \phi_n|) \tag{1}$$

This function represents the difference in partial charge, or site-specific reactivity (Δ_+), for each of the *para* and *meta* sites weighted by the direction and magnitude of the DM which maximize or minimize exposure of chlorine sites. Δ_{+n} is the normalized site-specific reactivity for a *para* or *meta* site n when removing one electron from the molecule. All six sites were summed regardless of chlorination where terms from chlorinated sites are positive (contribute to reactivity due to the electronegative Cl) and those from hydrogen sites are negative.

The site-specific reactivity was normalized to a value of 1.0 within the non-*ortho* and mono-*ortho* subgroups. $\left|\vec{P}_n\right|$ is the magnitude of the DM which is also normalized to 1 within the subgroups, such that it is of equivalent scale to reactivity. $\cos \phi_n$ represents the component of DM along the direction of the nth site's reactivity which is taken to act along the direction of the Cl-C (or H-C) bond for each *para* or *meta* site. Since all dioxin-like PCBs have an identical phenyl ring structure, the angular components ϕ_n are constant for all non-*ortho* or mono-*ortho* molecules (see **Figure 4**). The absolute value of the cosine was used because the DM acts as a weight for the reactivity only and should not produce negative values. That is, chlorinated sites always contribute positively to the DRF. The possible reduction of chemical reactivity due to dipole direction was taken into account by including hydrogen *para* and *meta* sites with normalized negative contribution.

This model, based on first-principles, was determined to work for non-*ortho* PCBs and mono-*ortho*-substituted, *or* dioxin-like PCBs (**Table 1**). The chlorination of PCB-169 is symmetric, and as such has no DM, but the chlorination pattern is such that all active sites (*para* and *meta*) are chlorinated and, thus can transfer electrons to, or take electrons from a neighboring receptor. In this case, toxic potency was determined by only reactivity of chlorinated sites. Because of its active surface, PCB-169 exhibits toxic potency that is lesser than that of congener 126, but greater than that of other congeners. Since these sites have already been established to be the most active, PCB-81 was classified as the third most potent congener. Finally PCB-77, which has a DM directed exactly perpendicular to the center axis does not preferentially expose *para* sites. However, it does expose the 3 and 3' *meta* sites, which are both chlorinated. Due to this DM direction, PCB-77 is classified as being less potent than the 3 most potent congeners, but because it is non-*ortho* it is still more potent than the other 8 dioxin-like PCBs.

All mono-*ortho* PCBs were ranked similarly with TEFs of 0.00003 (Van den Berg et al., 2006) (**Table 1**). PCB-123 (2',3,4,4',5-pentachlorobiphenyl) was predicted to be less potent than PCB-189 (2,3,3',4,4',5,5'-heptachlorobiphenyl) because the DM does not expose any *meta* sites, whereas PCB-189 does. An angle $\theta \sim 90°$ is least toxic, because it does not expose *para* sites. The model derived was able to rank this group without ambiguity. Due to its mono-*meta* nature, PCB-74 was ranked as the 13th most potent congener. The model produces a negative DRF value

for PCB-74 which indicates that the electron accepting nature of the hydrogenated *para* and *meta* sites outweighs the electron donating tendency of the chlorinated sites. In the DRF model, a value of 0 would be the threshold for dioxin-like toxicity, suggesting that PCB-74 has lesser toxic potency. Overall, the ranking of toxic potencies of PCB congeners suggested both DRF model and TEF were consistent. In addition, the DRF values of PCB congeners obtained from this study were well correlated with AhR binding affinity based on *in vitro* test ($r^2 = 0.63$, $p < 0.01$) (Safe et al., 1985; Mekenyan et al., 1996), indicating that the model was well-directed to predict the toxicity of PCB congeners. However, the DRF values were not correlated to the lipophilicity of PCB molecules such as log K_{ow} values (Svendsgaard et al., 1997). It is indicated that the log K_{ow} values of chemicals are not less explained the toxic potency in itself, but the values could be related to the bioaccessibility and bioavailability.

BIOACCUMULATION CHARACTERISTICS

Bioaccumulation of PCBs is a complex and multifaceted process since metabolism to hydroxylated PCBs is species-specific and dependent to the types of enzymes present (Kaminsky et al., 1981; WHO, 2003). As aforementioned, the HOMO-LUMO gap is one of the main parameters relating to the electron-mediated bonding. That is, if there is any increase in the HOMO-LUMO energy gap of a PCB molecule when absorbed into an organism, this molecule is stabilized by tight binding to the AhR and becomes more difficult to be metabolized. Bioaccumulation of chemicals is determined by their rates of depuration. In fact, hydroxylation makes molecules more polar and thus more likely to be excreted directly or as conjugates. Thus, monitoring of changes in HOMO-LUMO gap before and after exposure of animals to PCBs would allow prediction of bioaccumulation of the untransformed molecules.

Synchrotron-radiation X-ray spectroscopy was employed for empirical determination of HOMO-LUMO gaps. Near-Edge X-ray Absorption Fine Structure (NEXAFS) and X-ray Emission Spectroscopy (XES) probed unoccupied and occupied molecular orbitals of PCBs, respectively and determined the HOMO-LUMO energy gaps by superposing the C $1s$ NEXAFS and C $K\alpha$ XES spectra. To determine center peak locations corresponding to the HOMO and LUMO states more precisely, 2nd derivatives of spectra were used (Bazylewski et al., 2011). Spectroscopic measurements were obtained for four PCBs: PCB-101 (2,2',4,5,5'-pentachlorobiphenyl), PCB-105 (2,3,3',4,4'-pentachlorobiphenyl), PCB-118 (2,3',4,4',5-pentachlorobiphenyl), and PCB-153 (2,2'4, 4',5,5'-hexachlorobiphenyl). Superposition of non-resonant C $K\alpha$ XES and C $1s$ NEXAFS spectra of PCB-118 after microsome exposure (a) and their second derivatives (b), is given (**Figure 5**) as a representative example. This also demonstrates how the HOMO-LUMO energy gap was determined. The HOMO and LUMO levels were determined to be at 281.94 and 285.12 eV, respectively, which results in a gap of 3.18 eV. Measurements were repeated for four PCB congeners before and after exposure (**Figure 6**). It should be noted that the measured HOMO-LUMO gap of PCB-118 is smaller than that obtained by *ab-initio* calculations (4.82 eV). This is not unexpected due to the approximated

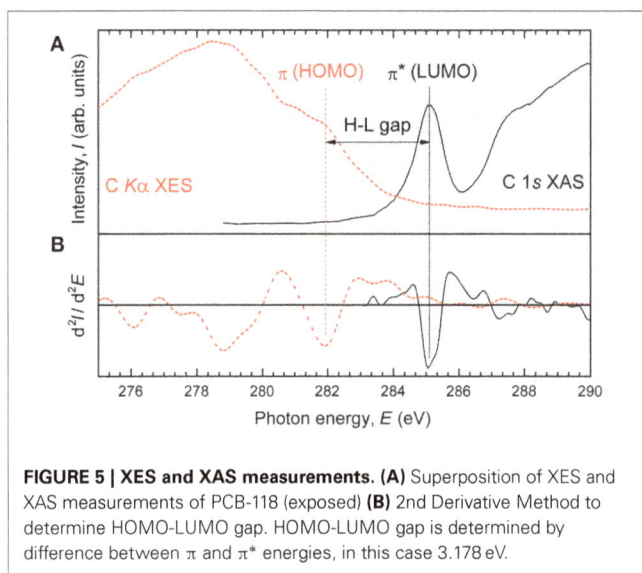

FIGURE 5 | XES and XAS measurements. (A) Superposition of XES and XAS measurements of PCB-118 (exposed) **(B)** 2nd Derivative Method to determine HOMO-LUMO gap. HOMO-LUMO gap is determined by difference between π and π* energies, in this case 3.178 eV.

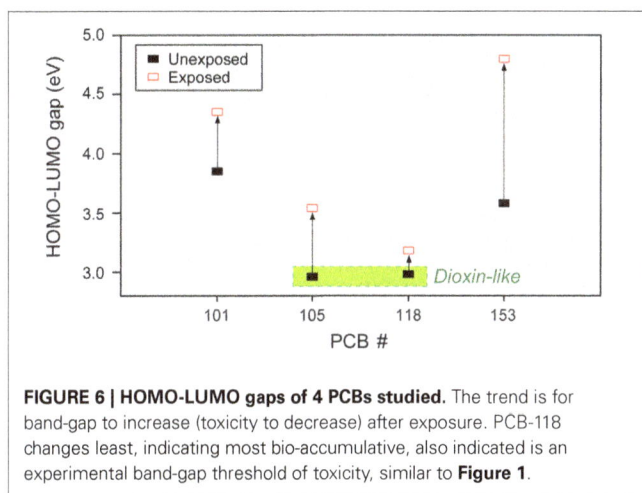

FIGURE 6 | HOMO-LUMO gaps of 4 PCBs studied. The trend is for band-gap to increase (toxicity to decrease) after exposure. PCB-118 changes least, indicating most bio-accumulative, also indicated is an experimental band-gap threshold of toxicity, similar to **Figure 1**.

be "band-gap." This separates more potent dioxin-like PCBs from less potent or non-dioxin-like PCB congeners. Dioxin-like PCBs were then examined more closely to determine "site-specific reactivity." This reactivity was shown to be greater for *para*, followed by *meta*, and finally *ortho* sites. Mono-*ortho*-substituted PCBs had, on average, lesser total reactivity than did non-*ortho* PCBs. This reactivity model is consistent with established purely structural models, and in fact explains the structure models in a more physical, non-empirical way. Ranking of toxic potency of the 12 known dioxin-like PCBs was accomplished by inclusion of a directional reactivity factor which allowed accurate prediction of TEF for and ranking of all 12 dioxin-like PCBs. Another possible dioxin-like PCB, PCB-74, was identified and ranked 13th. The model did not provide direct prediction of established TEF values since these are consensus values determined from a *meta*-analysis of a number of different endpoints for a wide range of species. Even though this model only ranks the dioxin-like PCBs, which mediate toxicity through the AhR, it could be used to predict relative toxicity of other AhR-active compounds. The methodology can also be applied to other receptor-mediated effects as long as the LBD of the receptor can be defined. The model is most powerful when the receptor structure, particularly the DM and position of likely attachment is known. At a minimum the model is useful for identifying other halogenated biphenyls, like polybrominated and mixed chloro-bromo analogs of the biphenyls, dioxins, furans and naphthalenes as well as alkylated polycyclic aromatic hydrocarbons that can bind to the AhR.

ACKNOWLEDGMENTS

We gratefully acknowledge support from the Natural Sciences and Engineering Research Council of Canada (NSERC) and Canada Foundation for Innovation. Research described in this paper was performed at the Canadian Light Source, which is supported by NSERC, the National Research Council Canada, the Canadian Institutes of Health Research, the Province of Saskatchewan, Western Economic Diversification Canada, and the University of Saskatchewan. This work was also supported by the project entitled "Development of Technology for CO_2 Marine Geological Storage" and "Oil Spill Environmental Impact Assessment and Environmental Restoration" funded by the Korean Ministry of Land, Transport, and Maritime Affairs given to Prof. Jong Seong Khim. Prof. John P. Giesy was supported by the Canada Research Chair program, a Visiting Distinguished Professorship in the Department of Biology and Chemistry and State Key Laboratory in Marine Pollution, City University of Hong Kong. He was also supported by the program of 2012 "High Level Foreign Experts" (#GDW20123200120) funded by the State Administration of Foreign Experts Affairs, P.R. China to Nanjing University and the Einstein Professor Program of the Chinese Academy of Sciences.

estimation of the exchange-correlation functional used in the DFT calculation. Despite this, there is a clear trend that the HOMO-LUMO energy gap increases for all PCBs after exposure.

Change in the HOMO-LUMO gap is different for each PCB. PCB-153 has a gap change of 1.2 eV, whereas PCB-118 shows a relatively small change of 0.2 eV. This result suggests that PCBs become more recalcitrant to chemical reaction after exposure to microsomes and that PCB-153 becomes more difficult to be metabolized and thus more bioaccumulative than other PCBs. Based on these results and the information from Section First Principles Model of Toxicity, it can be concluded that a small HOMO-LUMO gap is a necessary condition to estimate the toxic potency while the bioaccumulative behavior is closely associated with the change in the HOMO-LUMO gap rather than the magnitude of the gap (molecule's toxicity).

CONCLUDING REMARKS: ENVIRONMENTAL IMPLICATIONS

A model, based on first principles, was developed to predict the toxic potency of AhR-active compounds, including PCBs. The primary and necessary criterion for toxicity was determined to

REFERENCES

Alcock, R. E., Behnisch, P. A., Jones, K. C., and Hagenmaier, H. (1998). Dioxin-like PCBs in the environment—human exposure and the significance of sources. *Chemosphere* 37, 1457–1472. doi: 10.1016/S0045-6535(98)00136-2

Bazylewski, P. F., Kim, K. H., Forrest, J., Tada, H., Choi, D. H., and Chang, G. S. (2011). Side-chain effects on electronic structure and molecular stacking arrangement of PCBM spin-coated films. *Chem. Phys. Lett.* 508, 90–94. doi: 10.1016/j.cplett.2011.04.017

Birnbaum, L. S. (1994). Endocrine effects of prenatal exposure to PCBs, dioxins, and other xenobiotics: implications for policy and future research. *Environ. Health. Perspect.* 102, 676–679. doi: 10.1289/ehp.94102676

Chana, A., Concejero, M. A., de Frutos, M., González, M. J., and Herradón, B. (2002). Computational studies of biphenyl derivatives. Analysis of the conformational mobility, molecular electrostatic potential, and dipole moment of chlorinated biphenyl: searching for the rationalization of the selective toxicity of polychlorinated biphenyls (PCBs). *Chem. Res. Toxicol.* 15, 1514–1526. doi: 10.1021/tx025596d

DiGiovanni, J., Viaje, A., Berry, D. L., Slaga, T. J., and Juchau, M. R. (1987). Tumorinitiating ability of 2,3,7,8-tetrachlorodibenzo-p-dioxin (TCDD) and Arochlor 1254 in the two-stage system of mouse skin carcinogenesis. *Bull. Environ. Contam. Toxicol.* 18, 552–557. doi: 10.1007/BF01684000

Farmahin, R., Manning, G. E., Crump, D., Wu, D., Mundy, L. J., and Jones, S. P. (2013). Amino acid sequence of the ligan-binding domain of the aryl hydrocarbon receptor 1 predicts sensitivity of wild birds to effects of dioxin-like compounds. *Toxicol. Sci.* 131, 139–152. doi: 10.1093/toxsci/kfs259

Frisch, M. J., Trucks, G. W., Schlegel, H. B., Scuseria, G. E., Robb, M. A., and Cheeseman, J. R. (2004). *Gaussian 03, Revision C.02.* Wallingford, CT: Gaussian, Inc.

Giesy, J. P., and Kannan, K. (1998). Dioxin-like and non-dioxin like effects of polychlorinated biphenyls: implications for risk assessment. *Crit. Rev. Toxicol.* 28, 511–569. doi: 10.1080/10408449891344263

Giesy, J. P., Kannan, K., Jones, P. D., and Blankenship, A. L. (2006). "PCBs and related compounds," in *Endocrine Disruptors: Biological Basis for Health Effects in Wildlife and Humans*, ed A. Carr (New York, NY: Oxford University Press), 245–331.

Giesy, J. P., Ludwig, J. P., and Tillitt, D. E. (1994a). "Dioxins, dibenzofurans, PCBs and colonial, fish-eating water birds," in *Dioxin and Health*, ed A. Schecter (New York, NY: Plenum Press), 254–307.

Giesy, J. P., Ludwig, J. P., and Tillitt, D. E. (1994b). Embryolethality and deformities in colonial, fish-eating, water birds of the Great Lakes region: assessing causality. *Environ. Sci. Technol.* 28, 128A–135A.

Jia, J. J., Callcott, T. A., Yurkas, J., Ellis, A. W., Himpsel, F. J., and Samant, M. G. (1995). First experimental results from IBM/TENN/TULANE/LLNL/LBL undulator beamline at the advanced light source. *Rev. Sci. Instrum.* 66, 1394–1397. doi: 10.1063/1.1145985

Kaminsky, L. S., Kennedy, M. W., Adams, S. M., and Guengerich, F. P. (1981). Metabolism of dichlorobiphenyls by highly purified isozymes of rat liver cytochrome P450. *Biochemistry* 20, 7379–7384. doi: 10.1021/bi00529a009

Lynam, M., Kuty, M., Damborsky, J., Koca, J., and Adriaens, P. (1998). Molecular orbital calculations to describe microbial reductive dechlorination of polychlorinated dioxins. *Environ. Toxicol. Chem.* 17, 988–997. doi: 10.1002/etc.5620170603

Mekenyan, O. G., Veith, G. D., Call, D. J., and Ankley, G. T. (1996). A QSAR evaluation of Ah receptor binding of halogenated aromatic xenobiotics. *Environ. Health Perspect.* 104, 1302–1310. doi: 10.1289/ehp.961041302

Newsted, J. L., and Giesy, J. P. (1987). Predictive models for photoinduced acute toxicity of polycylic aromatic hydrocarbons to *Daphnia magna* Strauss (Cladocera, Crustacea). *Environ. Toxicol. Chem.* 6, 445–461. doi: 10.1002/etc.5620060605

Regier, T., Krochak, J., Sham, T. K., Hu, Y. F., Thompson, J., and Blyth, R. I. R. (2007). Performance and capabilities of the Canadian Dragon: the SGM beamline at the canadian light source. *Nucl. Instrum. Meth. A* 582, 93–95. doi: 10.1016/j.nima.2007.08.071

Safe, S. (1993). Toxicology, structure-function relationship, and human and environmental health impacts of polychlorinated biphenyls: progress and problems. *Environ. Health Perspect.* 100, 259–268. doi: 10.1289/ehp.93100259

Safe, S., Bandiera, S., Sawyer, T., Zmudzka, B., Mason, G., and Romkes, M. (1985). Effects of structure on binding to the 2,3,7,8-TCDD receptor protein and AHH induction-Halogenated biphenyls. *Environ. Health Perspect.* 61, 21–33.

Seegal, R. F. (1996). Epidemiological and laboratory evidence of PCB-induced neurotoxicity. *Crit. Rev. Toxicol.* 26, 709–737. doi: 10.3109/10408449609037481

Svendsgaard, D. J., Ward, T. R., Tilson, H. A., and Kodavanti, P. R. S. (1997). Empirical modeling of an *in vitro* activity of polychlorinated biphenyl congeners and mixtures. *Environ. Health Perspect.* 105, 1106–1115. doi: 10.1289/ehp.971051106

Van den Berg, M., Birnbaum, L. S., Denison, M., De Vito, M., Farland, W., and Freeley, M. (2006). The 2005 World Health Organization reevaluation of human and mammalian toxic equivalency factors for dioxins and dioxin-like compounds. *Toxicol. Sci.* 93, 223–241. doi: 10.1093/toxsci/kfl055

WHO (World Health Organization). (2003). *Polychlorinated Biphenyls: Human Health Aspects.* Geneva: World Health Organization, 1–64.

Yang, W., Liu, X., Liu, H., Wu, Y., Giesy, J. P., and Yu, H. (2010). Molecular docking and comparative molecular similarity indices analysis of estrogenicity of polybrominated diphenyl ethers and their analogues. *Environ. Toxicol. Chem.* 29, 660–668. doi: 10.1002/etc.70

Yang, W., Mu, Y., Giesy, J. P., Zhang, A., and Yu, H. (2009). Antiandrogen activity of polybrominated diphenyl ethers determined by comparative molecular similarity indices and molecular docking. *Chemosphere* 75, 1159–1164. doi: 10.1016/j.chemosphere.2009.02.047s

Conflict of Interest Statement: The authors declare that the research was conducted in the absence of any commercial or financial relationships that could be construed as a potential conflict of interest.

18

Need for monitoring and maintaining sustainable marine ecosystem services

Jacob Carstensen *

Department of Bioscience, Aarhus University, Roskilde, Denmark

Edited by:
Susana Agusti, The University of Western Australia, Australia

Reviewed by:
Jesus M. Arrieta, Instituto Mediterraneo de Estudios Avanzados, Spain
Juan-Carlos Molinero, GEOMAR Helmholtz Centre for Ocean Research Kiel, Germany
Dolors Vaque, Marine Sciences Institut (Consejo Superior de Investigaciones Científicas), Spain

*Correspondence:
Jacob Carstensen, Department of Bioscience, Aarhus University, Frederiksborgvej 399, DK-4000 Roskilde, Denmark
e-mail: jac@dmu.dk

Increases in human population and their resource use have drastically intensified pressures on marine ecosystem services. The oceans have partly managed to buffer these multiple pressures, but every single area of the oceans is now affected to some degree by human activities. Chemical properties, biogeochemical cycles and food-webs have been altered with consequences for all marine living organisms. Knowledge on these pressures and associated responses mainly originate from analyses of a few long-term monitoring time series as well as spatially scattered data from various sources. Although the interpretation of these data can be improved by models, there is still a fundamental lack of information and knowledge if scientists are to predict more accurately the effects of human activities. Scientists provide expert advices to society about marine system governance, but such advices should rest on a solid base of observations. Nevertheless, many monitoring programs around the world are currently facing financial reduction. Marine ecosystem services are already overexploited in some areas and sustainable use of these services can only be devised on a solid scientific basis, which requires more observations than presently available.

Keywords: biodiversity, ecosystem trends, eutrophication, food-webs, global change, ocean acidification, ocean governance, overfishing

INTRODUCTION

The last 10,000 years, known as the Holocene, have been a relatively stable period in earth's climate history (Petit et al., 1999), but recently human activities have become the main driver of environmental change at the local as well as global scale (Rockström et al., 2009). Humans have significantly altered the biogeochemical cycles on earth (Vitousek et al., 1997); something thought impossible just a few decades ago. Burning of fossil fuels, deforestation, mining, and other activities have increased the concentration of CO_2 in the atmosphere and ocean, elevating the greenhouse effect with rising temperatures as consequence. So far, the oceans have managed to store three times as much heat as the atmosphere (Levitus et al., 2001) and absorb about one third of the human-induced CO_2 emitted into the atmosphere (Steffen et al., 2007). However, recent studies suggest that the ocean's buffer capacity might decrease with further warming (Gruber et al., 2004).

Industrial nitrogen fixation and phosphate mining as well as fossil fuel burning have mobilized nitrogen and phosphorus (Vitousek et al., 1997). Humans have almost doubled the supply of nitrogen from the atmosphere to land, leading to an increased release of the greenhouse gas N_2O (Gruber and Galloway, 2008). Phosphate demands for agriculture have increased phosphorus inputs to the biosphere by factor of almost four (Falkowski et al., 2000). Nutrients applied to land as fertilizers are partly lost to the aquatic environment, eventually the ocean, where they stimulate production of organic matter, a process known as eutrophication (Nixon, 1995). One of the most deleterious effects of

eutrophication is the development of hypoxia (Carstensen et al., 2014), having strong ramifications on nutrient biogeochemical processes (Diaz and Rosenberg, 2008; Conley et al., 2009).

Human demand on fish has significantly reduced populations of marine top predators (Pauly et al., 1998), altering the flow of energy through food-webs and eventually leading to ecosystem collapses (Jackson et al., 2001). Fisheries landings have increased by more than 50% from 1970 to 2005 (Duarte et al., 2009) and the number of unsustainable fisheries is growing (Vitousek et al., 1997). In addition to reducing the overall population of marine top predators, overfishing has also selected toward smaller populations by removing the largest individuals (Jackson et al., 2001). It is possible that overfishing may exacerbate effects of eutrophication through trophic cascades, disrupting the normal flow of energy through marine food-webs (Scheffer et al., 2005). Another facet of altered energy flows is the global loss of biodiversity caused by overfishing, pollution, and habitat destruction reducing ocean ecosystem services (Worm et al., 2006).

Human pressures on marine ecosystems have increased recently to an extent where every area of the oceans is affected to some degree, although the human footprint is largest in the coastal zones with a high population density (Halpern et al., 2008). The multiple pressures of human activities have eroded the capacity of marine ecosystems to provide services benefitting humans. The oceans no longer constitute an infinite reservoir of natural resources that humans can exploit unconcerned. Therefore, science has an important role in identifying problems as well as their solutions, and conveying this knowledge broadly to the public and particularly, decision makers (Levin et al., 2009).

ASSESSING HUMAN IMPACTS ON MARINE ECOSYSTEMS

Our knowledge on human impacts on marine ecosystems has mainly been driven by observations supported by models for extrapolation. However, there is a significant lack of data on human pressures and marine effects, particularly in the open ocean. Data are often scattered in time and space, because they mostly arise from various research cruises and ships-of-opportunity; uncoordinated activities not aimed at assessing changes over time. Therefore, models are needed to integrate these data (e.g., Boyce et al., 2010; Halpern et al., 2012), but for many components of ocean health such models do not exist or they are so coarse that the reliability of the output may be disputable (Mackas, 2010; McQuatters-Gollop et al., 2010; Rykaczewski and Dunne, 2011).

Remote sensing data from satellites overcome the problem of spatial and temporal sampling heterogeneity and can be used for assessing changes in sea surface temperature and ocean color from which proxies for phytoplankton biomass and productivity can be derived (Behrenfeld et al., 2006), but they also have their limitations. Remote sensing applies to the upper surface layer only, and satellites cannot assess processes taking place at deeper depths. Algorithms for processing remote sensing data have mainly been developed for the open ocean, and the algorithms produce biases in shallower coastal waters. The proxy information obtained from satellite imagery provides only a small fraction of information needed to assess human impact on marine ecosystems.

Autonomous sensors typically placed on fixed buoys or floatable undulating devices such as Argo floats complement remote sensing by providing subsurface information on salinity, temperature, oxygen, and bio-optical properties (Roemmich et al., 2009). For instance, Argo float data with the support of global climate models revealed that the deep ocean (>300 m) was taking up more heat during the recent surface-temperature hiatus period (Meehl et al., 2011). At present, only the most basic physical-chemical variables are measured using these autonomous devices, since other measurements of interest (e.g., nutrient concentrations) typically require more regular maintenance, increasing the operating costs substantially.

Monitoring programs providing more consistent time series across a wide range of different physical, chemical and biological variables are found in certain coastal areas, e.g., the Chesapeake Bay and the Baltic Sea. These were typically initiated in the 1970s and 1980s, when pollution effects became clearly visible, to assess the efficiency of management actions to alleviate human pressure on overstressed marine ecosystems (Carstensen et al., 2006). In addition to assessing physical-chemical status, different organism groups from phytoplankton to top predators in the marine ecosystems were monitored. These monitoring programs have contributed substantially to our present understanding of trophic interactions in coastal areas and the disturbance of these imposed by human activities.

Understanding of long-term variations in ocean waters has so far been based on a few observatories, some of these organized

within the Long Term Ecological Research (LTER) Network (www.ilternet.edu). Long-term decreases in pH and aragonite saturation from the Hawaiian Ocean Time-series (HOT) and Bermuda Atlantic Time Series (BATS) have highlighted another problem associated with increased emission of CO_2, namely ocean acidification (Doney et al., 2009), which may alter ocean biogeochemistry (Beman et al., 2011). Long-term time series in coastal waters have revealed that pH is governed by changes in inputs from land rather than CO_2 in the atmosphere (Duarte et al., 2013). The Continuous Plankton Recorder (CPR) survey has been in operation since 1931 and has provided valuable insights into how climate oscillations affect plankton communities (Edwards et al., 2009). Since 1949 the California Cooperative Oceanic Fisheries Investigations (CalFOCI) program has investigated distributions of phytoplankton, zooplankton and fish distributions off Southern California and showed how changes in the Pacific Decadal Oscillation (PDO) can precipitate sudden shifts in these distributions (McGowan et al., 2003). Nevertheless, despite the value of these unique time series there is a need to establish and maintain ocean time series of high research quality, particularly in subtropical and tropical waters that are severely understudied at present.

DIRECTIONS FOR THE FUTURE

"We know more about the surface of the Moon and about Mars than we do about the deep sea floor, despite the fact that we have yet to extract a gram of food, a breath of oxygen or a drop of water from those bodies." This statement by Dr. Paul Snelgrove clearly articulates the need for improving our understanding of how marine ecosystems function, particularly as they provide essential ecosystem services to humans and because expanding human activities are putting these services under threat.

Our current understanding of marine ecosystem responses to human activities is limited by the availability of data, particularly long-term time series of physical and chemical conditions as well as biological properties. Moreover, efforts should be made to improve the accessibility and comparability of existing time series. Further development of models integrating monitoring data is needed to better assess changes over time and predict future trends, but models cannot stand alone without data. The lack of data is partly technical, as current measurement techniques may not necessarily provide the needed information, and partly financial, as costs of ocean sampling are indeed excessively expensive. Technological developments are expected to contribute more accurate, precise and cost-effective measurements over time. However, many marine monitoring programs are facing budget reductions, which have led to discontinuation of monitoring stations and abandoning sampling of biological components as well as decreasing monitoring frequencies. A possible consequence is loss of invested capital for establishing such long-term time series, simply because their value has to be written down. There is a growing discrepancy between the need for better understanding of human impact on marine ecosystems and the basis for addressing these scientific questions.

Ducklow et al. (2009) have identified seven key elements that will help science address critical issues on marine ecosystem services in times when human pressures on these are intensifying: (1) maintain existing monitoring programs and expand these with additional biological components, (2) establish new monitoring programs in under-sampled regions, (3) increase the use of remote sensing and autonomous monitoring devices, (4) establish targeted research program (process studies) in connection to long-term monitoring sites, (5) improve the integration of monitoring activities with ships-of-opportunity, (6) modify current funding for ecological research to balance consistent long-term research and short-term targeted studies, and (7) improve data access and synthesis using models. If these are recommendations are pursued we may eventually know more about our oceans than the surface of the Moon and Mars. The growing human imprint on marine ecosystems may, if left unmonitored and unattended, result in significant losses of ecosystem services that are crucial to support a globally growing population.

ACKNOWLEDGMENTS

This manuscript is a contribution from the DEVOTES project (DEVelopment Of Innovative Tools for understanding marine biodiversity and assessing good Environmental Status; www.devotes-project.eu), funded by the European Union under the 7th Framework Programme (grant agreement no.308392) and the WATERS project (Waterbody Assessment Tools for Ecological Reference conditions and status in Sweden).

REFERENCES

Behrenfeld, M. J., O'Malley, R. T., Siegel, D. A., McClain, C. R., Sarmiento, J. L., Feldman, G. C., et al. (2006). Climate-driven trends in contemporary ocean productivity. *Nature* 444, 752–755. doi: 10.1038/nature05317

Beman, J. M., Chow, C.-E., King, A. L., Feng, Y., Fuhrman, J. A., Andersson, A., et al. (2011). Global declines in oceanic nitrification rates as a consequence of ocean acidification. *Proc. Nat. Acad. Sci. U.S.A.* 108, 208–213. doi: 10.1073/pnas.1011053108

Boyce, D. G., Lewis, M. R., and Worm, B. (2010). Global phytoplankton decline over the past century. *Nature* 466, 591–596. doi: 10.1038/nature09268

Carstensen, J., Andersen, J. H., Gustafsson, B. G., and Conley, D. J. (2014). Deoxygenation of the Baltic Sea during the last century. *Proc. Nat. Acad. Sci. U.S.A.* 111, 5628–5633. doi: 10.1073/pnas.1323156111

Carstensen, J., Conley, D. J., and Andersen, J. H., Ærtebjerg, G. (2006). Coastal eutrophication and trend reversal: a danish case study. *Limnol. Oceanogr.* 51, 398–408. doi: 10.4319/lo.2006.51.1_part_2.0398

Conley, D. J., Björck, S., Bonsdorff, E., Carstensen, J., Destouni, G., and Gustafsson, B. G. (2009). Hypoxia-related processes in the Baltic Sea. *Environ. Sci. Technol.* 43, 3412–3420. doi: 10.1021/es802762a

Diaz, R. J., and Rosenberg, R. (2008). Spreading dead zones and consequences for marine ecosystems. *Science* 321, 926–929. doi: 10.1126/science.1156401

Doney, S. C., Fabry, V. J., Feely, R. A., and Kleypas, J. A. (2009). Ocean acidification: the other CO_2 problem. *Annu. Rev. Mar. Sci.* 1, 169–192. doi: 10.1146/annurev.marine.010908.163834

Duarte, C. M., Conley, D. J., Carstensen, J., and Sánchez-Camacho, M. (2009). Return to Neverland: shifting baselines affect eutrophication restoration targets. *Estuar. Coasts* 32, 29–36. doi: 10.1007/s12237-008-9111-2

Duarte, C. M., Hendriks, I. E., Moore, T. S., Olsen, Y. S., Steckbauer, A., Ramajo, L., et al. (2013). Is ocean acidification an open-ocean syndrome? Understanding anthropogenic impacts on seawater pH. *Estuar. Coasts.* 36, 221–236. doi: 10.1007/s12237-013-9594-3

Ducklow, H. W., Doney, S. C., and Steinberg, D. K. (2009). Contributions of long-term research and time-series observations to marine ecology and biogeochemistry. *Annu. Rev. Mar. Sci.* 1, 279–302. doi: 10.1146/annurev.marine.010908.163801

Edwards, M., Beaugrand, G., Hays, G. C., Koslow, A., and Richardson, A. J. (2009). Multi-decadal oceanic ecological datasets and their application in marine policy and management. *Trends Ecol. Evol.* 25, 602–610. doi: 10.1016/j.tree.2010.07.007

Falkowski, P., Scholes, R. J., Boyle, E., Canadell, J., Canᵭeld, D., Elser, J., et al. (2000). The global carbon cycle: a test of our knowledge of earth as a system. *Science* 290, 291–296. doi: 10.1126/science.290.5490.291

Gruber, N., Friedlingstein, P., Field, C. B., Valentini, R., Heimann, M., Richey, J. E., et al. (2004). "The vulnerability of the carbon cycle in the 21st century: an assessment of carbon-climate-human interactions" in *The Global Carbon Cycle: Integrating Humans, Climate, and the Natural World*, eds C. B. Field and M. R. Raupach (Washington, DC: Island Press), 45–76.

Gruber, N., and Galloway, J. N. (2008). An Earth-system perspective of the global nitrogen cycle. *Nature* 451, 293–296. doi: 10.1038/nature06592

Halpern, B. S., Longo, C., Hardy, D., McLeod, K. L., Samhouri, J. F., Katona, S. K., et al. (2012). An index to assess the health and benefits of the global ocean. *Nature* 488, 615–621. doi: 10.1038/nature11397

Halpern, B. S., Walbridge, S., Selkoe, K. A., Kappel, C. V., Micheli, F., D'Agrosa, C., et al. (2008). A global map of human impact on marine ecosystems. *Science* 319, 948–952. doi: 10.1126/science.1149345

Jackson, J. B. C., Kirby, M. X., Berger, W. H., Bjorndal, K. A., Botsford, L. W., and Bourque, B. J. (2001). Historical overfishing and the recent collapse of coastal ecosystems. *Science* 293, 629–638. doi: 10.1126/science.1059199

Levin, P. S., Fogarty, M. J., Murawski, S. A., and Fluharty, D. (2009). Integrated dcosystem assessments: developing the scientific basis for ecosystem-based management of the ocean. *PLoS Biol.* 7:e1000014. doi: 10.1371/journal.pbio.1000014

Levitus, S., Antonov, J. L., Wang, J., Delworth, T. L., Dixon, K. W., and Broccoli, A. J. (2001). Anthropogenic warming of Earth's climate system. *Science* 292, 267–270. doi: 10.1126/science.1058154

Mackas, D. L. (2010). Does blending of chlorophyll data bias temporal trend? *Nature* 472, E4–E5. doi: 10.1038/nature09951

McGowan, J. A., Bograd, S. J., Lynn, R. J., and Miller, A. J. (2003). The biological response to the 1977 regime shift in the California Current. *Deep Sea Res.* 50(Pt. II), 2567–2582. doi: 10.1016/S0967-0645(03)00135-8

McQuatters-Gollop, A., Reid, P. C., Edwards, M., Burkill, P. H., Castellani, C., Batten, S., et al. (2010). Is there a decline in marine phytoplankton? *Nature* 472, E6–E7. doi: 10.1038/nature09950

Meehl, G. A., Arblaster, J. M., Fasullo, J. T., Hu, A., and Trenberth, K. E. (2011). Model-based evidence of deep-ocean heat uptake during surface-temperature hiatus periods. *Nature Clim. Change* 1, 360–364. doi: 10.1038/nclimate1229

Nixon, S. W. (1995). Coastal marine eutrophication: a definition, social causes, and future concerns. *Ophelia* 41, 199–219. doi: 10.1080/00785236.1995.10422044

Pauly, D., Christensen, V., Dalsgaard, J., Froese, R., and Torres, F. Jr. (1998). Fishing down marine food webs. *Science* 279, 860–863. doi: 10.1126/science.279.5352.860

Petit, J. R., Jouzel, J., Raynaud, D., Barkov, N. I., Barnola, J.-M., Basile, I., et al. (1999). Climate and atmospheric history of the past 420,000 years from the Vostok ice core, Antarctica. *Nature* 399, 429–436. doi: 10.1038/20859

Rockström, J., Steffen, W., Noone, K., Persson, Å., Chapin, F. S. 3rd., Lambin, E. F., et al. (2009). A safe operating space for humanity. *Nature* 461, 472–475. doi: 10.1038/461472a

Roemmich, D., Johnson, G. C., Riser, S., Davis, R., Gilson, J., Owens, W. B., et al. (2009). The Argo program: observing the global ocean with profiling floats. *Oceanography* 22, 34–43. doi: 10.5670/oceanog.2009.36

Rykaczewski, R. R., and Dunne, J. P. (2011). A measured look at ocean chlorophyll trends. *Nature* 472, E5–E6. doi: 10.1038/nature09952

Scheffer, M., Carpenter, S., and de Young, B. (2005). Cascading effects of overfishing marine systems. *Trends Ecol. Evol.* 20, 579–581. doi: 10.1016/j.tree.2005.08.018

Steffen, W., Crutzen, P. J., and McNeill, J. R. (2007). The Anthropocene: are humans now overwhelming the great forces of nature? *Ambio* 36, 614–621. doi: 10.1579/0044-7447(2007)36[614:TAAHNO]2.0.CO;2

Vitousek, P. M., Mooney, H. A., Lubchenco, J., and Melillo, J. M. (1997). Human domination of Earth's ecosystems. *Science* 277, 494–499. doi: 10.1126/science.277.5325.494

Worm, B., Barbier, E. B., Beaumont, N., Duffy, J. E., Folke, C., Halpern, B. S., et al. (2006). Impacts of biodiversity loss on ocean ecosystem services. *Science* 314, 787–790. doi: 10.1126/science.1132294

Conflict of Interest Statement: The author declares that the research was conducted in the absence of any commercial or financial relationships that could be construed as a potential conflict of interest.

19

Broad distribution and high proportion of protein synthesis active marine bacteria revealed by click chemistry at the single cell level

Ty J. Samo*, Steven Smriga†, Francesca Malfatti†, Byron P. Sherwood† and Farooq Azam

Marine Biology Research Division, Scripps Institution of Oceanography, University of California, San Diego, La Jolla, CA, USA

Edited by:
Hongyue Dang, Xiamen University, China

Reviewed by:
Gerhard Josef Herndl, University of Vienna, Austria
Craig E. Nelson, University of Hawaii at Manoa, USA

*Correspondence:
Ty J. Samo, Department of Oceanography, Center for Microbial Oceanography: Research and Education, School of Ocean and Earth Science and Technology, University of Hawaii at Manoa, 1950 East-West Rd., Honolulu, HI 96822, USA
e-mail: tsamo@hawaii.edu

†Present address:
Steven Smriga, Ralph M. Parsons Laboratory, Department of Civil and Environmental Engineering, Massachusetts Institute of Technology, Cambridge, USA
Francesca Malfatti, Oceanography Section (OCE), National Institute of Oceanography and Experimental Geophysics (OGS), Trieste, Italy
Byron P. Sherwood, Department of Oceanography, Center for Microbial Oceanography: Research and Education, School of Ocean and Earth Science and Technology, University of Hawai'i at Mânoa, Honolulu, USA

Marine bacterial and archaeal communities control global biogeochemical cycles through nutrient acquisition processes that are ultimately dictated by the metabolic requirements of individual cells. Currently lacking, however, is a sensitive, quick, and quantitative measurement of activity in these single cells. We tested the applicability of copper (I)-catalyzed cycloaddition, or "click," chemistry to observe and estimate single-cell protein synthesis activity in natural assemblages and isolates of heterotrophic marine bacteria. Incorporation rates of the non-canonical methionine bioortholog L-homopropargylglycine (HPG) were quantified within individual cells by measuring fluorescence of alkyne-conjugated Alexa Fluor®488 using epifluorescence microscopy. The method's high sensitivity, along with a conversion factor derived from two *Alteromonas spp.* isolates, revealed a broad range of cell-specific protein synthesis within natural microbial populations. Comparison with ^{35}S-methionine microautoradiography showed that a large fraction of the natural marine bacterial assemblage (15–100%), previously considered inactive by autoradiography, were actively synthesizing protein. Data pooled from 21 samples showed that cell-specific activity scaled logarithmically with cell volume. Activity distributions of each sample were fit to power-law functions, providing an illustrative and quantitative comparison of assemblages that demonstrate individual protein synthesis rates were commonly partitioned between cells in low- and high-metabolic states in our samples. The HPG method offers a simple approach to link individual cell physiology to the ecology and biogeochemistry of bacterial (micro)environments in the ocean.

Keywords: marine bacteria, microscopy, click chemistry, HPG, single-cell protein production, ecology, oceanography, biogeochemistry

INTRODUCTION

The metabolism and growth responses of marine bacteria and archaea communities significantly affect global ocean ecology and biogeochemical cycles. Diverse microhabitats are an important component of microbial ecosystems, existing within an organic matter continuum typified by a variety and abundance of colloids, gels, and particles of varying chemical and physical attributes (Azam and Worden, 2004; Verdugo et al., 2004). Characterizing the adaptive responses, particularly growth, of individual bacteria and archaea (collectively here called "bacteria") may assist in quantifying the microenvironmental regulation of microbial communities and constrain estimates of their biogeochemical effects. A quantitative individual cell approach should help test the significance of microscale heterogeneity for the maintenance of bacterial genomic and functional diversity in the ocean.

Previous studies have shown the existence of microscale variation in bacterial community composition and activity (Long and Azam, 2001b; Barbara and Mitchell, 2003; Seymour et al., 2005). The challenge now is to develop sensitive and quantitative methods for measuring individual cell activities within these microscale patches. Currently, individual cell growth is studied using a variety of approaches. This includes ^{3}H-based microautoradiography (Fuhrman and Azam, 1982; Cottrell and Kirchman, 2004) and the use of a non-radioisotopic, fluorescence-based method detecting incorporation of the thymidine analog bromodeoxyuridine (BrdU) (Pernthaler et al., 2002; Hamasaki et al., 2004). Other studies have documented mRNA fluorescent *in situ* hybridization (mRNA FISH; Pernthaler and Amann, 2004) or single-cell analyses of activity using nano secondary ion mass spectrometry (nanoSIMS) to quantify metabolic fluxes of symbioses and inter-species interactions (Orphan et al., 2002; Foster et al., 2011; Musat et al., 2012). While microautoradiography has been used extensively, the use of radioisotopes can be problematic or even prohibited in many field settings. The BrdU method overcomes some of these challenges and has been applied to individual cell growth measurements (Tada et al., 2010; Galand et al., 2013). FISH methodologies are robust and focus on taxonomic and/or functional potential, but provide little in rate characterizations. Meanwhile, the high spatial resolution combined with sensitive quantification of incorporated, stable isotope-labeled compounds makes nanoSIMS the most sensitive technique for single cell taxonomic and metabolic quantification. But measurements are relatively low-throughput and dependent upon instrument availability.

Over the past several years, cell biology researchers have adopted the use of bioorthogonal amino acids (Dieterich et al., 2006; tom Dieck et al., 2012) coupled with copper catalyzed azide-alkyne cycloaddition chemistry (one type of many "click" methodologies) for proteomic and protein expression studies of mammalian cells *in vivo* (Best, 2009; below). Recently, a group has described a method using the azide-bearing methionine surrogate azidohomoalanine (AHA) on microbial assemblages to identify translationally active cells in combination with the taxonomic identity via FISH (Hatzenpichler et al., 2014). We have developed a separate method to measure protein synthesis rates in natural planktonic microbial assemblages using the alkyne-bearing methionine analog 2-Amino-5-hexynoic acid, commonly known as L-homopropargylglycine (HPG).

A significant technical advantage of click chemistry is use of a fluorophore bound to a small molecule, thus eliminating the need to permeabilize cells (as used in the BrdU method) and resulting in simple, rapid sample processing (Smriga et al., 2014). We found that the high sensitivity and low signal background of the HPG method allowed us to detect a broad spectrum of individual cell activity within natural marine assemblages. This new capability increases the capacity to directly address the long-standing question of what proportion of bacterial communities are active and their relative cell-specific activity, as well as estimated cell-specific growth rates. In this study, we describe the advantages and limitations of the method via comparisons with bulk radioisotope incorporation and microautoradiography. We then present a comprehensive set of experiments conducted in the laboratory and field that test the validity and limitations of bacterial protein synthesis rate measurements in both simple cultures and complex heterogeneous environmental samples.

RESULTS

ADAPTING CLICK CHEMISTRY FOR MARINE BACTERIA

Several important factors were identified and adopted into the HPG click chemistry protocol including: (i) immobilization of cells onto filter membranes rather than centrifugation; (ii) use of the filter-transfer-freeze technique to reduce background fluorescence; and (iii) use of glass coverslips during the click reaction to limit oxygen exposure and thereby maintain reduced conditions for conversion of copper (II) to copper (I) with subsequent bonding of the fluorescent azide (Alexa Fluor 488) to the incorporated alkyne (HPG) (**Figure 1A**).

SIGNIFICANT BACKGROUND REDUCTION BY THE FILTER-TRANSFER-FREEZE TECHNIQUE

When measured on polycarbonate filters, HPG-labeling ranged 9.8–47% of total DAPI-stained bacteria among five samples (**Figure S1**). Application of the FTF technique to five parallel samples resulted in a 10-fold reduction in background fluorescence (**Figure 1B**) and increased the proportion of HPG-labeled cells to 14.6–100% (**Figure S1**; example microautoradiography image in **Figure S2**). In parallel incubations of two separate seawater assemblages, percent labeling with the HPG method (using the FTF technique) was ~5- to 8-fold higher than with ^{35}S-methionine microautoradiography (**Table 1**). FTF bacterial transfer efficiency ranged 69–200% because some cells remained stuck to the filter (underestimation) and transferred cells aggregated in condensation pools on the coverslip (overestimation). The average labeling percentages calculated from each field of view were positively correlated with the absolute percentages calculated from all fields ($R = 0.99$, $p < 0.0001$), and they exhibited a nearly 1:1 ratio, as shown by a model II linear regression yielding a significantly non-zero slope of 0.98 ($p < 0.0001$; 95% confidence intervals = 0.95–1.017). Thus, assuming equal transfer efficiency of labeled and unlabeled cells, individual cell signal intensities and labeling percentages (**Table 2**) should not have been affected by FTF.

VALIDATING USE OF HPG

HPG incorporation specificity and efficacy as methionine analog

We determined that HPG labeling did not occur in the presence of chloramphenicol, and that the presence of methionine at 1 or 2 µM reduced or eliminated HPG labeling percentages in four separate tests of natural assemblages (**Figure S3**).

An experiment that tested the competition of HPG with ^{35}S-methionine incorporation suggested the presence of HPG (8 or 18 nM) did not significantly change ^{35}S-methionine incorporation while the presence of cold-methionine (8 or 18 nM) decreased ^{35}S-methionine incorporation by 2.3- and 3.7-fold, respectively ($p < 0.05$; **Figure S4**). In a second experiment, presence of HPG at 20 nM resulted in an insignificant reduction of ^{35}S-methionine incorporation (1.3-fold; $p = 0.06$), but 200 nM and 2 µM HPG elicited significant reductions (2.8-fold; $p < 0.05$ and 12.5-fold; $p < 0.05$, respectively; **Figure S4**).

Additional experiments examined the potential for HPG to competitively inhibit methionine incorporation. Enzyme kinetic curve fit calculations indicated that K_m for ^{35}S-methionine decreased with higher HPG concentrations but there was no effect on V_{max}. The calculated inhibition constant (K_i) for HPG was 373 nM (numerically equal to K_m for HPG incorporation) while the K_m for methionine incorporation was an order of magnitude lower at 31 nM (**Figure 2**). The results support the conclusion that HPG competes with methionine for incorporation into cells within an enzyme-substrate binding framework (Copeland, 2005).

Conversion factor for HPG fluorescence to protein production rate

We calculated a conversion factor using 3 different approaches and adopted the "top 10% isolate mean" calculation (see Materials and Methods). The individual cell HPG signal

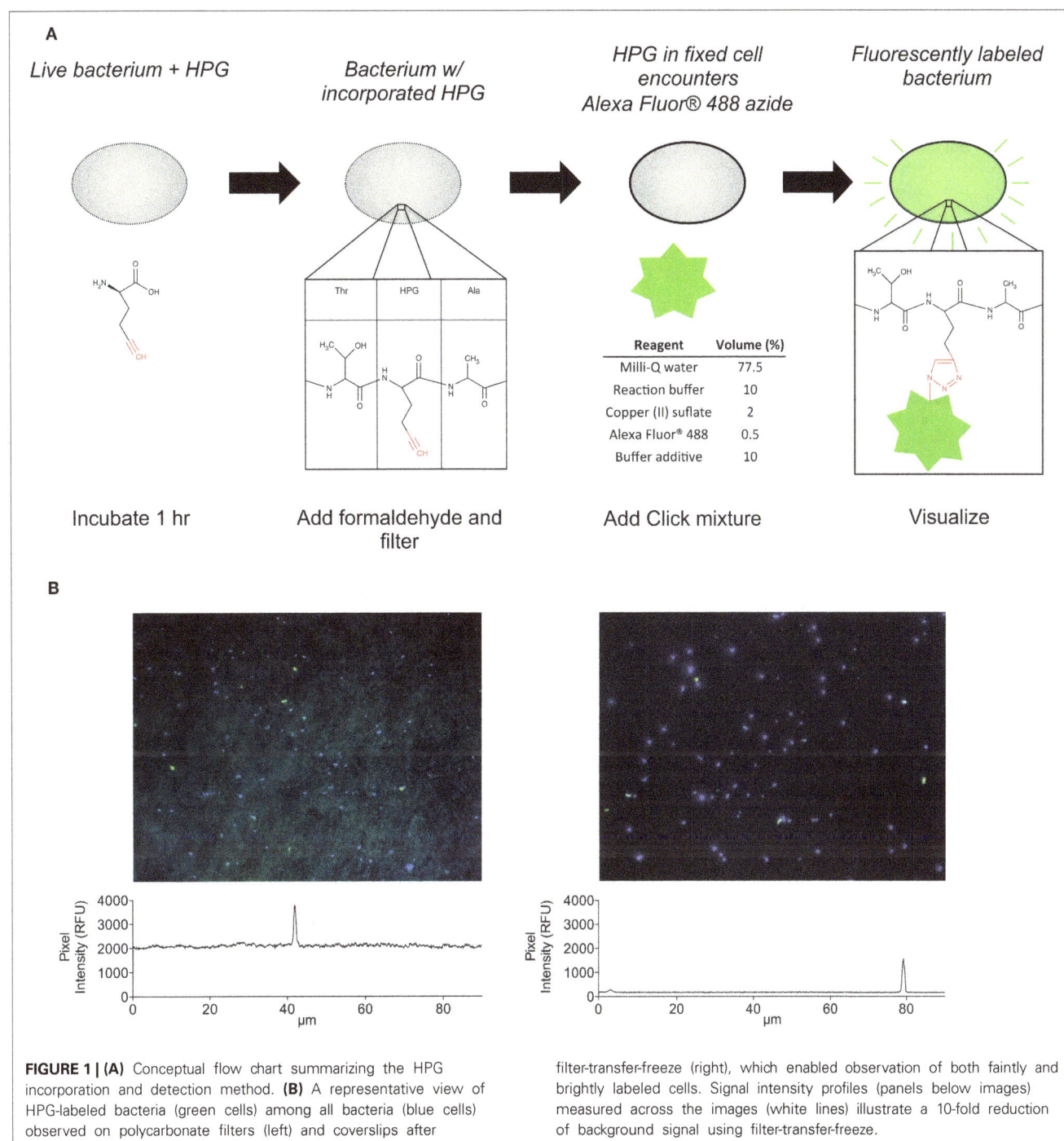

FIGURE 1 | (A) Conceptual flow chart summarizing the HPG incorporation and detection method. **(B)** A representative view of HPG-labeled bacteria (green cells) among all bacteria (blue cells) observed on polycarbonate filters (left) and coverslips after filter-transfer-freeze (right), which enabled observation of both faintly and brightly labeled cells. Signal intensity profiles (panels below images) measured across the images (white lines) illustrate a 10-fold reduction of background signal using filter-transfer-freeze.

intensities of the *Alteromonas* spp. AltSIO and As1 used for the conversion factors ranged 1.795–5.617 $\log_{10}$ RFU cell^{-1} ($n = 1697$). The range in protein for the top 10% cells was 18.2–136.1 fg cell^{-1} and the resulting conversion factor was 5.358 $\log_{10}$ RFU fg^{-1} protein; this value was used to convert a single cell HPG signal to a single cell protein production rate (scPP) for all labeled cells in natural communities. The detection limit of the method was determined to be 0.07 fg cell^{-1} ($SE = 6.1 \times 10^{-4}$, $n = 7136$) and was calculated by taking the average value of all control cells from all samples after applying the conversion factor.

Protein mass determined from all cell volumes ranged 8.1–136.1 fg cell^{-1}. The two remaining calculations, while not adopted, provided the following conversion values: (i) "isolates regression," 5.620 $\log_{10}$ RFU fg^{-1} protein ($R^2 = 0.41$, $p > 0.0001$) and (ii) "community sum ^{35}S-methionine," 5.166 $\log_{10}$ RFU fg^{-1} protein. The "isolates regression" calculation assumed all labeled cells, including low intensity cells, harbored

Table 1 | Percent labeling in two natural assemblages as quantified via three methods: (i) ^{35}S-methionine microautoradiography, (ii) HPG on polycarbonate filters, and (iii) HPG with the freeze-transfer-freeze method.

Date	^{35}S-methionine positive (%)	HPG positive on filter (%)	HPG positive with FTF (%)
8/17/10	16.1 ± 1.0	12.8 ± 0.5	25.7 ± 2.0
10/11/10	17.4 ± 1.9	14.3 ± 0.7	93.8 ± 7.6

Seawater was incubated for 1 h. Substrate concentrations were 2 nM ^{35}S-methionine + 18 nM cold methionine or 20 nM HPG.

Table 2 | Overview of HPG percent labeling for 21 sampling dates at Scripps Pier.

Date and time	Number of cells analyzed	Abundance (cells ml^{-1} × 10^6)	Percent labeled (±SE)
11/18/09 10:00	412	1.21	14.9 ± 1.6
4/30/10 10:00	1416	2.38	49.3 ± 2.3
8/17/10 16:00	1492	2.41	25.7 ± 2.0
10/6/10 10:00	1676	3.25	69.3 ± 2.8
10/11/10 14:00	1504	2.87	93.8 ± 7.6
10/20/10 14:00	1136	1.62	40.4 ± 2.8
12/15/10 10:00	399	1.18	16.3 ± 1.8
3/30/11 12:00	769	0.82	100 ± 0
3/30/11 16:00	1111	0.74	90.2 ± 10.2
3/30/11 21:00	804	0.56	100 ± 0
3/31/11 12:00	1034	1.20	63.2 ± 4.6
3/31/11 16:00	979	2.71	54.7 ± 12.1
3/31/11 20:00	1631	0.78	79.0 ± 8.6
4/1/11 12:00	1164	3.82	66.1 ± 6.6
4/1/11 16:00	1487	1.24	96.6 ± 6.0
4/1/11 21:00	817	0.70	64.6 ± 7.7
4/12/11 9:00	645	3.55	14.6 ± 1.6
9/30/11 9:00	646	2.24	64.8 ± 6.8
10/3/11 9:00	575	3.52	70.9 ± 6.5
12/9/11 14:30	726	1.70	78.6 ± 8.4
4/16/12 16:30	391	0.67	100 ± 0

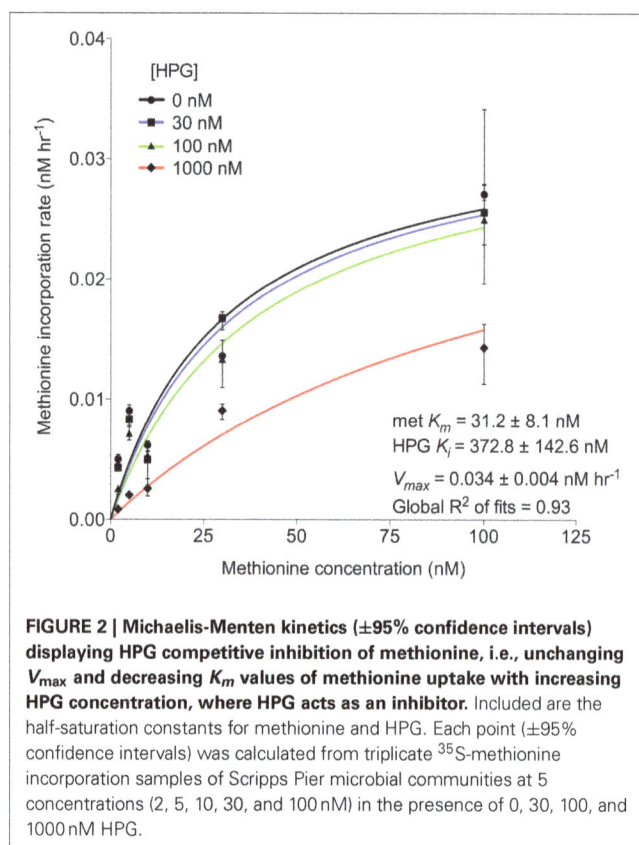

FIGURE 2 | Michaelis-Menten kinetics (±95% confidence intervals) displaying HPG competitive inhibition of methionine, i.e., unchanging V_{max} and decreasing K_m values of methionine uptake with increasing HPG concentration, where HPG acts as an inhibitor. Included are the half-saturation constants for methionine and HPG. Each point (±95% confidence intervals) was calculated from triplicate ^{35}S-methionine incorporation samples of Scripps Pier microbial communities at 5 concentrations (2, 5, 10, 30, and 100 nM) in the presence of 0, 30, 100, and 1000 nM HPG.

the sum HPG signal intensity for all cells in each community was calculated. The percentage of rank-ordered cells that comprised 10% of this "community sum" ranged 0.2–1.6% (**Table 3**). The percentage of "community sum" comprised by the top 10% of active cells ranged 34.7–77.8% while the bottom 50% of active cells ranged 1.9–13.7% (**Table 3**). Upon application of the conversion factor to each cell's HPG signal intensity, scPPs ranged 3.1×10^{-6} to 3.9 fg protein cell^{-1} h^{-1} ($n = 13645$; **Table 3** and **Figure S6**).

Cell volume and HPG signal intensity

For all labeled natural assemblage cells in the 21 Scripps Pier samples ($n = 13645$), a model II linear regression and correlation analysis of cell volume and log-transformed HPG signal intensity were significantly non-zero and correlated, respectively ($p < 0.0001$; Spearman $R = 0.306$; **Figure 3**). There was also a significantly non-zero model II linear regression and correlation between cell volume and size-normalized cell signal intensity ($p < 0.008$; Spearman $R = 0.063$; data not shown).

Distribution of single-cell protein production

The scPPs within each community exhibited non-Gaussian distributions, likely caused by a high number of cells with low rates. Neither log-normal nor exponential regressions of the frequency distributions were significant for any of the 21 communities. However, the distributions fit well to power law regressions (**Figure 4**) with $R^2 > 0.99$ in 9 of the 21 seawater samples while

protein where most methionine had been replaced by HPG. This regression fit calculation was sensitive to disproportionate influence by low intensity cells. The "community sum ^{35}S-methionine" factor assumed: (i) the sum crystal area for the ~900 cells we measured could be normalized to all cells per mL, allowing us to correlate crystal area with bulk growth rate and (ii) the bulk HPG community sum intensity could be accurately divided into the bulk ^{35}S-methionine protein synthesis rate.

USING HPG TO QUANTIFY SINGLE CELL PROTEIN PRODUCTION
Natural assemblages

HPG labeling across 21 seawater samples (i.e., communities) ranged 14.6–100% (**Table 2**). HPG signal intensities in labeled cells ranged six orders of magnitude (-0.155 to 6.949 log$_{10}$ RFU cell^{-1}; raw data shown in **Figure S5** and **Table S2**). The labeled cells were rank-ordered by intensity within each community, and

Table 3 | Overview of protein synthesis rates for 21 sampling dates at Scripps Pier incubated with 20 nM HPG for 1 h.

Date and time	scPP (fg protein cell^{-1} h^{-1})		Median	Percent cells comprising 10% of community sum signal	Percent of community sum signal by	
	Range	Mean (±95% confidence interval)			Top 10% labeled cells	Bottom 50% labeled cells
11/18/09 10:00	0.001–1.555	0.202 ± 0.082	0.071	1.6	48.2	6.8
4/30/10 10:00	<0.000–2.966	0.127 ± 0.014	0.076	1.0	38.8	12.3
8/17/10 16:00	<0.000–1.421	0.132 ± 0.112	0.051	1.1	46.6	6.8
10/6/10 10:00	<0.000–1.836	0.147 ± 0.135	0.056	1.5	42.1	6.4
10/11/10 14:00	<0.000–2.980	0.056 ± 0.048	0.012	0.3	65.2	4.8
10/20/10 14:00	<0.000–0.981	0.112 ± 0.098	0.044	1.2	46.3	6.8
12/15/10 10:00	<0.000–1.823	0.277 ± 0.204	0.187	1.4	34.7	13.3
3/30/11 12:00	0.004–2.024	0.183 ± 0.163	0.058	1.2	47.1	7.8
3/30/11 16:00	<0.000–1.537	0.079 ± 0.069	0.015	0.6	59.5	3.5
3/30/11 21:00	0.006–1.545	0.102 ± 0.090	0.047	0.7	48.8	13.7
3/31/11 12:00	<0.000–2.254	0.149 ± 0.129	0.041	0.8	52.4	4.4
3/31/11 16:00	<0.000–2.128	0.209 ± 0.177	0.044	1.0	55.1	3.5
3/31/11 20:00	<0.000–2.861	0.125 ± 0.112	0.031	0.8	56.5	3.6
4/1/11 12:00	<0.000–2.077	0.135 ± 0.063	0.042	0.8	52.2	5.3
4/1/11 16:00	<0.000–1.910	0.107 ± 0.052	0.031	0.7	57.1	5.3
4/1/11 21:00	<0.000–1.288	0.084 ± 0.038	0.024	0.8	56.3	5.4
4/12/11 9:00	0.001–0.890	0.107 ± 0.075	0.038	1.1	47.1	8.8
9/30/11 9:00	<0.000–0.498	0.035 ± 0.029	0.008	0.7	55.1	3.5
10/3/11 9:00	<0.000–1.743	0.184 ± 0.153	0.028	1.0	55.4	1.9
12/9/11 14:30	<0.000–3.893	0.067 ± 0.041	0.013	0.2	77.8	4.1
4/16/12 16:30	0.001–2.391	0.113 ± 0.084	0.015	0.3	73.6	3.8

Photopigmented cells are not included. Partitioning of community signal is shown in the rightmost three columns.

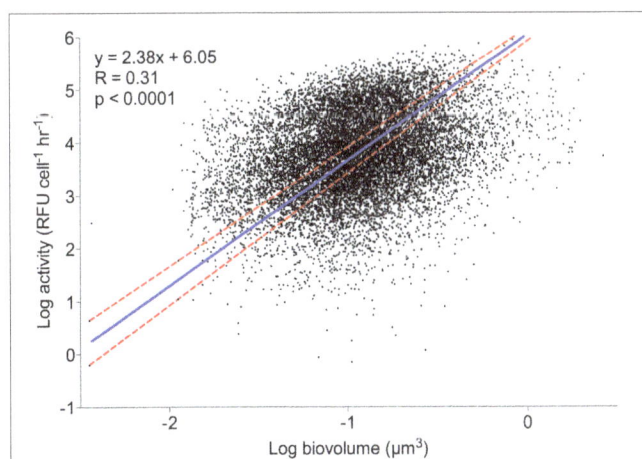

FIGURE 3 | Individual cell HPG signal intensity relative to cell size.
A model II regression (blue line) shows log-transformed sum intensity scaled significantly with log-transformed cell volume. Red dashed lines are 95% confidence intervals. $n = 13645$.

the remaining had $R^2 > 0.75$. The power law fits displayed similar visual trends in that all had negative slopes and were relatively confined to the same x-y space. However, the scaling factors (variable A) and exponents (variable B) were wide ranging (**Figure 4** inset). A consensus regression model fit to the global dataset did not account for all of the variation (extra sum of squares F test,

$p < 0.0001$; Akaike's informative criteria, 99.99% probability). Notably, the innermost and outermost regressions (15-Dec-2010 and 9-Dec-2011) originated from water samples that differed greatly with regard to sampling depth and chlorophyll concentration.

Particle-attached bacteria

When we applied the standard concentration and incubation protocol to seawater particles (20 nM HPG for 1 h), 80% of DAPI-stained attached bacteria were labeled, with a median scPP of 0.025 fg protein cell^{-1} h^{-1} (**Figure 5**; example images in **Figure S7**). Extended incubation times (up to 2 d) and higher HPG concentration (2 μM) increased labeling characteristics for particle-associated cells. Median signal intensity in 20 nM HPG increased by 2.8-fold from 1 h to 2 d then remained approximately constant through 8 d (**Figure 5**). In contrast, median intensity in 2 μM HPG increased by 4.8-fold from 1 h to 2 d, then another 6.6-fold from 2–8 d. Signal intensities for 20 nM HPG treatments were lower than 2 μM HPG treatments at each time point (**Figure 5**). The same pattern occurred for free-living cells in the same vessels (data not shown). Signal intensities for particle-attached cells followed non-Gaussian distributions (D'Agostino and Pearson omnibus and Shapiro-Wilk normality tests, $p < 0.001$). Methodologically, omission of filter-transfer-freeze for 8 μm filters had little effect on the particle attached cell measurements because high background fluorescence, as occurs on 0.2 μm filters, was due to small, closely-spaced pores. Background

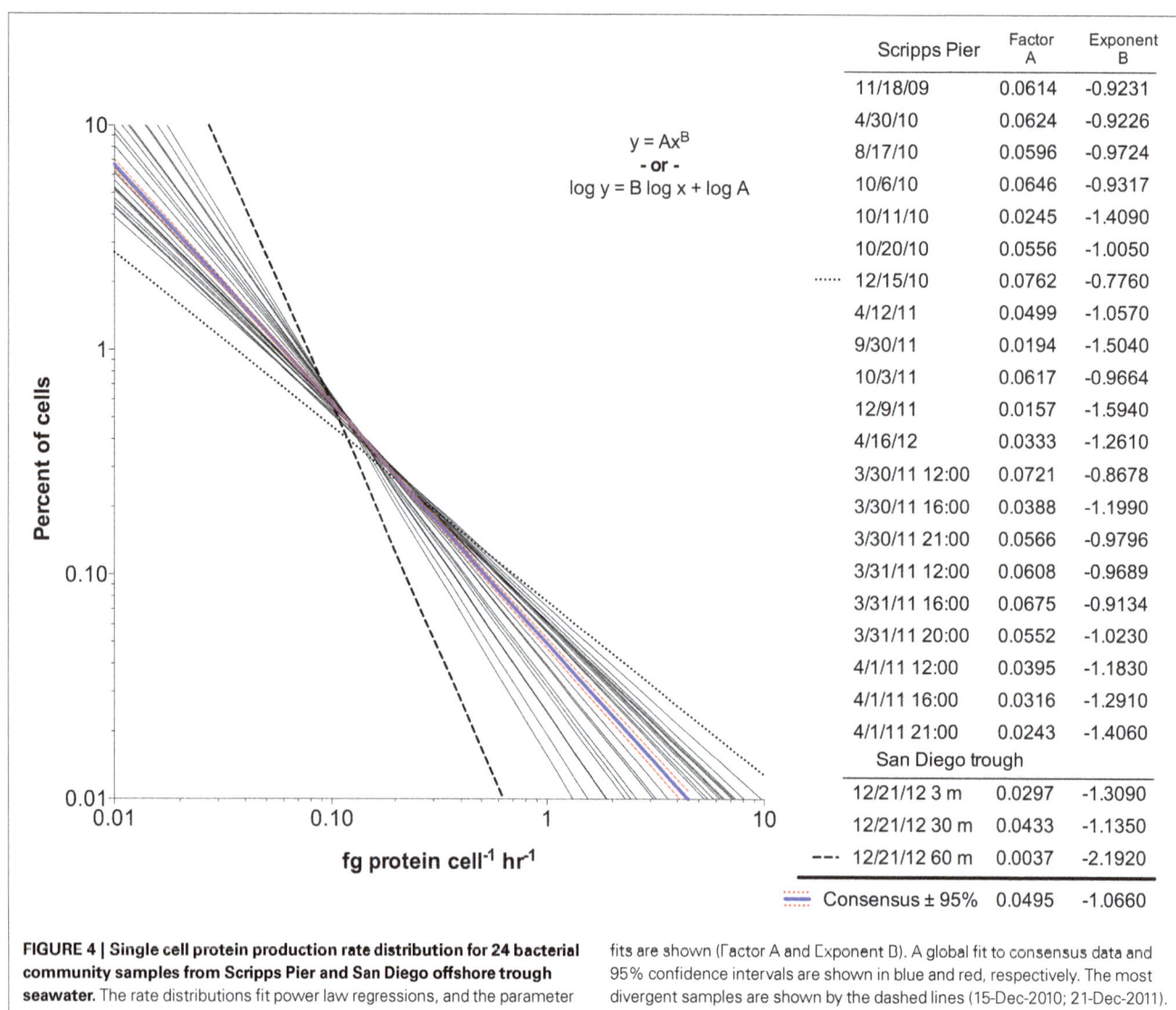

$$y = Ax^B$$
- or -
$$\log y = B \log x + \log A$$

Scripps Pier	Factor A	Exponent B
11/18/09	0.0614	-0.9231
4/30/10	0.0624	-0.9226
8/17/10	0.0596	-0.9724
10/6/10	0.0646	-0.9317
10/11/10	0.0245	-1.4090
10/20/10	0.0556	-1.0050
12/15/10	0.0762	-0.7760
4/12/11	0.0499	-1.0570
9/30/11	0.0194	-1.5040
10/3/11	0.0617	-0.9664
12/9/11	0.0157	-1.5940
4/16/12	0.0333	-1.2610
3/30/11 12:00	0.0721	-0.8678
3/30/11 16:00	0.0388	-1.1990
3/30/11 21:00	0.0566	-0.9796
3/31/11 12:00	0.0608	-0.9689
3/31/11 16:00	0.0675	-0.9134
3/31/11 20:00	0.0552	-1.0230
4/1/11 12:00	0.0395	-1.1830
4/1/11 16:00	0.0316	-1.2910
4/1/11 21:00	0.0243	-1.4060
San Diego trough		
12/21/12 3 m	0.0297	-1.3090
12/21/12 30 m	0.0433	-1.1350
12/21/12 60 m	0.0037	-2.1920
Consensus ± 95%	0.0495	-1.0660

FIGURE 4 | Single cell protein production rate distribution for 24 bacterial community samples from Scripps Pier and San Diego offshore trough seawater. The rate distributions fit power law regressions, and the parameter fits are shown (Factor A and Exponent B). A global fit to consensus data and 95% confidence intervals are shown in blue and red, respectively. The most divergent samples are shown by the dashed lines (15-Dec-2010; 21-Dec-2011).

intensities measured from circular regions (1 μm diameter) were only slightly higher on 8 μm polycarbonate filters relative to cover glass following filter-transfer-freeze (1.25-fold in 20 nM samples; 1.09-fold in 2 μM samples).

Clonal isolates

We hypothesized that limited variation in HPG labeling intensity would characterize marine bacterial monocultures. To address this, isolates AltSIO and As1 were grown in separate experiments to examine patterns and distribution of HPG incorporation in clonal cells. Labeling percentages were 100% for both isolates as determined from aliquots tested each hour from 16 to 22 h. Single-cell signal intensities ranged approximately three orders of magnitude at each timepoint, with the brightest cells of AltSIO and As1, respectively, at 6.176 and 6.279 $\log_{10}$ RFU (**Figure S8**). By comparison, the most fluorescent single cell measured among all Scripps Pier samples was 5.949 $\log_{10}$ RFU. Population signal intensities for both isolates followed non-Gaussian distributions at all timepoints except $T = 18$ h (D'Agostino and Pearson omnibus and Shapiro-Wilk normality tests, $0.03 < p < 0.0001$).

DISCUSSION

The HPG method we developed measures the percentage of bacteria engaged in protein synthesis and the protein synthesis rate of those single cells. After confirming the method's utility and specificity, we used it to quantify single cell protein synthesis rates in free-living natural seawater assemblages, particle-associated natural assemblages, and two cultured clonal bacterial strains.

METHOD VALIDATION

Prerequisites for HPG as a quantitative measurement

HPG is specifically incorporated into de novo synthesized protein. Incorporation of HPG into new protein has been previously demonstrated for *E. coli* (Beatty et al., 2005; Wang et al., 2008). We used chloramphenicol inhibition of protein synthesis to test whether this held true for natural assemblages of marine bacteria. Chloramphenicol blocks peptidyl-transferase and prevents protein chain elongation. The presence of chloramphenicol eliminated HPG labeling, indicating that HPG is very likely incorporated into newly synthesized protein in natural bacteria. We did not directly address the interaction of HPG and/or methionine

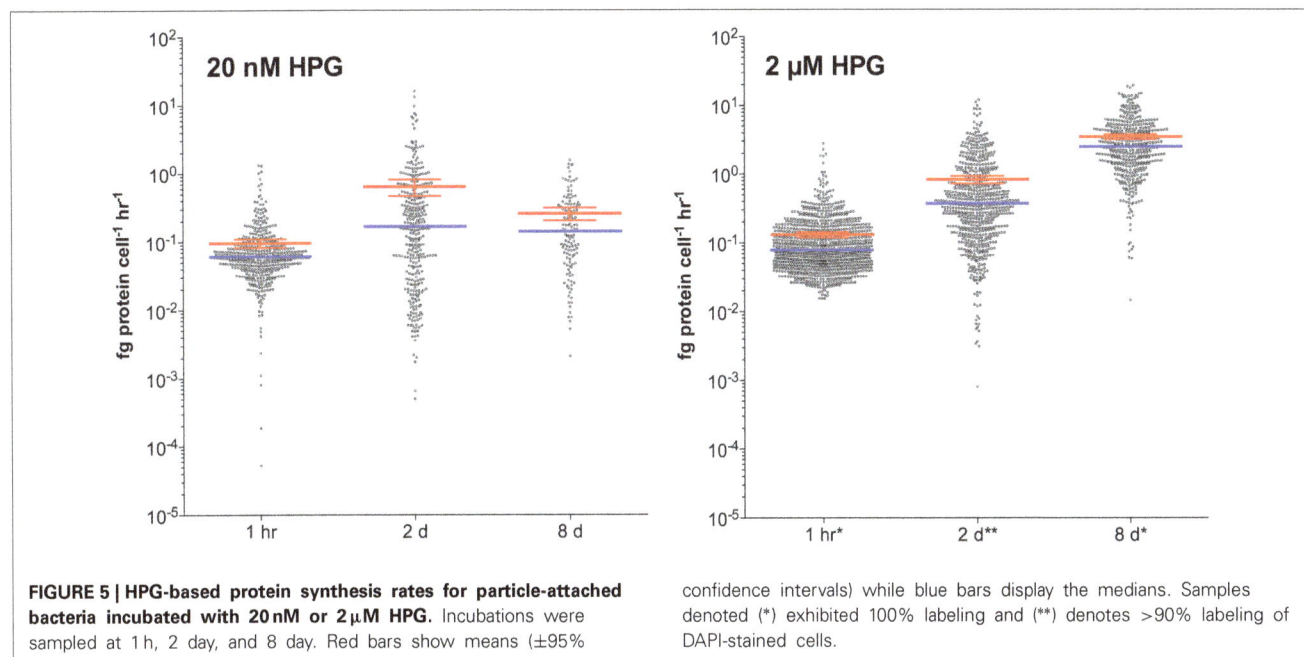

FIGURE 5 | HPG-based protein synthesis rates for particle-attached bacteria incubated with 20 nM or 2 μM HPG. Incubations were sampled at 1 h, 2 day, and 8 day. Red bars show means (±95% confidence intervals) while blue bars display the medians. Samples denoted (*) exhibited 100% labeling and (**) denotes >90% labeling of DAPI-stained cells.

at the level of membrane transport and translation, nor did we directly test whether HPG and/or methionine was converted to other compounds or brought into intracellular inclusion bodies rather than become incorporated into proteins.

HPG is a competitive inhibitor of methionine. Methionine addition reduced or eliminated HPG percent labeling in a dose-dependent manner, supporting the basic premise that HPG can be used as a substitute for methionine in protein (**Figure S3**). Given the non-zero, albeit significantly reduced, HPG labeling percentages in the presence of methionine on 1-Oct-2009, the experiment was repeated three times. The 16-May-2010 and 6-Oct-2010 samples showed no labeling, as expected, while 11-Oct-2010 maintained a low amount of labeling (**Figure S3**). We speculate that community acclimation to high exogenous methionine concentrations within the particular water masses resulted in preferential uptake of HPG. These results may also be due to naturally-occurring alkynes in the cells of these samples (Udwary et al., 2007). While such cells should have appeared in many samples, it may be that there was a higher abundance of such cells on 1-Oct-2009 and 11-Oct-2010. This result clearly defies expectation and is without a well-defined explanation.

Bulk methionine incorporation was reduced by the presence of HPG (**Figure S4**). Michaelis-Menten kinetics of HPG-methionine interaction showed competitive inhibition at 1000 nM, but overlap of the 95% confidence intervals for the non-linear uptake kinetic fits in the presence of 0, 30, and 100 nM HPG suggested insignificant competitive inhibition at these concentrations despite exhibiting similar V_{max} rates alongside increasing K_m concentrations with increasing inhibitor (i.e., HPG; **Figure 2**). This may partially be explained by: (i) the bacterial community exhibiting ~10-fold lower affinity for incorporation of HPG based on its high K_i and (ii) metabolic variation in the complex assemblage of cells within the triplicate measurements.

Additionally, this affinity may explain why 8 or 18 nM HPG did not significantly reduce ^{35}S-methionine incorporation while as little as 10 nM cold methionine was sufficient (**Figure S4**). Nevertheless, since typical methionine concentrations in surface seawater are ~200 pM (Zubkov et al., 2004), availability of 20 nM HPG should overcome competitive inhibition by methionine.

HPG percent labeling is comparable to ^{35}S-methionine microautoradiography. For parallel seawater incubations, the HPG method (without FTF) and ^{35}S-methionine microautoradiography yielded similar percent labeling (**Table 1**). Thus, the HPG method performs at least as well as microradiography for quantifying proportions of protein-synthesizing bacteria in natural assemblages. We attempted to directly compare the two methods by co-incubating seawater with ^{35}S-methionine and HPG, transferring co-labeled cells to photographic emulsion on a single slide, then sequentially processing the slide for microautoradiography or click. We found that AlexaFluor-488 azide bound non-specifically to the microradiography emulsion, rendering HPG-labeled cells indistinguishable from background. Application of the click reaction on the filters prior to microautoradiography processing was also unsuccessful as no HPG-labeled cells were observed, possibly due to fluorescence fading during 3 day exposure at 4°C.

Short incubations and high sensitivity. These prerequisites are desirable in order to limit time-dependent changes in protein synthesis rate. We adopted 1 h incubation time as standard protocol because the results of a time course experiment demonstrated that percent cell labeling increased 1.6-fold as incubation time increased from 30 min to 1 h, with no further increase up to 4 h incubation (data not shown). The 1 h incubation time, combined with use of the FTF technique, permitted detection of very low individual cell HPG fluorescence intensities. The scPP estimates were calculated by first acquiring accurate and precise measurements of fluorescence intensities and bacterial cell dimensions using the Nikon NIS Elements software. The HPG intensities were divided by the conversion factor to yield scPP. Using this individualized measurement approach, the lowest detected scPP was 3.1×10^{-7} fg protein $cell^{-1} h^{-1}$.

From data acquired in the particle experiment, we found that increasing HPG concentration from 20 nM to 2 μM yielded mean incorporation rates that were 1.5-fold higher while median incorporation rates increased 3.8-fold. As a standard protocol we chose to use 20 nM HPG to yield conservative estimates of the incorporation rate; even at this conservative HPG concentration bacterial communities on 3 out of 21 sampling dates exhibited 100% labeling.

METHOD LIMITATIONS

While the method can be used to determine percent protein synthesis active bacteria in communities and single cell protein production rates, a potential limitation is that HPG uptake affinity and subsequent incorporation may be species variable, which could underestimate percentages and influence scPP rates. The limitation may be overcome by using high HPG concentrations provided there is no effect of HPG concentration on protein synthesis rate, though greater HPG availability would not increase labeling for species that do not uptake HPG. Another possible limitation is that click chemistry labeling component(s) may not permeate through fixed membranes of all taxa. Though this topic was not specifically tested here, we saw no evidence for impermeability, and a separate click chemistry study demonstrated no improvement in cell labeling with enzymatic permeabilization (Smriga et al., 2014). Lastly, we note the inability to extrapolate single cell protein synthesis rates to bulk rates due to the FTF technique potentially biasing the relative proportion of zero, low, or high activity cells. This limitation may be ameliorated by developing *in vivo* bulk fluorometric detection of HPG in whole seawater using "strain promoted," copper-free click labeling (Agard et al., 2004; Baskin et al., 2007; Hatzenpichler et al., 2014). In addition to these three aspects, future work examining the dynamics between uptake and protein incorporation would further constrain the "top 10% isolate mean" conversion factor and supplement testing on oligotrophic seawater samples.

CONVERTING FLUORESCENCE TO PROTEIN PRODUCTION

We experimentally determined a numerical factor for converting single cell fluorescence intensity into single cell protein synthesis rate. Ideally this factor would have been determined from a direct, cell-specific comparison of microautoradiography and HPG, i.e., co-incubation of natural seawater samples with ^{35}S-methionine and HPG followed by simultaneous measurement of silver grain cluster area and Alexa Fluor 488 intensity in each cell. Experimental attempts at this were problematic (see above).

We instead opted for a parallel incubation approach to determine a conversion factor, i.e., experimental replicates amended only with ^{35}S-methionine or only with HPG. For this "community sum ^{35}S-methionine" approach we determined the bulk growth rate for the samples amended with ^{35}S-methionine and separately did both microradiography to quantify silver grain cluster areas and click labeling on HPG incubated samples. Percent labeling in the HPG replicates was much higher than in the ^{35}S-methionine replicates, but the distributions of single cell HPG signal intensities were similar to those for single cell crystal areas in two samples (**Figure 6**). While we cannot directly correlate the two signal types to get a conversion factor, the distributions imply that HPG and ^{35}S-methionine were comparable in their utility for estimating relative incorporation at the single cell level.

As an independent approach to determine a conversion factor, we quantified HPG incorporation in two bacterial isolates (AltSIO and As1) that grew well in unamended seawater and used the HPG signal intensity data for two calculations (see Materials and Methods).

The "top 10% isolate mean" calculation resulted in a conversion factor ($5.358 \log_{10} RFU fg^{-1}$ protein) that we assumed to be most accurate for transforming fluorescence signal to HPG content. It took into account the skewed distribution of labeling observed in AltSIO and As1 (described below) as evidence that a

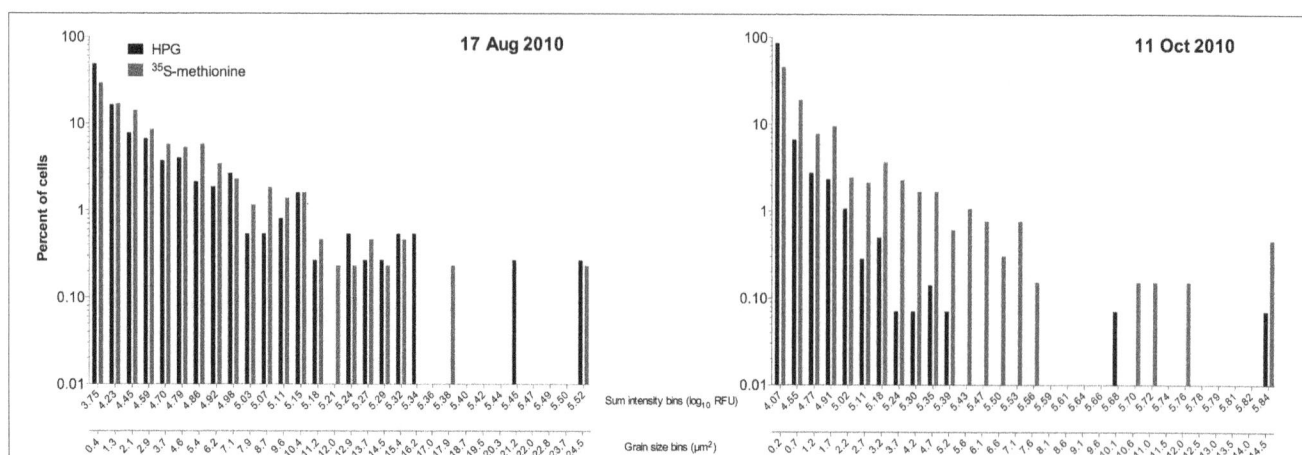

FIGURE 6 | Distributions of individual cell signal intensities for click chemistry (HPG) and micro radiography (^{35}S-methionine) determined via parallel incubations (i.e., separate 2 mL samples from the same flask) on two sample dates at Scripps Pier.

greater fraction of methionine was replaced by HPG in the protein content of the top 10% brightest cells. The rationale for the "top 10%" calculation arose from: (i) the large range and large coefficient of variation ($cv = 0.86$) in labeling intensities for AltSIO and As1, potentially due to heterogeneous activity patterns and/or departure from logarithmic growth, (ii) the lower variance among the top 10% of these cells ($cv = 0.49$) implying that this subpopulation was nearly saturated with HPG, and (iii) the fact that nearly saturated cells yielded a higher conversion factor which translated into more conservative scPP rates than the use of a lower conversion factor.

Our use of strains AltSIO and As1 assumed that their growth in unamended seawater reflected the growth conditions of natural communities (in the absence of significant viral or protistan mortality). We recognized that two closely related strains might only partially represent the vast taxonomic and physiological diversity of natural bacterial assemblages. We therefore attempted to grow other isolates in unamended seawater to determine a conversion factor: Tw3 (Vibrionaceae), Tw7 (Alteromonadaceae), and BBFL7 (Flavobacteriaceae). However, these isolates did not grow in unenriched seawater.

Variance in the methionine content per cell could effect the conversion factor calculation, but evidence suggests low variance in methionine content relative to other amino acids including leucine, the amino acid often used to estimate bulk bacterial production in seawater. First, methionine content of marine bacteria is low (2.2 mol%) relative to other amino acids including leucine (7.3 mol%) as quantified experimentally (Simon and Azam, 1989). Second, the average methionine content of all aquatic bacterial genomes in the IMG database (http://img.jgi.doe.gov) is 2.4 mol% ($SE = 0.02$, $n = 279$) while leucine is 10.4 mol% ($SE = 0.05$, $n = 279$; Moura et al., 2013). Similarly, methionine is the least frequently incorporated amino acid in *E. coli* proteins ($\sim 2\%$ of residues) while leucine is among the most (Saier, 2008). While total cell methionine content variance is low, this may be underpinned by high variance in the methionine residues per protein. Among three *Alteromonas* spp. genomes, most proteins (74–77%) encode 1–10 residues, but a small percentage require either >25 or none at all. For context, the genome of AltSIO (unpublished) contains an average of 8.4 methionine residues per protein ($SE = 0.095$; $n = 3941$). It is possible that in natural assemblages, expression of a few "high-methionine" proteins will cause an individual cell's total methionine content to be disproportionately high relative to other individuals in a clonal population. However, we predict this is limited since proteins with many methionine residues are also large proteins, and the space constraints of cells do not permit unregulated production of these molecules unless cell size is increased (Simon and Azam, 1989).

The adopted "top 10% isolate mean" conversion factor is currently a best estimate based on labeling response of two strains exhibiting growth characteristics that reflect elevated protein synthesis rates in heterotrophic bacteria encountering favorable conditions within the marine water column. Importantly, while we chose to adopt the "top 10% isolate mean" calculation, all three approaches gave conversion factors within the same order of magnitude (~ 5.2 to ~ 5.6 $\log_{10}$ RFU fg^{-1} protein), which provides confidence in the level of accuracy of the adopted value. Future experience with the method will likely improve accuracy of the conversion factor.

CHARACTERISTICS OF SINGLE CELL PROTEIN PRODUCTION IN MARINE BACTERIA

Low activity bacteria in natural assemblages

When we applied the conversion factor to natural assemblages in 21 samples, the resulting scPP ranged $\sim 10^{-6}$ to 3.9 fg protein cell^{-1} h^{-1}. This broad range is in agreement with Sintes and Herndl (2006) in which values ranged 1–100 amol Leu cell^{-1} d^{-1}, or equivalent to 0.15 and 15.1 fg protein cell^{-1} h^{-1}, respectively, assuming 3.6 kg protein per mole of leucine incorporated (Simon and Azam, 1989). Comparatively, our minimum rate detected with HPG is very low. The FTF technique enabled a high signal-to-background ratio for HPG-labeled cells and extended the lower threshold for detecting active cells. This is one potential reason why percent labeling was higher for HPG than for microradiography in parallel samples (**Table 1**). It may also be influenced by the dynamic range obtained from pixels vs. exposed silver grains. Using fluorescence, the collection of photons from a single cell occurs over hundreds of pixels comprising its dimensions, and not based on the tens of silver grains in its vicinity.

Several different methods for quantifying cell activity have shown that a majority of cells are inactive or "ghosts" or dormant in seawater (Fuhrman and Azam, 1982; Zweifel and Hagstrom, 1995; Lebaron et al., 1998; Schumann et al., 2003; Hamasaki et al., 2004). By contrast, the majority of cells were actively synthesizing protein in all samples we tested, albeit many cells were doing so at a low rate. The large proportion of low activity cells points to the existence of a background microbial community with low metabolic activity potentially primed to respond advantageously. It is known that bacterial communities harbor multiphasic uptake systems (Azam and Hodson, 1981). Low activity cells with high K_m permeases may maintain a minimal metabolic state until favorable conditions arise (Roszak and Colwell, 1987; Lauro et al., 2009). These low-activity populations may thus represent a substantial portion of the "microbial seed bank" suggested to exist throughout the ocean (Gibbons et al., 2013).

Relationship of HPG incorporation to cell volume

A positive linear relationship was observed between cell volume and HPG incorporation (i.e., $\log_{10}$ RFU intensity), consistent with previous studies of marine bacteria (Gasol et al., 1995; Lebaron et al., 2002). Larger volume cells tended to have higher protein synthesis rates (**Figure 3**). These relationships held when incorporation rates were normalized by volume, i.e., larger cells exhibited higher incorporation even when each cell's rate was divided by its volume. This suggests that carbon and nutrient incorporation is not similar for all cells per cell mass, showing that cell size is a significant parameter affecting anabolic processes derived from ambient material. This lends further support that size is an important factor governing microbially-mediated nutrient dynamics, along with substrate degradation specificity and growth efficiency (del Giorgio and Cole, 1998).

Power law distribution

The strong skew in fluorescence intensities within each sample led us to use log-log plots for viewing and comparing the spectrum of scPP across samples after fitting the data distributions to power law regressions. The visual clustering of regressions (and the global fit; **Figure 4**) indicates there is stability in the distribution of growth rates across individuals in the communities at Scripps Pier as well as in the offshore ecosystem of the San Diego trough. The consensus fit, however, could not explain all of the variation seen in the data, indicating each sampled water mass harbored microbial communities with unique continuums of single cell protein synthesis profiles that: (i) exerted differential influence over elemental cycling and (ii) reflected fluctuating nutrient availability supporting individual cell growth. Indeed, the two most divergent regression lines (15-Dec-2010 and 21-Dec-2011; **Figure 4**, bold lines) reflected the two most divergent sample types, specifically offshore waters at 60 m (overall lower community activity) and an intense bloom of the dinoflagellate *Lingulodinium polydrum* (overall higher community activity). Phytoplankton-derived organic carbon availability is among the factors that may explain these two divergent regressions (Mague et al., 1980; Fuhrman et al., 1985; Nelson and Carlson, 2012; Sarmento and Gasol, 2012).

The power law scaling values are useful to consider alongside means and medians when describing non-Gaussian distributions of single cell microbial activity. These values define the allocation of activity within the community (**Figure S9**). Increasing the scaling factor A (analogous to the y-intercept) results in an overall increase in community protein production rate. Changing the scaling exponent B (analogous to the slope) affects the relative proportion of higher or lower activity cells while also controlling the distribution of community activity. We found no significant relationship between chlorophyll and scPP mean or median, scaling factor, or scaling exponent. However, we did find significant positive correlations between: (i) the scPP means and scaling factors ($r = 0.61$, $p = 0.003$) or scaling exponents ($r = 0.65$, $p = 0.002$) and (ii) scPP medians and scaling factors ($r = 0.93$, $p < 0.0001$) or scaling exponents ($r = 0.95$, $p < 0.0001$).

The power law distributions are in general agreement with previous work examining single-cell incorporation of various substrates in the Arctic Ocean using microautoradiography (Nikrad et al., 2012). In one of the few examples of this distribution, the authors noted a negative slope of percent active cells vs. silver grain area (data were plotted on log-linear axes), similar to the decreasing trend between percent cells and the activity bins shown in our HPG study using log-log axes. One implication is that "cell-specific rate," i.e., bulk community rate divided by the total abundance of cells, does not accurately reflect activity distribution at the single cell level. This calculation assumes Gaussian rate distribution within the natural community (Smith and del Giorgio, 2003; and references therein) but this assumption is clearly not valid for HPG-based protein synthesis rates given their fit to a power law regression.

Activity of particle-attached bacteria

We applied the method to particles because these microenvironments are ubiquitous in coastal seawater and may regulate

bacterial community activity at larger scales. The standard protocol for free-living bacteria (20 nM HPG, 1 h incubation; **Figure 1**) also worked for particle-attached bacteria, and additional treatments in the particle incubations gave insight into the nature of protein production on particles. The experiment included parallel treatments of 2 µM HPG since we recognized *a priori* that high methionine concentrations could originate from particles, e.g., due to bacteria-mediated proteolysis, which could reduce HPG incorporation and thereby underestimate growth rates for bacteria closely-associated with particles (Smith et al., 1992; Azam and Long, 2001; Kiørboe and Jackson, 2001). This assumption may have been accurate since signal intensities were higher in 2 µM than in 20 nM HPG (**Figure 5**). Also, the dramatic increase in HPG signal intensities over 8 day for particle-associated cells suggests that methionine concentrations in the particles were initially high and decreased over time. Possible explanations include chemical diffusion away from the particles or bacterial incorporation and conversion into biomass. As a result, the removal of methionine relieved competition with HPG, which led to increased incorporation of HPG. An alternative explanation for low HPG signal intensities at 1 and 48 h is the occurrence of antagonistic interactions within the attached consortia that constrained cell growth (Long and Azam, 2001a).

The observations are consistent with previous work on marine particles that demonstrated particle-attached bacteria are active. One study used fluorescence microscopy to detect incorporation of the thymidine analog BrdU (Hamasaki et al., 2004). BrdU percent labeling was higher on particles than among free-living cells (56 vs. 35%) but signal intensities were not quantified for particle-attached bacteria. A few studies have used microautoradiography and showed that attached bacteria were active and exhibited higher percent labeling than free-living bacteria (Kirchman and Mitchell, 1982; Paerl, 1984). Specifically, size-fractionated samples demonstrated that particle-associated cells accounted for disproportionately high incorporation relative to free-living cells; the tested substrates included ^{14}C-glucose and ^{14}C-glutamate (Kirchman and Mitchell, 1982), or an ^{3}H-amino acid mixture and ^{33}PO$_4$ (Paerl, 1984).

Our results suggest that percent labeling on particles is influenced by incubation times and concentrations of both a compound and its analog in the surrounding environment. Single cell protein production rates on a particle can vary widely which may reflect microscale heterogeneity of resources on the particle itself, e.g., protein-depleted microscale patches may have enhanced HPG incorporation. The HPG method may enable cells to act as bioindicators for the "nutritional status" of a particle, simultaneously providing information on particle lability and single cell growth rate.

Growth heterogeneity among cells in cultured, clonal populations

When we applied the conversion factor to cultures of *Alteromonas* AltSIO and As1, protein synthesis rates varied by greater than 3 orders of magnitude. Here cells were grown ~21 h in non-enriched seawater in contrast to >40 h growth used in the conversion factor experiments. The large variance was unexpected but indicates that growth rate heterogeneity can occur among clonal cells undergoing logarithmic growth (**Figure S8**). The pattern is consistent with the adaptive strategy of bacterial aging

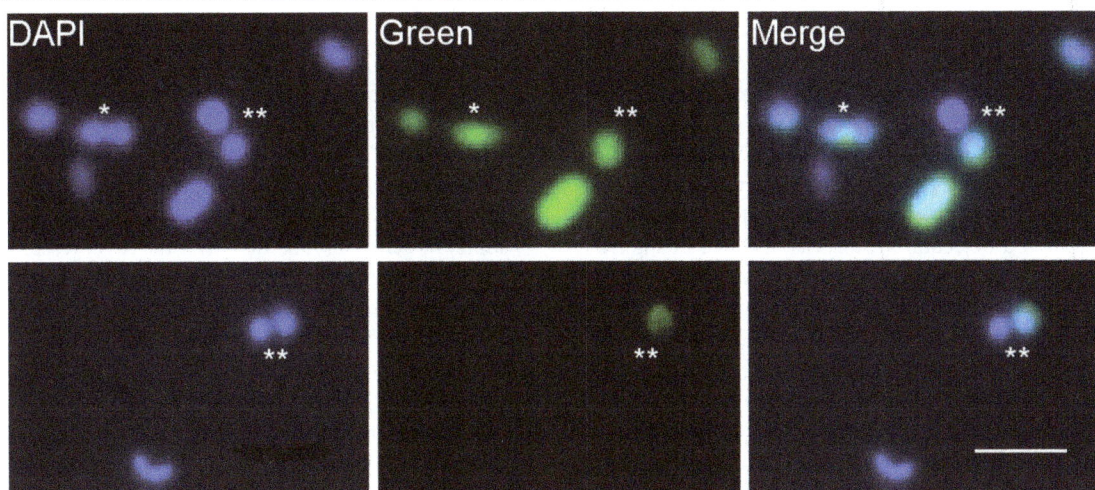

FIGURE 7 | HPG labeling patterns of putatively dividing cells. Each row shows a separate field of view. In most cases, both mother and daughter cells were labeled (*). Occasionally, one cell (either mother or daughter) was labeled and the other was not (**). Scale bar for all images = 2 μm.

in *E. coli* (Rang et al., 2011) in that heterogeneity may indicate repeated accumulation of oxidative damage to mother cells (lower intensity cells) that serve to rejuvenate daughter cells (higher intensity cells). If growth heterogeneity in AltSIO and As1 reflects similar behavior in natural marine communities, then individuals of specific taxa may exhibit a broad range of growth rates that depend on the overall health of their predecessors. This may partially explain the existence of dividing cells containing one HPG-labeled and one non-labeled cell (**Figure 7**). Such asymmetrical growth coupled with the prevalence of very slow growing cells could have important consequences for the rank-abundance of microbial taxa within the rare biosphere (Sogin et al., 2006).

CONCLUSIONS

The HPG click chemistry method provides a highly sensitive approach to estimate single cell bacterial protein synthesis and growth, and complements other single-cell quantifiable DNA synthesis viability/activity probes, e.g., click EdU, CTC, live/dead, CellTracker, Redox Sensor Green (del Giorgio et al., 1997; Grégori et al., 2001; Schumann et al., 2003; Kalyuzhnaya et al., 2008; Smriga et al., 2014). We used it to measure activity in both rapidly growing cells as well as slow growers previously considered inactive (Zweifel and Hagstrom, 1995; Smith and del Giorgio, 2003). Low abundances of very active cells suggest they were preferentially removed from the communities, possibly via protistan grazing or viral lysis, i.e., "kill the winner" dynamics (Thingstad, 2000). The method also provides a new capability to examine protein synthesis patterns among individual cells on the important microhabitats of planktonic particles. Future goals toward method improvement include development of conversion factors for open ocean regimes. Our results, as well as those of future studies that utilize the method, should help inform conceptual and numerical models that seek to predict the effects of single cell activities on ecosystem-scale biogeochemical processes.

MATERIALS AND METHODS

CLICK REACTION, SAMPLE PROCESSING, AND IMAGE ANALYSES

The click reaction is a conjugation of an azide to an alkyne group in the presence of copper (I) via the formation of a stable triazole (Rostovtsev et al., 2002; Tornøe et al., 2002). The sample is first incubated with an azide- or alkyne-functionalized bioorthogonal molecule. This study employed the alkyne-modified methionine analog L-homopropargylglycine (HPG). Incubation of the analog with copper (II) and an azide-linked fluorophore under reducing conditions enables copper (I) to catalyze binding of the azide group to the alkyne, thus fluorescently labeling the analog (Beatty et al., 2005; Salic and Mitchison, 2008). Labeled single-cell fluorescence is then measured in thousands of individual cells using an epifluorescence microscope and CCD camera.

GENERAL PROTOCOL

Stock solutions of HPG were prepared in dimethyl sulfoxide (DMSO, 200–2000 μM final; Smriga et al., 2014). Experiments were performed with 0.2 μm filtered 2–200 μM working solutions diluted in Milli-Q water (pH 7.0). Final concentrations were 1 to 2000 nM HPG incubated for 30 min to 8 d. Unless otherwise noted, most incubation durations were 1 h. Field samples were incubated at *in situ* temperature and protected from light. Experiments with bacterial isolates were incubated at 21°C. Control samples were added to vials containing HPG and 0.2 μm-filtered formaldehyde (2–4% w/v final) to measure non-specific labeling. Statistical tests, data plotting, and kinetics analyses were performed with Graphpad Prism 6 (GraphPad Software, Inc., La Jolla, CA).

Following incubation, 0.2 μm-filtered formaldehyde (2–4% w/v final) was added to each sample and allowed to fix for 15 min to 1 h. The volumes were then filtered onto white 0.2 μm pore size polycarbonate membrane filters (Nuclepore, GE Whatman, Piscataway, NJ). Each filter was cut into 1/8 or 1/6 slices using a sterile razor blade or surgical scissors and placed onto a microscope slide. The remaining portion was stored (dark; $-20°C$). Filters were processed with the Click-iT® reagent kit (Life Technologies, Grand Island, NY). The kit contained 5 mg of freeze-dried HPG, 0.5 mg Alexa Fluor 488 azide, 20 mL of reaction buffer (an unknown compound, likely ascorbic acid, in $1\times$ tris-buffered saline), 4 mL of 100 mM copper (II) sulfate, and 20 mL of buffer additive (also unknown, likely tris[(1-hydroxypropyl-1H-1,2,3-triazol-4-yl)methyl]amine or similar; Hong et al., 2009; Hatzenpichler et al., 2014). The Click-iT mixture and detection protocol is outlined in **Figure 1** and detailed below. The solution of Alexa Fluor® 488 azide was prepared in DMSO at 1 mM with a final concentration of 5 μM in the mixture. Processing steps were completed in darkened conditions to minimize fluorophore photobleaching and azide group decomposition (Abbenante et al., 2007). DNA counterstaining of bacterial cells with $4',6$-diamidino-2-phenylindole (DAPI) was achieved per Glöckner et al. (1996) using antifade mounting medium of Patel et al. (2007).

FILTER-TRANSFER-FREEZE

To reduce background signal, we adapted the filter-transfer-freeze technique (Hewes and Holm-Hansen, 1983). Briefly, the wet filter is placed cell side down on a glass slide, quickly frozen (Rapid Freeze Spray, Decon Laboratories, Inc., King of Prussia, PA), warmed with a gloved finger, frozen again, and peeled away, leaving a layer of cells on the slide. The remaining condensation is dried at room temperature (RT).

STEP-BY-STEP PROTOCOL

Preparation:

(1) Remove frozen Alexa 488-azide and buffer additive from the freezer; thaw at RT in the dark.
(2) Clean coverslips and slides with 70% ethanol.
(3) Cut polycarbonate filter into 1/8–1/6 slice and place cell-side up onto a glass slide.
(4) Label filters with a pen.
(5) Add sterile deionized (DI) water to tightly folded absorbent paper placed in a 50 mL falcon tube.

Click reaction steps:

(1) Prepare the Click reaction cocktail with the following components. Use higher volumes to process more than 8 filters: 155 μL sterile DI water, 20 μL 10× reaction buffer, 4 μL copper solution, 1 μL Alexa-488, and 20 μL buffer additive.
Note: be sure to add the buffer additive last and use the reaction cocktail immediately after adding the buffer additive.
(2) Apply 25 μL of reaction cocktail to a cleaned glass coverslip. Turn the coverslip over and place it reagent-side down onto the filter piece.
Note: when multiple filter pieces are processed at once, up to four pieces can be processed on one glass slide. Two 25 μL drops of reaction cocktail are placed onto each 25×25 mm coverslip, or four 25 μL drops are placed on a 24×50 mm coverslip. Each coverslip is inverted onto the 2–4 filter pieces.
(3) Place the slide into the 50 mL Falcon tube. Seal the Falcon tube and incubate in the dark at RT for 30 min.

Counterstaining:

(1) Prepare DAPI solution at 5 μg mL^{-1} final concentration. Each filter will need 25 μL.
(2) Two min before the incubation is complete, set up material for DAPI staining: fill 2 Petri dishes with ~20 mL sterile DI water ("wash"), and add 25 μL drops of DAPI per filter piece onto a EtOH-cleaned microscope slide (up to 4 per slide).
(3) Remove the slide from the Falcon tube. Remove the coverslip(s) by gently sliding them over the edge, grasping with the tweezers or gently with gloved fingers, and lifting up. The filters may stick to the coverslip or remain on the slide so be sure to remember which side is face up.
(4) Place the filter piece in the first DI wash, then place in the second DI wash. Dab the filter onto absorbent paper then place cell-side up on top of the DAPI drop (stain will permeate through filter) and incubate 5–15 min in the dark.

Filter-transfer-freeze and slide mounting:

(1) Add 25 μL drops of sterile DI water to EtOH-cleaned slides for as many filters as will be processed.
(2) Remove the filter from DAPI, wash each filter piece as in step 12, dab on absorbent paper, and place on the sterile DI water to maintain hydration.
(3) One at a time, place a filter cell-side down on the coverslip, mark the location & orientation with a sharpie, apply freeze spray for ~5 s, lightly press and warm filter with gloved finger, refreeze for ~5 s, and quickly peel away filter with forceps.
(4) Allow the coverslip to dry completely at RT. Prepare antifade mounting media by diluting a stock solution of 1,4-Phenylenediamine dihydrochloride (10% w/v; Sigma-Aldrich) to 0.1% in 1:1 PBS:Glycerol and keep dark on ice. The stock can be re-frozen and used as many times as the solution remains clear and not purple or dark purple-red.
(5) Add 15 μL of mounting medium to the slide and invert the coverslip on top of it. Let settle for 5–30 min.
(6) Immediately acquire images (as opposed to storing at $-20°C$ for later processing) to reduce fluorescence fading.

MICROSCOPY

We used a Nikon TE2000-U inverted microscope with a Plan Apochromat VC 100×, 1.4 NA oil immersion objective (Nikon Instruments, Japan) and illumination from an Exfo Xcite 120 Hg lamp (Lumen Dynamics, Canada). Images were acquired and analyzed using NIS-Elements 3.0 (Nikon Instruments). Each image (10–20 fields/filter) was haphazardly acquired at: (i) 350 nm excitation 457 nm emission (blue channel for DAPI), (ii) 490 nm excitation 528 nm emission (green channel for Alexa Fluor® 488 signal of incorporated HPG), and (iii) 555 nm excitation 617 nm emission (red channel for photopigment autofluorescence). Exposure ranged from 10 ms to a few seconds, depending on sample. The typical exposure for HPG conjugated Alexa Fluor 488 fluorescence was 1 s. Exposures of DAPI and photopigments were optimized for each sample set. It was critical to quantify between-sample variation in Hg lamp illumination and to inter-calibrate fluorescence intensity at different exposure times. We performed calibrations using the InSpeck Green Microscope Image Intensity Calibration Kit (Life Technologies, NY). Fluorescent beads (0.3% relative intensity) were imaged on three separate dates at 2, 10, 25, 100, and 200 ms exposure times. Additionally, the 0, 0.3, 1, 3, 10, 30, and 100% relative intensity beads were imaged at a constant 30 ms exposure time.

SINGLE-CELL SIGNAL MEASUREMENTS

Following filter-transfer-freeze, samples were imaged and processed to quantify single-cell intensities. Binary thresholding in Elements software was used to select individual cells in the DAPI channel. Cell area, width, and length data were automatically measured in the software and exported to Microsoft Excel. Cell volumes (V) were calculated from cell lengths (L) and widths (W) according to Bratbak (1985) with the equation $(V = L - W \div 3) \times (\pi \div 4 \times W^2)$. Protein content (P) was calculated from cell volume according to Simon and Azam (1989) with the equation $P = 88.6 \times V^{0.59}$. The conversion of 2-dimensional measurements into cell volumes and protein is robust for marine bacteria and has been used repeatedly in marine bacterial assemblages (Bratbak, 1985; Lee and Fuhrman, 1987; Gundersen et al., 2002; Terrado et al., 2008; Malfatti et al., 2009; Pedler et al., 2014).

For each image the DAPI binary layer was overlaid onto the corresponding green channel. Data in this channel was exported as two separate parameters: sum intensity (total fluorescence from each cell) and cell size normalized intensity (sum intensity $\div$ #pixels; referred to as "size normalized intensity"). Formaldehyde-fixed controls revealed faint fluorescence likely due to slight non-specific binding of HPG and/or Alexa Fluor® 488 to bacterial cell surfaces. The glass slide contributed little to overall background fluorescence. In some cases the background area normalized intensity of the slide was greater than cell size normalized intensity in control samples. Therefore, to calculate background signals, we measured the mean size normalized intensity of all cells from 5 haphazardly chosen fields (>200 cells) in formaldehyde-fixed controls for each sample. This average cell background intensity was multiplied by the pixel number (cell area divided by the square of pixel length) of each cell in HPG-amended treatments. We reasoned that background subtraction using the highest intensity control cell would give conservative corrected signal intensities in the natural samples, but this approach translated to very low percent labeling and very liberal scPP rates. Using the lowest intensity control cell for background subtraction yielded labeling and scPP rates similar to those obtained using average background approach. Thus, we adopted the average background calculation. It provided a liberal, unique, and cell volume-specific background sum intensity that was subtracted from the sum intensity of each respective cell, the difference being the HPG signal. Intensity values that were zero or negative were considered unlabeled and not analyzed further, but included in calculations of percent-labeled cells.

PROPORTION OF HPG-LABELED CELLS

The percent of HPG-labeled cells on filters was calculated as [(# HPG$^+$ cells) − (# red channel$^+$ cells)] $\div$ (# DAPI$^+$ cells)

from 10 haphazardly chosen fields and $\geq$20 cells per field. Red channel$^+$ cells were excluded since samples used for cell counts displayed seemingly HPG+ cells in the green channel due to carryover of photopigment autofluorescence from cyanobacteria and/or picoeukaryotes. They were manually removed from the thresholding-based selections using Elements software.

HPG-labeled cells in samples processed by filter-transfer-freeze were analyzed identically. After background subtraction, sum intensities >0 were used to calculate percentages of HPG-labeled cells: (i) as averages of 10 fields of view (arcsine transformed prior to calculation of means and standard error) and (ii) as absolute percentages (total # HPG$^+$ cells from all fields $\div$ total # DAPI$^+$ cells from all fields). This was done in order to determine whether cell abundance aggregation and variance in each field (caused by filter-transfer-freeze) skewed the results.

METHOD VALIDATION

Specificity of HPG incorporation into newly synthesized protein

Chloramphenicol inhibition of protein synthesis was used to determine the specificity of HPG incorporation. Samples were incubated with 50 μg mL^{-1} chloramphenicol + 20 nM HPG for 1 and 6 h, processed, and compared to unamended controls.

HPG as analog to and competitive inhibitor of methionine

E. coli uses methionyl-tRNA synthetase to incorporate HPG into protein in place of methionine (Kiick et al., 2002; Wang et al., 2008). We conducted four incorporation kinetics experiments to test that natural marine microbial communities also incorporated HPG in place of methionine.

Methionine inhibition of HPG incorporation. We incubated seawater samples with 20 nM HPG plus 0, 1, or 2 μM methionine for 0.5, 1, 4, and 6 h. Samples were click-processed (as above, without filter-transfer-freeze), imaged, and analyzed to quantify methionine inhibition of HPG incorporation.

HPG inhibition of ^{35}S-methionine incorporation. Scripps Pier seawater was incubated for 1 h with 2 nM ^{35}S-methionine alone and with 8 nM, 18 nM, 20 nM, 200 nM, 2 μM HPG, or 8 nM HPG + 10 nM cold methionine. ^{35}S incorporation was quantified by the centrifugation method (triplicate samples with duplicate controls; Smith and Azam, 1992).

HPG inhibition of methionine incorporation: Competitive or non-competitive?

We evaluated uptake kinetics to test whether HPG is a competitive, non-competitive, or mixed inhibitor of methionine (Michaelis and Menten, 1913). We used natural coastal seawater bacterial assemblages collected from the Scripps Pier, and analyzed the concentration-dependence of ^{35}S-methionine incorporation (K_m) at 5 concentrations (2, 5, 10, 30, and 100 nM). Deviation of ^{35}S-methionine incorporation kinetics due to the presence of 0, 30, 100, and 1000 nM HPG was analyzed to determine the K_i for HPG, which is numerically equivalent to HPG K_m (Experimental details in **Table S1**).

^{35}S-methionine microautoradiography. Parallel incubations were conducted to compare microautoradiography and click chemistry for detection of the proportion of protein synthesizing cells from natural assemblages. Slides were prepared for

microautoradiographic analysis from samples incubated with ^{35}S-methionine in the presence (200 nM and 2 μM) or absence of HPG. Simultaneously, samples were incubated with HPG and click-processed for visualization: (i) directly on filters and (ii) on glass slides after applying the filter-transfer-freeze technique. The objective was to compare the proportion of positively labeled cells by the three approaches, as well as to quantify the potential inhibitory effect of the presence of HPG on cellular protein synthesis.

Microautoradiography

Seawater was incubated for 1 h in the dark unless otherwise noted. Following incubation with 2 nM ^{35}S-methionine + 18 nM methionine, samples were formaldehyde fixed (2% final concentration) for >15 min. Killed controls received a mixture of radiolabeled amino acid and formaldehyde. Samples were filtered onto 0.2 μm pore size polycarbonate filters backed with 0.45 μm pore size mixed cellulose ester filters (Isopore membrane filters and MF membrane filters, Millipore, Billerica, MA). Filters were rinsed with 0.2 μm filtered 1× PBS or Milli-Q water, dried on absorbent tissue, and stored at −20°C.

Filters were cut into 1/8 to 1/4 pieces. Preparations for microautoradiography were performed under darkroom conditions using a Kodak GBX-2 safelight with a 15 W bulb placed ∼5 m away from the working area. This extremely faint light ensured low background, but necessitated the use of night vision monocular goggles (D-112MG; Night Optics USA, Inc., Huntington Beach, CA). We modified an emulsion coating and cell transferring protocol based on those from previous publications (Fuhrman and Azam, 1982; Teira et al., 2004; Longnecker et al., 2010). Ammersham LM-1 emulsion was melted at 43°C for 15 min to 1 h. Slides were dipped in emulsion for 5 s, allowed to drip for 5 s, wiped to remove emulsion on the back of the slide, and immediately placed onto a flat sheet of aluminum foil on ice. After 5–10 min, a filter slice was placed cell side down onto the emulsion. When this was completed for all filters, each slide was placed in a slide box with desiccant. The slide box was sealed with black tape, wrapped in aluminum foil, placed in a cardboard box, kept at RT for 1 h, and then transferred to 4°C for 72 h exposure. We did not test longer exposure times than 72 h, but note that this duration is longer than published studies examining both %-active and silver-grain sizes using ^{3}H-labeled compounds (Cottrell and Kirchman, 2003; Sintes and Herndl, 2006; Longnecker et al., 2010). Since ^{35}S emits nearly 10x more energy than ^{3}H, our results likely reflect the upper bound of %-active cells as revealed using ^{35}S-methionine.

Slides were developed with Kodak D19. In 50 mL tubes, 1.6 g of developer was mixed with 40 mL water and dissolved at 43°C for 15 min. Nine grams of fixer was mixed with 40 mL water at room temperature until fully dissolved. Four separate slide mailers (Fisher Scientific, Pittsburgh, PA, USA) were used as reagent containers and filled with 20 mL of developer, developer wash, fixer, and fixer, respectively. Under darkroom conditions, the slides were removed from the darkened box, and the filters were gently removed from each slide using forceps, thereby transferring the cells from the filters (now embedded in exposed emulsion).

Slides were submerged in developer for 4 min, washed for 10 s, then submerged in fixer for 5 min, washed again for 5 min,

and then allowed to dry on absorbent paper. Once the slide was dry, 50–100 μL of 1 μg mL^{-1} DAPI was added to the location of the cells and incubated for 10 min. The slide was then washed in autoclaved filtered water and dried again. A coverslip (#1.5 thickness) was mounted to the slide with 15–20 μL of antifade mounting medium.

Microscopy visualization was performed as described above with omission of the green channel. Transmitted light was used to observe the presence or absence of exposed emulsion grains. Percent labeling was calculated as the number of DAPI-labeled cells located within exposed emulsion divided by the total number of DAPI-labeled cells in each field of view for both live and control samples. The data were arcsine transformed, averaged, blank subtracted, and then back transformed. Autofluorescent cyanobacteria and/or eukaryotic cells were excluded from analyses.

Signal intensity conversion to single cell carbon production

We tested 3 approaches to calculate a valid factor converting sum intensity (SI) to fg protein, called "top 10% isolate mean," "isolates regression," and "community sum ^{35}S-methionine."

Alteromonas species As1 and AltSIO (Pedler et al., 2014) were used to quantify the "top 10% isolate mean." These strains were acclimated to and maintained in GF/F-filtered autoclaved seawater (FASW) to simulate natural seawater conditions. Exponentially growing cells of each isolate were inoculated into triplicate flasks containing 100 ml FASW at a starting concentration of 10^2 cells ml^{-1}. Flasks were incubated at room temperature for 16 h, at which point concentrations were ∼10^3 cells mL^{-1}, corresponding to growth by 4 doublings (confirmed via DAPI counts). HPG (20 nM) was added to one flask of each isolate and the other remained unamended as a control. Assuming each cell had ≤ 4.0 ×10^6 methionine residues (an upper limit), 20 nM HPG would be an ample supply for up to ∼10^9 cells and was not limited in the incubation volumes. After 24 and 48 h additional incubation (T40 and T64 h total growth time), 20 mL sub-samples were formaldehyde fixed (2%) and filtered. Each sample was click processed, imaged, counted and measured. Images for the green channel were acquired with 50 ms exposure due to bright labeling of cells. Cell volume (V) and protein content (P) was calculated as above. After fluorescence background subtraction, sum intensities for the cells (of both isolates) were divided by volume-derived protein content (P) to obtain a conversion factor of fluorescence in units of sum intensity per fg protein. This value was multiplied by 20 to scale the 50 ms exposure to images acquired from the standard 1 s exposure used in natural assemblages. We concluded this intensity quantification was valid after measuring the intensity linearity of calibrated fluorescent beads. The average intensity:protein value of the top 10% most intensely labeled cells at T40 and T64 was calculated as a conversion factor. Here it was assumed that only the top 10% most intensely labeled cells were "saturated" with HPG.

The "isolates regression" approach used the cell signal intensities (from the experiment described above) and plotted these against the cell volume-derived protein content (P). A linear regression was fit to the data, and the inverse of the slope in terms of log$_{10}$ RFU fg protein^{-1} was the calculated conversion factor. Here it was assumed that all cells, both faintly and brightly labeled, were saturated with HPG.

Lastly, the "community sum ^{35}S-methionine" approach was calculated from two experiments that quantified parallel HPG labeling and ^{35}S-methionine bulk incorporation. On two dates, 1 h incubations of Scripps Pier seawater with 2 nM ^{35}S-methionine + 18 nM methionine were performed alongside separate 1 h incubations with 20 nM HPG. Bulk protein synthesis was calculated from ^{35}S-methionine incorporation assuming zero isotope dilution and 2.2 mol%. The bulk assemblage HPG community sum (i.e., of $\sim 1 \times 10^6$ cells mL^{-1}) was extrapolated from directly measured cells. First, the HPG labeling percentage (calculated via microscopy) was multiplied by the DAPI-based cell density to provide the number of HPG$^+$ cells in the bulk assemblage. This was divided by the number of microscopy-based HPG$^+$ cells to yield a proportional value that was multiplied by the community sum to yield the bulk assemblage community sum. It was then divided by the ^{35}S-methionine bulk protein synthesis rate to provide a conversion factor in units of log$_{10}$ RFU fg protein^{-1}.

The "top 10% isolates mean" factor was applied to individual cell HPG intensities acquired from natural samples and other experiments to obtain single-cell protein production (scPP; fg cell^{-1} h^{-1}).

USING THE METHOD TO QUANTIFY SINGLE CELL ACTIVITY
Sampling locations
Seawater was collected from the Ellen Browning Scripps Memorial Pier (32° 52.02′ N, 117° 15.43′ W; "Scripps Pier"). Measurement of environmental parameters including temperature, salinity, and chlorophyll fluorescence were conducted as part of the Southern California Coastal Ocean Observing System (SCCOOS; www.sccoos.org). In total, 21 samples were collected and processed.

Offshore samples were collected at 3, 30, and 60 m depth from the San Diego trough (32° 38.022′ N, 117° 34.034′ W) via Niskin bottles attached to a CTD rosette (Sea-Bird Electronics, Bellevue, WA).

Individual cell activity distribution
Comparison of HPG-based activity profiles among all the sampling dates was performed by plotting the frequency distributions of cell-specific protein production on log$_{10}$-log$_{10}$ axes. The percent cell labeling vs. fg cell^{-1} h^{-1} within 0.05 fg bins were fit to power-law functions for each sample, and the resulting equations were used to generate a linear regression by reorganizing the power law equation (y = Ax^B) into logarithmic form (log y = B log x + log A). This enabled B, the scaling exponent, to be used as the regression slope alongside A, the scaling factor, as the y-intercept.

Particle-attached bacteria
Duplicate 40 mL seawater samples were treated as follows: (i) 20 nM HPG, (ii) 20 nM HPG + 2% formaldehyde, (iii) 2 μM HPG, (iv) 2 μM HPG + 2% formaldehyde, and (v) no addition. Each was incubated for 1 h, 48 h, and 8 day. At each time point, 30 mL was removed, formaldehyde-fixed, and filtered onto 8 μm white polycarbonate filters (30 mL). Filters were click-processed, but filter-transfer-freeze was not applied to 8 μm filters (to minimize potential damage to particles and/or alteration of bacterial associations). Ten particles (>10 μm in diameter and containing

at least 20 cells) were imaged on the 8 μm filters. Particle-attached cells were only measured and quantified if they were in focus, a characteristic of most particles observed in this study. Particles with high proportions of unfocused cells were not visualized, and those with few unfocused cells were visualized, but the cells were not analyzed.

Clonal isolates
Alteromonas spp. strains As1 and AltSIO were inoculated at $\sim 10^3$ cells mL^{-1} into GF/F filtered and autoclaved seawater. Through preliminary growth experiments in the same growth medium, it was determined that sampling would begin 16 h after inoculation to target the onset of exponential growth phase. One hour incubations amended with 20 nM HPG were carried out every hour from 16 to 22 h (duplicate with one control). Samples were fixed, filtered, stored, and then click- and filter-transfer-freeze processed as above. Images were acquired at 200 ms to maintain unsaturated pixels and signal intensities measured for each cell were multiplied by 5 for congruence with all samples.

SUPPLEMENTARY MATERIAL
The Supplementary Material for this article can be found online at: http://www.frontiersin.org/journal/10.3389/fmars.2014.00048/abstract

Table S1 | Setup of 20 incubations to measure ^{35}S-methionine incorporation in the presence of homopropargylglycine (HPG).

Table S2 | Raw data of all HPG-labeled cells analyzed in this study.
Measurements include background subtracted sum intensity, cell dimensions (area, length, width, volume), calculated protein content, and conversion factor-based protein synthesis rate.

Figure S1 | Percent HPG-labeling for cells visualized directly (filter) or after filter-transfer-freeze (FTF). Seawater samples were incubated with 20 nM HPG for 1 h. Bars represent *SE*.

Figure S2 | Example images of DAPI-stained cells within radiographic emulsion for parallel HPG and ^{35}S-methionine incubation experiments performed 17-Aug-2010 (A) or 11-Oct-2010 (B). Scale bars = 10 μm.

Figure S3 | HPG percent labeling in bacterial assemblages collected on four dates from Scripps Pier under different amendment conditions and incubation times. No labeling occurred in HPG plus 1 μM methionine for two dates, as indicated by asterisks. HPG plus 2 μM methionine was tested only on 11-Oct-2010. Bars represent *SE*.

Figure S4 | Disintegrations per minute (DPM) in seawater samples incubated with 2 nM ^{35}S-methionine alone and with varying concentrations of HPG and/or cold methionine. Mean values for two replicates are shown; error bars represent *SE* of the mean.

Figure S5 | Rank ordered plot of raw sum intensity values of all single cells measured in this study without background subtraction or protein conversion. Note: the "fixed" cells' values are not the size-normalized numbers used for background subtractions.

Figure S6 | Distribution of HPG-based single-cell activity rates for all dates. Each point is calculated from one bacterium. Red bars represent the mean ± 95% confidence intervals. Blue bars are median values. Green bars are single-cell protein production rates calculated from bulk ^{35}S-methionine incorporation (two samples).

**Figure S7 | Composite example images of HPG labeled (green) and
DAPI-stained cells (blue) of the free-living (A) and particle-associated (B)
communities following incubation for 1 h in 20 nM HPG.** For comparison,
the bottom row of images show fixed controls of free-living **(C)** and
particle-associated **(D)** cells. Insets depict signal in the green channel only.
Scale bars = 10 μm.

**Figure S8 | Individual cell HPG incorporation in clonal cultures of
Alteromonas strains AltSIO (left) and As1 (right).** Cultures were initiated in
seawater at $T = 0$ h, and at each timepoint shown an aliquot was
incubated for 1 h with HPG. Means are shown by red bars ± 95%
confidence intervals. Blue dashed bars show medians. Note the different
intervals between the top and bottom y-axis segments.

**Figure S9 | Simulated power law regressions showing effects of changes
in scaling values on HPG activity structures.**

REFERENCES

Abbenante, G., Le, G. T., and Fairlie, D. P. (2007). Unexpected photolytic decomposition of alkyl azides under mild conditions. *Chem. Commun.* 4501–4503. doi: 10.1039/b708134k. Available online at: http://pubs.rsc.org/en/content/articlelanding/2007/cc/b708134k#!divAbstract

Agard, N. J., Prescher, J. A., and Bertozzi, C. R. (2004). A strain-promoted [3 + 2] azide-alkyne cycloaddition for covalent modification of biomolecules in living systems. *J. Am. Chem. Soc.* 126, 15046–15047. doi: 10.1021/ja044996f

Azam, F., and Hodson, R. E. (1981). Multiphasic kinetics for D-glucose uptake by assemblages of natural marine bacteria. *Mar. Ecol. Prog. Ser.* 6, 213–222. doi: 10.3354/meps006213

Azam, F., and Long, R. A. (2001). Oceanography: sea snow microcosms. *Nature* 414, 495–498. doi: 10.1038/35107174

Azam, F., and Worden, A. Z. (2004). Microbes, molecules, and marine ecosystems. *Science* 303, 1622–1624. doi: 10.1126/science.1093892

Barbara, G. M., and Mitchell, J. G. (2003). Marine bacterial organisation around point-like sources of amino acids. *FEMS Microbiol. Ecol.* 43, 99–109. doi: 10.1111/j.1574-6941.2003.tb01049.x

Baskin, J. M., Prescher, J. A., Laughlin, S. T., Agard, N. J., Chang, P. V., Miller, I. A., et al. (2007). Copper-free click chemistry for dynamic *in vivo* imaging. *Proc. Natl. Acad. Sci. U.S.A.* 104, 16793–16797. doi: 10.1073/pnas.0707090104

Beatty, K. E., Xie, F., Wang, Q., and Tirrell, D. A. (2005). Selective dye-labeling of newly synthesized proteins in bacterial cells. *J. Am. Chem. Soc.* 127, 14150–14151. doi: 10.1021/ja054643w

Best, M. D. (2009). Click chemistry and bioorthogonal reactions: unprecedented selectivity in the labeling of biological molecules. *Biochemistry* 48, 6571–6584. doi: 10.1021/bi9007726

Bratbak, G. (1985). Bacterial biovolume and biomass estimations. *Appl. Environ. Microbiol.* 49, 1488–1493.

Copeland, R. A. (2005). *Evaluation of Enzyme Inhibitors in Drug Discovery: A Guide for Medicinal Chemists and Pharmacologists.* 1st Edn. Hoboken, NJ: Wiley-Interscience.

Cottrell, M. T., and Kirchman, D. L. (2003). Contribution of major bacterial groups to bacterial biomass production (thymidine and leucine incorporation) in the Delaware estuary. *Limnol. Oceanogr.* 48, 168–178. doi: 10.4319/lo.2003.48.1.0168

Cottrell, M. T., and Kirchman, D. L. (2004). Single-cell analysis of bacterial growth, cell size, and community structure in the Delaware estuary. *Aquat. Microb. Ecol.* 34, 139–149. doi: 10.3354/ame034139

del Giorgio, P. A., and Cole, J. J. (1998). Bacterial growth efficiency in natural aquatic systems. *Annu. Rev. Ecol. Syst.* 29, 503–541. doi: 10.1146/annurev.ecolsys.29.1.503

del Giorgio, P. A., Prairie, Y. T., and Bird, D. F. (1997). Coupling between rates of bacterial production and the abundance of metabolically active bacteria in lakes, enumerated using CTC reduction and flow cytometry. *Microb. Ecol.* 34, 144–154. doi: 10.1007/s002489900044

Dieterich, D. C., Link, A. J., Graumann, J., Tirrell, D. A., and Schuman, E. M. (2006). Selective identification of newly synthesized proteins in mammalian

cells using bioorthogonal noncanonical amino acid tagging (BONCAT). *Proc. Natl. Acad. Sci. U.S.A.* 103, 9482–9487. doi: 10.1073/pnas.0601637103

Foster, R. A., Kuypers, M. M. M., Vagner, T., Paerl, R. W., Musat, N., and Zehr, J. P. (2011). Nitrogen fixation and transfer in open ocean diatom–cyanobacterial symbioses. *ISME J.* 5, 1484–1493. doi: 10.1038/ismej.2011.26

Fuhrman, J. A., and Azam, F. (1982). Thymidine incorporation as a measure of heterotrophic bacterioplankton production in marine surface waters: evaluation and field results. *Mar. Biol.* 66, 109–120. doi: 10.1007/BF00397184

Fuhrman, J. A., Eppley, R. W., Hagström, Å., and Azam, F. (1985). Diel variations in bacterioplankton, phytoplankton, and related parameters in the Southern California Bight. *Mar. Ecol. Prog. Ser.* 27, 9–20. doi: 10.3354/meps027009

Galand, P. E., Alonso-Saez, L., Bertilsson, S., Lovejoy, C., and Casamayor, E. O. (2013). Contrasting activity patterns determined by BrdU incorporation in bacterial ribotypes from the Arctic Ocean in winter. *Front. Microbiol.* 4:118. doi: 10.3389/fmicb.2013.00118.

Gasol, J. M., Del Giorgio, P. A., Massana, R., and Duarte, C. M. (1995). Active versus inactive bacteria: size-dependence in a coastal marine plankton community. *Mar. Ecol. Prog. Ser. Oldendorf* 128, 91–97. doi: 10.3354/meps128091

Gibbons, S. M., Caporaso, J. G., Pirrung, M., Field, D., Knight, R., and Gilbert, J. A. (2013). Evidence for a persistent microbial seed bank throughout the global ocean. *Proc. Natl. Acad. Sci. U.S.A.* 110, 4651–4655. doi: 10.1073/pnas.1217767110

Glöckner, F. O., Amann, R., Alfreider, A., Pernthaler, J., Psenner, R., Trebesius, K., et al. (1996). An *in situ* hybridization protocol for detection and identification of planktonic bacteria. *Syst. Appl. Microbiol.* 19, 403–406. doi: 10.1016/S0723-2020(96)80069-5

Grégori, G., Citterio, S., Ghiani, A., Labra, M., Sgorbati, S., Brown, S., et al. (2001). Resolution of viable and membrane-compromised bacteria in freshwater and marine waters based on analytical flow cytometry and nucleic acid double staining. *Appl. Environ. Microbiol.* 67, 4662–4670. doi: 10.1128/AEM.67.10.4662-4670.2001

Gundersen, K., Heldal, M., Norland, S., Purdie, D. A., and Knap, A. H. (2002). Elemental C, N, and P cell content of individual bacteria collected at the Bermuda Atlantic Time-Series Study (BATS) site. *Limnol. Oceanogr.* 47, 1525–1530. doi: 10.4319/lo.2002.47.5.1525

Hamasaki, K., Long, R. A., and Azam, F. (2004). Individual cell growth rates of marine bacteria, measured by bromodeoxyuridine incorporation. *Aquat. Microb. Ecol.* 35:217. doi: 10.3354/ame035217

Hatzenpichler, R., Scheller, S., Tavormina, P. L., Babin, B. M., Tirrell, D. A., and Orphan, V. J. (2014). *In situ* visualization of newly synthesized proteins in environmental microbes using amino acid tagging and click chemistry. *Environ. Microbiol.* 16, 2568–2590. doi: 10.1111/1462-2920.12436

Hewes, C. D., and Holm-Hansen, O. (1983). A method for recovering nanoplankton from filters for identification with the microscope: the filter-transfer-freeze (FTF) technique. *Limnol. Oceanogr.* 28, 389–394. doi: 10.4319/lo.1983.28.2.0389

Hong, V., Presolski, S. I., Ma, C., and Finn, M. G. (2009). Analysis and optimization of copper-catalyzed azide–alkyne cycloaddition for bioconjugation. *Angew. Chem. Int. Ed. Engl.* 48, 9879–9883. doi: 10.1002/anie.200905087

Kalyuzhnaya, M. G., Lidstrom, M. E., and Chistoserdova, L. (2008). Real-time detection of actively metabolizing microbes by redox sensing as applied to methylotroph populations in Lake Washington. *ISME J.* 2, 696–706. doi: 10.1038/ismej.2008.32

Kiick, K. L., Saxon, E., Tirrell, D. A., and Bertozzi, C. R. (2002). Incorporation of azides into recombinant proteins for chemoselective modification by the Staudinger ligation. *Proc. Natl. Acad. Sci. U.S.A.* 99, 19–24. doi: 10.1073/pnas.012583299

Kiørboe, T., and Jackson, G. A. (2001). Marine snow, organic solute plumes, and optimal chemosensory behavior of bacteria. *Limnol. Oceanogr.* 46, 1309–1318. doi: 10.4319/lo.2001.46.6.1309

Kirchman, D., and Mitchell, R. (1982). Contribution of particle-bound bacteria to total microheterotrophic activity in five ponds and two marshes. *Appl. Environ. Microbiol.* 43, 200–209.

Lauro, F. M., McDougald, D., Thomas, T., Williams, T. J., Egan, S., Rice, S., et al. (2009). The genomic basis of trophic strategy in marine bacteria. *Proc. Natl. Acad. Sci. U.S.A.* 106, 15527–15533. doi: 10.1073/pnas.0903507106

Lebaron, P., Catala, P., and Parthuisot, N. (1998). Effectiveness of SYTOX green stain for bacterial viability assessment. *Appl. Environ. Microbiol.* 64, 2697–2700.

Lebaron, P., Servais, P., Baudoux, A.-C., Bourrain, M., Courties, C., and Parthuisot, N. (2002). Variations of bacterial-specific activity with cell size and nucleic acid content assessed by flow cytometry. *Aquat. Microb. Ecol.* 28, 131–140. doi: 10.3354/ame028131

Lee, S., and Fuhrman, J. A. (1987). Relationships between biovolume and biomass of naturally derived marine bacterioplankton. *Appl. Environ. Microbiol.* 53, 1298–1303.

Long, R. A., and Azam, F. (2001a). Antagonistic interactions among marine pelagic bacteria. *Appl. Environ. Microbiol.* 67, 4975–4983. doi: 10.1128/AEM.67.11.4975-4983.2001

Long, R. A., and Azam, F. (2001b). Microscale patchiness of bacterioplankton assemblage richness in seawater. *Aquat. Microb. Ecol.* 26:103. doi: 10.3354/ame026103

Longnecker, K., Wilson, M., Sherr, E., and Sherr, B. (2010). Effect of top-down control on cell-specific activity and diversity of active marine bacterioplankton. *Aquat. Microb. Ecol.* 58, 153–165. doi: 10.3354/ame01366

Mague, T. H., Friberg, E., Hughes, D. J., and Morris, I. (1980). Extracellular release of carbon by marine phytoplankton; a physiological approach. *Limnol. Oceanogr.* 25, 262–279. doi: 10.4319/lo.1980.25.2.0262

Malfatti, F., Samo, T. J., and Azam, F. (2009). High-resolution imaging of pelagic bacteria by Atomic Force Microscopy and implications for carbon cycling. *ISME J.* 4, 427–439. doi: 10.1038/ismej.2009.116

Michaelis, L., and Menten, M. L. (1913). Die kinetik der invertinwirkung. *Biochem. Z.* 49, 352.

Moura, A., Savageau, M. A., and Alves, R. (2013). Relative amino acid composition signatures of organisms and environments. *PLoS ONE* 8:e77319. doi: 10.1371/journal.pone.0077319

Musat, N., Foster, R., Vagner, T., Adam, B., and Kuypers, M. M. M. (2012). Detecting metabolic activities in single cells, with emphasis on nanoSIMS. *FEMS Microbiol. Rev.* 36, 486–511. doi: 10.1111/j.1574-6976.2011.00303.x

Nelson, C. E., and Carlson, C. A. (2012). Tracking differential incorporation of dissolved organic carbon types among diverse lineages of Sargasso Sea bacterioplankton. *Environ. Microbiol.* 14, 1500–1516. doi: 10.1111/j.1462-2920.2012.02738.x

Nikrad, M. P., Cottrell, M. T., and Kirchman, D. L. (2012). Abundance and single-cell activity of heterotrophic bacterial groups in the Western Arctic Ocean in summer and winter. *Appl. Environ. Microbiol.* 78, 2402–2409. doi: 10.1128/AEM.07130-11

Orphan, V. J., House, C. H., Hinrichs, K.-U., McKeegan, K. D., and DeLong, E. F. (2002). Multiple archaeal groups mediate methane oxidation in anoxic cold seep sediments. *Proc. Natl. Acad. Sci. U.S.A.* 99, 7663–7668. doi: 10.1073/pnas.072210299

Paerl, H. W. (1984). Alteration of microbial metabolic activities in association with detritus. *Bull. Mar. Sci.* 35, 393–408.

Patel, A., Noble, R. T., Steele, J. A., Schwalbach, M. S., Hewson, I., and Fuhrman, J. A. (2007). Virus and prokaryote enumeration from planktonic aquatic environments by epifluorescence microscopy with SYBR Green I. *Nat. Protoc.* 2, 269–276. doi: 10.1038/nprot.2007.6

Pedler, B. E., Aluwihare, L. I., and Azam, F. (2014). Single bacterial strain capable of significant contribution to carbon cycling in the surface ocean. *Proc. Natl. Acad. Sci. U.S.A.* 111, 7202–7207. doi: 10.1073/pnas.1401887111

Pernthaler, A., and Amann, R. (2004). Simultaneous fluorescence in situ hybridization of mRNA and rRNA in environmental bacteria. *Appl. Environ. Microbiol.* 70, 5426–5433. doi: 10.1128/AEM.70.9.5426-5433.2004

Pernthaler, A., Pernthaler, J., Schattenhofer, M., and Amann, R. (2002). Identification of DNA-synthesizing bacterial cells in coastal north sea plankton. *Appl. Environ. Microbiol.* 68, 5728–5736. doi: 10.1128/AEM.68.11.5728-5736.2002

Rang, C. U., Peng, A. Y., and Chao, L. (2011). Temporal dynamics of bacterial aging and rejuvenation. *Curr. Biol.* 21, 1813–1816. doi: 10.1016/j.cub.2011.09.018

Rostovtsev, V. V., Green, L. G., Fokin, V. V., and Sharpless, K. B. (2002). A stepwise huisgen cycloaddition process: copper(I)-catalyzed regioselective "ligation" of azides and terminal alkynes. *Angew. Chem. Int. Ed. Engl.* 41, 2596–2599. doi: 10.1002/1521-3773(20020715)41:14<2596::AID-ANIE2596>3.0.CO;2-4

Roszak, D. B., and Colwell, R. R. (1987). Survival strategies of bacteria in the natural environment. *Microbiol. Rev.* 51, 365–379.

Saier, M. H. (2008). The bacterial chromosome. *Crit. Rev. Biochem. Mol. Biol.* 43, 89–134. doi: 10.1080/10409230801921262

Salic, A., and Mitchison, T. J. (2008). A chemical method for fast and sensitive detection of DNA synthesis in vivo. *Proc. Natl. Acad. Sci. U.S.A.* 105, 2415–2420. doi: 10.1073/pnas.0712168105

Sarmento, H., and Gasol, J. M. (2012). Use of phytoplankton-derived dissolved organic carbon by different types of bacterioplankton. *Environ. Microbiol.* 14, 2348–2360. doi: 10.1111/j.1462-2920.2012.02787.x

Schumann, R., Schiewer, U., Karsten, U., and Rieling, T. (2003). Viability of bacteria from different aquatic habitats. II. Cellular fluorescent markers for membrane integrity and metabolic activity. *Aquat. Microb. Ecol.* 32, 137–150. doi: 10.3354/ame032137

Seymour, J. R., Seuront, L., and Mitchell, J. G. (2005). Microscale and small-scale temporal dynamics of a coastal planktonic microbial community. *Mar. Ecol. Prog. Ser.* 300, 21–37. doi: 10.3354/meps300021

Simon, M., and Azam, F. (1989). Protein content and protein synthesis rates of planktonic marine bacteria. *Mar. Ecol. Prog. Ser.* 51, 201–213. doi: 10.3354/meps051201

Sintes, E., and Herndl, G. J. (2006). Quantifying substrate uptake by individual cells of marine bacterioplankton by catalyzed reporter deposition fluorescence in situ hybridization combined with microautoradiography. *Appl. Environ. Microbiol.* 72, 7022–7028. doi: 10.1128/AEM.00763-06

Smith, D. C., and Azam, F. (1992). A simple, economical method for measuring bacterial protein synthesis rates in seawater using 3H-leucine. *Mar. Microb. Food Webs* 6, 107–114.

Smith, D. C., Simon, M., Alldredge, A. L., and Azam, F. (1992). Intense hydrolytic enzyme activity on marine aggregates and implications for rapid particle dissolution. *Nature* 359, 139–142. doi: 10.1038/359139a0

Smith, E. M., and del Giorgio, P. A. (2003). Low fractions of active bacteria in natural aquatic communities? *Aquat. Microb. Ecol.* 31, 203–208. doi: 10.3354/ame031203

Smriga, S., Samo, T., Malfatti, F., Villareal, J., and Azam, F. (2014). Individual cell DNA synthesis within natural marine bacterial assemblages as detected by "click" chemistry. *Aquat. Microb. Ecol.* 72, 269–280. doi: 10.3354/ame01698

Sogin, M. L., Morrison, H. G., Huber, J. A., Welch, D. M., Huse, S. M., Neal, P. R., et al. (2006). Microbial diversity in the deep sea and the underexplored "rare biosphere." *Proc. Natl. Acad. Sci. U.S.A.* 103, 12115–12120. doi: 10.1073/pnas.0605127103

Tada, Y., Taniguchi, A., and Hamasaki, K. (2010). Phylotype-specific growth rates of marine bacteria measured by bromodeoxyuridine immunocytochemistry and fluorescence in situ hybridization. *Aquat. Microb. Ecol.* 59, 229–238. doi: 10.3354/ame01412

Teira, E., Reinthaler, T., Pernthaler, A., Pernthaler, J., and Herndl, G. J. (2004). Combining catalyzed reporter deposition-fluorescence in situ hybridization and microautoradiography to detect substrate utilization by bacteria and archaea in the deep ocean. *Appl. Environ. Microbiol.* 70, 4411–4414. doi: 10.1128/AEM.70.7.4411-4414.2004

Terrado, R., Lovejoy, C., Massana, R., and Vincent, W. F. (2008). Microbial food web responses to light and nutrients beneath the coastal Arctic Ocean sea ice during the winter–spring transition. *J. Mar. Syst.* 74, 964–977. doi: 10.1016/j.jmarsys.2007.11.001

Thingstad, T. F. (2000). Elements of a theory for the mechanisms controlling abundance, diversity, and biogeochemical role of lytic bacterial viruses in aquatic systems. *Limnol. Oceanogr.* 45, 1320–1328. doi: 10.4319/lo.2000.45.6.1320

tom Dieck, S., Müller, A., Nehring, A., Hinz, F. I., Bartnik, I., Schuman, E. M., et al. (2012). Metabolic labeling with noncanonical amino acids and visualization by chemoselective fluorescent tagging. *Curr. Protoc. Cell Biol.* 56, 7.11.1–7.11.29. doi: 10.1002/0471143030.cb0711s56

Tornøe, C. W., Christensen, C., and Meldal, M. (2002). Peptidotriazoles on solid phase: [1,2,3]-triazoles by regiospecific copper(I)-catalyzed 1,3-dipolar cycloadditions of terminal alkynes to azides. *J. Org. Chem.* 67, 3057–3064. doi: 10.1021/jo011148j

Udwary, D. W., Zeigler, L., Asolkar, R. N., Singan, V., Lapidus, A., Fenical, W., et al. (2007). Genome sequencing reveals complex secondary metabolome in the marine actinomycete *Salinispora tropica. Proc. Natl. Acad. Sci. U.S.A.* 104, 10376–10381. doi: 10.1073/pnas.0700962104

Verdugo, P., Alldredge, A. L., Azam, F., Kirchman, D. L., Passow, U., and Santschi, P. H. (2004). The oceanic gel phase: a bridge in the

DOM–POM continuum. *Mar. Chem.* 92, 67–85. doi: 10.1016/j.marchem.2004. 06.017

Wang, A., Winblade Nairn, N., Johnson, R. S., Tirrell, D. A., and Grabstein, K. (2008). Processing of N-terminal unnatural amino acids in recombinant human interferon-β in *Escherichia coli. Chembiochem* 9, 324–330. doi: 10.1002/cbic.200700379

Zubkov, M. V., Tarran, G. A., and Fuchs, B. M. (2004). Depth related amino acid uptake by *Prochlorococcus cyanobacteria* in the Southern Atlantic tropical gyre. *FEMS Microbiol. Ecol.* 50, 153–161. doi: 10.1016/j.femsec.2004. 06.009

Zweifel, U. L., and Hagstrom, A. (1995). Total counts of marine bacteria include a large fraction of non-nucleoid-containing bacteria (ghosts). *Appl. Environ. Microbiol.* 61, 2180–2185.

Conflict of Interest Statement: The Reviewer Craig E. Nelson declares that, despite being affiliated to the same institution as the authors Ty J. Samo and Byron P. Sherwood, the review process was handled objectively and no conflict of interest exists. The authors declare that the research was conducted in the absence of any commercial or financial relationships that could be construed as a potential conflict of interest.

20

Invading the Mediterranean Sea: biodiversity patterns shaped by human activities

Stelios Katsanevakis[1]*, Marta Coll[2], Chiara Piroddi[1], Jeroen Steenbeek[3], Frida Ben Rais Lasram[4], Argyro Zenetos[5] and Ana Cristina Cardoso[1]

[1] Water Resources Unit, Institute for Environment and Sustainability, Joint Research Centre, Ispra, Italy
[2] Institut de Recherche pour le Développement, UMR EME 212, Centre de Recherche Halieutique Méditerranéenne et Tropicale, Sète, France
[3] Ecopath International Initiative Research Association, Barcelona, Spain
[4] Unité de Recherche Ecosystèmes et Ressources Aquatiques UR03AGRO1, Institut National Agronomique de Tunisie, Tunis, Tunisia
[5] Institute of Marine Biological Resources and Inland Waters, Hellenic Centre for Marine Research, Agios Kosmas, Greece

Edited by:
Christos Dimitrios Arvanitidis, Hellenic Centre for Marine Research, Greece

Reviewed by:
Melih Ertan Çinar, Ege University, Turkey
Salud Deudero, Instituto Español de Oceanografía, Spain
Christos Dimitrios Arvanitidis, Hellenic Centre for Marine Research, Greece
Theodoros Tzomos, Aristotle University of Thessaloniki, Greece

*Correspondence:
Stelios Katsanevakis, Water Resources Unit, Institute for Environment and Sustainability, Joint Research Centre, Via E. Fermi 2749, Building 46 (TP 460), Ispra I-21027, Italy
e-mail: stelios@katsanevakis.com

Human activities, such as shipping, aquaculture, and the opening of the Suez Canal, have led to the introduction of nearly 1000 alien species into the Mediterranean Sea. We investigated how human activities, by providing pathways for the introduction of alien species, may shape the biodiversity patterns in the Mediterranean Sea. Richness of Red Sea species introduced through the Suez Canal (Lessepsian species) is very high along the eastern Mediterranean coastline, reaching a maximum of 129 species per 100 km², and declines toward the north and west. The distribution of species introduced by shipping is strikingly different, with several hotspot areas occurring throughout the Mediterranean basin. Two main hotspots for aquaculture-introduced species are observed (the Thau and Venice lagoons). Certain taxonomic groups were mostly introduced through specific pathways—fish through the Suez Canal, macrophytes by aquaculture, and invertebrates through the Suez Canal and by shipping. Hence, the local taxonomic identity of the alien species was greatly dependent on the dominant maritime activities/interventions and the related pathways of introduction. The composition of alien species differs among Mediterranean ecoregions; such differences are greater for Lessepsian and aquaculture-introduced species. The spatial pattern of native species biodiversity differs from that of alien species: the overall richness of native species declines from the north-western to the south-eastern regions, while the opposite trend is observed for alien species. The biodiversity of the Mediterranean Sea is changing, and further research is needed to better understand how the new biodiversity patterns shaped by human activities will affect the Mediterranean food webs, ecosystem functioning, and the provision of ecosystem services.

Keywords: alien species, biological invasions, Lessepsian migrants, aquaculture, shipping, pathways, biodiversity patterns

INTRODUCTION

The Mediterranean Sea is a hotspot of marine biodiversity with >17,000 reported marine species, of which approximately one fifth are considered to be endemic (Coll et al., 2010). Such increased endemism and high species richness makes the Mediterranean Sea one of the world's biodiversity hotspots (Lejeusne et al., 2010). However, Mediterranean marine ecoregions are amongst the most impacted ecoregions globally (Halpern et al., 2008; Costello et al., 2010), due to increasing levels of human threats that affect all levels of biodiversity (Mouillot et al., 2011; Coll et al., 2012; Micheli et al., 2013), severe impacts from climate change (Lejeusne et al., 2010), and biological invasions (Zenetos et al., 2012; Katsanevakis et al., 2013).

Introduction of marine alien species in the Mediterranean Sea has been fostered by the opening of the Suez Canal, fouling and ballast transportation along shipping routes, aquaculture, and aquarium trade (Zenetos et al., 2012; Katsanevakis et al.,

2013). Nearly 1000 marine alien species have been introduced in the Mediterranean up to now, of which more than half are considered to be established and spreading (Zenetos et al., 2010, 2012). Marine alien species may become invasive and displace native species, cause the loss of native genotypes, modify habitats, change community structure, affect food-web properties and ecosystem processes, impede the provision of ecosystem services, impact human health, and cause substantial economic losses (Grosholz, 2002; Wallentinus and Nyberg, 2007; Molnar et al., 2008; Vilà et al., 2010; Katsanevakis et al., in press). On the other hand many alien species have positive impacts on ecosystem services and biodiversity, e.g., by acting as ecosystem engineers and creating novel habitats, controlling other invasive species, providing food, and supporting ecosystem functioning in stressed or degraded ecosystems (Schlaepfer et al., 2011; Simberloff et al., 2013; Katsanevakis et al., in press). Understanding the role of biological invasions in modifying biodiversity patterns and the functionality of ecosystems is a major challenge for marine ecosystems ecology (Borja, 2014).

In the Mediterranean Sea, despite the variability in monitoring and reporting effort among countries and the gaps in our knowledge of alien species distribution, there is an enormous amount of information scattered in various databases, institutional repositories, and the literature (including single- or multi-species reviews). By harmonizing and integrating information that has often been collected based on different protocols and is distributed in various sources (Gatto et al., 2013), the needed knowledge basis to assess the distribution and status of marine alien species can be built. Recently, the European Alien Species Information Network (EASIN; Katsanevakis et al., 2012) increased the accessibility to alien species spatial information by creating a network of interoperable web services through which data in distributed sources is accessed. Integrated distribution maps of single species or species aggregations can be easily produced with EASIN's freely available mapping tools.

Here, we utilize information on alien species distribution from EASIN to investigate the distribution patterns of marine alien species in the Mediterranean Sea, in relation to the main pathways of introduction. We investigate how specific human activities (opening of the Suez Canal, shipping, aquaculture) may shape the patterns of alien species distribution and consequently the overall biodiversity patterns in the Mediterranean Sea. We also compare the distributions of alien species with those of native ones to investigate differences in their patterns, and thus induced changes in pre-existing distribution patterns of native biodiversity.

MATERIALS AND METHODS

The EASIN alien species inventory was used (available online in: http://easin.jrc.ec.europa.eu/use-easin/species-search/combined-criteria-search) as of January 2014 (version 3.2 of the EASIN catalog). The link between alien species and pathways was based on Zenetos et al. (2012) and Katsanevakis et al. (2013); the pathway classification proposed by the latter authors was used herein. For each species one of the following uncertainty categories on the pathway(s) of introduction was adopted:

(1) There is direct evidence of a pathway/vector: The species was clearly associated to a specific pathway/vector at the time of introduction to a particular locality. This is the case e.g., in all intentional introductions (i.e., aquaculture/commodity) and in many cases of Lessepsian immigrants, when there was direct evidence of a gradual expansion along the Suez Canal and then in the localities around the exit of the Canal in the Mediterranean).

(2) A most likely pathway/vector can be inferred: The species appears for the first time in a locality where a single pathway/vector(s) is known to operate and there is no other rational explanation for its presence except by this pathway/vector(s). This applies e.g., to many species introduced by shipping or as aquaculture contaminants. In many cases inference is based on known examples of introductions elsewhere for the same or similar species, the biology and ecology of the species, the habitats it occupies in both the native and introduced range, and its pattern of dispersal (if known),

e.g., for a fouling species frequently recorded in/near ports, shipping has been assumed to be the most probable vector.

(3) One or more possible pathways/vectors can be inferred: The species cannot be convincingly ascribed to a single pathway/vector. Inference is based on the activities in the locality where the species was found and may include evidence on similarly behaving species reported elsewhere.

(4) Unknown: Where there is doubt as to any specific pathway explaining the arrival of the species.

Of the 986 species reported from the Mediterranean, 799 have been assigned to a single pathway, 114 have been assigned to two or more possible pathways, and the remaining 73 species have been classified as "unknown" (Zenetos et al., 2012). In the present analysis of spatial distribution by pathway only species linked to a single pathway were included (i.e., species of uncertainty categories 1 and 2). Species of uncertainty category 3 were excluded from any pathway-specific analysis to avoid the distortion of pathway-related spatial patterns by erroneously including species that might actually have been introduced through another pathway. In the absence of a permanent monitoring network and of a biased effort favoring specific locations (e.g., ports, marinas, and aquaculture facilities), some uncertainty remains (especially for category 2 species). All alien species were included in all other analyses (non-pathway-specific).

The "Species Search/Mapping By Multiple Criteria" tool of EASIN was used to select and map species introduced in the Mediterranean by the three major pathways of introduction, i.e., (1) through the Suez Canal; (2) by shipping; and (3) by aquaculture (**Figure 1**). The spatial data used herein through EASIN originate from the following sources: (1) the CIESM Atlas of Exotic Species (http://www.ciesm.org/online/atlas/index.htm); (2) the Global Biodiversity Information Facility (GBIF; http://www.gbif.org/); (3) the Global Invasive Species Information Network (GISIN; http://www.gisin.org); (4) the Regional Euro-Asian Biological Invasions Centre (REABIC; http://www.reabic.net/); (5) the Hellenic Network on Aquatic Invasive species (ELNAIS: https://services.ath.hcmr.gr/); and (6) EASIN-Lit (http://easin.jrc.ec.europa.eu/About/EASIN-Lit). EASIN-Lit is an EASIN product providing georeferenced records as retrieved from published literature (Trombetti et al., 2013). For the present work, we used the current (as of January 2014) version of EASIN-Lit, including 227 publications (L00001–L00227; full references in http://easin.jrc.ec.europa.eu/About/EASIN-Lit).

To investigate spatial patterns in the composition of alien communities, we randomly "sampled" five sites in each of the seven Mediterranean ecoregions (sensu Spalding et al., 2007, i.e., Levantine, Aegean, Ionian, Adriatic, Tunisian plateau and Gulf of Sidra, western Mediterranean, Alboran Sea); only in the western Mediterranean ecoregion, seven sites were "sampled" due to its relatively larger size. In derogation of the random sampling approach, we included the Venice and the Thau lagoons, as these sites have been well-studied and highlighted in the literature as hotspots of alien species (Occhipinti Ambrogi, 2000; Boudouresque et al., 2011). For each site (considered to be a 10 × 10 km quadrat), the list of recorded alien species was retrieved from EASIN (presence/absence data).

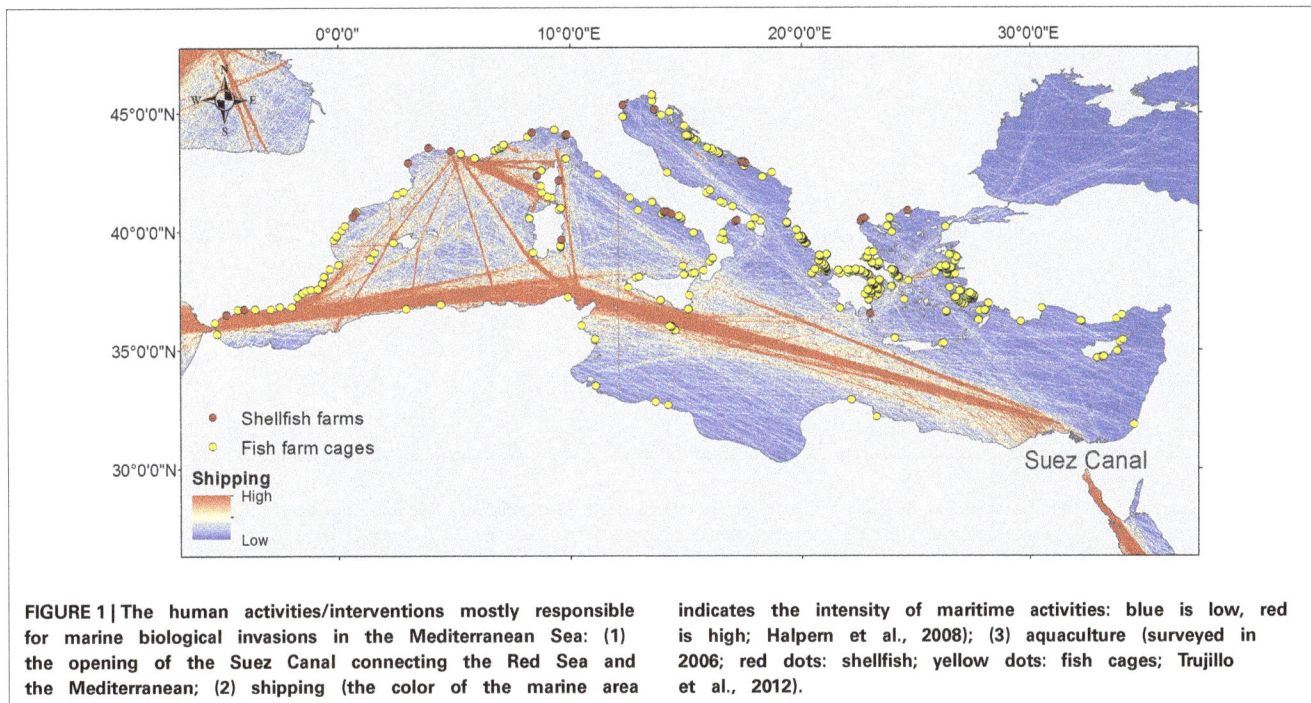

FIGURE 1 | The human activities/interventions mostly responsible for marine biological invasions in the Mediterranean Sea: (1) the opening of the Suez Canal connecting the Red Sea and the Mediterranean; (2) shipping (the color of the marine area indicates the intensity of maritime activities: blue is low, red is high; Halpern et al., 2008); (3) aquaculture (surveyed in 2006; red dots: shellfish; yellow dots: fish cages; Trujillo et al., 2012).

Similarity patterns were explored through non-metric multidimensional scaling (nMDS; Kruskal, 1964), based on a similarity matrix constructed using the Jaccard coefficient (Jaccard, 1901). Permutational multivariate analysis of variance (Permanova; Anderson, 2001) was used to test for differences among ecoregions, using type III sum of squares and 999 random permutations of the appropriate units. The software Primer 6 was used for multivariate analysis (Clarke and Warwick, 2001) and Permanova+ v.1.0.3 for the PERMANOVA analysis.

We also included available data regarding the spatial distribution of native fish and invertebrate species described in the Mediterranean Sea (Coll et al., 2010, 2012) to compare the spatial patterns of species richness between native and alien species. For fish species we used data available from the "Fishes of the Northern Atlantic and Mediterranean" (FNAM atlas; Whitehead et al., 1986) updated and integrated by Ben Rais Lasram and Mouillot (2009) and Coll et al. (2012). Data on invertebrates were compiled from the Food and Agriculture Organization of the United Nations (FAO: www.fao.org/fishery/species/distribution) and the Sea Around Us (www.seaaroundus.org) databases (Coll et al., 2012). To estimate the distribution of native species richness, we grouped all the species as the sum of the species co-occurring by overlapping distribution maps at fine-scale resolution (10 × 10 km).

We assessed the spatial congruence of native and alien species by calculating the correlation coefficient between the native and alien raster layers, i.e., the ratio of the covariance between the two layers divided by the product of their standard deviations. Only the cells adjacent to the coastline were included in this analysis, as alien species are generally concentrated in coastal and shelf waters (otherwise the overabundance of zero values in the offshore cells would mask any significant correlation). For this estimation, we used the Band Collection Statistics tool in ArcGis 10. The ratio of alien to native species richness was also estimated for each 10 × 10 km cell, as an indicator of the environmental impact of alien species (EC, 2010).

RESULTS

A total of 420 species of uncertainty levels 1 and 2 have been introduced in the Mediterranean Sea through the Suez Canal. An aggregated map of these Lessepsian species (**Figure 2**) shows a characteristic pattern of high species richness in the south-eastern Levantine Sea, which declines anticlockwise along the coastline of the Levantine Sea and further westwards and northwards along the northern Mediterranean coast, and also westwards along the north-African coastline. In the Israeli coastline, species richness reaches a maximum of 129 species per 10 × 10 km cell (in the Haifa coastal area), while it is markedly lower in the Ionian Sea, the Adriatic Sea, and the western Mediterranean basin.

Shipping, through ballast waters and hull-fouling, was the most probable pathway for the introduction of 308 species (uncertainty levels 1 and 2). The distribution of these species (**Figure 3**) is strikingly different to the one of Lessepsian species. Hotspot areas include the north-western Mediterranean coastline from Martigues and Marseille (France) to Genova (Italy), eastern Sicily (Italy), the Saronikos, Thermaikos and Evvoikos Gulfs (Greece), and the coastlines of the eastern Levantine (SE Turkey, Syria, Israel, and Lebanon).

Through aquaculture, either as commodities or as contaminants, 64 species have been introduced in the Mediterranean Sea (uncertainty levels 1 and 2). Two main hotspot areas were identified, the Thau lagoon (Gulf of Lion, France), and the Venice lagoon (northern Adriatic, Italy) (**Figure 4**). Most of species introduced through aquaculture are macrophytes (41 species)

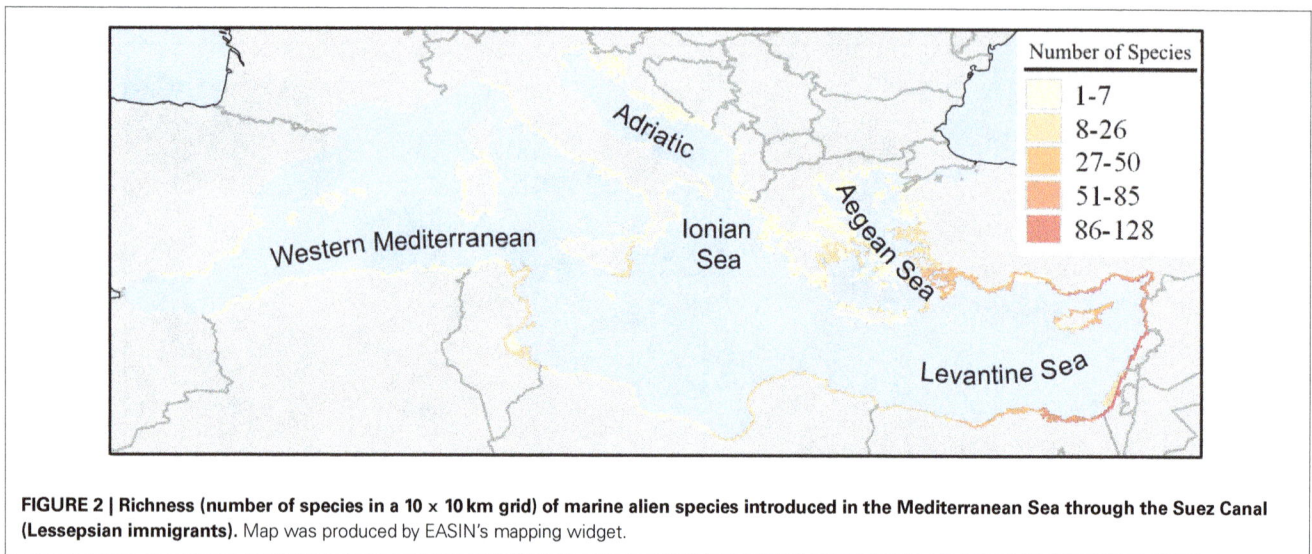

FIGURE 2 | Richness (number of species in a 10 × 10 km grid) of marine alien species introduced in the Mediterranean Sea through the Suez Canal (Lessepsian immigrants). Map was produced by EASIN's mapping widget.

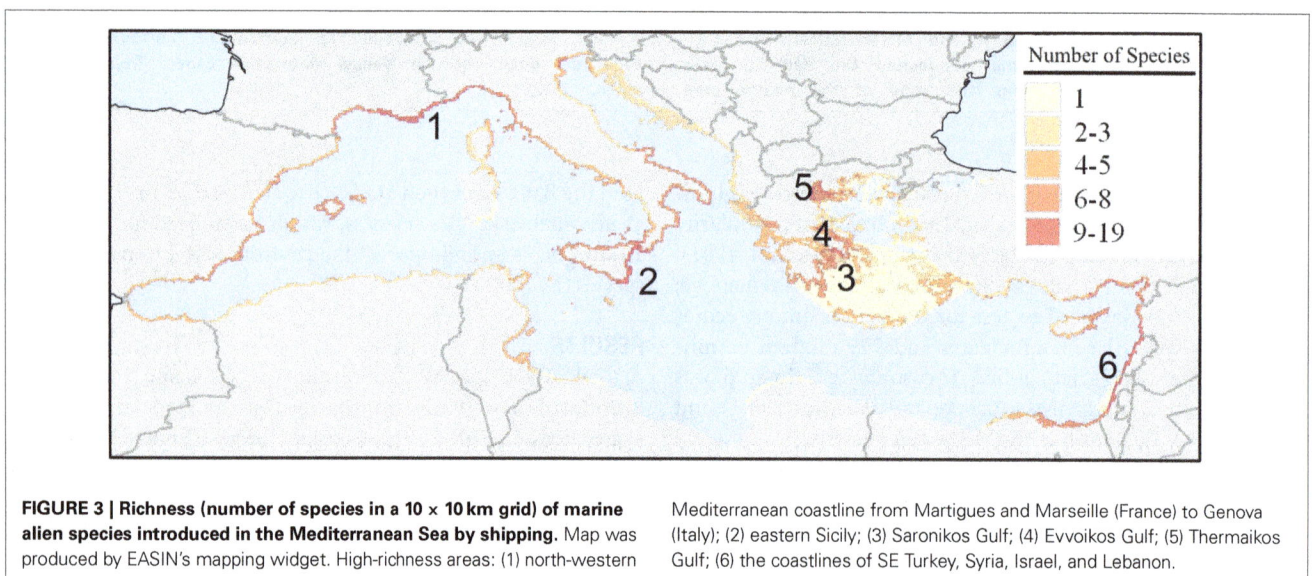

FIGURE 3 | Richness (number of species in a 10 × 10 km grid) of marine alien species introduced in the Mediterranean Sea by shipping. Map was produced by EASIN's mapping widget. High-richness areas: (1) north-western Mediterranean coastline from Martigues and Marseille (France) to Genova (Italy); (2) eastern Sicily; (3) Saronikos Gulf; (4) Evvoikos Gulf; (5) Thermaikos Gulf; (6) the coastlines of SE Turkey, Syria, Israel, and Lebanon.

and invertebrates (14 species) that arrived as contaminants of shellfish. Richness of species introduced by aquaculture is quite low in the Near East and northern African coastlines, with the exception of northern Tunisia (**Figure 4**).

There is a difference in the magnitude of species richness among the species introduced through the Suez Canal, shipping, and aquaculture (**Figures 2–4**). Much higher maximum values of species richness per 10 ×10 km cell are reached for Lessepsian species than for species introduced through aquaculture and shipping, although the total number of species introduced via shipping is not much lower than those introduced through the Suez Canal. This indicates the higher contribution of Lessepsian species in the overall spatial pattern of species richness of all alien species.

Besides differences in the spatial patterns of species richness by pathway, varying patterns among the main taxonomic groups are also observed (**Figure 5**). Alien fish richness is the highest in the Levantine and the southeastern part of the Aegean Sea and the lowest in the western and northern regions of the Mediterranean. For alien invertebrates, the spatial pattern of species richness is similar but there are some additional areas of increased richness such as the French coastline around the Thau lagoon, northern Adriatic, and eastern Sicily. Richness of alien macrophytes has a quite different spatial pattern, with increased richness in the western Mediterranean. These patterns are linked to the dominant pathways of introduction for each group, i.e., the Suez Canal for fish, aquaculture for macrophytes, and the Suez Canal but also shipping for invertebrates (**Figure 5**).

Alien species composition differs among ecoregions (**Figure 6**). With the exception of the Venice lagoon (site 18) and the Thau lagoon (site 24) that appear more similar to each other than to other sites of the same ecoregions (western Mediterranean and Adriatic, respectively), sites from the same ecoregion appear close together in the nMDS plot. Excluding

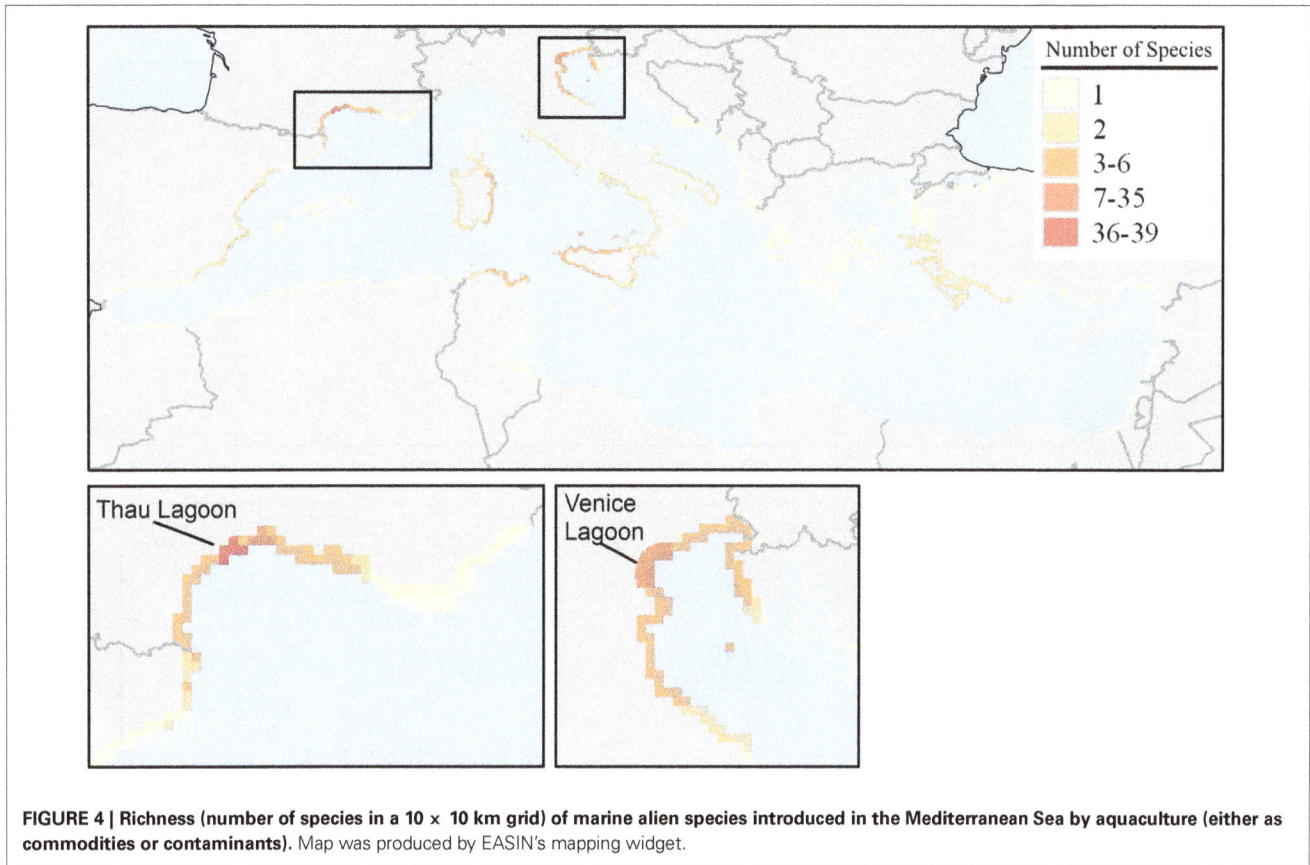

FIGURE 4 | Richness (number of species in a 10 × 10 km grid) of marine alien species introduced in the Mediterranean Sea by aquaculture (either as commodities or contaminants). Map was produced by EASIN's mapping widget.

these two outliers (sites 18, 24), PERMANOVA showed significant differences among ecoregions ($p < 0.001$). Pairwise tests gave p-values <0.05 for all pairs of ecoregions except for Ionian-Adriatic ($p = 0.06$) and Alboran-western Mediterranean ($p = 0.06$). Similarity appears well-correlated to geographical distance, with sites of the Levantine being more similar to sites of the Tunisian plateau and Gulf of Sidra and the Aegean Sea than to sites of the Adriatic, western Mediterranean, and Alboran Sea. Sites of the latter ecoregions are grouped closely in the nMDS plot, while the Ionian Sea is in the middle of all other ecoregions, in conformity to its geographical location.

The biodiversity spatial pattern of native species (fish and invertebrates) differs to that of alien species (**Figure 7**). The highest richness is observed in the Western Mediterranean Sea with a maximum of 391 species in a 10 ×10 km cell. Native species richness decreases from the north-western to the south-eastern regions of the basin, where a minimum value of 84 species is mapped. Native species richness is also higher in coastal and shelf areas, and decreases with depth. The correlation coefficient between alien and native species richness (in coastal areas) was −0.25 (significant, $p < 0.001$), thus the two distributions are negatively correlated.

The highest estimated values of the ratio of alien to native species richness are observed in the eastern Mediterranean (especially in the Levantine and the south-eastern Aegean Sea), with a maximum value of 0.69 (**Figure 8**). In the central and eastern Mediterranean, the alien to native species ratio is much lower.

DISCUSSION

The evidence herein provided demonstrates how human activities and interventions (shipping, aquaculture, opening of the Suez Canal) modify large-scale biodiversity patterns in the Mediterranean Sea by assisting biological invasions. In the Mediterranean Sea, a northwestern-to-southeastern gradient of native species richness is observed, although this could be, at least partly, due to gaps in our knowledge of the biota along the southern and eastern rims (**Figure 7**; Coll et al., 2010, 2012; Bianchi et al., 2012). Native biodiversity is generally higher in coastal and shelf waters in most groups of both vertebrates and invertebrates, with some local exceptions. Similarly, alien species are concentrated in coastal and shelf waters. Very few alien species have been reported in offshore areas, which may be explained by the thriving of shallow-water thermophilic demersal aliens, or because important vectors of alien species (ships and aquaculture) operate in shallow waters, but also due to the reduced sampling effort off-shore (Danovaro et al., 2010). However, the opposite (in relation to native biodiversity) basin-wide trend of alien species richness is observed, decreasing from southeast to northwest. Biodiversity patterns are substantially modified, and locally the induced change in species composition, abundance and richness can be even more marked. For example, in the Thau Lagoon (**Figure 4**) at least 58 introduced macrophytes have been identified, representing 32% of the species diversity and 48–99% of the macrophyte biomass on hard substrates (Boudouresque et al., 2011).

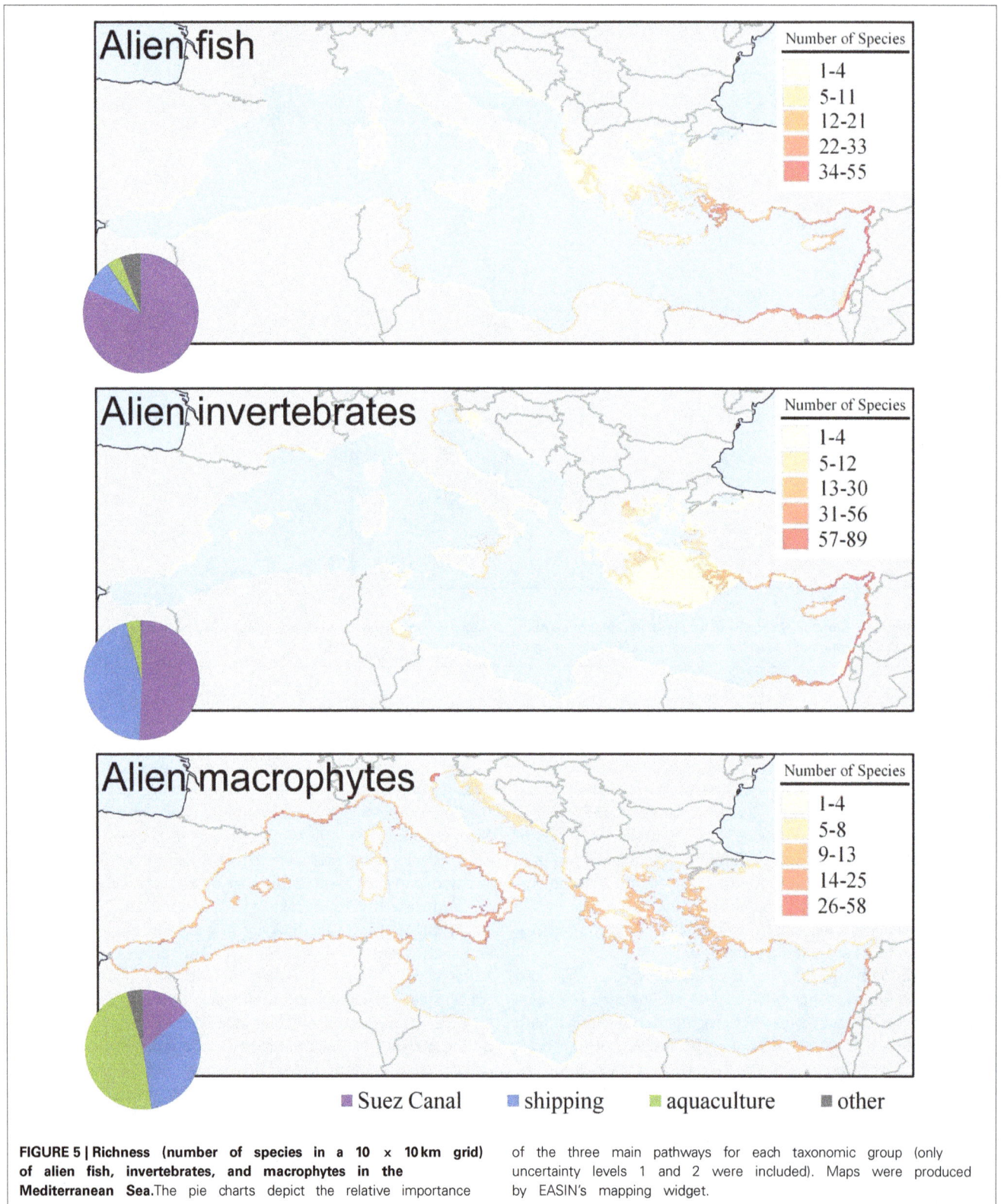

FIGURE 5 | Richness (number of species in a 10 × 10 km grid) of alien fish, invertebrates, and macrophytes in the Mediterranean Sea. The pie charts depict the relative importance of the three main pathways for each taxonomic group (only uncertainty levels 1 and 2 were included). Maps were produced by EASIN's mapping widget.

In several hotspot areas, alien species now constitute a substantial part of the communities and have in many cases caused a shift to novel habitats, with an entirely modified ecosystem functioning (Katsanevakis et al., in press). Species richness per 10 × 10 km cell is generally markedly higher for Lessepsian species than for species introduced by shipping or aquaculture. In the eastern Mediterranean (Levantine Sea), this high richness of alien species is highly reflected also in terms of total biomass and community

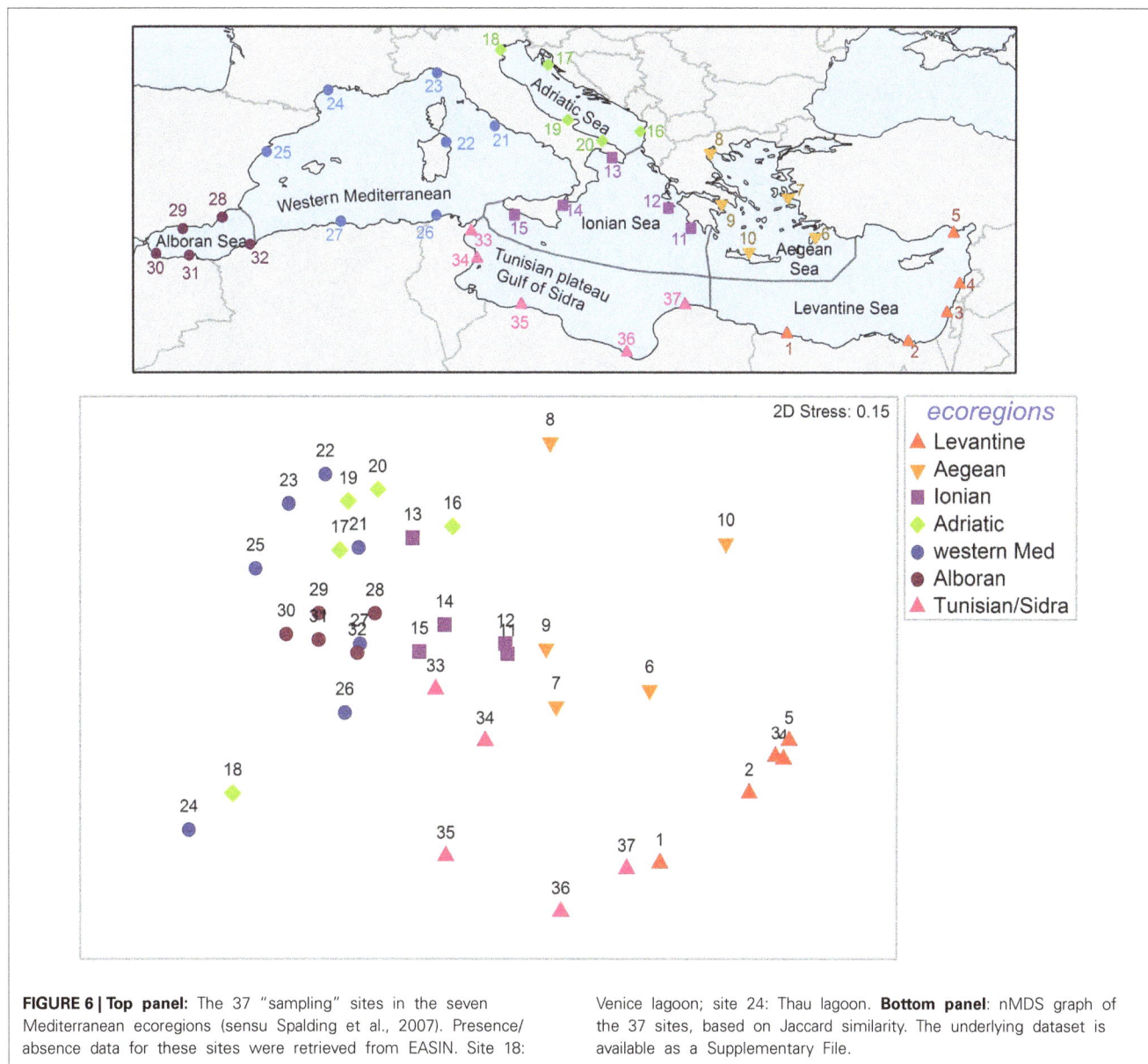

FIGURE 6 | Top panel: The 37 "sampling" sites in the seven Mediterranean ecoregions (sensu Spalding et al., 2007). Presence/absence data for these sites were retrieved from EASIN. Site 18: Venice lagoon; site 24: Thau lagoon. **Bottom panel**: nMDS graph of the 37 sites, based on Jaccard similarity. The underlying dataset is available as a Supplementary File.

structure. The proportion of alien fish in trawl catches along the Levantine continental shelf has been increasing, reaching 54% in abundance and 55% in biomass (84 and 66%, respectively for the 15–30 m depth stratum) (Edelist et al., 2013).

Most of the alien species that are established in the Mediterranean Sea were introduced in the last decades. Less than 200 alien species were introduced in the Mediterranean before 1950, while >800 species have been introduced after that date (Zenetos et al., 2012). Hence, the observed large-scale change of biodiversity patterns in the Mediterranean is a phenomenon that has been evolving mainly during the last century. This unprecedented change has been greatly driven by the opening of the Suez Canal in 1869 and its continuous enlargement, but also by the increasing seaborne trade, responsible for many shipping-mediated introductions, and the intentional introduction of alien commodity species (and, unintentionally, of contaminant species) for aquaculture (Katsanevakis et al., 2013; Nunes et al., 2014).

Herein, we focused on species richness as an indicator of biodiversity, as is common in the ecological literature (May, 1995; Bianchi and Morri, 2000). Alien species richness and the ratio between alien and native species (**Figure 8**) were used as indicators of biodiversity change and impact. Another indicator that has been previously used is the change in the intensity of spatial congruence between alien and endemic fauna (Ben Rais Lasram and Mouillot, 2009). However, in many cases these indicators are not the best to indicate biodiversity change or impact. Some individual keystone and high-impact alien species can have a much more severe impact than dozens of other non-invasive aliens. For example, the two herbivore rabbitfish *Siganus luridus* and *S. rivulatus* have radically altered the community structure and the native food web of the rocky infralittoral zone in the

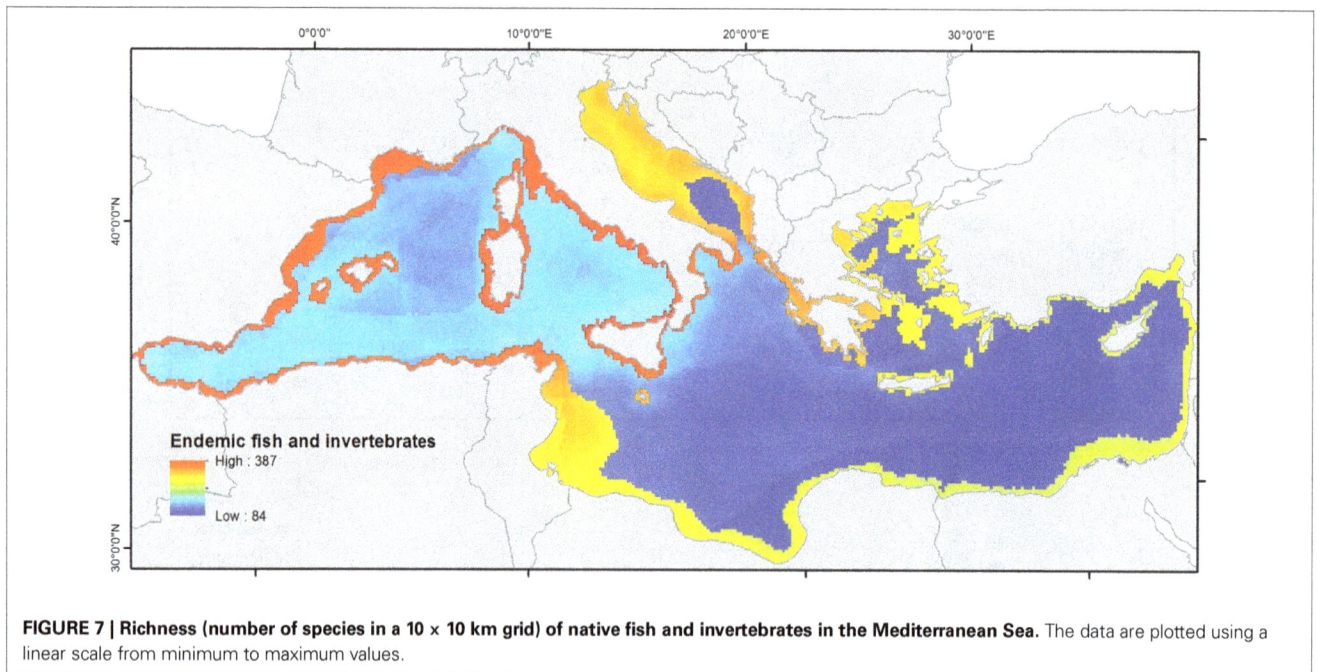

FIGURE 7 | Richness (number of species in a 10 × 10 km grid) of native fish and invertebrates in the Mediterranean Sea. The data are plotted using a linear scale from minimum to maximum values.

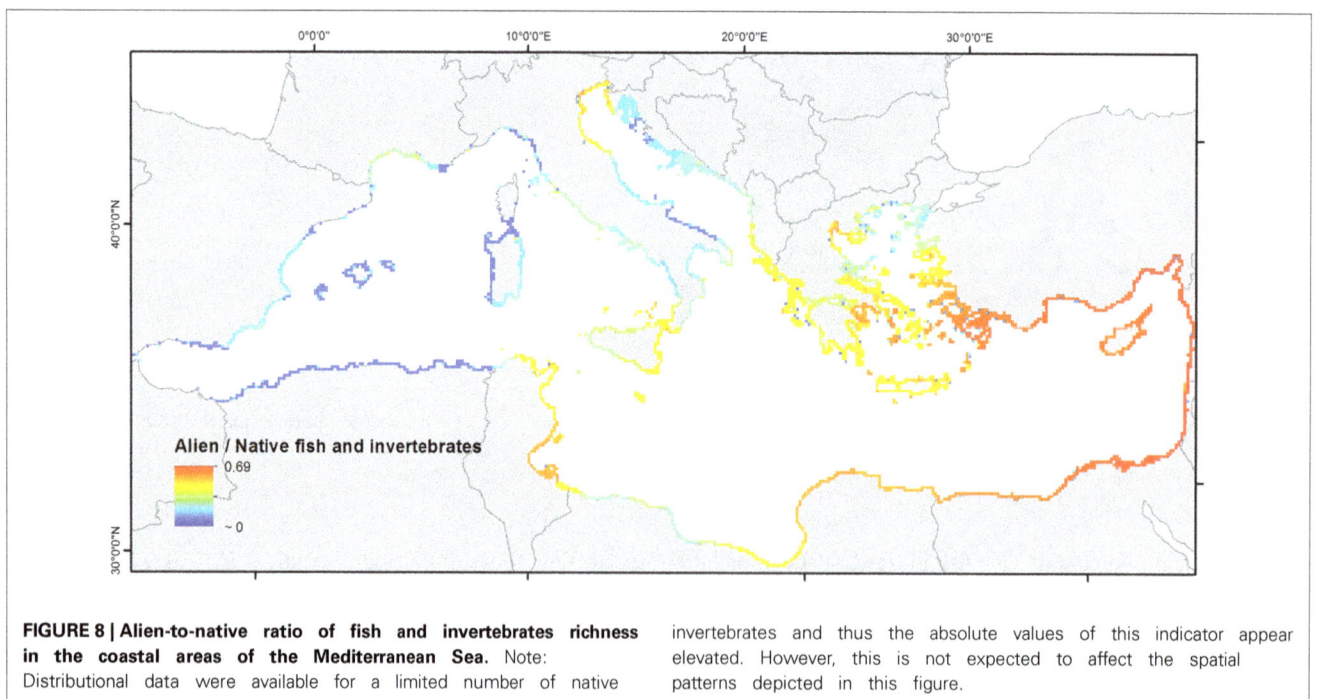

FIGURE 8 | Alien-to-native ratio of fish and invertebrates richness in the coastal areas of the Mediterranean Sea. Note: Distributional data were available for a limited number of native invertebrates and thus the absolute values of this indicator appear elevated. However, this is not expected to affect the spatial patterns depicted in this figure.

eastern Mediterranean, through overgrazing. They are able to create and maintain barrens (rocky areas almost devoid of erect algae) and contribute to the transformation of the ecosystem from one dominated by lush and diverse brown algal forests to a degraded one dominated by bare rock and patches of crustose coralline algae (Sala et al., 2011; Giakoumi, 2014). The large-scale and severe impact of these two species in the shallow rocky shores of the eastern Mediterranean is probably greater than that of the other alien fish in the Mediterranean altogether, which

is not depicted by a species richness indicator. Similarly, a few very invasive macroalgae can dominate algal assemblages creating homogenized microhabitats, greatly impacting native communities. This is the case of the invasive green alga *Caulerpa cylindracea*, which can easily overgrow and eliminate other macroalgal or invertebrate species and may form compact multilayered mats up to 15 cm thick that trap sediment and may create an anoxic layer underneath (Klein and Verlaque, 2008; Katsanevakis et al., in press). Several biotopes, such as Mediterranean communities of sublittoral algae and coralligenous communities, are affected by *C. cylindracea*, which smothers indigenous populations, outcompetes native communities, and diminishes the structural complexity and species richness. This species alone can have much greater impact than dozens of other non-invasive alien macrophytes, which again is not evident in an alien species richness indicator.

The ecosystems of many regions of the Mediterranean Sea have been substantially modified (Boudouresque et al., 2011; Sala et al., 2011; Edelist et al., 2013). Especially Lessepsian migration is considered as the most significant biogeographic change currently underway worldwide (Bianchi et al., 2012). Although so far there are no recorded basin-wide extinctions of native marine species in the Mediterranean, there are many examples of local extirpations and range shifts concurrent with alien invasions (Galil, 2007). Hence, γ-diversity has increased in the Mediterranean by >5% due to the overall increase of species richness, while α-diversity has locally decreased in some cases (see the *Siganus* spp. example above) and increased in others because of the habitat-specific increase of species richness. However, there is no evidence so far of extensive basin-wide taxonomic homogenization of the Mediterranean biota due to biological invasions. There are marked differences in the introduced biota among ecoregions (**Figure 6**), which is more intense for Lessepsian and aquaculture-introduced species. Hence, at a Mediterranean scale, communities are continuously changing but there is no sign of a reduced degree of heterogeneity across ecoregions. This may not be the case at smaller scales (e.g., among habitats within an ecoregion), possibly leading to an important decrease of β-diversity within ecoregions. The effect of biological invasions on β-diversity is greatly dependent on scale (Olden, 2006) and needs further investigation.

The future of the Mediterranean Sea biota is difficult to predict. During the past two decades, Mediterranean waters have been warming at a rather high rate, especially in the eastern region, and this trend is predicted to continue in the long-term influencing biogeochemical cycles and ecosystem functioning (Durrieu de Madron et al., 2011; Macias et al., 2013). The sea surface temperature contours are shifted northwards (Coll et al., 2010) and the boundaries of Mediterranean ecoregions are expected to change substantially. Warming of the Mediterranean Sea favors the establishment and spread of thermophilic species, such as most of the Lessepsian migrants (Bianchi, 2007; Bianchi et al., 2013). The high incidence of alien species of tropical affinity and origin is driving the eastern Mediterranean biota toward a phase of "tropicalization" (Bianchi and Morri, 2003). At least in the Levant Basin, environmental conditions are favorable for communities of Indo-Pacific hermatypic corals, and the arrival and establishment of the first reef builders and a great diversity of associated fish and invertebrates is probably only a matter of time (Por, 2009).

A better understanding of how the human-shaped new biodiversity patterns will affect the Mediterranean food webs, ecosystem functioning, and the provision of ecosystem services for the benefit of humans is challenging (Borja, 2014) but urgently needed. A possible way to assess this is through the employment of ecosystem models, which in the last decades have been increasingly used worldwide to evaluate ecosystem structure and functions and the impacts of human activities on marine systems (e.g., Christensen and Walters, 2004; Shin et al., 2004; Fulton, 2010). Despite the ability of some ecosystem models to provide useful indicators to address biological invasions, gaps still remain in relation to the understanding of role/impact of alien species in the food web (Piroddi et al., under review). Thus, future studies should be set up and carried out to assess and better understand alien species in an ecosystem context.

Alien species often benefit some components of native biodiversity and can enhance or provide new ecosystem services (Katsanevakis et al., in press). In marine regions subject to rapid change, such as the Mediterranean Sea, introduced species may even secure ecosystem processes and functioning (Walther et al., 2009). It is unknown if the future Mediterranean ecosystems will be more resilient, and may continue to provide the same ecosystem services, but it is likely they will be very different than the past ecosystems before the major wave of biological invasions of the last century.

AUTHOR CONTRIBUTIONS

Stelios Katsanevakis and Ana Cristina Cardoso conceived the study. Data on native species distributions were provided by Frida Ben Rais Lasram and Marta Coll. Analysis of pathways of introduction was conducted by Argyro Zenetos and Stelios Katsanevakis. The mapping of human activities was provided by Chiara Piroddi. Stelios Katsanevakis created the alien species maps and conducted the MDS analysis of **Figure 6**. Mapping of native species distribution, estimation and mapping of alien-to-native ratios, and related spatial analyses were conducted by Marta Coll, Jeroen Steenbeek, and Frida Ben Rais Lasram. Stelios Katsanevakis prepared a first draft of the manuscript and all coauthors contributed to the final version.

REFERENCES

Anderson, M. J. (2001). A new method for non-parametric multivariate analysis of variance. *Aust. Ecol.* 26, 32–46. doi: 10.1111/j.1442-9993.2001.01070.pp.x

Ben Rais Lasram, F., and Mouillot, D. (2009). Increasing southern invasion enhances congruence between endemic and exotic Mediterranean fish fauna. *Biol. Invasions* 11, 697–711. doi: 10.1007/s10530-008-9284-4

Bianchi, C. N. (2007). Biodiversity issues for the forthcoming tropical Mediterranean Sea. *Hydrobiologia* 580, 7–21. doi: 10.1007/s10750-006-0469-5

Bianchi, C. N., Boudouresque, C. F., Francour, P., Morri, C., Parravicini, V., Templado, J., et al. (2013). The changing biogeography of the Mediterranean Sea: from the old frontiers to the new gradients. *Boll. Mus. Ist. Biol. Univ. Genova* 75, 81–84.

Bianchi, C. N., and Morri, C. (2000). Marine biodiversity of the Mediterranean Sea: situation, problems and prospects for future research. *Mar. Pollut. Bull.* 40, 367–376. doi: 10.1016/S0025-326X(00)00027-8

Bianchi, C. N., and Morri, C. (2003). Global sea warming and 'tropicalization' of the Mediterranean Sea: biogeographic and ecological aspects. *Biogeographia* 24, 319–327.

Bianchi, C. N., Morri, C., Chiantore, M., Montefalcone, M., Parravicini, V., and Rovere, A. (2012). "Mediterranean Sea biodiversity between the legacy from the past and a future of change," in *Life in the Mediterranean Sea: a Look at Habitat Changes*, ed N. Stambler (New York, NY: Nova Science Publishers), 1–60.

Borja, A. (2014). Grand challenges in marine ecosystems ecology. *Front. Mar. Sci.* 1:1. doi: 10.3389/fmars.2014.00001

Boudouresque, C. F., Klein, J., Ruitton, S., and Verlaque, M. (2011). "Biological invasion: the Thau Lagoon, a Japanese biological island in the Mediterranean Sea," in *Global Change: Mankind-Marine Environment Interactions, Proceedings of the 13th French-Japanese Oceanography Symposium*, eds H. J. Ceccaldi, I. Dekeyser, M. Girault, and G. Stora (Dordrecht: Springer), 151–156.

Christensen, V., and Walters, C. (2004). Ecopath with ecosim: methods, capabilities and limitations. *Ecol. Model.* 72, 109–139. doi: 10.1016/j.ecolmodel.2003.09.003

Clarke, K. R., and Warwick, R. M. (2001). *Change in Marine Communities: an Approach to Statistical Analysis and Interpretation, 2nd Edn.* Plymouth: PRIMER-E.

Coll, M., Piroddi, C., Albouy, C., Ben Rais Lasram, F., Cheung, W. W. L., Christensen, V., et al. (2012). The Mediterranean under siege: spatial overlap between marine biodiversity, cumulative threats and marine reserves. *Glob. Ecol. Biogeogr.* 21, 465–481. doi: 10.1111/j.1466-8238.2011. 00697.x

Coll, M., Piroddi, C., Steenbeek, J., Kaschner, K., Ben Rais Lasram, F., Aguzzi, J., et al. (2010). The biodiversity of the Mediterranean Sea: estimates, patterns and threats. *PLoS ONE* 5:e11842. doi: 10.1371/journal.pone. 0011842

Costello, M. J., Coll, M., Danovaro, R., Halpin, P., Ojaveer, H., and Miloslavich, P. (2010). A census of marine biodiversity knowledge, resources and future challenges *PLoS ONE* 5:e12110. doi: 10.1371/journal.pone. 0012110

Danovaro, R., Company, B. J., Corinaldesi, C., D'Onghia, G., Galil, B. S., Gambi, C., et al. (2010). Deep-Sea biodiversity in the Mediterranean Sea: the known the unknown and the unknowable. *PLoS ONE* 5:e11832. doi: 10.1371/journal.pone.0011832

de Madron, X. D., Guieu, C., Sempéré, R., Conan, P., Cossa, D., D'Ortenzio, F., et al. (2011). Marine ecosystems' responses to climatic and anthropogenic forcings in the Mediterranean. *Prog. Oceanogr.* 91, 97–166. doi: 10.1016/j.pocean.2011.02.003

EC. (2010). European Commission Decision 2010/477/EU on criteria and methodological standards on good environmental status of marine waters. *J. Eur. Union L* 53, 232/14. doi: 10.3000/17252555.L_2010.232.eng

Edelist, D., Rilov, G., Golani, D., Carlton, J. T., and Spanier, E. (2013). Restructuring the Sea: profound shifts in the world's most invaded marine ecosystem. *Divers. Distrib.* 19, 69–77. doi: 10.1111/ddi.12002

Fulton, E. A. (2010). Approaches to end-to-end ecosystem models. *J. Marine Syst.* 81, 171–183. doi: 10.1016/j.jmarsys.2009.12.012

Galil, B. S. (2007). Loss or gain? Invasive aliens and biodiversity in the Mediterranean Sea. *Mar. Pollut. Bull.* 55, 314–322. doi: 10.1016/j.marpolbul.2006.11.008

Gatto, F., Katsanevakis, S., Vandekerkhove, J., Zenetos, A., and Cardoso, A. C. (2013). Evaluation of online information sources on alien species in Europe – the need of harmonization and integration. *Environ. Manag.* 51, 1137–1146. doi: 10.1007/s00267-013-0042-8

Giakoumi, S. (2014). Distribution patterns of the invasive herbivore *Siganus luridus* (Rüppell, 1829) and its relation to native benthic communities in the central Aegean Sea, Northeastern Mediterranean. *Mar. Ecol.* 35, 96–105. doi: 10.1111/maec.12059

Grosholz, E. (2002). Ecological and evolutionary consequences of coastal invasions. *Trends Ecol. Evol.* 17, 22–27. doi: 10.1016/S0169-5347(01)02358-8

Halpern, B. S., Walbridge, S., Selkoe, K. A., Kappel, C. V., Micheli, F., D'Agrosa, C., et al. (2008). A global map of human impact on marine ecosystems. *Science* 319, 948–952. doi: 10.1126/science.1149345

Jaccard, P. (1901). Étude comparative de la distribution florale dans une portion des Alpes et des Jura. *Bull. Soc. Vaudoise Sci. Nat.* 37: 547–579.

Katsanevakis, S., Bogucarskis, K., Gatto, F., Vandekerkhove, J., Deriu, I., and Cardoso, A. C. (2012). Building the European Alien Species Information Network (EASIN): a novel approach for the exploration of distributed alien species data. *Bioinvasions Rec.* 1, 235–245. doi: 10.3391/bir.2012. 1.4.01

Katsanevakis, S., Wallentinus, I., Zenetos, A., Leppäkoski, E., Çinar, M. E., Oztürk, B., et al. (in press). Impacts of marine invasive alien species on ecosystem

services and biodiversity: a pan-European critical review. *Aquat. Invasions.* doi: 10.3391/ai.2014.9.4.01

Katsanevakis, S., Zenetos, A., Belchior, C., and Cardoso, A. C. (2013). Invading European Seas: assessing pathways of introduction of marine aliens. *Ocean Coast. Manag.* 76, 64–74. doi: 10.1016/j.ocecoaman.2013.02.024

Klein, J., and Verlaque, M. (2008). The *Caulerpa racemosa* invasion: a critical review. *Mar. Pollut. Bull.* 56, 205–225. doi: 10.1016/j.marpolbul.2007.09.043

Kruskal, J. B. (1964). Multidimensional scaling by optimizing goodness of fit to a nonmetric hypothesis. *Psychometrika* 29, 1–27. doi: 10.1007/BF02289565

Lejeusne, C., Chevaldonne, P., Pergent-Martini, C., Boudouresque, C. F., and Perez, T. (2010). Climate change effects on a miniature ocean: the highly diverse, highly impacted Mediterranean Sea. *Trends Ecol. Evol.* 25, 250–260. doi: 10.1016/j.tree.2009.10.009

Macias, D., Garcia-Gorriz, E., and Stips, A. (2013). Understanding the causes of recent warming of Mediterranean waters. How much could be attributed to climate change? *PLoS ONE* 8:e81591. doi: 10.1371/journal.pone. 0081591

May, R. M. (1995). Conceptual aspects of the quantification of the extent of biological diversity. *Philos. Trans. R Soc. Lond. B Biol. Sci.* 345, 13–20. doi: 10.1098/rstb.1994.0082

Micheli, F., Halpern, B. S., Walbridge, S., Ciriaco, S., Ferretti, F., Fraschetti, S., et al. (2013). Cumulative human impacts on Mediterranean and Black Sea marine ecosystems: assessing current pressures and opportunities. *PLoS ONE* 8:e79889. doi: 10.1371/journal.pone.0079889

Molnar, J. L., Gamboa, R. L., Revenga, C., and Spalding, M. D. (2008). Assessing the global threat of invasive species to marine biodiversity. *Front. Ecol. Environ.* 6, 458–492. doi: 10.1890/070064

Mouillot, D., Albouy, C., Guilhaumon, F., Ben Rais Lasram, F., Coll, M., Devictor, V., et al. (2011). Protected and threatened components of fish biodiversity in the Mediterranean Sea. *Curr. Biol.* 21, 1044–1050. doi: 10.1016/j.cub.2011. 05.005

Nunes, A. L., Katsanevakis, S., Zenetos, A., and Cardoso, A. C., (2014). Gateways to alien invasions in the European Seas. *Aquat. Invasions.* 9, 133–144. doi: 10.3391/ ai.2014.9.2.02

Occhipinti Ambrogi, A. (2000). Biotic invasions in a Mediterranean Lagoon. *Biol. Invasions* 2, 165–176. doi: 10.1023/A:1010004926405

Olden, J. D. (2006). Biotic homogenization: a new research agenda for conservation biogeography. *J. Biogeogr.* 33, 2027–2039. doi: 10.1111/j.1365-2699.2006.01572.x

Por, F. D. (2009). Tethys returns to the Mediterranean: success and limits of tropical re-colonization. *Biorisk* 3:5e19. doi: 10.3897/biorisk.3.30

Sala, E., Kizilkaya, Z., Yildirim, D., and Ballesteros, E. (2011). Alien marine fishes deplete algal biomass in the eastern Mediterranean. *PLoS ONE* 6:e17356. doi: 10.1371/journal.pone.0017356

Schlaepfer, M. A., Sax, D. F., and Olden, J. D. (2011). The potential conservation value of non-native species. *Conserv. Biol.* 25, 428–437. doi: 10.1111/j.1523-1739.2010.01646.x

Shin, Y. J., Shannon, L. J., and Cury, P. M. (2004). Simulations of fishing effects on the southern Benguela fish community using an individual-based model: learning from a comparison with ECOSIM. *Afr. J. Mar. Sci.* 26, 95–114. doi: 10.2989/18142320409504052

Simberloff, D., Martin, J. L., Genovesi, P., Maris, V., Wardle, D. A., Aronson, J., et al. (2013). Impacts of biological invasions: what's what and the way forward. *Trends Ecol. Evol.* 28, 58–66. doi: 10.1016/j.tree.2012.07.013

Spalding, M. D., Fox, H. E., Allen, G. R., Davidson, N., Ferdaña, Z. A., Finlayson, M., et al. (2007). Marine ecoregions of the world: a bioregionalization of coastal and shelf areas. *Bioscience* 57, 573–583. doi: 10.1641/ B570707

Trombetti, M., Katsanevakis, S., Deriu, I., and Cardoso, A. C. (2013). EASIN-Lit: a geo-database of published alien species records. *Manag. Biol. Invasions* 4, 261–264. doi: 10.3391/mbi.2013.4.3.08

Trujillo, P., Piroddi, C., and Jacquet, J. (2012). Fish farms at sea: the ground truth from Google Earth. *PLoS ONE* 7:e30546. doi: 10.1371/journal.pone.0030546

Vilà, M., Basnou, C., Pysek, P., Josefsson, M., Genovesi, P., Gollasch, S., et al. (2010). How well do we understand the impacts of alien species on ecosystem services? A pan-European, crosstaxa assessment. *Front. Ecol. Environ.* 8:135–144. doi: 10.1890/080083

Wallentinus, I., and Nyberg, C. D. (2007). Introduced marine organisms as habitat modifiers. *Mar. Pollut. Bull.* 55, 323–332. doi: 10.1016/j.marpolbul.2006.11.010

Walther, G.-R., Roques, A., Hulme, P. E., Sykes, M. T., Pyšek, P., and Kühn, I. (2009). Alien species in a warmer world: risks and opportunities. *Trends Ecol. Evol.* 24, 686–693. doi: 10.1016/j.tree.2009.06.008

Whitehead, P., Bauchot, L., Hureau, J., Nielsen, J., and Tortonese, E. (1986). *Fishes of the North-Eastern Atlantic and the Mediterranean.* Paris: UNESCO.

Zenetos, A., Gofas, S., Morri, C., Rosso, A., Violanti, D., García Raso, J. E., et al. (2012). Alien species in the Mediterranean Sea by 2012.A contribution to the application of European Union's Marine Strategy Framework Directive (MSFD). Part 2. Introduction trends and pathways. *Mediterr. Mar. Sci.* 13, 328–352. doi: 10.12681/mms.327

Zenetos, A., Gofas, S., Verlaque, M., Çinar, M. E., García Raso, E., Azzurro, E., et al. (2010). Alien species in the Mediterranean by 2010. A contribution to the application of European Union's Marine Strategy Framework Directive (MSFD). Part I. Spatial distribution. *Mediterr. Mar. Sci.* 11, 381–493. doi: 10. 12681/mms.87

Conflict of Interest Statement: The Associate Editor Christos Dimitrios Arvanitidis declares that, despite being affiliated to the same institution as author Argyro Zenetos, the review process was handled objectively and no conflict of interest exists. The authors declare that the research was conducted in the absence of any commercial or financial relationships that could be construed as a potential conflict of interest.

21

Climatic and ecological drivers of euphausiid community structure vary spatially in the Barents Sea: relationships from a long time series (1952–2009)

Emma L. Orlova[1†], Andrey V. Dolgov[1], Paul E. Renaud[2,3], Michael Greenacre[2,4], Claudia Halsband[2]*
and Victor A. Ivshin[1,5]

[1] *Laboratory of Trophology, Polar Research Institute of Marine Fisheries and Oceanography, Murmansk, Russia*
[2] *Akvaplan-niva, Fram Centre for Climate and Environment, Tromsø, Norway*
[3] *Department of Arctic Biology, University Centre in Svalbard, Longyearbyen, Norway*
[4] *Barcelona Graduate School of Economics, Universitat Pompeu Fabra, Barcelona, Spain*
[5] *Laboratory of Fisheries Oceanography, Polar Research Institute of Marine Fisheries and Oceanography, Murmansk, Russia*

Edited by:
Paul F. J. Wassmann, University of Tromsø - Norway's Arctic University, Norway

Reviewed by:
Jan Marcin Weslawski, Institute of Oceanology Polish Academy of Sciences, Poland
Rolf Gradinger, University of Alaska Fairbanks, USA

Correspondence:
Andrey V. Dolgov, Laboratory of Trophology, Polar Research Institute of Marine Fisheries and Oceanography, Knipovich-St., 6, Murmansk 183038, Russia
e-mail: dolgov@pinro.ru

[†]*Deceased.*

Euphausiids play an important role in transferring energy from ephemeral primary producers to fish, seabirds, and marine mammals in the Barents Sea ecosystem. Climatic impacts have been suggested to occur at all levels of the Barents Sea food-web, but adequate exploration of these phenomena on ecologically relevant spatial scales has not been integrated sufficiently. We used a time-series of euphausiid abundance data spanning 58 years, one of the longest biological time-series in the Arctic, to explore qualitative and quantitative relationships among climate, euphausiids, and their predators, and how these parameters vary spatially in the Barents Sea. We detected four main hydrographic regions, each with distinct patterns of interannual variability in euphausiid abundance and community structure. Assemblages varied primarily in the relative abundance of *Thysanoessa inermis* vs. *T. raschii*, or *T. inermis* vs. *T. longicaudata*, and *Meganyctiphanes norvegica*. Climate proxies and the abundance of capelin or cod explained 30–60% of the variability in euphausiid abundance in each region. Climate also influenced patterns of variability in euphausiid community structure, but correlations were generally weaker. Advection of boreal euphausiid taxa from the Norwegian Sea is clearly more prominent in warmer years than in colder years, and interacts with seasonal fish migrations to help explain spatial differences in primary drivers of euphausiid community structure. Non-linear effects of predators were common, and must be considered more carefully if a mechanistic understanding of the ecosystem is to be achieved. Quantitative relationships among euphausiid abundance, climate proxies, and predator stock-sizes derived from these time series are valuable for ecological models being used to predict impacts of climate change on the Barents Sea ecosystem, and how the system should be managed.

Keywords: Arctic, capelin, cod, krill, pelagic food web

INTRODUCTION

Euphausiids in many high latitude ecosystems are important food for fish, seabirds, and mammals (Nilssen et al., 1995; Mehlum, 2001; Lindstrøm et al., 2013). In the Barents Sea, these lipid-rich zooplankton may represent up to 60% of the diet of capelin when centers of abundance of the two taxa overlap (e.g., Dalpadado and Mowbray, 2013). The share of lipid-rich food in capelin diets can determine their overwintering success and reproductive output during the next year (Orlova et al., 2010a,b). In addition, early year-classes of cod (*Gadus morhua*) and haddock (*Melanogrammus aegelfinus*) may feed heavily upon euphausiids in some regions of the Barents Sea (Ponomarenko, 1973; Ponomarenko and Yaragina, 2003; Dalpadado et al., 2009; Renaud et al., 2012), as will adult cod and haddock (Kovtsova et al., 1989; Drobysheva, 1994; Orlova et al., 2001). Due to these trophic relationships, euphausiid biomass and abundance have been the subject of monitoring efforts in the Barents Sea for almost six decades.

The Polar Research Institute of Marine Fisheries and Oceanography (PINRO, Murmansk, Russia) has conducted annual plankton surveys in most regions of the Barents Sea since 1952. Some of the most relevant findings from these studies include the coupling of population cycles of euphausiids with decadal climatic oscillations, links between capelin and euphausiid biomass, and the share of euphausiids in capelin (*Mallotus villosus*) and cod diets (Drobysheva, 1994; Orlova et al., 2001, 2010a, 2013; Dalpadado et al., 2003). The entire data series, however, has not been investigated in detail in a single study, and identification of responses to long-term trends in climate requires such long data series. Climatic shifts are implicated in impacting the distribution, development and composition of euphausiid communities (Gómez-Gutiérrez et al., 1995; Brinton and Townsend, 2003; Dorman et al., 2011), as well as predatory fish and seabird populations (Abraham and Sydeman, 2004; Sydeman et al., 2006; Coyle et al., 2011), in other areas of the world's oceans. Since climatic warming is already being felt in the Barents region, understanding and quantifying long-term trends in euphausiid response to warming and different predator stocks is important to build both conceptual and mathematical models of the future Barents Sea food web.

The Barents Sea ecosystem has been described as being structured by top-down processes (Dalpadado and Skjoldal, 1996), bottom-up processes (Drinkwater, 2006), or "wasp-waist" regulation, i.e., controlled by planktivorous fish populations (Yaragina and Dolgov, 2009). Different conclusions, however, may be a consequence of the particular species receiving the most focus or the time period studied. Climate oscillations have been implicated in determining the alternating top-down and bottom-up structuring of the Bering Sea pelagic food web (Hunt et al., 2002), and a similar mechanism may operate in the Barents Sea. Recently, a 40-y time series was used to suggest that the krill-planktivorous fish relationship alternated between top-down and bottom-up structuring in the Barents Sea (Johannesen et al., 2012). The recent warming, however, appears to overwhelm typical climate oscillation patterns (e.g., Eriksen and Dalpadado, 2011; Dalpadado et al., 2012; Johannesen et al., 2012; Orlova et al., 2013), indicating the need for longer time series and more complex modeling techniques.

There are four main species of euphausiids in the Barents Sea. *Thysanoessa raschii* is a neritic resident and reproduces in the colder waters of the Barents Sea (Drobysheva, 1994). *T. inermis* is the most abundant species in the Barents Sea, consisting both of locally reproducing populations and populations advected into the region from the Norwegian Sea (Drobysheva, 1982). These two species have historically made up the majority of the euphausiid community in the Barents Sea (e.g., Drobysheva, 1994). *T. longicaudata* and *Meganyctiphanes norvegica* are boreal species that spawn in the Norwegian Sea and arrive in the Barents Sea via warm ocean currents (e.g., Drobysheva, 1994; Zhukova et al., 2009). Only within the last 10 years have these two species been common inhabitants in this region (Dalpadado and Skjoldal, 1996; Buchholz et al., 2010). *T. inermis* and *T. raschii* reproduce in the southern Barents Sea (e.g., Zelikman, 1958, 1964), and *T. raschii* has recently been observed to spawn in Spitsbergen/Svalbard fjords (Buchholz et al., 2012). All four species complete their life cycles in 2–4 years, and all undergo dramatic seasonal vertical migrations where nearly the entire populations spend the winter months in near-bottom waters (e.g., Mauchline, 1980). Whereas there are some similarities in biology and overlap in distribution, the different species have distinct affinities for specific water masses, advective processes determine their population cycles differently, and they vary (in time and space) in terms of their contribution to fish diets. These traits indicate the importance of conducting studies that consider how species compositions vary among regions. This is an approach that has been little used in the Barents Sea region (but see Orlova et al., 2013), where primary questions until now have

centered on euphausiid biomass in general, and on interannual variability in prey resources for cod and capelin (e.g., Eriksen and Dalpadado, 2011). Only through this regional approach, however, can mechanisms responsible for determining community structure be identified, and environmental and biological drivers of these processes be explored.

In order to identify environmental and ecological impacts of euphausiids in the pelagic ecosystem, and ultimately their support of commercial fishes, seabirds, and mammal stocks, we analyzed a unique long-term data series within a multi-species and spatially-defined framework. Specifically, we asked: (1) how has euphausiid community structure varied over the duration of the time series within different hydrographically-defined regions in the Barents Sea? (2) what are the qualitative and quantitative effects of different climate indicators and fish stock sizes on populations of overwintering euphausiids? and (3) how can this new understanding be used to predict pelagic ecosystem structure and function in response to expected changes in oceanographic and biological drivers?

METHODS
THE EUPHAUSIID DATA SERIES
The Barents Sea is a relatively shallow-water sea, most of the area being less than 300–400 m. Between 1952 and 2009, data on the distribution, abundance, and species composition of euphausiids in the Barents Sea (Table S1, **Figure 1**) were collected annually during autumn-winter (October-March). This represents one of the most extensive biological time-series available in the Arctic in time and area covered, and data on species composition and abundance have been published for some of the years (Drobysheva, 1982, 1994; Orlova et al., 2001; Zhukova et al., 2009). The winter period was chosen because: (1) most planktonic species have reduced diurnal vertical migration and occur in a more limited range of depths than during the spring and summer, and (2) this season follows the period when fishes feed intensively on euphausiid stocks. Consequently, the estimates of euphausiid distribution and abundance in this season are consistent, reflect the state of euphausiids stock at the end of feeding season, and are useful for forecasting euphausiid stock size and food-resource state for commercial fish in the next year.

To collect euphausiids, a net (diameter 50 cm, 0.2 m² opening area, 564 μm mesh size) was attached to the headline of a bottom trawl. Euphausiids were sampled at 6–10 m above the bottom (the upper edge of the vertical opening of the bottom trawl). Individuals with body length >6–7 mm were in the catches. Euphausiids from most samples were identified to the species level and abundances were expressed as number of individuals per 1000 m³. The season and the sampling net were chosen in the 1950s after some experiments with timing of sampling and net shape and orientation (Orlova et al., 2008). This approach has been the standard at PINRO for estimation of euphausiid stock in the Barents Sea since the 1960s (Drobysheva, 1994), greatly enhancing data comparability. Community structure for a given year is represented by catches of the overwintering community during the winter months following the main growing season (i.e., euphausiid data for 1970 are from samples collected October 1970-March 1971). This allows matching of climatic conditions

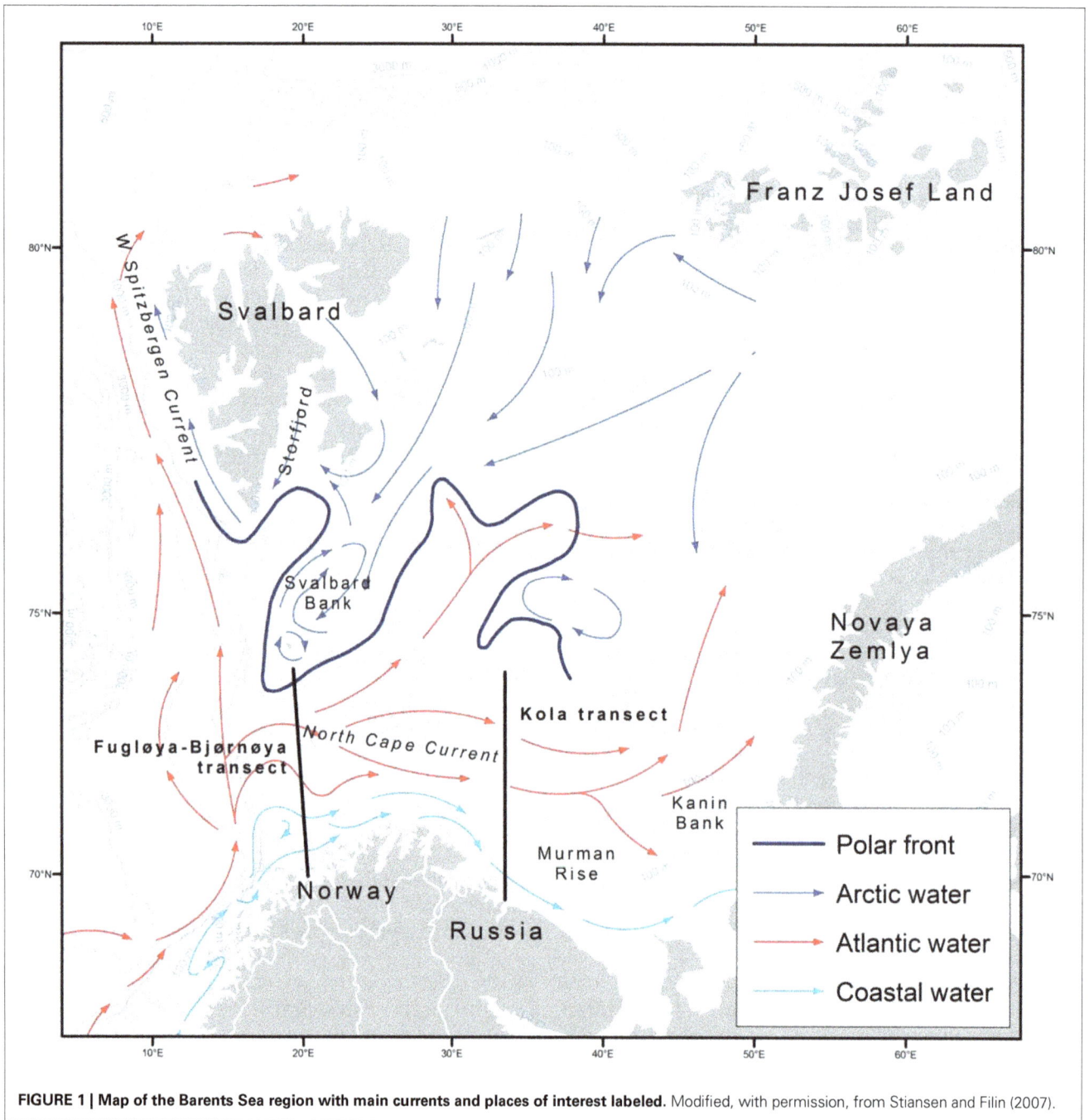

FIGURE 1 | Map of the Barents Sea region with main currents and places of interest labeled. Modified, with permission, from Stiansen and Filin (2007).

and fish stocks with the structure of the overwintering euphausiid assemblages.

DEFINITION OF BARENTS SEA SUB-REGIONS

In order to define regions with similar oceanographic charac-teristics, four variables (temperature and salinity values at 10 and 125 m) were extracted from the NOAA climatic atlas of the Barents Sea (Matishov et al., 1998). These depths were chosen to include as many stations as possible, but also to represent areas below the surface and Arctic water layers. A total of 19,270 data points were available for the period of this study and the four

variables in question were standardized to have means equal to 0 and variances equal to 1. Each pair of stations had an oceano-graphic distance (inter-station similarity value) based on their standardized values of these four oceanographic variables, as well as a geographic distance, which was similarly standardized. An overall inter-station distance was then computed, combining the oceanographic and geographic distances so that the resultant groups would be more contiguous geographically. A k-means cluster analysis of these combined distances, using 20 random starting points, yielded five distinct groups, consisting of four main groups with a clear interpretation and a fifth group split between the far north and far south containing only a few stations. The k-means clustering algorithm is suitable for large data sets like this one, but different numbers of groups need to be considered and in each case many random starting points—for example, see Greenacre and Primicerio (2013; Chapter 8). To assign our 10,357 krill sampling stations to one of the groups, each of the stations was assigned to the most common region of the 150 geographically nearest oceanographic stations.

LONG-TERM PATTERNS IN COMMUNITY STRUCTURE

Euphausiid sampling was uneven over the regions across the years (See Supplementary Material). This sampling bias was partly corrected for by reweighting the data to agree with the overall proportion of samples in each region across the whole time period. For example, 23.8% of the samples were collected in region 1. In 2009 only 15.9% of the samples were obtained in region 1, indicating the region was "under-sampled" relative to the long-term percentage. The total abundance values in this region for this year, therefore, were scaled up by a factor of (23.8/15.9). Each value in each year and each region were reweighted so that values across years were comparable. Unfortunately, nothing could be done about years in which no samples were taken in a given region, so a certain level of sampling bias still exists.

To investigate long-term patterns in community structure, data were aggregated to form a table of yearly abundances for the four species for the study period. Correspondence analysis (CA— see, for example, Greenacre, 2007; Greenacre and Primicerio, 2013) was used to visualize the data in the form of a biplot, showing differences in relative abundances of the four species among regions and among years. Each year of sampling was also plotted on the biplot for each region.

ECOLOGICAL DRIVERS OF EUPHAUSIID ABUNDANCE

General additive model (GAM) analyses were conducted to find the best descriptors of total euphausiid population size by region from the time series and to diagnose possible nonlinear relationships. The models were then parameterized and estimated by multiple regression (MR). The descriptor variables entering the model were the North Atlantic Oscillation (NAO) index [winter and summer principal component (PC)-based, Climate Analysis Section, NCAR, Boulder, USA, Hurrell et al. (2003)], the temperature anomaly from the Kola Transect (Stations 3–7, an oceanographic and biological transect run continuously for more than 100 years from the Russian coast to 74°N latitude roughly along the 33° 30′ E meridian, Orlova et al., 2010b), and stock sizes (for the entire Barents Sea) of capelin (Orlova et al., 2013), herring (Toresen and Østvedt, 2000), and cod, as well as 0+ year cod, 0+ year haddock (both only from 1980 onwards), and 1–3+ year cod (ICES, 2012). Fish and euphausiid abundances were natural-log(x+1)-transformed before entering the statistical analyses. The best combinations of variables in the model were selected for each region based on the Akaike Information Criterion (AIC). Relationships between the variable and log-euphausiid abundance were tested with GAM, and those that were not linear appeared to be quadratic. In this case both the log-variable and its square were entered into the subsequent MR analysis to estimate the parametric model. Total variance explained was calculated,

and, when relationships were linear, a quantitative relationship between the variable and euphausiid densities was determined (i.e., what percentage change in euphausiid abundance was associated with a standardized unit change in temperature or fish abundance).

Because of the significance of advection in general in the Barents Sea (Hunt et al., 2013), and the suggestion that boreal krill species have a larger role in the pelagic food web in the last two decades (Dalpadado et al., 2012; Orlova et al., 2013), we developed a euphausiid "advection index," which was in fact suggested by the second ordination axis of the CA. This index is expressed as:

$$(M.\ norvegica + T.\ longicaudata)\ /(\text{Total euphausiid abundance})$$

We conducted similar GAM and MR modeling as described above in an attempt to explain variation in the log of the advection index by climatic and predator variables. In addition, since CA results suggested that the primary axis of discrimination among years was explained by abundance of *T. inermis* vs. *T. raschii*, we ran the same GAM and MR regression analyses on the log of the ratio *T. inermis/T. raschii*.

RESULTS

SAMPLE DOMAIN AND IDENTIFICATION OF HYDROGRAPHICAL SUB-REGIONS IN THE BARENTS SEA

Between 1952 and 2009, up to 414 samples per year (mean 194 y^{-1}) were collected and processed, and a total of 10,357 euphausiid samples were analyzed for total abundance (Table S1). Of these 3912 contained data on species composition of euphausiids (Table S2). No samples were available for 1958, 1963, and 1991. The k-means cluster analysis of oceanographic data (temperature and salinity at 10 and 125 m, i.e., four data points per station) combined with geographical proximity among data points revealed five distinct zones (**Figure 2**). These regions conform roughly to water mass locations and definitions from Ingvaldsen and Loeng (2009) (Figure S1), although local mixing, warming, and cooling processes must be considered. Region 1 is the coastal region from approximately the Norwegian-Russian border and eastward, bordered on the north by the Atlantic inflow. This sector is characterized by the Norwegian Coastal Current, and is largely restricted to the shallow waters over the Murman Rise and North Kanin Bank to the east. Region 2 is the area north of the average position of the Polar Front, and waters here are of Atlantic origin, after it is cooled and mixed in the southern Barents Sea (southern area), or of Arctic origin but underlain by warmer transformed Atlantic Water (northern area). It is bounded on the south by the inflowing Atlantic water, and also includes the Svalbard Bank and Storfjord areas in the west. Region 3 is a highly variable, disjoint collection of sample locations in the far south and far north. These areas are characterized by similar temperature and salinity values, but the fresher water at the surface is probably from different sources, most likely from either riverine input (south) or ice melt (north). Since it is not a natural region and contained few euphausiid sampling stations (**Table 1**), we conducted few analyses of biological data from this zone. Region 4 represents the Atlantic Water inflow and is restricted to deeper

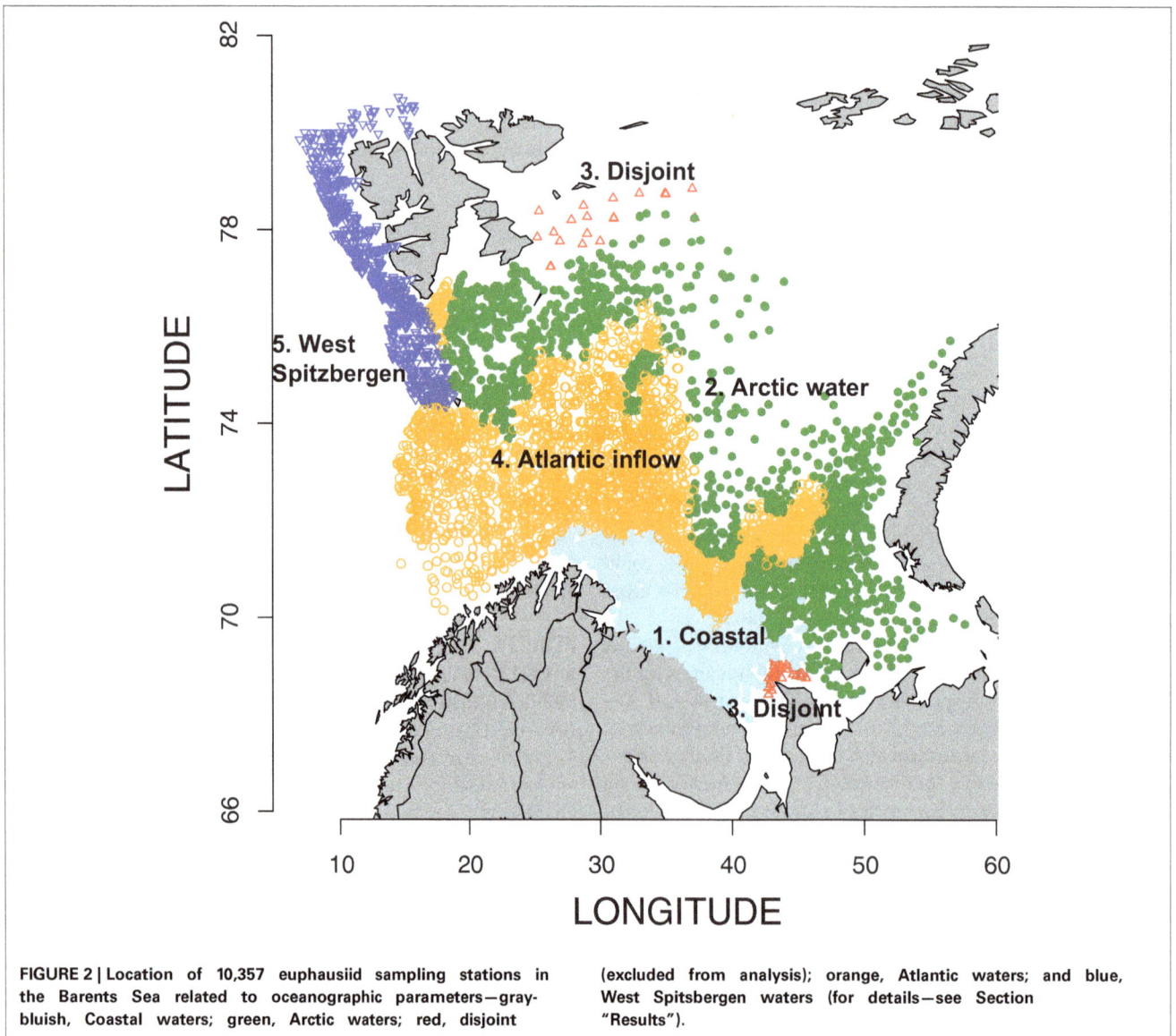

FIGURE 2 | Location of 10,357 euphausiid sampling stations in the Barents Sea related to oceanographic parameters—gray-bluish, Coastal waters; green, Arctic waters; red, disjoint (excluded from analysis); orange, Atlantic waters; and blue, West Spitsbergen waters (for details—see Section "Results").

Table 1 | Number of stations (percentage of total in parentheses) for each region where total euphausiid abundance was determined (Samples for total) and where species composition was determined (Samples for species).

Region	Samples for total	Samples for species	Temperature (°C)	Salinity
Coastal	2468 (23.8%)	1073 (27.4%)	3.7 (1.2)/3.3 (1.0)	34.5 (0.17)/34.7 (0.11)
Arctic Water	2695 (26.0%)	1251 (32.0%)	1.6 (1.8)/1.6 (1.3)	34.4 (0.32)/34.8 (0.10)
Disjoint	80 (0.8%)	36 (0.9%)	2.3 (2.4)/2.5 (1.9)	33.4 (0.5)/34.5 (0.12)
Atlantic inflow	4053 (39.1%)	1222 (31.2%)	3.4 (1.5)/3.0 (1.3)	34.8 (0.23)/34.9 (0.1)
W Spitzbergen	1061 (10.2%)	330 (8.4%)	0.2 (1.6)/1.2 (1.2)	34.2 (0.3)/34.7 (0.09)
Total	10,357	3912		

See Supplementary Material for information on distribution of this sampling by region for each year of the time series. Region assignment was made following k-means cluster analysis on oceanographic data and geographic proximity. Mean (standard deviation) temperature and salinity of the 10/125 m depths from each zone are also included.

areas in the north and east, bounded on the north by the Polar Front and in the south by the Coastal Current and shallow waters. Region 5 is the West Spitsbergen Current area, characterized by cooled Atlantic Water north of Bear Island and along the west and northwest of the Svalbard/Spitsbergen Archipelago (**Figure 2**).

LONG-TERM PATTERNS IN COMMUNITY STRUCTURE

Community compositions of euphausiids were separated by the CA along two axes explaining over 99% of the inter-regional variation (**Figure 3**). Most of the variability (94%) was explained by the continuum between *Thysanoessa inermis* and *T. raschii* (horizontal axis), whereas the relative abundance of *T. longicaudata* and *Meganyctiphanes norvegica* described the second axis. The centers of gravity for each region, when plotted on these axes, revealed two regions (1 and 2) most strongly influenced by *T. raschii*, two regions (3 and 5) by *T. inermis*, and one (Region 4) that typically had a relatively strong contribution of *T. longicaudata* and *M. norvegica* (**Figure 3**).

When factors defining "euphausiid space" are examined in greater detail for each region (**Figure 4**), interannual variability in species composition becomes clear. The coastally influenced Region 1 fluctuates among years with high *T. raschii* abundances (1964, 1968–1982, and 2005), and several years in the early 1960s and late 2000s that were dominated by *T. inermis* (**Figure 4**). Warmer years (indicated by seawater-temperature anomalies, Figure 1.1.1 in Orlova et al., 2010a) were usually mapped in the top part of the figure, indicating the influence of boreal taxa *T. longicaudata* and *M. norvegica*. In the modified Arctic waters of Region 2, years were described nearly exclusively by the relative abundance of *T. inermis* and *T. raschii*, with little influence of the year's climatology. The only years with significant contributions of *T. longicaudata* and *M. norvegica* were the warm 1954, and 1983, and moderate to cold 1953 and 1956 (**Figure 4**). The Atlantic inflow region of the Barents Sea (Region 4) was dominated by the two advected boreal taxa and *T. inermis* in nearly all years. In the warmest years (1954, 1989, and most of the 2000s) there is a strong component of *T. longicaudata* and *M. norvegica* evident (**Figure 4**). A similar pattern was observed in Region 5 along the West Spitsbergen Current, although this region was sampled in fewer years when *T. longicaudata* and *M. norvegica* were, again, relatively less abundant during cold years (**Figure 4**).

ECOLOGICAL DRIVERS OF EUPHAUSIID ABUNDANCE

GAM, followed by MR, identified the two factors best accounting for variation in euphausiid abundance in each region (except for Region 3, which was discontinuous and had few samplings). The quadratic of the NAO winter PC-values and the quadratic of total cod densities taken together explained 37.4% (adjusted R^2) of the total variability in euphausiid abundance in Region 1. Both quadratic relationships were concave, with an increase in euphausiid abundance with increasing values of NAO (winter) or cod, up to values just around 0 and 1400×10^9 g (k tons), respectively, beyond which euphausiid abundance decreased with increasing values. In Region 2, linear effects of the Kola temperature anomaly and the log-abundance of capelin explained 64.4% of the total euphausiid variability (**Table 2**). Since the relationships were linear, it was possible to calculate the quantitative

responses of euphausiid abundances to each factor. Each 0.1 increase in the Kola anomaly resulted in a 9.7% increase in euphausiid abundance; and euphausiids declined by 2.9% per 10% increase in capelin biomass.

As in Region 2, euphausiid abundance in Region 4 was linearly related to the Kola temperature anomaly and capelin. These two factors explained over 30% of the total variability in euphausiid abundance, which increased by 6.5% for each 0.1 increase in the Kola anomaly, and decreased by 2% for each 10% increase in capelin biomass. The combination of the Kola anomaly and the quadratic relationship with total cod abundance provided similar explanatory power for euphausiid abundance (29.8%), again with a concave relationship between euphausiid abundance and cod biomass. A similar trend also observed for Region 5, where the linear effect of the Kola temperature anomaly (6.5% for each 0.1 increase) and the quadratic relationship with total cod biomass explained 35.3% of total variability. Here, euphausiid abundance increased linearly with cod stock up to approximately 1400×10^9 g, followed by a negative relationship at higher cod biomasses (**Table 2**).

Modeling efforts using species composition instead of total abundance also yielded significant effects of climate and/or predator populations. The factors explaining significant variation, however, were different in different regions. Where they were significant, temperature was positively related to euphausiid abundance and predator abundance was negatively related to euphausiid abundance. The Kola temperature anomaly explained only a modest 8.8% of the variability in the log-ratio of *T. inermis:T. raschii* in Region 1, but in combination with cod abundance (log) explained nearly half the variability in the ratio in Region 4 (**Table 3**). In Region 5, capelin abundance (log) alone explained nearly 40% of the variability in the log ratio (**Table 3**). Effects on the ratio were positive for the Kola temperature anomaly (increase in anomaly resulted in an increase in *T. inermis/T. raschii* log ratio), but the association was negative for both species of predators. No significant relationships were found in Region 2.

Both the Kola temperature anomaly and capelin abundance strongly influenced the advection index in Region 1, explaining over 19% of the variability (**Table 3**). Climatic effects were statistically significant, but small, in Regions 2, 4, and 5 (explaining between 6.5 and 10.5% of the variability), and here it was either the summer or winter component of the NAO that had the greatest explanatory power (**Table 3**). Predators had no statistically significant impact on the advection index in these regions. Again, the significant climatic anomalies were positively related to the advective index, whereas capelin abundance in Region 1 was negatively related.

DISCUSSION

The Barents Sea is widely acknowledged to exhibit considerably regional variability in hydrography (e.g., water mass properties, ice cover), community structure of biotic components, and impacts of advection (e.g., Sakshaug et al., 2009; Stiansen et al., 2009). Despite such variability, the Barents Sea (including its Arctic and sub-Arctic parts) is often considered as a single ecosystem when exploring impacts of predation and climatic variability on zooplankton and fish (e.g., Hjermann et al.,

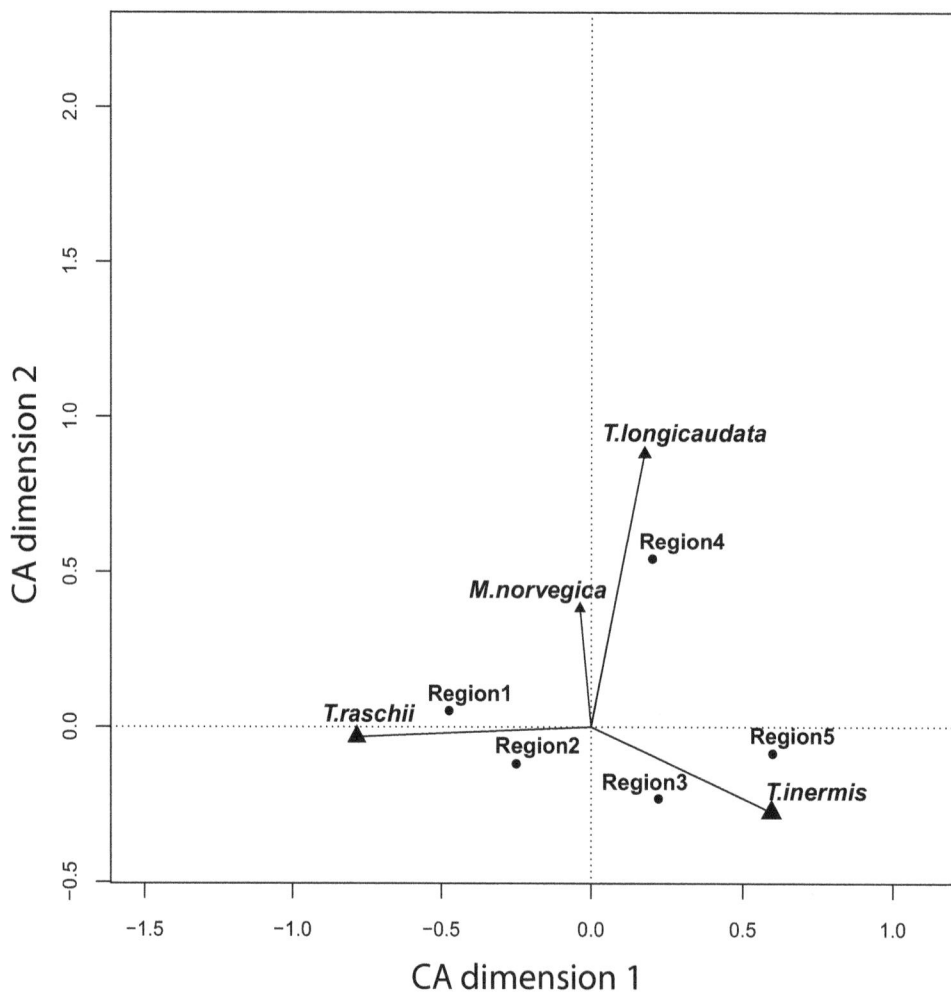

FIGURE 3 | Correspondence Analysis (CA) biplot for all stations and years. The two axes explain more than 99% of the total variability among regions. Black points represent the center of mass for each region, and the vectors indicate the strength (length) and direction of positive loadings of these species (T., *Thysanoessa*; M, *Meganctiphanes*). The horizontal axis is represented by a continuum between *T. raschii* (left) and *T. inermis* (right), whereas the vertical axis is largely described by the contribution of the *M. norvegica* and *T. longicaudata*. Whereas Region 5 is strongly characterized by *T. inermis*, *T. raschii* is more dominant in Regions 1 and 2. Region 4 has a high component of the taxa contributing to the vertical axis.

2004, 2007; Eriksen and Dalpadado, 2011; Dalpadado et al., 2012; Johannesen et al., 2012). Our results indicate that, mechanistically, the Barents Sea should be viewed regionally and not as one ecosystem. We provide estimates of quantitative links between climate, predation, and the structure of euphausiid assemblages, which are valuable for evaluation of ecosystem resilience and the development of management strategies in a changing Arctic.

LONG-TERM AND SPATIAL CHANGES IN EUPHAUSIID COMMUNITY

Thysanoessa inermis and *T. raschii* comprise the majority of the euphausiid stocks in the Barents Sea, and there is spatial segregation in their distributions with *T. inermis* being more abundant in the west and south, an *T. raschii* more restricted to colder waters of the north and east (e.g., Drobysheva, 1994). Further, *T. raschii* populations expand westward during cooler years. An increase in *T. inermis* during warm years is responsible for an overall increase

in euphausiid abundance in the Barents Sea (Drobysheva, 1994; Orlova et al., 2010b, 2013; Dalpadado et al., 2012). We found that Coastal and Arctic regions (1 and 2) fluctuated primarily between dominance by *T. inermis* (warmer years) and *T. raschii* (colder years) (**Figure 4**). Furthermore, *T. inermis* was always the predominant euphausiid in the Atlantic inflow and West Spitzbergen regions (4 and 5), with variable importance of advected *T. longicaudata* and *M. norvegica* (**Figure 3**). This finding is already more complex than the east-west segregation discussed above. Northward advection of boreal species, *T. inermis* as well as *T. longicaudata* and *Meganctiphanes norvegica*, is more intense during warm years (e.g., Drobysheva et al., 2003; Zhukova et al., 2009; Buchholz et al., 2010, 2012), an effect particularly evident since 2000 in Regions 1, 4, and 5 (**Figure 4, Table 3**). In these recent years, warmer conditions (e.g., Walczowski and Piechura, 2006) have led to a more reliable input of these taxa to some areas of the Barents Sea and west Spitzbergen, perhaps also

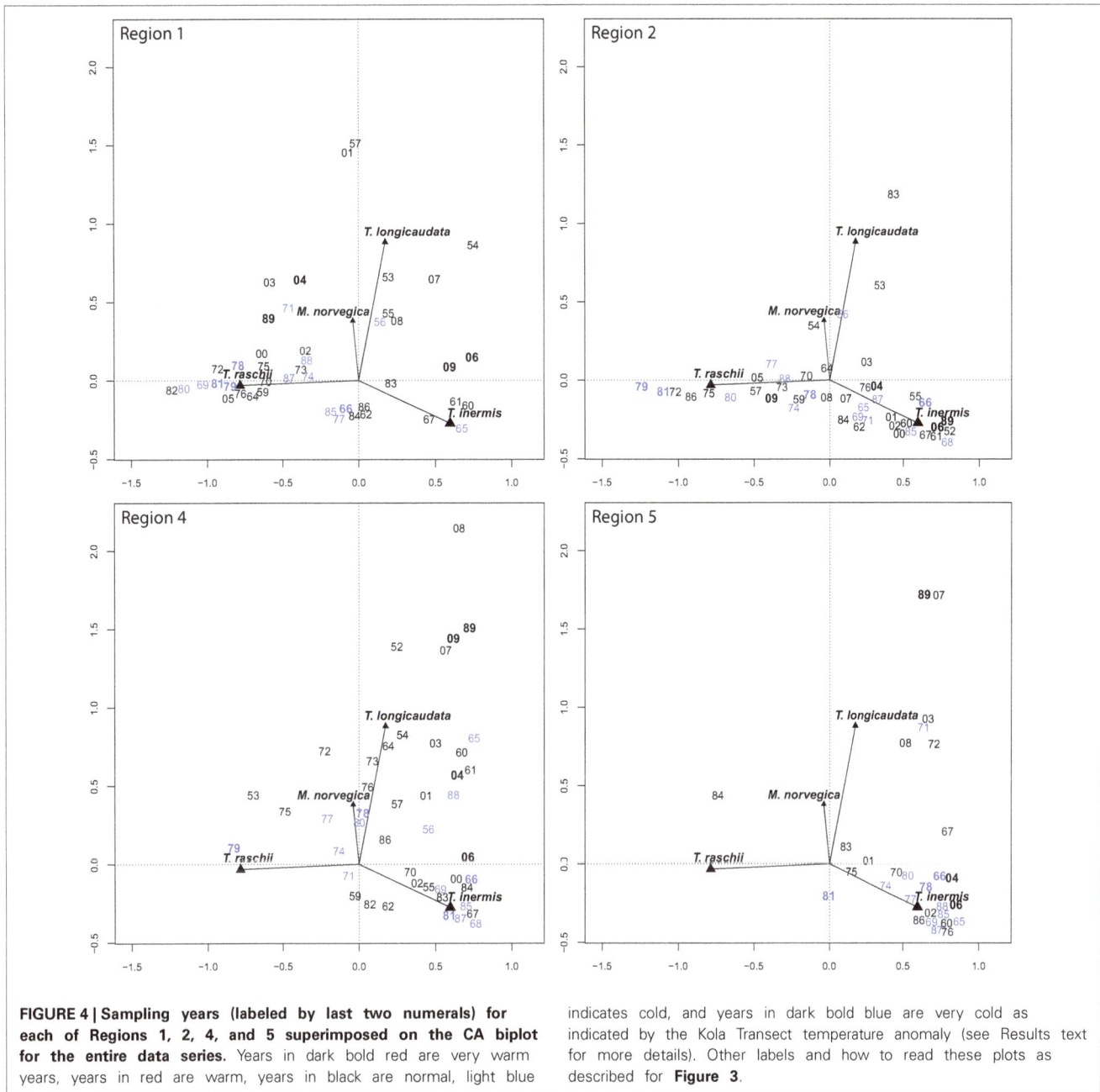

FIGURE 4 | Sampling years (labeled by last two numerals) for each of Regions 1, 2, 4, and 5 superimposed on the CA biplot for the entire data series. Years in dark bold red are very warm years, years in red are warm, years in black are normal, light blue indicates cold, and years in dark bold blue are very cold as indicated by the Kola Transect temperature anomaly (see Results text for more details). Other labels and how to read these plots as described for **Figure 3**.

contributing to high stock abundance and broad distribution of Atlantic cod in the Barents Sea (Orlova et al., 2013). Thus, examining patterns in hydrographically defined regions allowed us to better understand variability in spatial patterns in euphausiids over time.

These findings indicate greater complexity than the generally accepted pattern (see above) of euphausiid distributions in the Barents Sea, and the influence of thermal conditions on species composition. Climatic state certainly influences advection of *T. longicaudata* and *M. norvegica* into Regions 4 and 5, but strong advection years based on appearance of boreal migrants in Region 1 (e.g., 1953–1957, 1971, 2000) are not well explained by climatology. The dual nature of *T. inermis*, as a resident reproducing

in the Barents Sea and as a boreal migrant imported from the Norwegian Sea, also complicates matters, and likely limits the value of the advective index tested here. Climatic conditions and predator abundances generally had low explanatory power (low R^2 values) for this index, except in Region 1 (**Table 3**). Even in a region most frequently influenced by advection (Region 4), relative amounts of *T. raschii* and *T. inermis* vary more or less independently of climatic condition, but are strongly negatively related to cod abundance (**Tables 2, 3**). It is possible that different amounts of resident *T. inermis* may buffer the system against low advection years (high resident fraction), or may enhance the apparent impact of low advection years (low resident fraction), in determining where on the *T. raschii*- *T. inermis* continuum a

Table 2 | Results of the best multiple regressions and general additive models (GAM) describing impact of climatic and predator abundances [log(total Barents Sea stock size)] on euphausiid abundances [log(abundance m^{-3})] by region.

Region (adjusted R^2 as %) sample size N	Constant	Variables in regression coefficients (standard errors) p-values		GAM relationships
Region 1 (37.4%) $N = 30$	**3.10**	quadratic in *NAOw* *NAOw* **0.210** (0.129) $p = 0.12$ *NAOw*^{2} **−0.336** (0.096) $p = 0.002$	quadratic in log(*cod*) *lcod* **46.7** (17.3) $p = 0.01$ *lcod*^{2} **−3.27** (1.20) $p = 0.01$	
Region 2 (64.4%) $N = 34$	**7.17**	linear in *anomaly* **0.928** (0.142) $p < 0.0001$	linear in log(*capelin*) **−0.302** (0.075) $p = 0.0003$	
Region 4 (30.2%) $N = 34$	**6.44**	linear in *anomaly* **0.629** (0.181) $p = 0.002$	linear in log(*capelin*) **−0.209** (0.075) $p = 0.03$	
alternative Region 4 (29.8%) $N = 31$	**−194.9**	linear in *anomaly* **0.570** (0.214) $p = 0.01$	quadratic in log(*cod*) *lcod* **55.6** (18.0) $p = 0.005$ *lcod*^{2} **−3.86** (1.26) $p = 0.005$	
Region 5 (35.3%) $N = 27$	**−225.7**	linear in *anomaly* **0.617** (0.288) $p = 0.04$	quadratic in log(*cod*) *lcod* **63.1** (23.2) $p = 0.01$ *lcod*^{2} **−4.31** (1.62) $p = 0.01$	

Two models with similar explanatory power are presented for Region 4. The left column contains total sample size (number of years) and the adjusted R^2 of the multiple regression model. Constants and coefficients along with p-values are presented in the third column, whereas GAM relationships with 95% confidence intervals (shaded) are presented on the right. "anomaly" = Kola Transect temperature anomaly; "NAOw" = North Atlantic Oscillation Index winter value.

particular year will lie. But for now, there is no way of differentiating resident from advective fractions. So while climatic conditions are clearly important in determining euphausiid community structure, yet-to-be-determined interacting effects of advection, redistribution by local processes, and predator pressure also play important roles.

IMPACTS OF CLIMATE AND PREDATION ON EUPHAUSIID ABUNDANCE

Climate, and temperature in particular, has been linked with euphausiid abundance and biomass in the Barents Sea. Specifically, increased temperatures such as those observed since the mid-1990s, have a positive correlation with euphausiid biomass (Dalpadado et al., 2003, 2012; Zhukova et al., 2009; Orlova et al., 2010b). Certainly much of the impact of climate on euphausiids is through supply of both juvenile and adult stages to different regions of the Barents Sea. Climate variability, however, may also affect the timing and quantity for food for euphausiids during the growing season, thus contributing to growth and overwintering success via bottom-up forcing. We have no data

on food availability, however. Our results show a strong coupling of climatic conditions with euphausiid abundances on a regional basis, although these results vary qualitatively and quantitatively among regions. The temperature anomaly at the Kola Transect was positively and linearly correlated with euphausiid abundance in Regions 2, 4, and 5 (**Table 2**). The quantitative effect was strongest in the Arctic Water region (Region 2), where a 9.7% increase in abundance was associated with an increase of 0.1 in the anomaly. Similar 6.5 and 6.4% increases were recorded in Regions 4 and 5, respectively, but surprisingly, no relationship was observed between the Kola anomaly and euphausiids in Region 1, which contains part of the Transect (**Table 2**). Instead, euphausiid abundance in Region 1 showed a quadratic relationship with the NAO, a positive correlation when the winter NAO index was below 0.2 and a sharp decline in abundance when the NAO index was above 1.

The Kola Transect generally reflects conditions over the entire Barents Sea, and has been correlated with the NAO and regional ecology (Ottersen and Stenseth, 2001), at least until recent years

Table 3 | Results of the best multiple linear regressions describing impact of climatic and predator abundances (log) on the log of the ratio T. inermis/T. raschii (upper panel) and the log of the "advective index" (lower panel) by region.

Region (adjusted R^2 as %) sample size N	Constant	MODELING log(Th_in/Th_ra) Variables in regression coefficients (standard errors) p-values		Interpretation
Region 1 (8.8%) $N = 42$	**−0.335**	linear in *anomaly* **0.803** (0.360) $p = 0.03$		8.4% increase in Th_in/Th_ra per 0.1 increase in *anomaly*
Region 2	–	–		–
Region 4 (48.1%) $N = 22$	**23.44**	linear in *anomaly* **0.942** (0.403) $p = 0.03$	linear in log(*cod*) **−3.08** (0.69) $p = 0.0003$	9.9% increase in Th_in/Th_ra per 0.1 increase in *anomaly* 34.1% decrease in Th_in/Th_ra per 10% increase in *cod*
Region 5 (38.6%) $N = 22$	**−11.53**	linear in log(*capelin*) **−1.14** (0.30) $p = 0.001$		10.3% decrease Th_in/Th_ra per 10% increase in *capelin*

Region (adjusted R^2 as %) sample N	Constant	MODELING log($Advec.Index$) Variables in regression coefficients (standard errors) p-values		Interpretation
Region 1 (19.2%) $N = 24$	**0.630**	linear in *anomaly* **0.769** (0.372) $p = 0.05$	linear in log(*capelin*) **−0.501** (0.251) $p = 0.05$	8.0% increase in *Advec.Index* per 0.1 increase in *anomaly* 4.7% decrease in *Advec.Index* per 10% increase in *capelin*
Region 2 (9.8%) $N = 43$	**−4.030**	linear in *NAOs* **0.474** (0.224) $p = 0.04$		4.8% increase in *Advec.Index* per 0.1 increase in *NAOs*
Region 4 (6.5%) $N = 43$	**−2.123**	linear in *NAOw* **0.265** (0.134) $p = 0.05$		2.7% increase in *Advec.Index* per 0.1 unit increase in *NAOw*
Region 5 (10.5%) $N = 26$	**−3.072**	linear in *NAOw* **0.526** (0.265) $p = 0.05$		5.4% increase in *Advec.Index* per 0.1 increase in *NAOw*

["Advec. Index" = (M. norvegica + Ih. longicaudata)/(total euphausiids)]. The left column contains total sample size (number of years) and the adjusted R^2 of the model. Constants and coefficients along with p-values are presented in the third column. Numerical interpretation is included in the right column. No significant regression was found for the log of T. inermis/T. raschii in Region 2. "anomaly" = Kola Transect temperature anomaly; "NAOs" and "NAOw"= North Atlantic Oscillation Index summer and winter values, respectively.

(Johannesen et al., 2012). Each year, Regions 4 and 5 receive direct influx of warm waters and the organisms advected with it, including euphausiids. It is, therefore, no surprise that warmer climatic conditions, indicating greater advection and inflow into the Barents Sea, result in a strong increase in euphausiids in these regions (Orlova et al., 2013). The strong response of euphausiid communities to increased advection in Region 2 may be due to greater inflow of *T. inermis* from the Atlantic into areas traditionally characterized by Arctic Water, which in warm years can support high growth of both Arctic and boreal taxa. Region 1's complex (quadratic) relationship with the NAO is difficult to explain, but may be related to how inflowing water is distributed among the three branches of the Atlantic inflow (North Cape Current, **Figure 1**) in years of different climatic condition. Region 1 is also affected by the Norwegian Coastal Current, and mixing with locally-produced waters over several shallow banks.

The length of our time-series (1952–2009) spans multiple climatic cycles, providing increased confidence that relationships between euphausiid densities and climatic state are well-founded.

The first decade of the 21st century, however, appears to be quite different from previous decades. Since 1996, the historically strong correlation between the NAO and the temperatures along the Fugløya-Bjørnøya Transect, at the southwestern opening to the Barents Sea (**Figure 1**), has disappeared (Johannesen et al., 2012). Further, there has been a decrease in the variability of euphausiid standing stocks, despite an increase in advection of capelin from boreal waters (Orlova et al., 2013). Euphausiid biomass was high between 2007 and 2010 despite high capelin populations (Eriksen and Dalpadado, 2011), whereas during the previous two decades both euphausiids (Eriksen and Dalpadado, 2011) and zooplankton (mostly copepod) biomass varied inversely with capelin abundance (Dalpadado et al., 2003). It is unclear whether this change in the euphausiid-capelin relationship constitutes a regime shift, but these differences can have significant impact on trophic relationships throughout the food web.

Planktivorous fish reduce the size of zooplankton populations, including those of euphausiids in the Barents Sea (e.g., Dalpadado and Skjoldal, 1996). This can overwhelm climatic effects when fish abundances are high (Stige et al., 2009). We found predation effects varied significantly among regions, with capelin having strong linear (negative) effects on euphausiid abundance in Regions 2 and 4 (2–3% decrease in euphausiid abundance per 10% increase in capelin biomass), and cod having a greater role in Regions 1 and 5. Capelin are known to have strong impacts on euphausiid populations in the Barents Sea (e.g., Drobysheva, 1994; Orlova, 2002; Orlova et al., 2013), but we also found strong (high explanatory power) and complex (non-linear) relationships between cod standing stock and euphausiid abundance. Regions 1, 4, and 5 exhibited a threshold-type response curve, where increased cod biomass was positively correlated with euphausiid stock size until approximately 1400×10^9 g cod, when the relationship became inverse. An analogous threshold effect was noted for capelin feeding on copepods in the southern Barents Sea (Stige et al., 2009). One possible mechanism for such an effect could be the following—cod feed on capelin, which then decreases predation pressure of capelin on euphausiids (c.f. Fiksen et al., 2005). But beyond a certain level of cod biomass (1400×10^9 g cod in our data) capelin abundance is no longer high enough to support the cod, and cod begin feeding on euphausiids. Climate, through its impact of advection, may also have strong, positive relationships with both cod and euphausiids. Such a response is consistent with the observations of Drobysheva (1994). The very similar cod-threshold value for each of these three regions is striking, but we so far have no mechanism explaining why this value is so consistent.

Our finding of regional differences in which fish species had the larger effect on euphausiids can be linked with seasonality in both advection and capelin feeding-migrations. Stige et al. (2009) found effects of capelin feeding along the Fugløya-Bjørnøya Transect within Region 4 in spring, whereas the effect along the Kola Transect (partly in Region 1, but also 2 and 4, **Figure 1**) was not felt until summer. Capelin seasonal feeding migrations begin in spring from the mainland coast and proceed in a more or less clock-wise direction through the Barents Sea (Gjøsæter, 1998), whereas advection of euphausiids (and other zooplankton) into the Barents Sea is strongest in spring (Drobysheva, 1994). Since capelin will have the strongest effects where their feeding overlaps with concentrations of advected (or resident) euphausiids, it is not to be expected that they will have significant impact in Region 5, and perhaps only limited impact in Region 4. Here, resident Atlantic cod along the Norwegian coast, and advected cod from the Norwegian Sea and up toward Svalbard would be expected to have more impact over a longer period (not just during spring peaks in inflow). This scenario is suggested by the strong negative impact of cod on euphausiid community structure in Regions 4 and 5 (**Table 3**). Effects of cod (and lack of strong capelin effects) in Region 1 can be explained by the different seasonality of feeding by these two fish species in this Region. Capelin only feed on euphausiids in Region 1 during their pre-spawning migrations in February-April, so will have no statistical impact on euphausiid abundance in autumn-winter. In contrast, cod consume euphausiids intensively in this area in summer (so-called "euphausiid feeding"), with consequences during the sampling period (Zatsepin and Petrova, 1939; Orlova et al., 2001).

Surprisingly, we found no strong impact of herring, or of 0+, 1+, and 2+ cod and haddock. Stige et al. (2009) also found little effect of predation by herring in their studies, but the young cod and haddock stages are all known to feed on euphausiids (e.g., Ponomarenko and Yaragina, 2003; Dalpadado et al., 2009), and in fact cod largely stop feeding on euphausiids after 3 years. Our data on 0+ age classes of cod and haddock only go back to 1980, so this may have limited our ability to find effects during this period of strong climatic shifts.

IMPLICATIONS FOR THE FUTURE BARENTS SEA PELAGIC ECOSYSTEM

Combined impacts of climate and predatory fish explain from 30 to over 60% of the variation in euphausiid *abundance* on a regional basis, and these relationships were quantified in our regression models. Furthermore, impacts of these two factors on euphausiid *community structure* can be highly significant, and also vary regionally. Although mechanistic understanding of these relationships is not complete, these results and the growing body of literature on the topic suggest the system is dynamic, spatially segregated, and susceptible to variability in advection, fish stock size, and their interactions with climatic forcing. Thus, predictions for a future Barents Sea pelagic food-web need to reflect these complexities and the non-linear relationships inherent in the system.

Predicted shifts in capelin populations to the eastern Barents Sea (Hop and Gjøsæter, 2013) could enhance the capelin effect in the Arctic Water region (Region 2) and decrease it in Region 4 in the future. And if cod continue to increase, the impact of this (quadratic) threshold response could result in enhanced top-down control by cod over euphausiid abundance. This may make the pelagic ecosystem – now buffered by high advection of euphausiids (Orlova et al., 2013) –more susceptible to periodic crashes, particularly if other boreal planktivores (mackerel, blue whiting) also increase in abundance (Anon, 2010).

Quantification of variability in the standing stocks of zooplankton and fishes is important for understanding and management of ecosystems (Hjermann et al., 2007). Regional differences in prey/predator interactions can provide insight into ecological processes driving variability in these standing stocks. Significant quantitative effects of both climate and predation on euphausiid densities, integrated on an annual basis by sampling during winter months, can help parameterize ecosystem models testing climate and ecological change scenarios. Our results indicate that these models should be both spatially discrete and consider both annual routines of predators and their prey as well as possible lagged effects.

ACKNOWLEDGMENTS

The authors wish to thank PINRO and Russian/Soviet scientists who started and supported euphausiid investigations in the Barents Sea, making it one of the longest biological time-series anywhere. Boris P. Manteifel initiated euphausiid investigations as food supply for commercially important fish in the Barents Sea; Tamara S. Berger and Elena A. Pavshtiks provided the methodological base for euphausiid surveys by testing appropriate sampling tools and seasons; and Svetlana S. Drobysheva and Valentina N. Nesterova organized and supported these investigations in 1950–1990s. We also gratefully acknowledge Chris Emblow for graphics support, and financial support from Statoil and Akvaplan-niva. Michael Greenacre's research is partially supported by grant MTM2012-37195 of the Spanish Ministry of Economy and Competitiveness. This manuscript was improved by comments from F. Buchholz.

REFERENCES

Abraham, C. L., and Sydeman, W. J. (2004). Ocean climate, euphausiids and auklet nesting: inter-annual trends and variation in phenology, diet and growth of a planktivorous seabird, *Ptychoramphus aleuticus*. *Mar. Ecol. Prog. Ser.* 274, 235–250. doi: 10.3354/meps274235

Anon. (2010). "Survey report from the joint Norwegian/Russian ecosystem survey in the Barents Sea August-September 2010," in *IMR/PINRO Joint Report Ser, No. 4/2010* (Bergen).

Brinton, E., and Townsend, A. (2003). Decadal variability in abundances of the dominant euphausiid species in southern sectors of the California Current. *Deep Sea Res II* 50, 2449–2472. doi: 10.1016/S0967-0645(03)00126-7

Buchholz, F., Buchholz, C., and Weslawski, J. M. (2010). Ten years after: krill as indicator of changes in the macro-zooplankton communities of two Arctic fjords. *Polar Biol.* 33, 101–113. doi: 10.1007/s00300-009-0688-0

Buchholz, F., Werner, T., and Buchholz, C. (2012). First observation of krill spawning in the high Arctic Kongsfjorden, west Spitsbergen. *Polar Biol.* 35, 1273–1279. doi: 10.1007/s00300-012-1186-3

Coyle, K. O., Eisner, L. B., Mueter, F. J., Pinchuk, A. I., Janout, M. A., Cieciel, K. D., et al. (2011). Climate change in the southeastern Bering Sea: impacts on pollock stocks and implications for the oscillating control hypothesis. *Fisheries Oceanogr.* 20, 139–156. doi: 10.1111/j.1365-2419.2011.00574.x

Dalpadado, P., Bogstad, B., Eriksen, E., and Rey, L. (2009). Distribution and diet of 0-group cod (*Gadus morhua*) and haddock (*Melanogrammus aeglefinus*) in the Barents Sea in relation to food availability and temperature. *Polar Biol.* 32, 1583–1596. doi: 10.1007/s00300-009-0657-7

Dalpadado, P., Ingvaldsen, R. B., Stige, L. C., Bogstad, B., Knutsen, T., Ottersen, G., et al. (2012). Climate effects on Barents Sea ecosystem dynamics. *ICES J. Mar. Sci.* 69, 1303–1316. doi: 10.1093/icesjms/fss063

Dalpadado, P., Ingvaldsen, R., and Hassel, A. (2003). Zooplankton biomass variation in relation to climatic conditions in the Barents Sea. *Polar Biol.* 26, 233–241. doi: 10.1007/s00300-002-0470-z

Dalpadado, P., and Mowbray, F. (2013). Comparative analysis of feeding ecology of capelin from two shelf ecosystems, off Newfoundland and in the Barents Sea. *Progr. Oceanogr.* 114, 97–105. doi: 10.1016/j.pocean.2013.05.007

Dalpadado, P., and Skjoldal, H. R. (1996). Abundance, maturity and growth of the krill species *Thysanoessa inermis* and *T. longicaudata* in the Barents Sea. *Mar. Ecol. Progr. Ser.* 144, 175–183. doi: 10.3354/meps144175

Dorman, J. G., Powell, T. M., Sydeman, W. J., and Bograd, S. J. (2011). Advection and starvation cause krill (*Euphausia pacifica*) decreases in 2005 Northern California coastal populations: implications from a model study. *Geophys. Res. Lett.* 38:L04605. doi: 10.1029/2010GL046245

Drinkwater, K. F. (2006). The regime shift of the 1920s and 1930s in the North Atlantic. *Progr. Oceanogr.* 68, 134–151. doi: 10.1016/j.pocean.2006.02.011

Drobysheva, S. S. (1982). *Degree of Isolation of Thysanoessa inermis (Krøyer) and T. raschii (M. Sars, 1864) (Crustacea, Euphausiacea) Populations in the Southern Barents Sea.* ICES C.M.1982/L:19. (Copenhagen: International Council for the Exploration of the Sea (ICES)), 21.

Drobysheva, S. S. (1994). *Euphausiidae of the Barents Sea and their Role in the Formation of Commercial Bioproduction.* Murmansk: PINRO Press (in Russian)

Drobysheva, S. S., Nesterova, V. N., Nikiforov, A. G., and Zhukova, N. G. (2003). Role of warm water component in formation of euphausiids local aggregations in the southern Barents Sea. *Voprosy Rybolovstva* 4, 209–216 (in Russian).

Eriksen, E., and Dalpadado, P. (2011). Long-term changes in krill biomass and distribution in the Barents Sea. Are the changes mainly related to capelin stock size

and temperature conditions? *Polar Biol.* 34, 1399–1409. doi: 10.1007/s00300-011-0995-0

Fiksen, Ø., Melle, W., Torgersen, T., Breien, M. T., and Klevjer, T. A. (2005). Piscivorous fish patrol krill swarms. *Mar. Ecol. Progr. Ser.* 299, 1–5. doi: 10.3354/meps299001

Gjøsæter, H. (1998). The population biology and exploitation of capelin (*Mallotus villosus*) in the Barents Sea. *Sarsia* 88, 261–273.

Gómez-Gutiérrez, J., Palomares-García, R., and Gendron, D. (1995). Community structure of the euphausiid populations along the west coast of Baja California, Mexico, during the weak ENSO 986-1987. *Mar. Ecol. Prog. Ser.* 120, 41–51. doi: 10.3354/meps120041

Greenacre, M. (2007). *Correspondence Analysis in Practice, 2nd Edn.* Boca Raton, FL: Chapman & Hall/CRC. doi: 10.1201/9781420011234

Greenacre, M., and Primicerio, R. (2013). *Multivariate Analysis of Ecological Data.* Madrid: BBVA Foundation. Available online at: www.multivariatestatistics.org

Hjermann, D. Ø., Bogstad, B., Eikeset, A. M., Ottersen, G., Gjøsæter, H., and Stenseth, N. C. (2007). Food web dynamics affect Northeast Arctic cod recruitment. *Proc. R. Soc. Lond. B* 274, 661–669. doi: 10.1098/rspb.2006.0069

Hjermann, D. Ø., Ottersen, G., and Stenseth, N. C. (2004). Competition among fishermen and fish causes the collapse of the Barents Sea capelin. *Proc. Natl. Acad. Sci. U.S.A.* 101, 11679–11684. doi: 10.1073/pnas.0402904101

Hop, H., and Gjøsæter, H. (2013). Polar cod (*Boreogadus saida*) and capelin (*Mallotus villosus*) as key species in marine food webs of the Arctic and the Barents Sea. *Mar. Biol. Res.* 9, 878–894. doi: 10.1080/17451000.2013.775458

Hunt, G. L., Blanchard, A. L., Boveng, P., Dalpadado, P., Drinkwater, K., Eisner, L., et al. (2013). The Barents and Chukchi Seas: comparison of two Arctic shelf ecosystems. *J. Mar. Syst.* 109/110, 43–68. doi: 10.1016/j.jmarsys.2012.08.003

Hunt, G. L., Stabeno, P., Walters, G., Sinclair, E., Brodeur, R. D., Napp, J. M., et al. (2002). Climate change and control of the southeastern Bering Sea pelagic ecosystem. *Deep Sea Res. II* 49, 5821–5853. doi: 10.1016/S0967-0645(02)00321-1

Hurrell, J. W., Kushnir, Y., Ottersen, G., and Visbeck, M. (eds.). (2003). The North Atlantic oscillation: climate significance and environmental impact. *Geophys. Monogr. Ser.* 134, 279.

ICES (International Council for Exploration of the Seas). (2012). *Report of the Arctic Fisheries Working Group 2012 (AFWG).* ICES CM 2012/ACOM:05, ICES Headquarters, Copenhagen.

Ingvaldsen, R., and Loeng, H. (2009). "Physical oceanography," in *Ecosystem Barents Sea*, eds E. Sakshaug, G. Johnsen, and K. Kovacs (Trondheim: Tapir Academic Press), 33–64.

Johannesen, E., Ingvaldsen, R. B., Bogstad, B., Dalpadado, P., Eriksen, E., Gjøsæter, H., et al. (2012). Changes in Barents Sea ecosystem state, 1970-2009: climate fluctuations, human impact, and trophic interactions. *ICES J. Mar. Sci.* 69, 880–889. doi: 10.1093/icesjms/fss046

Kovtsova, M. V., Antonov, S. G., and Orlova, E. M. (1989). 'Peculiarities of feeding and fatness dynamics of haddock *Melanogrammus aeglefinus* (L.) in the Barents Sea," in *Trophic Relations of Benthic Organisms and Demersal Fishes in the Barents Sea*, ed A. D. Chinarina (Apatity: Kola Science Center for USSR Academy of Science), 27–36 (in Russian).

Lindstrøm, U., Nilssen, K. T., Pettersen, L. M. S., and Haug, T. (2013). Harp seal foraging behaviour during summer around Svalbard in the northern Barents Sea: diet composition and the selection of prey. *Polar Biol.* 36, 305–320. doi: 10.1007/s00300-012-1260-x

Matishov, G., Zuyev, A., Golubev, A., Adrov, N., Slobodin, V., Levitus, S., et al. (1998). *Climatic Atlas of the Barents Sea 1998: Temperature, Salinity, Oxygen. NOAA Atlas NESDIS 26, 1-24 + Appendices and CD.* Washington, DC: US Department of Commerce.

Mehlum, F. (2001). Crustaceans in the diet of adult common and Brünnich's guillemots (*Uria aalga* and *U. lomvia*) in the Barents Sea during the breeding period. *Mar. Ornithol.* 29, 19–22.

Mauchline, J. (1980). The biology of mysids and euphausiids. *Adv. Mar. Biol.* 18, 1–677.

Nilssen, K. T., Haug, T., Potelov, V., and Timoshenko, Y. K. (1995). Feeding habits of harp seals (*Phoca groenlandica*) during early summer and autumn in the northern Barents Sea. *Pol. Biol.* 15, 485–493. doi: 10.1007/BF00237462

Orlova, E. L., Boitsov, V. D., and Nesterova, V. N. (2010a). *The Influence of Hydrographic Conditions on the Structure and Functioning of the Trophic Complex Plankton-Pelagic Fishes-Cod.* PINRO. Murmansk: Murmansk Printing Company.

Orlova, E. L., Dolgov, A. V., Renaud, P. E., Boitsov, V. D., Prokopchuk, I. P., and Zashihina, M. V. (2013). Structure of the macroplankton–pelagic fish–cod trophic complex in a warmer Barents Sea. *Mar. Biol. Res.* 9, 851–866. doi: 10.1080/17451000.2013.775453

Orlova, E. L., Nesterova, V. N., and Dolgov, A. V. (2001). Euphausiids and their role in feeding of Arcto-Norwegian cod (80-90th). *Voprosy Rybolovstva* 2, 86–103 (in Russian).

Orlova, E. L., Ushakov, N. G., Nesterova, V. N., and Boitsov, V. D. (2002). Food supply and feeding of capelin (Mallotus villosus) of different size in the central latitudinal zone of the Barents Sea during intermediate and warm years. *ICES J. Mar. Sci.* 59, 968–975. doi : 10.1006/jmsc.2002.1255

Orlova, E. L., Rudneva, G. B., Nesterova, V. N., and Yurko, A. S. (2008). On the quantitative estimation of macroplankton abundance in the Barents Sea. *Izvestia TINRO* 152, 186–200 (in Russian).

Orlova, E. L., Rudneva, G. B., Renaud, P. E., Eiane, K., Savinov, V. M., and Yurko, A. S. (2010b). Climate impacts on feeding and condition of capelin (*Mallotus villosus*) in the Barents Sea. Evidence and mechanisms from a 30-year data series. *Aquat. Biol.* 10, 105–118. doi: 10.3354/ab00265

Ottersen, G., and Stenseth, N. C. (2001). Atlantic climate governs oceanographic and ecological variability in the Barents Sea. *Limnol. Oceanogr.* 46, 1774–1780. doi: 10.4319/lo.2001.46.7.1774

Ponomarenko, I. Y. A. (1973). The effects of food and temperature conditions on the survival of young bottom-dwelling cod in the Barents Sea. *Rapp P-v Reun Cons. Int. Explor. Mer.* 164, 199–207.

Ponomarenko, I. Y. A., and Yaragina, N. A. (2003). "Feeding and trophic relations. Cod in the Barents Sea ecosystem," in *The Barents Sea Cod: Biology and Fisheries, 2nd Edn.*, ed V. N. Shleinik (Murmansk: PINRO Press), 62–113 (in Russian).

Renaud, P. E., Berge, J., Varpe, Ø., Lønne, O. J., Nahrgang, J., Ottesen, C., et al. (2012). Is the poleward expansion by Atlantic cod and haddock threatening native polar cod, *Boreogadus saida*? *Polar Biol.* 35, 401–412. doi: 10.1007/s00300-011-1085-z

Sakshaug, E., Johnsen, G., Kristiansen, S., von Quillfeldt, C. H., Rey, F., Slagstad, D., et al. (2009). "Phytoplankton and primary production," in *Ecosystem Barents Sea*, eds E. Sakshaug, G. Johnsen, and K. M. Kovacs (Trondheim: Tapir Academic Press), 167–208.

Stiansen, J. E., and Filin, A. A. (eds.). (2007). Joint PINRO/IMR report on the state of the Barents Sea ecosystem 2006, with expected situation and considerations for management. *IMR/PINRO Joint Report Series*, 209.

Stiansen, J. E., Korneev, O., Titov, O., Arneberg, P. (eds.) Filin, A., Hansen, J. R., et al. (co-eds.). (2009). "Joint Norwegian-Russian environmental status 2008," in *Report on Barents Sea Ecosystem Part II – Complete Report*. Bergen: IMR/PINRO Joint Report Series 2009, 375.

Stige, L. C., Lajus, D. L., Chan, K.-S., Dalpadado, P., Basedow, S., Berchenko, I., et al. (2009). Climatic forcing of zooplankton dynamics is stronger during low densities of planktivorous fish. *Limnol. Oceanogr.* 54, 1025–1036. doi: 10.4319/lo.2009.54.4.1025

Sydeman, W. J., Bradley, R. W., Warzybok, P., Abraham, C. L., Jahncke, J., Hyrenbach, K. D., et al. (2006). Planktivorous auklet *Ptychoramphus aleuticus* responses to ocean climate, 2005: Unusual atmospheric blocking? *Geophys. Res. Lett.* 33:L22S09. doi: 10.1029/2006GL026736

Toresen, R., and Østvedt, O. J. (2000). Variation in abundance of Norwegian spring-spawning herring (*Clupea harengus*, Clupeidae) throughout the 20th century and the influence of climatic variations. *Fish Fisheries* 1, 231–256. doi: 10.1046/j.1467-2979.2000.00022.x

Walczowski, W., and Piechura, J. (2006). New evidence of warming propagating toward the Arctic Ocean. *Geophys. Res. Lett.* 33:L12601. doi: 10.1029/2006GL025872

Yaragina, N. A., and Dolgov, A. V. (2009). Ecosystem structure and resilience – a comparison between the Norwegian and the Barents Sea. *Deep Sea Res. II* 56, 2141–2153. doi: 10.1016/j.dsr2.2008.11.025

Zatsepin, V. I., and Petrova, N. S. (1939). Feeding of fishery concentrations of cod in the southern Barents Sea (by observations in 1934-1938). *Trudy PINRO* 5, 1–170 (in Russian).

Zelikman, E. A. (1958). Materials on distribution and reproduction of euphausiids in the coastal zone of Murman. *Trudy Murmansk Biol. Stations* 4, 79–117 (in Russian).

Zelikman, E. A. (1964). On ecology of reproduction of the abundant species of Euphausiañåà in the south-eastern Barents Sea. *Trudy Murmansk Mar. Biol. Inst.* 6, 12–21. (in Russian).

Zhukova, N. G., Nesterova, V. N., Prokopchuk, I. P., and Redneva, G. B. (2009). Winter distributions of euphausiids (Euphausiacea) in the Barents Sea (2000-2005). *Deep Sea Res. II* 56, 1959–1967. doi: 10.1016/j.dsr2.2008.11.007

Conflict of Interest Statement: The authors declare that the research was conducted in the absence of any commercial or financial relationships that could be construed as a potential conflict of interest.

22

Revealing the regime of shallow coral reefs at patch scale by continuous spatial modeling

Antoine Collin[1,2]*, Philippe Archambault[3] and Serge Planes[2]

[1] Nadaoka Laboratory, Department of Mechanical and Environmental Informatics, Tokyo Institute of Technology, Tokyo, Japan
[2] USR 3278 CNRS-EPHE, Centre de Recherches Insulaires et Observatoire de l'Environnement (CRIOBE), Papetoai, French Polynesia
[3] Institut des Sciences de la Mer, Université du Québec à Rimouski, Rimouski, Canada

Edited by:
Alberto Basset, University of Salento, Italy

Reviewed by:
Simonetta Fraschetti, University of Salento and Consorzio Nazionale Interuniversitario per le Scienze del Mare, Italy
Alberto Basset, University of Salento, Italy

*Correspondence:
Antoine Collin, Nadaoka Laboratory, Department of Mechanical and Environmental Informatics, Tokyo Institute of Technology, O-okayama 2-12-1-W8-13, Meguro-ku, Tokyo, 152-8552, Japan
e-mail: antoinecollin1@gmail.com

Reliably translating real-world spatial patterns of ecosystems is critical for understanding processes susceptible to reinforce resilience. However, the great majority of studies in spatial ecology use thematic maps to describe habitats and species in a binary scheme. By discretizing the transitional areas and neglecting the gradual replacement across a given space, the thematic approach may suffer from substantial limitations when interpreting patterns created by many continuous variables. Here, local and regional spectral proxies were used to design and spatially map at very fine scale a continuous index dedicated to one of the most complex seascapes, the coral reefscape. Through a groundbreaking merge of bottom-up and top-down approach, we demonstrate that three to seven-habitat continuous indices can be modeled by nine, six, four, and three spectral proxies, respectively, at 0.5 m spatial resolution using hand- and spaceborne measurements. We map the seven-habitat continuous index, spanning major Indo-Pacific coral reef habitats through the far red-green normalized difference ratio over the entire lagoon of a low (Tetiaroa atoll) and a high volcanic (Moorea) island in French Polynesia with 84 and 82% accuracy, respectively. Further examinations of the two resulting spatial models using a customized histoscape (density function of model values distributed on a concentric strip across the reef crest-coastline distance) show that Tetiaroa exhibits a greater variety of coral reef habitats than Moorea. By designing such easy-to-implement, transferrable spectral proxies of coral reef regime, this study initiates a framework for spatial ecologists tackling coral reef biodiversity, responses to stresses, perturbations and shifts. We discuss the limitations and contributions of our findings toward the study of worldwide coral reef resilience following stochastic environmental change.

Keywords: non-thematic mapping, coral reefs, spectral proxy, reefscape ecology, Moorea, Tetiaroa, French Polynesia, resilience

INTRODUCTION

Elucidating emergent properties of complex adaptive ecosystems composed of interacting ecological patches (i.e., elements of a hierarchy) requires an innovative conceptual, theoretical, and methodological framework (Holling, 2001). An overarching component of this framework is to better understand the non-linear dynamics of the ecological patches across various organizational and spatial scales (Cumming, 2011). Identifying the spatial patterns of these patches is traditionally done by quantifying the surface area occupied by the targeted elements (Gotelli and Colwell, 2001; Rooney et al., 2004; Collin et al., 2008). The ecological patches are thereby spatially referred to as homogeneous discrete patches. Depending on the level of ecological organization studied, patches are composed of either individual organisms or assemblages as well as habitats (Forman, 1995). Classifying populations, communities or habitats into categorical types has traditionally been accepted by ecologists (Turner et al., 2001). Studies so far mainly strive to monitor evolution between alternate spatially-enclosed domains of attraction or regimes (Folke,

2006). Since the advent of digital data acquisition, computation and storage, ecological research has increasingly used images to outline ecological patches separated from each other by sharp boundaries. However, the detection of these boundaries is heavily dependent on the capabilities of sensors and statistical classifiers. In addition, the binary approach underlying these boundaries has sparked off a wide-ranging theoretical debate on the ecological representativeness between real world and digitalized information (Austin, 2002, 2007).

Investigating the spatial patterns of ecological patches in a continuous (or non-thematic) way has received relatively little attention so far. However, it is obvious that a continuous approach is better suited for describing natural pattern-process interactions that are emerging from a range of gradients and resources (Austin, 2007). The continuously ranging approach provides insights into the spatial modeling of species diversity (Harborne et al., 2008; Mellin et al., 2009) and can successfully map the variations in ecological processes by means of appropriate indicators such as the functional index (Borja et al., 2000), the resilience index (Rowlands et al., 2012), and the vegetation index (Tucker, 1979). Moreover, the latter two indices are of high importance in natural resource management as they may reveal or predict high biodiversity areas or hotspots (Myers et al., 2000).

Since the Normalized Difference Vegetation Index (NDVI) was designed by Tucker (1979), it has benefited from self-selection in terrestrial ecology work and come to the forefront of the study of global change in, for instance, shedding light on the increased plant growth in northernmost latitudes (Myneni et al., 1997; Sturm, 2010), and assessing the carbon balance and net primary production of terrestrial ecosystems (Piao et al., 2009; Zhao and Running, 2010). Based on the light interaction with vegetation, absorbing the red wavelengths and reflecting the near-infrared, the NDVI enables tree species, tree phenology and bare ground to be reliably discriminated over a vast range of spatial scales. Even though the use of the NDVI has become a key component for ecologists (Pettorelli et al., 2005), use of the index is strongly confined to terrestrial ecosystems. This is due to the fact that near-infrared wavelengths are substantially absorbed by water, incurring a great reduction in signal return/detection (Smith and Baker, 1981). Marine ecologists inevitably limit their investigation of ecological patch's spectral signatures to wavelengths constrained by the visible spectrum, i.e., 400–700 nm (Hochberg et al., 2003; Collin and Planes, 2012).

An in-depth examination of the water-attenuated gamut near both range boundaries shows that a larger spectral window has the potential of delivering meaningful information from particular ecological patches, hitherto neglected (Pegau et al., 2003; Leon et al., 2012). Specifically, insofar as the attenuation coefficient is moderately high in clear seawaters (Maritorena et al., 1994), the very near-infrared (between 700 and 750 nm) can reveal targets interacting with light at these wavelengths. The interaction can correspond to (1) the overall reflection by the chlorophyll patches (i.e., red-shift effect), or, conversely, (2) the absorption by the chlorophyll photosystem I (while the photosystem II absorbs the red wavelengths, Koning, 1994), or (3) the absorption by the recently discovered chlorophyll f, allowing reef building stromatolites to absorb light around a 706 nm centered peak (Chen et al., 2010). If the interplay of these sparse findings turns out to be fruitful, it would enhance the monitoring of shallow ecosystems, such as coral reefs, which are at the top marine biodiversity hotspots.

Remote surveying of coastal and marine ecosystem patterns at high (~10 m) and very high (~1 m) spatial resolution is increasingly required. The launch of the WorldView-2 (WV2) sensor in 2009 has been an incentive that greatly stimulates this study. This spaceborne sensor has doubled the spectral capabilities (notably a far red waveband) of all very high resolution counterparts whilst furnishing ecological information down to 0.5 m. Based on the WV2 synergistic spectral dataset, the coastal habitats and bathymetry can now be mapped at the submeter scale over ecosystems showing clear waters, typically over coral reefscapes (Collin and Planes, 2011; Collin and Hench, 2012). The shallow seascapes were thereby selected as the study cases by virtue of their current socio-ecological paramountcy. Coral reefscapes are regarded as the beacon seascape given the essential ecological services they provide and the increased number of abiotic and biotic factors

threatening them (Emanuel, 2005; Hoegh-Guldberg et al., 2007; Wilkinson, 2008; AMAP Expert Group, 2011). Rapidly monitoring the regime of coral reefscapes across various spatial scales, while using a reliable proxy with little ground-truthing, constitutes an inevitable issue to be addressed in the near future, owing to the accelerating pace of reef loss (De'ath et al., 2012). In this study, the regime of the coral reefscape refers to the set of the habitats that has the same essential structure, function and feedbacks (Walker et al., 2004).

The purpose of this study is to design, validate, and use a continuously ranging spatial model of coral reefscape regime for the first time. Co-evolving with cutting-edge imagery (ground based and remotely sensed), this model, dedicated to an ecologically complex seascape, must be able to (1) concur with comprehensive ground measurements, and (2) reliably map components (i.e., highlight live coral colonies) across various spatial scales while being decreasingly sensitive to various algae taxa, bleached coral and sediment. As a first use, (3) the spatial model is employed to portray the regime akin to two entirely-modeled high (volcano) and low islands (atoll) in French Polynesia, outlining its potential key role in assessing the resilience of shallow coral reefs to global environmental change.

MATERIALS AND METHODS
STUDY SITE
Indo-Pacific reefs exhibit diversified benthic communities, ranging from the sediment to healthy coral colonies through various genera of macroalgae. This high degree of seascape diversity seemed ideally suited to test and validate a continuous (nonthematic) coral index. The study was conducted around the islands of Tetiaroa (17°0'S, 149°33'W) and Moorea (17°47'S, 149°80'W) in the Archipelago of the Society Islands (French Polynesia) (**Figure 1**). Tetiaroa, located 53 km North of Tahiti, is a 34 km^2 atoll (crown of reef peaking at 3 m), comprising 13 islets (*motu*) and a 30 m deep lagoon with no reef pass. Benefiting from a relative anthropogenic-related protection due to its ownership history (sacred location of the Tahitian kingship and finally property of the Brando family), Tetiaroa shelters various coral colonies (Porites *sp.*, Synarea *sp.*, Acropora *sp.*, *Pavona cactus*) with attached lagoon organisms (pink whipray *Himanturafai*, black tip shark *Carcharhinus melanopterus*). Contrary to Tetiaroa, Moorea, as a 187 km^2 volcanic island, peaked at 1207 m and experiences an increase of demography pressure compounded with runoff disturbances (16,490 inhabitants, 2007 census population). Located 17 km North-West of Tahiti, Moorea is situated at the southeastern end of the Society volcanic chain, displacing to the northwest as the Pacific plate. Given the wide spectrum of habitats and species and the ease of sampling (clear, warm, and shallow waters), both islands fit the requirements for our study.

CORAL REEF HABITATS
A field campaign conducted over the two islands between 21 June 2010 and 16 November 2011, enabled a total of 897 sites to be characterized in respect to their geographic location, water depth and benthic habitats. Supporting the accuracy assessment of the spatial model and bridging the hand- and spaceborne measurements, these sites were classified in eight benthic habitats

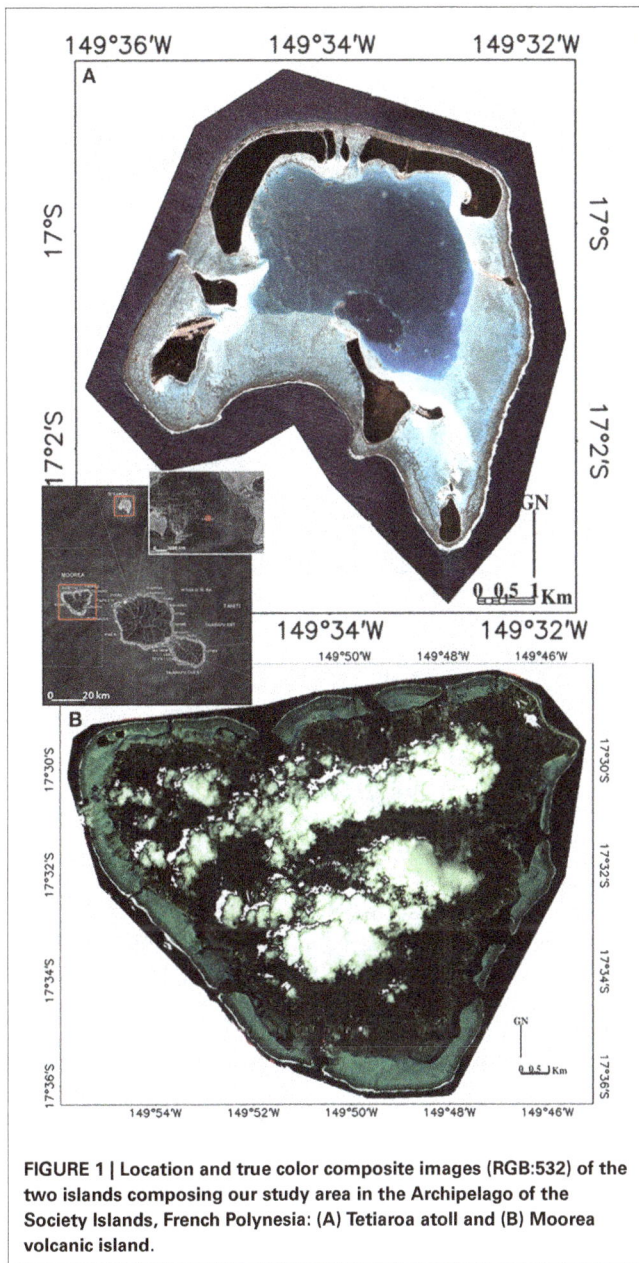

FIGURE 1 | Location and true color composite images (RGB:532) of the two islands composing our study area in the Archipelago of the Society Islands, French Polynesia: (A) Tetiaroa atoll and (B) Moorea volcanic island.

based on an influential classification scheme (Hochberg et al., 2003) deemed representative of main habitats in global coral reefs (Roff and Mumby, 2012): brown stony coral (*Acropora hyancinthus, Acropora pulchra, Pavona cactus, Pocillopora damicornis, Porites sp., Synarearus*), blue stony coral (*Montipora sp., Porites sp.*), bleached stony coral (all previous taxa), brown fleshy algae (*Padina boryana, Sargassum pacificum, Turbinaria ornata*), red fleshy algae (*Actinotrichia fragilis*), green fleshy algae (*Halimeda opuntia, Cladophora patentiramea*), red calcareous algae (Lithotamniae), and calcareous sand. Over every site, a georeferenced 0.5 × 0.5 m quadrat (matching the satellite spatial resolution) allowed the quantification of benthic habitats using a digitally-superimposed 6 × 6 grid. Since the quadrat was recorded using a high-resolution digital RGB camera, each of the

36 grid squares composing the grid were thoroughly examined (2.10^{-4} m resolution). The habitat dominating the grid composition (>18 sub-quadrats) was considered as being representative of the site. Since all grids were analyzed and classified by the same analyst, we assume that the vision-inherent bias is sufficiently buffered by the 36 sub-quadrats for an acceptable accuracy in quadrat assignment. Benthic habitats not included in our classification scheme (i.e., Echinodermata and Porifera phyla) were recorded when encountered. Further, there was no effect of including them as these habitats did not dominate any quadrat and hence, affect site classification.

In addition to the photograph-derived habitat reconnaissance, the water depth was recorded for each georeferenced site using a 0.1 m accuracy acoustic system. Whilst water depths greater than 1 m were surveyed with an aluminum boat and sea kayak, in Moorea and Tetiaroa, respectively, shallower sites were investigated by foot on and around both islands.

LOCAL SCALE MEASUREMENTS

Targeting a statistical robustness, a total of 2300 spectral signatures collected for the eight coral reef habitats were individually measured between 7 and 22 November 2011 in a coral nursery enclosed in the Moorea Intercontinental Resort. The proximity of the samples to a dry and powered pontoon enabled for a controlled recording, requiring calm conditions and a continuous connection between the portable spectrographic instrument and the software-supplied laptop. Provided with one nm wavelength accuracy (hyperspectral), each spectral measurement was recorded as a reflectance (R) signature (ratio situated between 0 and 1) by virtue of the white correction. Reflectance signatures were then averaged across replicates to provide a mean spectral signature representative of each benthic habitat (**Figure 2**). The spectral range (spanning 350–750 nm) recorded in this study overcomes the gamut traditionally covered (400–700 nm) by other studies (Hochberg et al., 2003; Leiper et al., 2009). The traditional limitations take the light attenuation in water into account, measured by the attenuation coefficient (Mobley et al., 2002). However, given (1) the magnitude of the coefficient encompassed in the new spectrum (Maritorena et al., 1994), and (2) the ground measured reflectance that remains exploitable (coefficient < 2 m^{-1} and R < 1, respectively), we point out upfront that the spectral information tied to wavelengths slightly greater (until 750 nm) than the arbitrary upper limit (i.e., 700 nm) has the potential for underpinning the design of the coral reef indices. This assumption may be confirmed by the amount of underwater habitats detectable by remotely-sensed wavelengths of interest.

Measurements of the reflectance hyperspectral signatures akin to the seven targeted habitats consist of the initial step related to the first approach employed in this study. Depicted as bottom-up (see green ring in **Figure 3**), the approach is dedicated to identify the coral reef indices based on their spectral signatures, further discernable at regional scale by remote sensing.

REGIONAL SCALE MEASUREMENTS

The WV2 is the remote sensor selected for its capabilities of spatial coverage and its high ratio linking the number of spectral

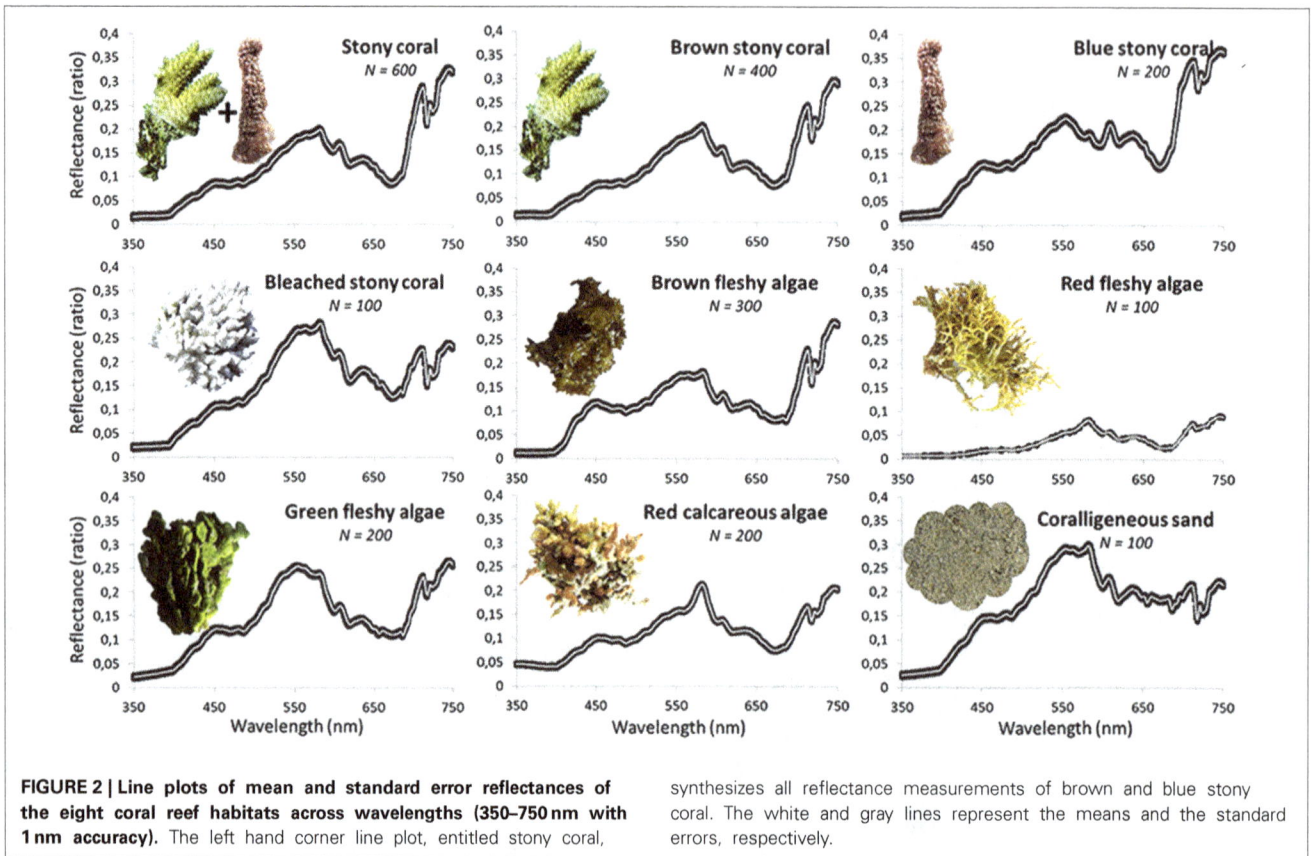

FIGURE 2 | Line plots of mean and standard error reflectances of the eight coral reef habitats across wavelengths (350–750 nm with 1 nm accuracy). The left hand corner line plot, entitled stony coral, synthesizes all reflectance measurements of brown and blue stony coral. The white and gray lines represent the means and the standard errors, respectively.

optical bands with the spatial resolution. Providing both six optical bands (including the "red edge" band, 705–745 nm) with a 2 m spatial resolution and one overarching band, called panchromatic, with a 0.5 m spatial resolution across a 16.4 km swath, WV2 currently outperforms any space- and airborne sensors for mapping coral reef habitats very finely over regional areas (Collin et al., 2013). The underlying premise of this sensor selection is that coral reef indices will be operational for both space- and airborne sensors fitted with a greater amount of optical bands, irrespective of the decay of their spatial resolution. A total of four WV2 images covering 388 km^2 (1.552 billion 0.5 m pixels) were collected over Moorea and Tetiaroa islands on 12 February and 14 October 2010, respectively. WV2 imagery was corrected for geometric distortions and atmospheric attenuation using the MODTRAN-4 algorithm (Matthew et al., 2000), resulting in six water-leaving reflectance matrices for each island (see blue ring in **Figure 3** and Collin et al., 2013 for further details). The innovative use of the "red edge" band is encouraged to the extent that the amount of reliable information, empirically measured, is significant up to 2 m deep (**Figure 4**). The pansharpening method is then applied to scale up the spatial resolution of the six bands (i.e., 2 m) to that of the panchromatic band (i.e., 0.5 m) using a reliable fine-tuning procedure (Collin et al., 2013). Focusing only on benthic habitats (possibly overshadowed by higher signals common to air and terrestrial features), clouds and land were masked out based on the probability density function of the strongly-attenuated eighth band.

Water depth was mapped using the 897 field measurements coupled with the ratio transform (Stumpf et al., 2003) applied to the first and third WV2 bands ("coastal" and green, respectively), selected based on a systematic analysis (Collin and Hench, 2012). Based on the differential attenuation of optical wavelengths by water, the transform "ratios" the natural logarithms of the two bands and furnishes a relative water depth model, suitable for obtaining actual water depths after calibration (**Figures 4A,D**). Given the increasing loss of benthic information in respect to the water depth, areas deeper than 2 m were masked out, namely 63.48% (32.37 km^2) and 88.31% (14.73 km^2) of Moorea and Tetiaroa lagoon parts, respectively (**Figure 4**).

The pre-processing steps leading to the reflectance multispectral images are required to initialize the second approach. Described here as top-down (see blue ring in **Figure 3**), this approach aims to test the regional scale applicability of the coral reef indices derived from local scale measurements.

CORAL REEF INDICES

The methodology revolves around the fusion of the bottom-up and top-down approaches. The success of the methodology lies in the bridging of outcomes derived from the two processes(see rings in **Figure 3**). Based on hyperspectral data, the local scale characterization requires conversion with the spatial tool which will allow interpolation at the regional scale (based on multispectral data). The WV2 spectral sensitivity of the first six bands illustrates the key function of joining the 500-bin hyperspectral

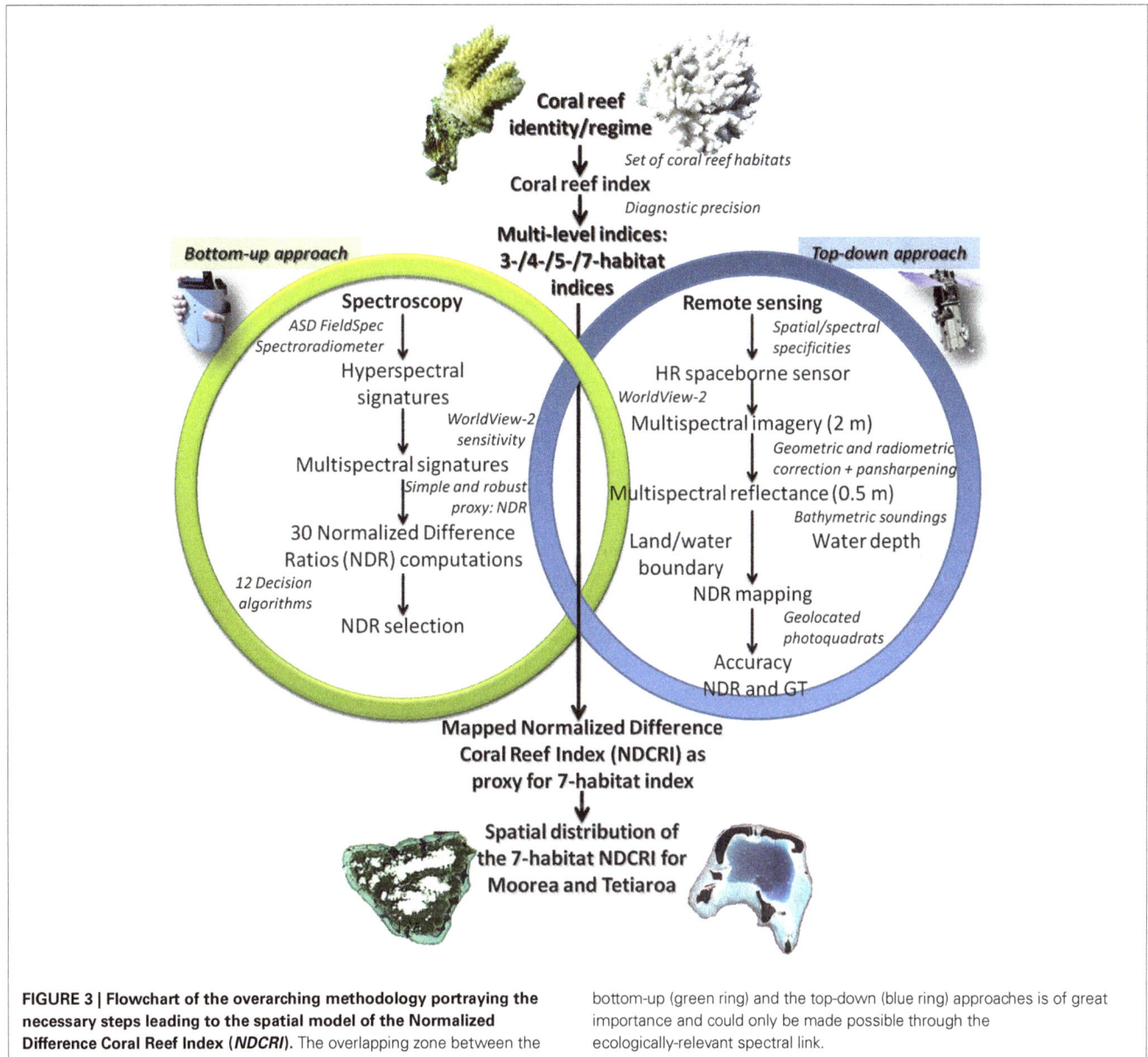

FIGURE 3 | Flowchart of the overarching methodology portraying the necessary steps leading to the spatial model of the Normalized Difference Coral Reef Index (NDCRI). The overlapping zone between the bottom-up (green ring) and the top-down (blue ring) approaches is of great importance and could only be made possible through the ecologically-relevant spectral link.

measurements to the six-bin multispectral pixel data (**Figure 5**). Once the conversion is done for each hyperspectral signature, reflectance multispectral signatures are averaged by benthic habitat so that coral reef indices can be designed.

The core principle of the design consists in establishing indices which allow attribution of the highest values for living coral, medium values for algae, and the lowest values for bleached coral and sediment. Beyond the traditional coral health index aiming at discriminating healthy and unhealthy corals in a dualistic (even simplistic) way, our bespoke indices have the potential to overarch the common Indo-Pacific shallow coral habitats in a rapid, spatially-fine and extensive way. Logical expressions of 12 coral reef indices, based on three, four, five and seven habitats, are built to surrogate four categories of diagnostic precision (**Figure 6**). Each diagnostic category is subdivided into three coral reef indices accounting for either one coral super habitat or two coral habitats

depending on their color, i.e., blue or brown (Hochberg et al., 2003) (see Index name column in **Figure 6**). For each coral reef index, the logical expressions between habitats are portrayed by strict inequality signs (like thresholds values) except for the three algae habitats for which there is no logical hierarchy (see < or > in **Figure 6**). Encompassing the three algae habitats, a algae super habitat is created since the far-reaching studies demonstrated the ecological differentiation occurring between general coral- and macroalgae-dominated reefs (Bellwood et al., 2004; Roff and Mumby, 2012).

Given the substantial scope of the NDVI spectral index in major land-based ecological studies (Myneni et al., 1997; Pettorelli et al., 2005; Piao et al., 2009; Sturm, 2010; Zhao and Running, 2010) and its facilitated use, the coral reef indices will be revealed in the form of spectral Normalized Difference Ratios (*NDR*), mathematically formulated as follows:

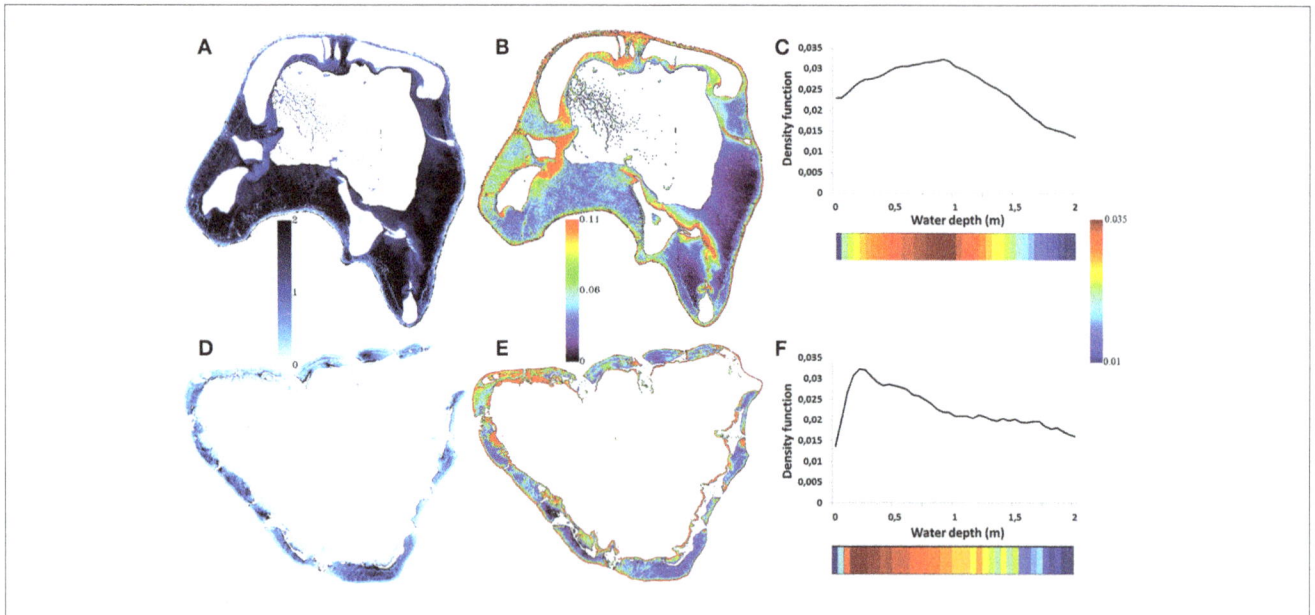

FIGURE 4 | Digital models of the water depth and reflectance of the "red edge" band compounded with the water depth in respect to the density function of the reflectance of the "red edge" band related to Tetiaroa (A–C), and to Moorea (D–F). The water depth is positively correlated with the darkness of the blue gradient (0–2 m). The reflectance of the "red edge" is here represented by a rainbow-scaled gradient (0–0.11). Electro-magnetic radiation akin to the 705–745 nm gamut (i.e., spectral range of the "red edge" band) is reflected from benthos as deep as 2 m, as revealed by the positive values of the density function values in both islands.

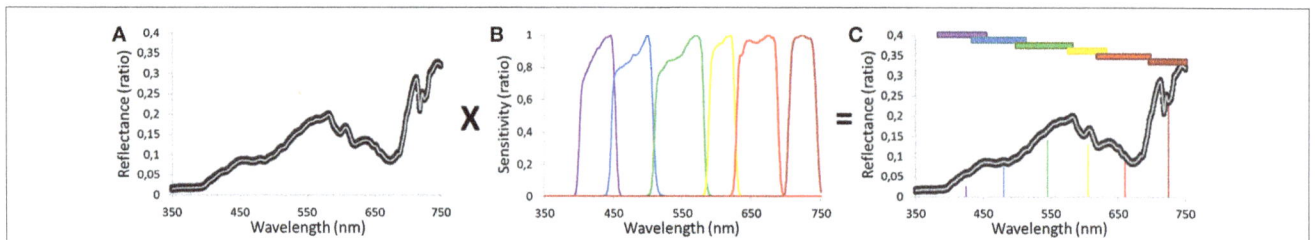

FIGURE 5 | Sequence of line plots describing (A) the local scale measurement of hyperspectral reflectance data along with (B) the spectral sensitivity of the regional scale sensor (WorldView-2) and **(C) the product of combining (A,B) to obtain the multispectral reflectance signature, suitable for regional scale spatial modeling.**

$$NDRij = \frac{i - j}{i + j} \qquad (1)$$

where $i, j \in \{$"coastal," blue, green, yellow, red, "red edge" two-dimensional matrices$\}$ and $i \neq j$. An exploratory dataset of 30 NDR was computed as a result of the pair combinations of the six optical bands. NDR are named according to the number index of the two bands involved, i.e., "coastal," blue, green, yellow, red, "red edge" related to 1, 2, 3, 4, 5, 6, respectively. An array of 30 potential NDR by 11 habitats (eight single habitats + three super habitats: coral, algae, fleshy algae) is then subject to the 12 logical expressions (see structure of **Table 1**). When one logical expression is successfully addressed by a NDR, the matching cell is highlighted in **Table 1**.

SPATIAL ANALYSIS

The spatial mapping of the retained NDR is realized in applying Equation (1) at each pixel of the two island imageries. Resulting spatial models are then corrected for the water depth using a non-linear least squares fit, called the Gaussfit function (Research Systems 2005), based on a six-termed linear combination of Gaussian and quadratic functions. NDR residual values of the spatial model need only be scaled to the actual NDR values; namely:

$$NDCRI = m \left\| X - \left(A0e^{\left(-\left[\frac{X - A1}{A2} \right]^2 / 2 \right)} + A3 + A4X + A5X^2 \right) \right\| \qquad (2)$$

in which $NDCRI$ means Normalized Difference Coral Reef Index, X refers to retained NDR, m is a tunable constant to scale the NDR residuals, and $A0, A1, A2, A3, A4,$ and $A5$ are the unknown parameters of the non-linear depth regression.

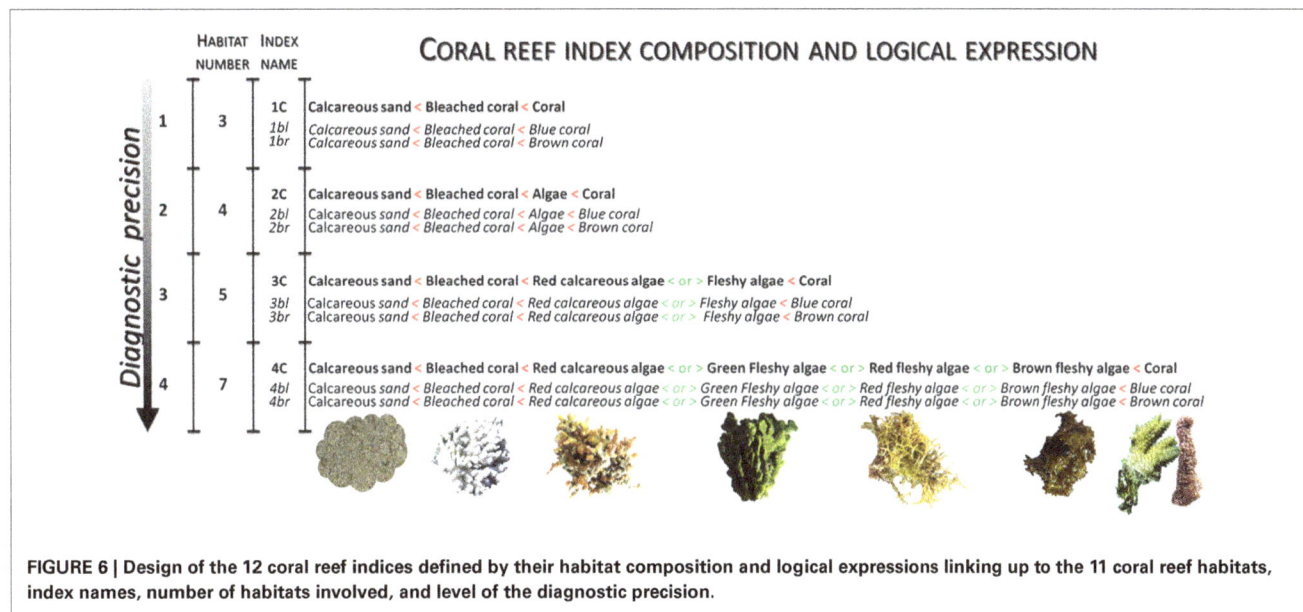

FIGURE 6 | Design of the 12 coral reef indices defined by their habitat composition and logical expressions linking up to the 11 coral reef habitats, index names, number of habitats involved, and level of the diagnostic precision.

The accuracy assessment of the processed spatial models is sourced from the statistical relationships manifesting between the georeferenced measurements and associated pixel values retrieved from the spatial models. Since the models do not need any training data, all 897 measurements are used in the accuracy assessment. Whilst the dataset of measurements appears in the form of classification, the models are made of continuous values, matching the purpose of this study. One way around this dilemma is to convert the continuous model into a discrete one through the data slicing of the model into intervals intrinsic to the outcome of the index design. Boundaries separating intervals result from the mean value between index values of two contiguous habitats. The consistency between the classified measurements and sliced model is appraised in computing the contingency matrix, for each island, summarized by the user's accuracy (UA), the producer's accuracy (PA) and the overall accuracy (OA). OA is the number of correct observations divided by the total number of observations; UA and PA detail errors of commission and omission, respectively (Congalton and Green, 1999). The spatial model associated with the most relevant NDR (highest diagnostic precision and data amplitude) is segmented into the number of classes corresponding to the diagnostic level. Note that the discretization of the spatial model is a necessary method to validate our models, but appears as a temporary result compounded with intrinsic limitations, mentioned in the introduction, that this study searches to overcome.

Once model validation is done, the ecological status of each island is determined using the NDCRI featuring both the highest diagnostic precision and amplitude. A novel quantitative method aiming at reliably and synoptically diagnosing the structure and likely near-future shifts of coral reef regime is designed by outlining the shape of the reef crest surrounding the island and by calculating the distance between coastline and reef crest. The gradient of the distance is then mapped across the island, producing 2640 contour lines ranging from 0 to 1320 m by 0.5 m increments.

A probability density function of the NDCRI is then computed for each contour line in the form of histogram. Plotting the 2640 histograms in respect to the distance from the coastline frames a quantitative tool, so-called histoscape, having the potential to help visualize the NDCRI-related basins of habitat and ecological distances between them, within the coral reefscape. Whilst the gradient of data values is tightly linked to the abundance of related coral reef habitats, the horizontal and vertical dimensions of a basin of habitat may be related to its spatial variability and landward sprawl, respectively. The distance between two basins may furthermore be associated with the habitat inertia, i.e., the inverse likelihood of shifting toward the neighbor habitat.

RESULTS

CORAL REEF INDICES

The nested bottom-up and top-down approaches provide insight into the dataset of NDR successfully responding to the constraints enacted by the logical sequences of designed indices (see **Figure 6**). The amount of successful NDR decreases with the diagnostic precision (accounting for the 10 successful NDR types across the 12 designed indices): 25 for three-habitat level, 17 for four-habitat level, 13 for five-habitat level, and 7 for seven-habitat level (**Table 1**). A substantial amount ($n = 49$ or 79%) of successful NDR include the "red edge" band. Combining the "red edge" band with the "coastal," blue and green bands satisfactorily respect the requirements related to the three-, four-, and five-habitat levels, respectively, regardless the coral habitat. Although the NDR based on the "red edge" band and the yellow or the red bands adequately translate the indices up to the five-habitat level for all coral habitats, the "red edge"-yellow does not correctly translate the seven-habitat level overarched by the Brown coral. Also, the "red edge"-red does not address the seven-habitat level headed by the Blue coral. The Coral and Brown coral show a greater amount of successful NDR ($n = 22$) than Blue coral ($n = 18$).

Table 1 | Boolean results of the 30 regional scale Normalized Difference Ratio (*NDR*) successfully responding to logical expressions associated with the 12 different coral reef indices.

	3-Habitat Index			4-Habitat Index			5-Habitat Index			7-Habitat Index		
	1C	1bl	1br	2C	2bl	2br	3C	3bl	3br	4C	4bl	4br
12	1	1	1	0	0	0	0	0	0	0	0	0
13	0	0	0	0	0	0	0	0	0	0	0	0
14	0	0	0	0	0	0	0	0	0	0	0	0
15	0	0	0	0	0	0	0	0	0	0	0	0
16	0	0	0	0	0	0	0	0	0	0	0	0
23	0	0	0	0	0	0	0	0	0	0	0	0
24	0	0	0	0	0	0	0	0	0	0	0	0
25	0	0	0	0	0	0	0	0	0	0	0	0
26	0	0	0	0	0	0	0	0	0	0	0	0
34	0	0	0	0	0	0	0	0	0	0	0	0
35	0	0	1	0	0	0	0	0	0	0	0	0
36	0	0	0	0	0	0	0	0	0	0	0	0
45	1	0	1	1	0	1	0	0	1	0	0	0
46	0	0	0	0	0	0	0	0	0	0	0	0
56	0	0	0	0	0	0	0	0	0	0	0	0
21	0	0	0	0	0	0	0	0	0	0	0	0
31	0	0	0	0	0	0	0	0	0	0	0	0
41	0	0	0	0	0	0	0	0	0	0	0	0
51	0	0	0	0	0	0	0	0	0	0	0	0
61	1	1	1	1	1	1	0	0	0	0	0	0
32	0	0	0	0	0	0	0	0	0	0	0	0
42	1	0	0	0	0	0	0	0	0	0	0	0
52	0	0	0	0	0	0	0	0	0	0	0	0
62	1	1	1	1	1	1	1	1	1	0	0	0
43	1	1	1	0	0	0	0	0	0	0	0	0
53	0	0	0	0	0	0	0	0	0	0	0	0
63	1	1	1	1	1	1	1	1	1	1	1	1
54	0	0	0	0	0	0	0	0	0	0	0	0
64	1	1	1	1	1	1	1	1	1	1	1	0
65	1	1	1	1	1	1	1	1	1	1	0	1

Notes: The Boolean results are coded in the form of bit. **Figures 1–6** *are respectively related to "coastal," blue, green, yellow, red, "red edge." Coded names C, bl, and br mean Coral, Blue coral, and Brown coral, respectively.*

The 10 successful *NDR* types are plotted for each of the 12 coral reef indices so that their evolution across and amplitude among habitats can be quantitatively embraced (**Figure 7**). Corroborating the trend shown in the **Table 1**, where the "red edge" band plays a key role in pivotal indices, the highest amplitudes (0.17–0.43 range) akin to retained *NDR* are intricately associated with the "red edge"-included *NDR*, as the degree of slope visually confirms it (**Figure 7**). One exception is the *NDR* combining red and yellow bands (0.1–0.2 range). It is the only *NDR* built from conventionally-limited visible spectrum (400–700 nm) displaying an amplitude greater than 0.1 (0.12 for 1br, 2br, and 3br in **Figure 7**). Striving to maximize the ecological information conveyed by the spectral tool, we choose the *NDR* type (1) able to translate the detailed seven-habitat level index (4C), (2) evidently crossing the zero value, and (3) endowed with the highest amplitude, i.e., "red edge"-green ratio (0.41 for 1C, 2C, 3C, and 4C in **Figure 7**). This "red edge"-green *NDR* neatly illustrates the outcome of the bottom-up approach enhanced by the top-down by means of the spectral sensitivity of the regional scale sensor.

SPATIAL MODELING OF THE CORAL REEF INDEX

Supported by the result originating from the bottom-up process, the top-down approach can be achieved in testing the validity of the spatial modeling of the *NDCRI* based on the most compelling *NDR* type, i.e., "red edge"-green. The accuracy summarizing the relationships between the seven-habitat field observations and partitioned "red edge"-green *NDCRI* is assessed using the 340 and 557 georeferenced quadrat-pixels related to Tetiaroa and Moorea, respectively (**Table 2**). Surpassing 80%, both accuracies are obviously high with a slight advantage to Tetiaroa (83.53 vs. 81.69%). Diagonal cells consistently highlight Coral, Bleached coral and Calcareous sand at the expense of Algae habitats across both islands, indicating correct matching between observed and modeled quadrat-pixels. Conversely, the off diagonal cells reveal that

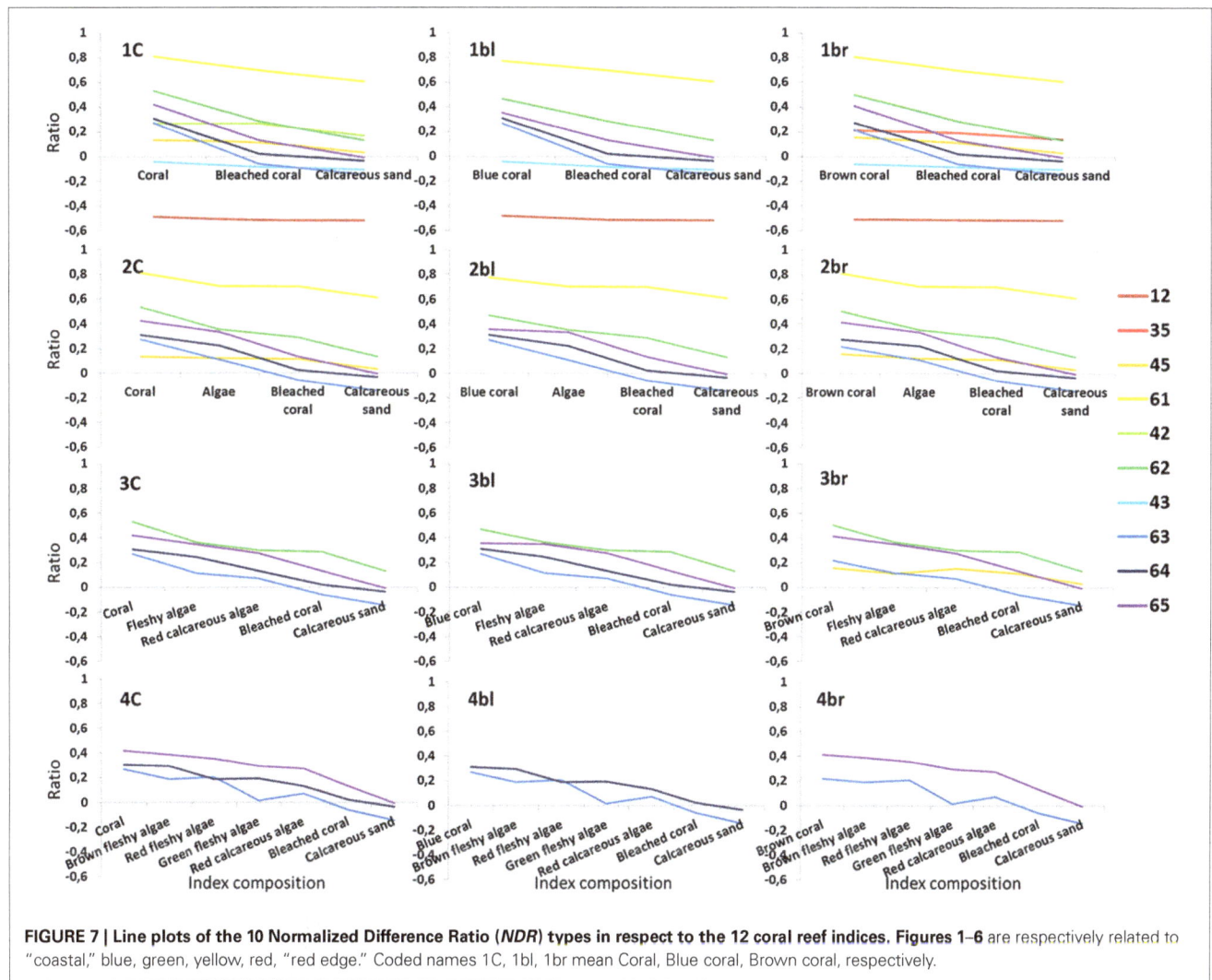

FIGURE 7 | Line plots of the 10 Normalized Difference Ratio (*NDR*) types in respect to the 12 coral reef indices. Figures 1–6 are respectively related to "coastal," blue, green, yellow, red, "red edge." Coded names 1C, 1bl, 1br mean Coral, Blue coral, Brown coral, respectively.

Algae habitats both self-undermine and undermine Coral habitat, and that Bleach coral and Calcareous sand mutually overshadow, attesting misclassified accuracy control points. Measuring how accurately (1) the slicing-based classification performs (rows) and (2) the analyst assigns classes (columns), the UA and PA interestingly demonstrate opposite patterns in respect to the island. UA are, on the one hand, greater than PA concerning Coral and Algae habitats, and on the other hand, lesser than PA regarding Bleached coral and Calcareous sand for Tetiaroa. Also, UA are lesser and greater than PA about the Coral-Algae habitats and the Bleached coral-Calcareous sand clusters, respectively, for Moorea.

The spatial modeling of the *NDCRI* shows marking similarities as well as dissimilarities in the structural patterns of the islands (**Figures 8A,C**). Resemblances lie in high values of the *NDCRI* in the barrier reef, medium values within the reef flat area and low values near the coastlines. Discrepancies emerge from specificities pertaining to each island, such as Tetiaroa eastern plain and lagoon reef network, characterized by low and high values, respectively, and Moorea outer slope of the fringing reef featured by high values. Driven by optimally exploiting the spatial information included in the *NDCRI* models, we embrace the histoscape

derived from a continuous variable spanning the entire study area rather than the traditional thematic classes distributed along discrete transects.

Histoscape results exacerbate differences occurring between the two islands (**Figures 8B,D**). Although the first dozens of meters of the histograms across the reef crest-coastline distance show a predominance of high *NDCRI* values (*x*-axis) in the two islands, the gap widens between them for the rest of the gradient. The Tetiaroa zone along the reef crest (close to 0 on the *y*-axis) is characterized by extensive *xyz* data values in a Coral-dominated, wide and deep basin within the histoscape. Conversely, Moorea's counterpart exhibits restrained *yz* dimensions, appearing in the form of a very surficial depression. Medium-high *x* values are observed leaving Tetiaroa reef crest toward the coastline, being relatively widespread and abundant around 0.05–0.1 *x* values, characterized by a Coral-mediated stream bed with an Algae-dominated channel and meanders in the histoscape. Otherwise, Moorea's counterpart of the histoscape is defined by low *x* values, confined to 0 with a marked decline close to the coastline, characterized by a Green fleshy algae-dominated trench ending by Bleached coral and a Calcareous sand basin. Two overall trends

Table 2 | Contingency matrices depicting Tetiaroa ($n = 340$) and Moorea ($n = 557$) field-observed coral reef habitats vs. their classification based on the seven-habitat slicing of the spatial model built from the Normalized Difference Ratio combining the "red edge" and the green WorldView-2 bands.

	Stony coral	Brown fleshy algae	Red fleshy algae	Green fleshy algae	Red calcareous algae	Bleached stony coral	Calcareous sand	Row Total	UA
TETIAROA									
Stony coral	**42**	2	0	1	2	1	1	49	*85.71*
Brown fleshy algae	1	**40**	0	2	2	1	2	48	*83.33*
Red fleshy algae	0	1	**40**	0	0	1	1	43	*93.02*
Green fleshy algae	2	0	0	**38**	4	0	0	44	*86.36*
Red calcareous algae	3	1	5	0	**39**	0	0	48	*81.25*
Bleached stony coral	2	3	0	3	0	**41**	4	53	*77.366*
Calcareous sand	0	2	0	2	3	4	**44**	55	*80*
Column Total	50	49	45	46	50	48	52	340	
PA	84	81.63	88.89	82.61	78	85.42	84.61		
MOOREA									
Stony coral	**70**	4	2	0	4	3	1	84	*83.33*
Brown fleshy algae	4	**64**	2	0	0	2	4	76	*84.21*
Red fleshy algae	0	0	**57**	6	7	1	0	71	*80.28*
Green fleshy algae	5	0	5	**65**	1	9	3	88	*73.86*
Red calcareous algae	0	3	1	3	**60**	0	3	70	*85.71*
Bleached stony coral	2	1	1	7	1	**68**	6	86	*79.07*
Calcareous sand	0	3	0	2	0	6	**71**	82	*86.59*
Column Total	81	75	68	83	73	89	88	557	
PA	86.42	85.33	83.82	78.31	82.19	76.40	80.68		

Notes: UA and PA refer to user's and producer's accuracy, respectively. Bold values correspond to correctly classified validation pixels.

unequivocally typify the histoscapes inherent to the atoll and volcanic island: a Coral/Algae-dominated wide stream bed with intertwined channels and a narrow Green fleshy algae-mediated channel, respectively.

DISCUSSION

Synergistically coupling bottom-up and top-down approaches, we find a continuous index discriminating and ranking the main coral reef habitats and we spatially model it across two ecogeomorphologically opposed coral reefscapes. The histoscapes explicitly highlight basins of habitat along the reef crest-coastline distance and show distinctions between the two islands.

ADVANCES AND LIMITATIONS OF THE CORAL REEF INDEX

Opting for a continuous spatial model to identify and quantify the structure of complex ecosystems like coral reefs presents substantial advantages compared to thematic models, also called classifications. Tailored to mimic subtle variations in spatial patterns akin to the real world (Austin, 2007), continuous modeling can also suitably document linear and non-linear dynamics (May, 2001; Harborne et al., 2006; Hastings and Wysham, 2010). Conversely, thematic modeling tends to perform well when defining non-linear processes endowed with thresholds around which regime shifts occur (Ives and Dakos, 2012), while aggregating

heterogeneous data into an average unit assigned to an overarching class. Specifically, spatial aggregation is very conducive in inducing misclassification when the related surface area encompasses the unit, and the variability is substantial between neighbor or within-aggregate units. Interestingly, an array of coral reef indices positively respond to the designed logical sequences against the diagnostic precision, i.e., the number of habitats involved. Although the amount of successful *NDR* declines with the precision, their variety shows great promise in addressing a broad list of purposes. In this study, the seven-habitat index is the only index examined in-depth, mostly because of its suitability to our detailed analysis but also for the challenge it represents. Notwithstanding, the proxies for three-, four-, and five-habitat indices have the obvious potential to help stakeholders examine coral bleaching, regime shifts occurring between coral and algae, and between coral and fleshy algae, respectively. A large body of literature seeking to demonstrate the regime shifts intrinsic to coral reefs focuses on the ratio of the abundance of the coral and algae habitats (macroalgae more specifically) (see in Hughes et al., 2010). However, a very recent study shows that the traditional coral-macroalgae ratio may have a low informational value (i.e., poor correlation) in terms of identifying the regime shift, but these limitations may be overcome using a ratio built on coral and fleshy macroalgae (Smith, 2012). Including the habitats commonly tackled in coral reef studies, such as soft coral, seagrass or

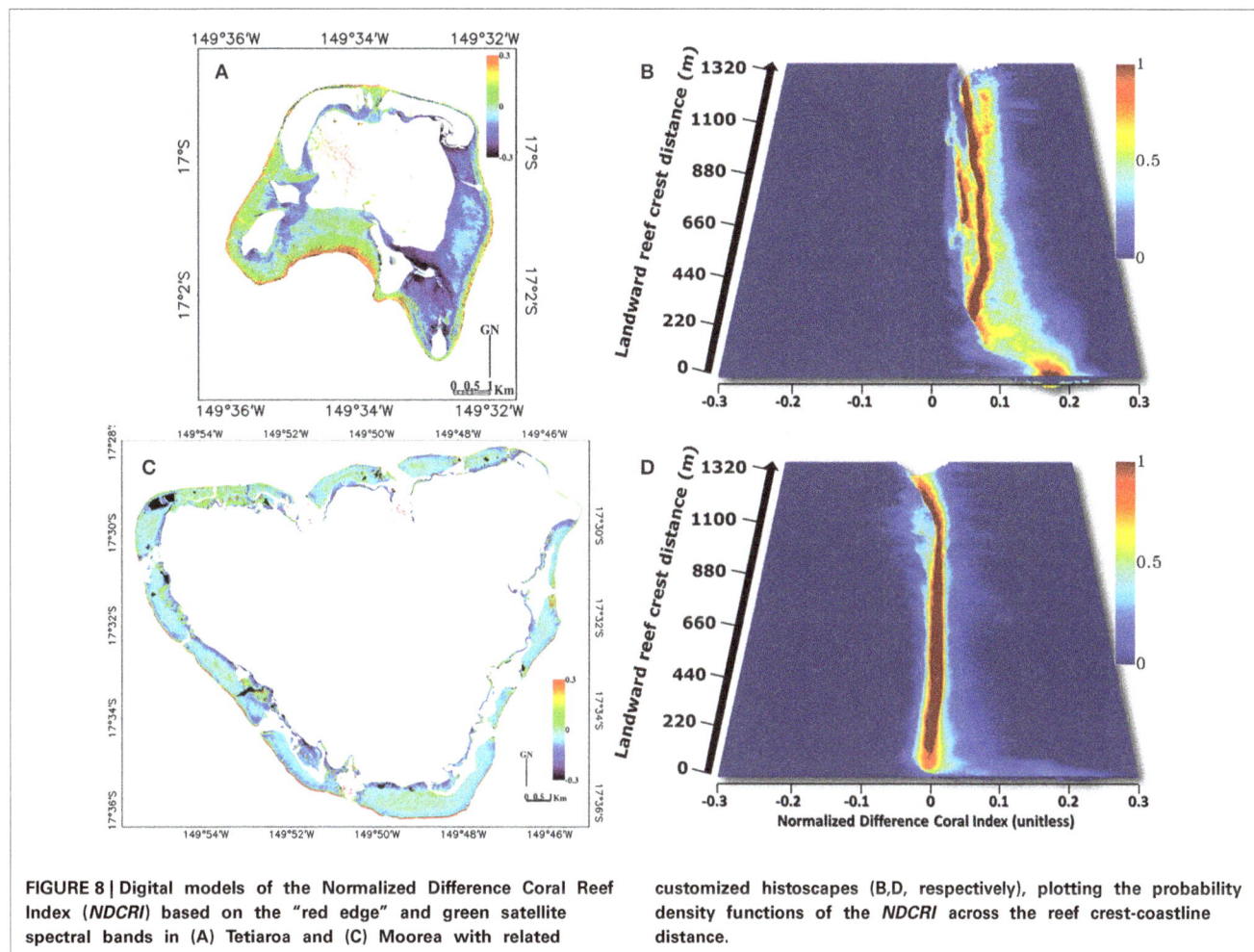

FIGURE 8 | Digital models of the Normalized Difference Coral Reef Index (*NDCRI*) based on the "red edge" and green satellite spectral bands in (A) Tetiaroa and (C) Moorea with related customized histoscapes (B,D, respectively), plotting the probability density functions of the *NDCRI* across the reef crest-coastline distance.

terrigeneous mud (Hochberg et al., 2003) would allow real world patterns to be better elucidated. Manifesting versatility against purposes, the developed continuous models turn out promising in quantifying the volume of primary coral reef habitats by associating the water depth as the vertical proxy.

Using the spectral tool as a proxy in characterizing the ecological regime of coral reefscapes facilitates the dissemination and promotion of the *NDCRI* given its ease of implementation and increasing commonness in reef ecologists' toolbox. Doing so ensures that coral reef habitats are reliably detected and discerned over large extents at very high spatial scales, especially when performed in clear oligotrophic waters. Inspired by research dedicated to coastal waters, coral reef scientists employing a spectrally-based methodology do not delve into expanding the conventional 400–700 nm gamut, owing to the quantity of information already provided (Hochberg et al., 2003; Leiper et al., 2009). The thorough examination of the 700–745 nm spectrum (i.e., "red edge") nevertheless testifies that shallow benthic habitats can be better detected and discerned (see **Figures 2, 4B,D**). This assessment concurs with the review of light attenuation by pure and oceanic water. It demonstrates that the visible wavelengths span 350–750 nm and the measurements precisely superimpose each other for wavelengths equal to and greater than 500 nm (White et al.,

2002, with a special emphasis on their **Figure 4**). We plotted the measurement of the absorption and penetration of the 350–750 gamut by oligotrophic oceanic water (Smith and Baker, 1981) so that the maximum depth of benthic investigation can be predicted and incorporated into future coral reef studies using spectral proxies (**Figure 9**). By integrating the spectral capabilities of various satellite sensors (i.e., regional-scale instruments), this figure will aid scientists and managers replicate and scale up the spectral proxies so that multiscale-reliant patterns and processes can be elucidated and enhanced (from any single sub- and tropical island to the Great Barrier Reef). It is crucial to bear in mind that the use of the spectrum beyond 700 nm is restricted to examine shallow coral reefs no deeper than 2 m. The shorter a proxies' wavelengths are, the deeper coral reef habitats will be detected. To know the maximum penetration of a proxy, we recommend referring to the penetration length tied to the mean wavelength of the spectral band with the highest wavelengths. For instance, the *NDCRI* based upon the "red edge" and the green spectral bands could be provided by the WV2, Hyperion, MERIS, MODIS, and SeaWiFS at 0.5, 10, 260, 1000, and 1100 m up to 0.4 m deep. By contrast the *NDCRI* based upon the blue and "coastal" spectral bands, liable to discriminate and rank three coral reef habitats (see **Table 1**), might be derived from all sensors presented in **Figure 9** up and to

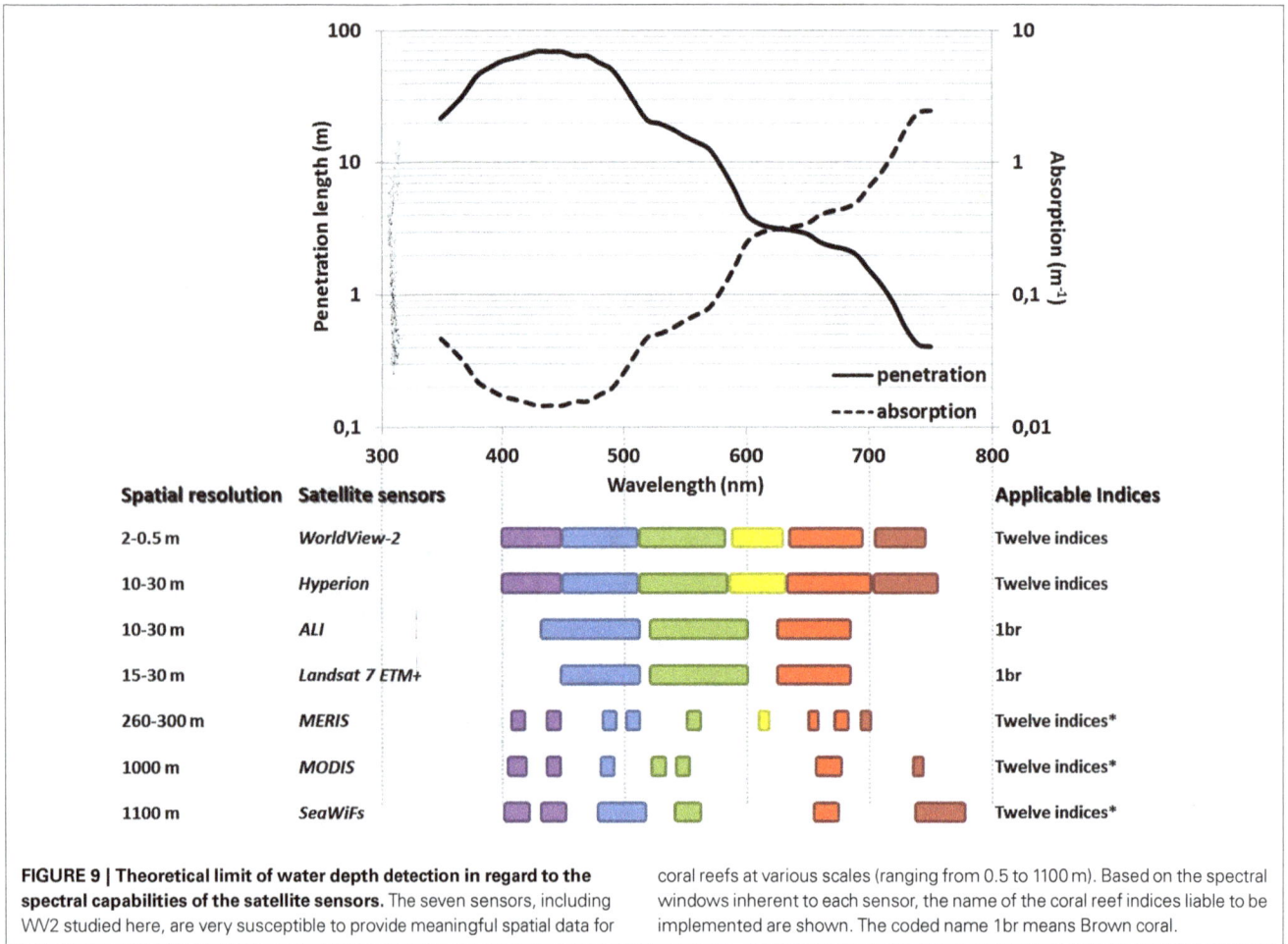

FIGURE 9 | Theoretical limit of water depth detection in regard to the spectral capabilities of the satellite sensors. The seven sensors, including WV2 studied here, are very susceptible to provide meaningful spatial data for coral reefs at various scales (ranging from 0.5 to 1100 m). Based on the spectral windows inherent to each sensor, the name of the coral reef indices liable to be implemented are shown. The coded name 1br means Brown coral.

30 m deep. However, the indicated water depths have to be considered as theoretical thresholds that may over- or under-estimate the actual limits. Even though the theoretical limit of the "red edge" band is near 0.7 m depth (**Figure 9**), the empirical relationships between the associated reflectance and water depth reveal a limit approximating 1.5 m depth (**Figures 4C,F**). Theoretical limits are deemed informational insofar as the magnitude of the sun glint and water clarity can be neglected. Integrating these two limitation factors into a decision-support tool will refine outcomes of the **Figure 9** in order to optimize the spatial data acquisition and selection.

CONTINUOUS SPATIAL MODELS

The spatial modeling of ecological indicators has proven to be excessively helpful in achieving conservation purposes during the process of area selection. From planning to outreach, any conservation projects to date are to use maps. The spatially-explicit quantification of ecologically-relevant proxies proposes a visual support that may facilitate understanding to many stakeholders, irrespective of their technical skills. By virtue of its versatility compared to sensors, and its ease to build through any open-source Geographical Information System and/or Remote Sensing software, the *NDCRI* lends strong support to managers tasking with coral reef sustainability and ecologists delving into understanding coral reef regime shifts. Initiating the latter aspect, we design, deepen and apply the histoscape concept to the Tetiaroa Pacific atoll and Moorea volcanic island. Summarizing the spatial distribution of the *NDCRI* across the reef crest-coastline distance, histoscapes tied to both islands evidently manifest multiple basins composed of Coral, Brown, Red, and Green algae and a single deep basin of Green algae, for the atoll and the high island, respectively. The differences in the topology (field of study of geometry and set theory) of both histoscapes may be caused by the sediment-, nutrient- and pollutant-laden runoff intrinsic to the populated high island. Specifically, the increased nutrient and pollutant discharge form urban and agricultural development are well known to drastically reduce the water quality required so that coral reef can thrive (Bellwood et al., 2004; Fabricius, 2005). The land-based anthropogenic inductions may thus explain why the fringing part depicted in Moorea's histoscape (1100–1300 m in **Figure 8D**) reveals lower *NDCRI* values, corresponding to Red calcareous algae and Bleached coral.

The histoscape tends to exemplify the probability density function of the *NDCRI* values in respect to the reef crest-coastline distance. The spatial gradient selected results from the decision process aimed at finding patterns that better suit the dynamics of coral reef ecosystems tied to islands (high and low), i.e., a centripetal growth starting from the surrounding reef crest pseudo-circle. Rather than investigating the coral reef area of an island monolithically, the histoscape may be insightful when applied to a set of areas segmented in respect to the spatial ecology. For instance, an East-West partition would be advocated to meet the explicative scale of *NDCRI* changes in response to the yearly-dominant Eastern winds and driven swell. In addition, to aid in addressing long-term issues, the histoscape may also finely diagnose mid- and short-term variations following anthropogenic and cyclone perturbations.

IMPLICATIONS FOR CORAL REEF RESILIENCE

Even though this study focuses on a single mapping of coral reef regime, precluding any considerations of stressors or temporal analyses, the *NDCRI* histoscape may bridge the empirical characterization of coral reef ecosystems and some theoretical aspects of the ecological resilience (Holling, 1973). The depth of a histoscape basin is very amenable to correlate with the magnitude of disturbance that a habitat can absorb while conserving its intrinsic topology. Defining the complexity of a geometrical figure, a comprehensive topology would require supplementary parameters such as the length and width of the surface of a basin, which can intuitively be associated with the area of the (1) dominance occupied by a habitat and (2) influence exerted by a habitat toward its neighborhood. The influence of a habitat stops at the edge of the basin (threshold), beyond it another habitat predominates. The depth, length and width of a histoscape basin obviously remind the Resistance and Latitude terms forged with the stability landscape (Walker et al., 2004). Since the histoscape represents the empirical basins of habitat across a spatial gradient, it does not have to be confounded with the stability landscape in which a single ecosystem is a point defined by tridimensional coordinates into a state space diagram constituting of putative basins of attraction (Walker et al., 2004). However, both landscapes can assist in weaving trends of the foreseeable dynamics. The histoscape, as regime or identity landscape, provides the actual habitat (basin) most likely to influence a concentric strip summarizing an evenly-habitated distribution (plain). The stability landscape indicates the position of the ecosystem within its basin of attraction bottomed by the equilibrium state (the attractor), thus influencing the ecosystem's trajectory. As a concise snapshot of the ecosystem's regime, the histoscape is strongly subject to changes due to external forces (e.g., runoff) and internal processes (e.g., coral succession, predator-prey cycles), such as changes in the number and topology of the basins. The decrease in the number of basins may be caused by a basin sprawl (increased surface area of a habitat) to the detriment of neighbors and explained by a linear growth followed by a non-linear phenomenon of coalescence, i.e., process by which two or more basins rapidly merge during contact to form a single daughter basin. In addition to taking the length and width of a basin into account, the topology will incorporate in further studies the overall and inherent structural complexity and diversity, such as the rugosity and the entropy of morphometric features, respectively, since these readily interrelate with the degree of self-organization that an ecosystem and nested habitats can demonstrate. From the glimpse of the dynamical aspect, we point out upfront that Tetiaroa, typified by multiple shallow basins, is more conducive

to exhibit ongoing adaptations to perturbations, comparatively to Moorea, characterized by a single deep channel. The parameters retrieved from the histoscape, such as depth, intrinsic topology and the overall complexity and diversity respectively suit with the three components of the definition of resilience commonly admitted (Carpenter et al., 2001): (1) the amount of disturbance that a system can absorb while still remaining within the same identity, (2) the degree to which the system is capable of self-organization, and (3) the degree to which the system can build and increase its capacity for adaptation and learning. Although learning is not explicitly documented here, some studies have revealed that corals that have experienced bleaching episodes may become thermally tolerant through adapted symbionts, suggesting an acclimation rather than genetic adaptation (Hoegh-Guldberg et al., 2002; Rowan, 2004). Importantly, the scale, as critical issues for understanding pattern-process interactions (Levin, 1992; Levin and Lubchenco, 2008), may be partly examined through the histoscape. Offering the possibility to modulate both the spatial (nested basins) and the diagnostic (habitat and super habitat) scale, the histoscape can tangibly investigate the scaling relationships of the *NDCRI*. The resilience of coral reefs can be acutely augmented in scaling up the organizational level, that is to say in encapsulating the sociological aspect into the system at stake. By networking coral reefs through multi-scale marine protected areas and stimulating ecological memory (Nystrom and Folke, 2001) through human incentives, scales of human demands and ecological supplies (Cumming, 2011) will be matched. The implementation of our easy-to-interpret spectral proxies of coral reef regime into decision-making process has the potential to reinforce positive feedback loops essential for growing a resilient coral reefscape coping with inevitable stochasticity.

ACKNOWLEDGMENTS

This study is part of the CREM (Coral Reefscape Ecology and Mapping) project, funded by the Marie Curie FP7-PEOPLE-2010-RG. The first author gratefully acknowledges the French Agency of Marine Protected Areas and the University of Quebec in Rimouski for the purchase of Moorea's imagery and the use of the handborne spectroradiometer, respectively. Ground-based hyperspectral measurements have been achieved with the CRIOBE's support, which has developed the coral nursery jointly with the Total Foundation for the Biodiversity, Te Mana o te Moana association and Moorea Intercontinental Resort. We are indebted to CRIOBE staff for collecting 897 georeferenced photographs and water depths.

REFERENCES

AMAP Expert Group. (2011). *Climate Change and POPS: Predicting the Impacts, Arctic Monitoring and Assessment Programme.* Oslo: Report of the UNEP/AMAP Expert Group.

Austin, M. P. (2002). Spatial prediction of species distribution: an interface between ecological theory and statistical modelling. *Ecol. Model.* 157, 101–118. doi: 10.1016/S0304-3800(02)00205-3

Austin, M. P. (2007). Species distribution models and ecological theory: a critical assessment and some possible new approaches. *Ecol. Modell.* 200, 1–19. doi: 10.1016/j.ecolmodel.2006.07.005

Bellwood, D. R., Hughes, T. P., Folke, C., and Nyström, M. (2004). Confronting the coral reef crisis. *Nature* 429, 827–833. doi: 10.1038/nature02691

Borja, A., Franco, J., and Pérez, V. (2000). A marine biotic index to establish the ecological quality of soft-bottom benthos within European estuarine and coastal environments. *Mar. Pollut. Bull.* 40, 1100–1114. doi: 10.1016/S0025-326X(00)00061-8

Carpenter, S. R., Walker, M., Anderies, J. M., and Abel, N. (2001). From metaphor to measurement: resilience of what to what? *Ecosystems* 4, 765–781. doi: 10.1007/s10021-001-0045-9

Chen, M., Schliep, M., Willows, Z. R., Cai, D.-L., Neilan, B. A., and Scheer, H. (2010). A red-shifted chlorophyll. *Science* 329, 1318–1319. doi: 10.1126/science.1191127

Collin, A., Archambault, P., and Planes, S. (2013). Bridging ridge-to-reef patches: seamless classification of the coast using very high resolution satellite. *Remote Sens.* 5, 3583–3610. doi: 10.3390/rs5073583

Collin, A., and Hench, J. (2012). Towards deeper measurements of tropical reefscape structure using the worldview-2 spaceborne sensor. *Remote Sens.* 4, 1425–1447. doi: 10.3390/rs4051425

Collin, A., Long, B., and Archambault, P. (2008). Mapping the shallow water seabed habitat with the SHOALS. *IEEE Trans. Geosci. Remote Sens.* 46, 2947–2955. doi: 10.1109/TGRS.2008.920020

Collin, A., and Planes, S. (2011). "What is the value added of 4 bands within the submetric remote sensing of tropical coastscape? Quickbird-2 vs WorldView-2," in *Proceedings of the 31st IGARSS* (Vancouver, BC).

Collin, A., and Planes, S. (2012). Enhancing coral health detection using spectral diversity indices from worldview-2 imagery and machine learners. *Remote Sens.* 4, 3244–3264. doi: 10.3390/rs4103244

Congalton, R., and Green, K. (1999). "Basic analysis techniques," in *Assessing the Accuracy of Remotely Sensed Data: Principles and Practices,* Chapter 5, eds R. Congalton and K. Green (Boca Raton, FL: CRC Press).

Cumming, G. S. (2011). Spatial resilience: integrating landscape ecology, resilience, and sustainability. *Landsc. Ecol.* 26, 899–909. doi: 10.1007/s10980-011-9623-1

De'ath, G., Fabricius, K. E., Sweatman, H., and Puotinen, M. (2012). The 27-year decline of coral cover on the GreatBarrier Reef and its causes. *Proc. Natl. Acad. Sci. U.S.A.* 109, 17995–17999. doi: 10.1073/pnas.1208909109

Emanuel, K. (2005). Increasing destructiveness of tropical cyclones over the past 30-years. *Nature* 436, 686–688. doi: 10.1038/nature03906

Fabricius, K. (2005). Effects of terrestrial runoff on the ecology of corals and coral reefs: review and synthesis. *Mar. Pollut. Bull.* 50, 125–146. doi: 10.1016/j.marpolbul.2004.11.028

Folke, C. (2006). Resilience: the emergence of aperspective for social-ecological system analyses. *Glob. Environ. Change* 16, 253–267. doi: 10.1016/j.gloenvcha.2006.04.002

Forman, R. T. T. (1995). *Land Mosaics: the Ecology of Landscapes and Regions.* Cambridge: Cambridge University Press.

Gotelli, N. J., and Colwell, R. K. (2001). Quantifying biodiversity: procedures and pitfalls in the measurement and comparison of species richness. *Ecol. Lett.* 4, 379–391. doi: 10.1046/j.1461-0248.2001.00230.x

Harborne, A. R., Mumby, P. J., Kappel, C. V., Dahlgren, C. P., Micheli, F., Holmes, K. E., et al. (2008). Tropical coastal habitats as surrogates of fish community structure, grazing, and fisheries value. *Ecol. Appl.* 18, 1689–1701. doi: 10.1890/07-0454.1

Harborne, A. R., Mumby, P. J., Zychaluk, K., Hedley, J. D., and Blackwell, P. G. (2006). Modeling the beta diversity of coral reefs. *Ecology* 87, 2871–2881. doi: 10.1890/0012-9658(2006)87[2871:MTBDOC]2.0.CO;2

Hastings, A., and Wysham, D. B. (2010). Regime shifts in ecological systems can occur with no warning. *Ecol. Lett.* 13, 464–472. doi: 10.1111/j.1461-0248.2010.01439.x

Hochberg, E. J., Atkinson, M. J., and Andréfouët, S. (2003). Spectral reflectance of coral reef bottom-types worldwide and implications for coral reef remote sensing. *Remote Sens. Environ.* 85, 159–173. doi: 10.1016/S0034-4257(02)00201-8

Hoegh-Guldberg, O., Jones, R. J., Ward, S., and Loh, W. K. (2002). Communication arising. Is coral bleaching really adaptive? *Nature* 415, 601–602. doi: 10.1038/415601a

Hoegh-Guldberg, O., Mumby, P. J., Hooten, A. J., Steneck, R. S., Greenfield, P., Gomez, E., et al. (2007). Coral reefs under rapid climate change and ocean acidification. *Science* 318, 1737–1742. doi: 10.1126/science.1152509

Holling, C. S. (1973). Resilience and stability of ecological systems. *Annu. Rev. Ecol. Syst.* 4, 1–23. doi: 10.1146/annurev.es.04.110173.000245

Holling, C. S. (2001). Understanding the complexity of economic, ecological, and social systems. *Ecosystems* 4, 390–405. doi: 10.1007/s10021-001-0101-5

Hughes, T. P., Graham, N. A. J., Jackson, J. B. C., Mumby, P. J., and Steneck, R. S. (2010). Rising to the challenge of sustaining coral reef resilience. *Trends Ecol. Evol.* 25, 633–642. doi: 10.1016/j.tree.2010.07.011

Ives, A. R., and Dakos, V. (2012). Detecting dynamical changes in nonlinear time series using locally linear state-space models. *Ecosphere* 3, 58. doi: 10.1890/ES11-00347.1

Koning, R. E. (1994). *Light. Plant Physiology Information Website.* Available online at: http://plantphys.info/plant_physiology/phytochrome.shtml (Accessed August 11, 2012).

Leiper, I. A., Siebeck, U. E., Marshall, N. J., and Phinn, S. R. (2009). Coral health monitoring: linking coral colour and remote sensing techniques. *Can. J. Remote Sens.* 35, 276–286. doi: 10.5589/m09-016

Leon, J. X., Phinn, S. R., Hamylton, S., and Saunders, M. I. (2012). Filling the 'white ribbon' - a seamless multisource digital elevation/depth model for Lizard Island, northern Great Barrier Reef. *Int. J. Remote Sens.* 34, 6337–6354. doi: 10.1080/01431161.2013.800659

Levin, S. A. (1992). The problem of pattern and scale in ecology. *Ecology* 73, 1943–1967. doi: 10.2307/1941447

Levin, S. A., and Lubchenco, J. (2008). Resilience, robustness, and marine ecosystem-based management. *Bioscience* 58, 1–6. doi: 10.1641/B580107

Maritorena, S., Morel, A., and Gentili, B. (1994). Diffuse reflectance of oceanic shallow waters: influence of water depth and bottom albedo. *Limnol. Oceanogr.* 39, 1689–1703. doi: 10.4319/lo.1994.39.7.1689

Matthew, M. W., Adler-Golden, S. M., Berk, A., Richtsmeier, S. C., Levine, R. Y., Bernstein, L. S., et al. (2000). Status of atmospheric correction using a MODTRAN4-based algorithm. *SPIE Proc.* 4049, 199–207. doi: 10.1117/12.410341

May, R. M. (2001). *Stability and Complexity in Model Ecosystems.* Princeton, NJ: Princeton University Press.

Mellin, C., Andréfouët, S., Kulbicki, M., Dalleau, M., and Vigliola, L. (2009). Remote sensing and fish–habitat relationships in coral reef ecosystems: review and pathways for systematic multi-scale hierarchical research. *Mar. Pollut. Bull.* 58, 11–19. doi: 10.1016/j.marpolbul.2008.10.010

Mobley, C. D., Sundman, L. K., and Boss, E. (2002). Phase function effects on oceanic light fields. *Appl. Opt.* 41, 1035–1050. doi: 10.1364/AO.41.001035

Myers, N., Mittermeier, R. A., Mittermeier, C., Da Fonseca, G. A. B., and Kent, J. (2000). Biodiversity hotspots for conservation priorities. *Nature* 403, 853–858. doi: 10.1038/35002501

Myneni, R. B., Keeling, C. D., Tucker, C. J., Asrar, G., and Nemani, R. R. (1997). Increased plant growth in the northern high latitudes from 1981 to 1991. *Nature* 386, 698–702. doi: 10.1038/386698a0

Nystrom, M., and Folke, C. (2001). Spatial resilience of coral reefs. *Ecosystems* 4, 406–417 doi: 10.1007/s10021-001-0019-y

Pegau, S., Zaneveld, J. R. V., Mitchell, B. G., Mueller, J. L., Kahru, M., Wieland, J., et al. (2003). *Inherent Optical Properties: Instruments, Characterizations, Field Measurements and Data Analysis Protocols. Ocean Optics Protocols for Satellite Ocean Color Sensor Validation.* NASA Tech Memo, 211621.

Pettorelli, N., Vik, J. O., Mysterud, A., Gaillard, J.-M., Tucker, C. J., and Stenseth, N. C. (2005). Using the satellite-derived NDVI to assess ecological responses to environmental change. *Trends Ecol. Evol.* 20, 503–510. doi: 10.1016/j.tree.2005.05.011

Piao, S., Fang, J., Ciais, P., Peylin, P., Huang, Y., Sitch, S., et al. (2009). The carbon balance of terrestrial ecosystems in China. *Nature* 458, 1009–1013. doi: 10.1038/nature07944

Roff, G., and Mumby, P. J. (2012). Global disparity in the resilience of coral reefs. *Trends Ecol. Evol.* 27, 404–413. doi: 10.1016/j.tree.2012.04.007

Rooney, T. P., Wiegmann, S. M., Rogers, D. A., and Waller, D. M. (2004). Biotic impoverishment and homogenization in unfragmented forest understory communities. *Conserv. Biol.* 18, 787–798. doi: 10.1111/j.1523-1739.2004.00515.x

Rowan, R. (2004). Coral bleaching: thermal adaptation in reef coral symbionts. *Nature* 430, 742. doi: 10.1038/430742a

Rowlands, G., Purkis, P., Riegl, B., Bruckner, B., and Renaud, P. (2012). Satellite imaging coral reef resilience at regional scale. *Mar. Pollut. Bull.* 64, 1222–1237. doi: 10.1016/j.marpolbul.2012.03.003

Smith, J. (2012). "Baselines and degradation of central Pacific benthic reef communities," in *Proceedings of the 12th International Coral Reef Symposium* (Cairns, QLD).

Smith, R. C., and Baker, K. S. (1981). Optical properties of the clearest naturalwaters (200-800nm). *Appl. Opt.* 20, 177–184. doi: 10.1364/AO.20.000177

Stumpf, R. P., Holderied, K., and Sinclair, M. (2003). Determination of water depth with high-resolution satellite imagery over variable bottom types. *Limnol. Oceanogr.* 48, 547–556. doi: 10.4319/lo.2003.48.1_part_2.0547

Sturm, M. (2010). Arctic plants feel the heat. *Sci. Am.* 302, 48–55. doi: 10.1038/scientificamerican0510-66

Tucker, C. J. (1979). Red and photographic infrared linear combinations for monitoring vegetation. *Remote Sens. Environ.* 8, 127–150. doi: 10.1016/0034-4257(79)90013-0

Turner, M. G., Gardner, R. H., and O'Neill, R. V. (2001). *Landscape Ecology in Theory and Practice*. New-York, NY: Springer-Verlage.

Walker, B., Holling, C. S., Carpenter, S. R., and Kinzig, A. (2004). Resilience, adaptability and transformability in social–ecological systems. *Ecol. Soc.* 9:5. Available online at: http://www.ecologyandsociety.org/vol9/iss2/art5/

White, S. N., Chave, A. D., and Reynolds, G. T. (2002). Investigations of ambient light emission at deep-sea hydrothermal vents. *J. Geophys. Res.* 107, B1. doi: 10.1029/2000JB000015

Wilkinson, C. (2008). *Status of Coral Reefs of the World: 2008*. Townsville, QLD: Global Coral Reef Monitoring Network and Reef and Rainforest Research Centre.

Zhao, M., and Running, S. W. (2010). Drought-induced reduction in global terrestrial net primary production from 2000 through 2009. *Science* 329, 940–943. doi: 10.1126/science.1192666

Conflict of Interest Statement: The authors declare that the research was conducted in the absence of any commercial or financial relationships that could be construed as a potential conflict of interest.

Permissions

The contributors of this book come from diverse backgrounds, making this book a truly international effort. This book will bring forth new frontiers with its revolutionizing research information and detailed analysis of the nascent developments around the world.

We would like to thank all the contributing authors for lending their expertise to make the book truly unique. They have played a crucial role in the development of this book. Without their invaluable contributions this book wouldn't have been possible. They have made vital efforts to compile up to date information on the varied aspects of this subject to make this book a valuable addition to the collection of many professionals and students.

This book was conceptualized with the vision of imparting up-to-date information and advanced data in this field. To ensure the same, a matchless editorial board was set up. Every individual on the board went through rigorous rounds of assessment to prove their worth. After which they invested a large part of their time researching and compiling the most relevant data for our readers.

The editorial board has been involved in producing this book since its inception. They have spent rigorous hours researching and exploring the diverse topics which have resulted in the successful publishing of this book. They have passed on their knowledge of decades through this book. To expedite this challenging task, the publisher supported the team at every step. A small team of assistant editors was also appointed to further simplify the editing procedure and attain best results for the readers.

Apart from the editorial board, the designing team has also invested a significant amount of their time in understanding the subject and creating the most relevant covers. They scrutinized every image to scout for the most suitable representation of the subject and create an appropriate cover for the book.

The publishing team has been an ardent support to the editorial, designing and production team. Their endless efforts to recruit the best for this project, has resulted in the accomplishment of this book. They are a veteran in the field of academics and their pool of knowledge is as vast as their experience in printing. Their expertise and guidance has proved useful at every step. Their uncompromising quality standards have made this book an exceptional effort. Their encouragement from time to time has been an inspiration for everyone.

The publisher and the editorial board hope that this book will prove to be a valuable piece of knowledge for researchers, students, practitioners and scholars across the globe.

List of Contributors

Nathalie Gypens
Laboratoire d'Ecologie des Systèmes Aquatiques, Ecole Interfacultaire de Bioingénieurs, Université Libre de Bruxelles, Brussels, Belgium

AlbertoV.Borges
Unité d'Océanographie Chimique, Department of Astrophysics Geophysics and Oceanography, Institutde Physique (B5), Université de Liège, Liège, Belgium

Ibon Galparsoro, Angel Borja and María C.Uyarra
Marine Research Division, AZTI-Tecnalia, Pasaia, Spain

William G. Ambrose Jr.
Department of Biology, Bates College, Lewiston, ME,USA
Akvaplan-niva, FRAM-High North Research Centre for Climate and the Environment, Tromsø, Norway

Lisa M. Clough
Division of Polar Programs, National Science Foundation, Arlington, VA, USA

Jeffrey C. Johnson
Department of Sociology and Institute for Coastal Science and Policy, East Carolina University, Greenville, NC, USA
Department of Anthropology, University of Florida, Gainesville, FL, USA

Michael Greenacre
Akvaplan-niva, FRAM-High North Research Centre for Climate and the Environment, Tromsø, Norway
Department of Economics and Business, Unıversıtat Pompeu Fabra, and Barcelona Graduate School of Economics, Barcelona, Spain

David C. Griffith
Department of Anthropology and Institute for Coastal Science and Policy, East Carolina University, Greenville, NC, USA

Michael L. Carroll
Akvaplan-niva, FRAM-High North Research Centre for Climate and the Environment, Tromsø, Norway

Alex Whiting
Native Village of Kotzebue, Kotzebue, AK, USA

Esteban Acevedo-Trejos and Agostino Merico
Systems Ecology Group, Leibniz Center for Tropical Marine Ecology, Bremen, Germany
School of Engineering and Science, Jacobs University Bremen, Bremen, Germany

Gunnar Brandt
Systems Ecology Group, Leibniz Center for Tropical Marine Ecology, Bremen, Germany

Marco Steinacher
Climate and Environmental Physics, Physics Institute, University of Bern, Bern, Switzerland
Oeschger Centre for Climate Change Research, University of Bern, Bern, Switzerland

Riaan van der Merwe, Sabine Lattemann and Gary L. Amy
Water Desalination and Reuse Center, Biological and Environmental Sciences and Engineering Division, King Abdullah University of Science and Technology, Thuwal, Saudi Arabia

Till Röthig and Christian R. Voolstra
Red Sea Research Center, Biological and Environmental Sciences and Engineering Division, King Abdullah University of Science and Technology, Thuwal, Saudi Arabia

Michael A. Ochsenkühn
Biological and Organometallic Catalysis Laboratories, Physical Sciences and Engineering Division, King Abdullah University of Science and Technology, Thuwal, Saudi Arabia

Jesper H. Andersen
NIVA Denmark Water Research, Copenhagen, Denmark
Marine Research Center, Finnish Environment Institute (SYKE), Helsinki, Finland

Karsten Dahl, Cordula Göke and Ciarán Murray
Department of Bioscience, Aarhus University, Roskilde, Denmark

Martin Hartvig
DTU Aqua, Section for Marine Ecosystem-Based Management, Technical University of Denmark, Charlottenlund, Denmark
Centre for Macroecology, Evolution and Climate, University of Copenhagen, Copenhagen, Denmark

Anna Rindorf and Morten Vinther
DTU Aqua, Section for Marine Ecosystem-Based Management, Technical University of Denmark, Charlottenlund, Denmark

Henrik Skov
DHI, Hørsholm, Denmark

Samuli Korpinen
Marine Research Center, Finnish Environment Institute (SYKE), Helsinki, Finland

Giorgos Chatzigeorgiou
Biology Department, University of Crete, Heraklion, Greece
Hellenic Centre for Marine Research, Institute of Marine Biology, Biotechnology and Aquaculture, Heraklion, Greece

Elena Sarropoulou, Katerina Vasileiadou, Sarah Faulwetter, Giorgos Kotoulas and Christos D. Arvanitidis
Hellenic Centre for Marine Research, Institute of Marine Biology, Biotechnology and Aquaculture, Heraklion, Greece

Christina Brown
Department of Biology, Chemistry and Pharmacology, Institute of Biology, Free University of Berlin, Berlin, Germany

Miquel Alcaraz
Institut de Ciències del Mar, CSIC, Barcelona, Spain

Rodrigo Almeda
Centre for Ocean Life, DTU Aqua, Technical University of Denmark, Charlottenlund, Denmark

Carlos M. Duarte and Susana Agustí
IMEDEA, CSIC-UIB, Esporles, Spain
UWA Oceans Institute, Crawley, WA, Australia

Burkhard Horstkotte and, Sebastien Lasternas
IMEDEA, CSIC-UIB, Esporles, Spain

Frédéric Gazeau and Samir Alliouane
Sorbonne Universités, Université Pierre et Marie Curie Univ Paris 06, Unité Mixte de Recherche 7093, Laboratoire d' Océanographie de Villefranche, Villefranche/Mer, France
Centre National de la Recherche Scientifique, Unité Mixte de Recherche 7093, Laboratoire d'Océanographie de Villefranche, Observatoire Océanologique, Villefranche/Mer, France

Christian Bock, Timo Hirse and Hans-Otto Pörtner
Alfred-Wegener-Institute Helmholtz Zentrum für Polar und Meeresforschung, Am Handelshafen 12,D-27570 Bremerhaven, Germany

Lorenzo Bramanti
Sorbonne Universités, Université Pierre et Marie Curie Univ Paris 06, Unité Mixte de Recherche 8222, Laboratorio de Ecofisiología para la Conservación de Bosques, Observatoire Océanologique, Banyuls/mer, France
Centre National de la Recherche Scientifique, Unité Mixte de Recherche 8222, Laboratorio de Ecofisiología para la Conservación de Bosques, Observatoire Océanologique, Banyuls/mer, France

Matthias López Correa
GeoZentrum Nordbayern, Universität Erlangen-Nürnberg, Erlangen, Germany
German University of Technology in Oman, Halban Campus, Muscat, Sultanate of Oman

Miriam Gentile
Consejo Superior de Investigaciones Científicas, Institut de Ciencies del Mar, Barcelona, Spain

Patrizia Ziveri
Universitat Autòno made Barcelona, Institutde Cienciai Tecnologia Ambientals,Barcelona, Spain
Institució Catalana de Recerca I Estudis Avançats, Barcelona, Spain

Sarah Q. Foster
Department of Earth and Environment, Boston University, Boston, MA, USA

Robinson W. Fulweiler
Department of Earth and Environment, Boston University, Boston, MA, USA
Department of Biology, Boston University, Boston, MA, USA

Christian Lindemann and Michael A. St. John
National Institute of Aquatic Resources, Technical University of Denmark, Charlottenlund, Denmark

Lisa-Marie K. Harrison and Robert Harcourt
Marine Predator Research Group, Department of Biological Sciences, Faculty of Science and Engineering, Macquarie University, North Ryde, NSW, Australia

Martin J.Cox
Marine Predator Research Group, Department of Biological Sciences, Faculty of Science and Engineering, Macquarie University, North Ryde, NSW, Australia
Australian Antarctic Division, Department of the Environment, Australian Government, Kingston, TAS, Australia

Georg Skaret
Institute of Marine Research, Bergen, Norway

Lucy C. Woodall and Gordon L .J. Paterson
Department of Life Sciences, The Natural History Museum, London, UK

Laura F. Robinson
School of Earth Sciences, University of Bristol, Bristol, UK

Alex D. Rogers
Department of Zoology, University of Oxford, Oxford, UK

Bhavani E. Narayanaswamy
The Scottish Association for Marine Science, Ecology Department, Scottish Marine Institute, Oban, UK

Jay J. Lunden, Conall G. McNicholl, Christopher R. Sears and Erik E. Cordes
Department of Biology, Temple University, Philadelphia, PA, USA

Cheryl L. Morrison
United States Geological Survey, Leetown Science Center, Kearneysville, WV, USA

Alex Rattray
Dipartimento di Biologia, Università di Pisa, Pisa, Italy
Centre for Integrative Ecology, School of Life and Environmental Sciences, Deakin University, Warrnambool, VIC, Australia

Daniel Ierodiaconou
Centre for Integrative Ecology, School of Life and Environmental Sciences, Deakin University, Warrnambool, VIC, Australia

Tim Womersley
DHI Water and Environment Pty Ltd., Perth, WA, Australia
Water Technology Pty Ltd., Melbourne, VIC, Australia

Sascha B. Sjollema, Charlotte D. Vavourakis, Harm G. van der Geest and Wim Admiraal
Department of Aquatic Ecology and Ecotoxicology, Institute for Biodiversity and Ecosystem Dynamics, University of Amsterdam, Amsterdam, Netherlands

A. Dick Vethaak
Deltares, Marine and Coastal Systems, Delft, Netherlands
Department Chemistry and Biology, Institute for Environmental Studies (IVM), VU University Amsterdam, Amsterdam, Netherlands

Jay Forrest, Paul Bazylewski, Robert Bauer and Gap Soo Chang
Department of Physics and Engineering Physics, University of Saskatchewan, Saskatoon, SK, Canada

Seongjin Hong and Jong Seong Khim
School of Earth and Environmental Sciences and Research Institutes of Oceanography, Seoul National University, Seoul, Republic of Korea

Chang Yong Kim
Canadian Light Source, Saskatoon, SK, Canada

JohnP.Giesy
Department of Veterinary Biomedical Sciences and Toxicology Centre, University of Saskatchewan, Saskatoon, SK, Canada
Department of Zoology, and Center for Integrative Toxicology, Michigan State University, East Lansing, MI, USA
Department of Biology and Chemistry and State Key Laboratory in Marine Pollution, City University of Hong Kong, Kowloon,Hong Kong, China
School of Biological Sciences, University of HongKong, HongKong, China

State Key Laboratory of Pollution Control and Resource Reuse, School of the Environment, Nanjing University, Nanjing, China

Jacob Carstensen
Department of Bioscience, Aarhus University, Roskilde, Denmark

Ty J. Samo, Steven Smriga, Francesca Malfatti, Byron P. Sherwood and Farooq Azam Marine
Biology Research Division, Scripps Institution of Oceanography, University of California, San Diego, La Jolla, CA, USA

Stelios Katsanevakis, Chiara Piroddi and Ana Cristina Cardoso
Water Resources Unit, Institute for Environment and Sustainability, Joint Research Centre, Ispra, Italy

Marta Coll
Institut de Recherche pourle Développement, UMREME212, Centre de Recherche Halieutique Méditerranéenneet Tropicale, Sète, France

Jeroen Steenbeek
Ecopath International Initiative Research Association, Barcelona, Spain

Frida Ben Rais Lasram
Unité de Recherche Ecosystèmes et Ressources Aquatiques UR03AGRO1, Institut National Agronomiquede Tunisie, Tunis, Tunisia

Argyro Zenetos
Institute of Marine Biological Resources and Inland Waters, Hellenic Centre for Marine Research, Agios Kosmas, Greece

Emma L. Orlova and Andrey V. Dolgov
Laboratory of Trophology, Polar Research Institute of Marine Fisheries and Oceanography, Murmansk, Russia

Paul E. Renaud
Akvaplan-niva, Fram Centre for Climate and Environment, Tromsø, Norway
Department of Arctic Biology, University Centrein Svalbard, Longyearbyen, Norway

Michael Greenacre
Akvaplan-niva, Fram Centre for Climate and Environment, Tromsø, Norway
Barcelona Graduate School of Economics, Universitat Pompeu Fabra, Barcelona, Spain

Claudia Halsband
Akvaplan-niva, Fram Centre for Climate and Environment, Tromsø, Norway

Victor A. Ivshin
Laboratory of Trophology, Polar Research Institute of Marine Fisheries and Oceanography, Murmansk, Russia Laboratory of Fisheries Oceanography, Polar Research Institute of Marine Fisheries and Oceanography, Murmansk, Russia

Antoine Collin
Nadaoka Laboratory, Department of Mechanical and Environmental Informatics, Tokyo Institute of Technology, Tokyo, Japan
USR 3278CNRS-EPHE, Centre de Recherches Insulaireset Observatoire de l'Environnement (CRIOBE), Papetoai, French Polynesia

Philippe Archambault
Institut des Sciences dela Mer, Université du Québec à Rimouski, Rimouski, Canada

Serge Planes
USR 3278CNRS-EPHE, Centre de Recherches Insulaireset Observatoire de l'Environnement (CRIOBE), Papetoai, French Polynesia